Reviews in Plasmonics

More information about this series at http://www.springer.com/series/7164

Chris D. Geddes
Editor

Reviews in Plasmonics 2015

Springer

Editor
Chris D. Geddes
Institute of Fluorescence
University of Maryland Baltimore County
Baltimore, MD, USA

ISSN 1555-7677
Reviews in Plasmonics
ISBN 978-3-319-24604-8 ISBN 978-3-319-24606-2 (eBook)
DOI 10.1007/978-3-319-24606-2

Library of Congress Control Number: 2015961024

Springer Cham Heidelberg New York Dordrecht London

Printed on acid-free paper

Springer International Publishing AG Switzerland is part of Springer Science+Business Media (www.springer.com)

Preface

This is the second volume in the Plasmonics series, *Reviews in Plasmonics*. The first volume was very well received by the Plasmonics community with several notable reviews of the volume.

In this 2015 volume, we are pleased again with the broad and timely Plasmonic content. We subsequently thank the authors for their very timely and exciting contributions again this year. We hope you all will find this volume as useful as the first volume.

In closing, I would like to thank both Tanja Koppejan and Meran Owen at Springer for their help in compiling this volume.

Chris D. Geddes
Institute of Fluorescence
University of Maryland Baltimore County
Baltimore, MD, USA
July 29 2015

Contents

Chapter 1
Surface Plasmon Polariton Assisted Optical Switching in Noble Metal Nanoparticle Systems: A Sub-Band Gap Approach

Sandip Dhara

Abstract Understanding the light-matter interaction at nanometre scale is a fundamental issue in optoelectronics and nanophotonics which are prerequisites for advanced sensor applications of optical switching. The electrical transport process in noble metal-insulator nanocomposite or dispersed noble metal nanocluster in dielectric matrix is discussed. Banking on high value of third-order nonlinear susceptibility, optical switching was reported in the percolation threshold of noble metals in dielectric matrices. The optical switching originating from the excitation of the surface plasmon was recorded for metal–oxide–metal tunnelling junctions. The surface plasma polariton (SPP) in the form of drifting hot electrons across the oxide barrier and tunnelling to the counter electrode in the evanescent field of surface plasmon resonance (SPR) was made responsible for electrical transport mechanism. These models, for the first time, are discussed in ambit of having a sub-band gap feature in the SPP assisted photoresponse where transport of electrical carriers may manifest either at the percolation threshold with enhanced electro-magnetic field, and in the form of tunnelling current through the potential barrier at the Fermi level or in the propagation of plasmon coupled electrons at SPR for metal-dielectric composites.

Keywords SPP • SPR • Optical switching • Noble metal • Nanoparticle

1.1 Introduction

Understanding the light-matter interaction at nanometre scale is a fundamental issue in optoelectronics and nanophotonics which are prerequisites for advanced sensor applications, namely, optical switching (abrupt change in resistance with exposed light signal) devises using all-optical signal processing [1, 2], two photon absorption (TPA) enhanced second harmonic generation (SHG) [3], renewable

---

S. Dhara (✉)
Surface and Nanoscience Division, Materials Science Group, Indira Gandhi Centre for Atomic Research, Kalpakkam 603 102, India
e-mail: dhara@igcar.gov.in

C.D. Geddes (ed.), *Reviews in Plasmonics 2015*,
DOI 10.1007/978-3-319-24606-2_1

energy resourcing by guiding and localizing light and considerable reduction in absorption layer thickness [4]. It is also used in the bio-organic sensing, in medical therapy encompassing early detection of beginning of life using shift in plasmonic resonance frequency due to the presence of molecular motors [5] and in destroying tumour by generating heat using the adiabatic heating of plasmonic vibration [6].

Optical switching using the enhanced third-order nonlinear susceptibility in noble metal systems [7–10], especially near the surface-plasmon-resonance (SPR) frequency evoked considerable interest among scientific researchers during last one and half decade. Coherent oscillation of conduction electrons, particularly in noble metal nanoclusters, with the excitation of visible light gives rise to surface plasmon polaritons (SPP) which propagate near the metal-dielectric interface (schematically shown in Fig. 1.1) [11–14]. The evanescent field allows the observation of interference [15–17], and sub-diffraction limited optical imaging of nanostructure [18–22]. The plasmon coupling within arrays of metal nanoparticles can lead to the formation of nanoscale hot spots in which the intensity of light from an incident beam can be concentrated by more than four orders of magnitude. The effect of light concentration by means of plasmon is most obvious in phenomena dealing nonlinearity in light intensity, as demonstrated recently by the on-chip generation of extreme-ultraviolet light by pulsed laser high harmonic generation [23]. This opens an affluence of prospects in lithography or imaging at the nanoscale through the use of soft x-rays. In fact, SPR induced localized SHG using Au nanoparticles are well studied system [24–26]. Among many such significant phenomena and applications, the most well known is the giant surface-enhanced Raman scattering (SERS) [27] or noble metal coated tip enhanced Raman scattering (TERS) [28], those allow both detection and spectroscopic imaging of a single molecule [29]. Electromagnetic energy transfer in chains of closely spaced metal nanoparticles was reported in the sub-diffraction limit by means of coupled plasmon modes [30, 31]. In a dispersion model for coupled plasmon modes considering equi-spaced metal nanoclusters is developed using an analytical

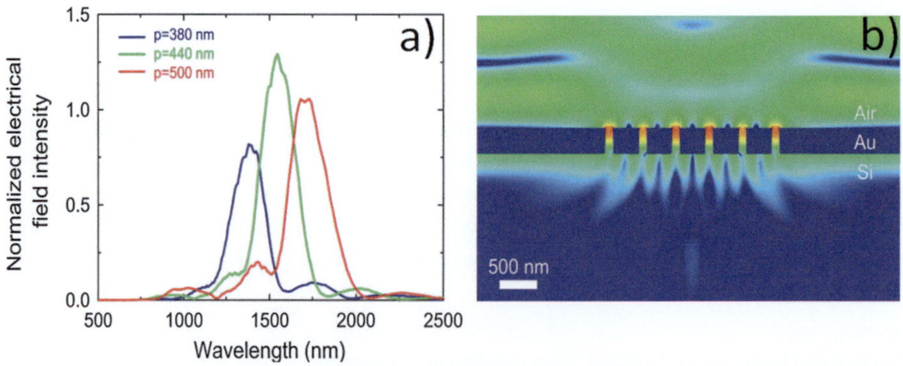

Fig. 1.1 (**a**) Incident-light wavelength dependencies of the electrical field intensity for different slit periods and (**b**) electrical field intensity distribution in a nano-slit grating for 1500 nm excitation (Ref. [14]; Copyright (2013) by the IEEE)

model that describes the near-field electromagnetic interaction between the particles in the dipole limit [30]. Coherent propagation with group velocities exceeding 0.1 *c* was calculated in straight wires of dimension less than 0.1λ (wavelength) and around sharp corners with bending radius less than wavelength of visible light. In another report using generalized Mie theory, the light-transport properties of Ag particles of 50-nm diameter, an optimum guiding conditions for an inter-particle spacing of 25 nm, and a corresponding 1/*e* signal-damping length of 900 nm was estimated [31]. These models of optical energy transport in a plasmonic chain are useful for sub-wavelength transmission lines within integrated optical circuits and for near-field optical microscopy in futuristic ultra fast photonic device applications.

Coinage elements of Au, Ag, Cu are extensively studied for plasmonic applications [32]. Ultrafast switching of 360 fs in Ag nanoclusters embedded in SiO_2 matrix, originating from self-diffraction of a pump pulse due to transient grating, was recorded in a femtosecond optical-Kerr-shutter experiment [7]. A very high optical nonlinearlity $\chi^{(3)} \sim 10^{-7}$ esu which was comparable [8] or one order lower [9] than that for Au nanoclusters system, was reported for Ag nanoclusters. An order lower value of $\chi^{(3)} \sim 10^{-8}$ esu was recorded on Cu nanoclusters dispersed in silica matrix with picosecond nonlinear optical response [10]. In a microreactor approach [33], Au NPs embedded in silica matrix showed optical switching with the exposure of 532 nm, which was close to the SPR absorption peak for Au NPs. In the similar approach, an optical switching was also reported for Au functionalized GaN (Au–GaN) nanowires with an excitation of 532 nm [34]. Role of band conduction in GaN is ruled out in the absence of optical switching using 325 nm (~3.8 eV) excitation (energy being higher than the band gap of GaN ~ 3.4 eV). The SPR, on the other hand, in bimetallic nanocluster of noble metals of Au and Ag is also useful for various applications with the ability of tuning the absorption peak position [35–37]. The SHG in the Pt/Cu [38] and Ag/Cu [39] systems drew lot of attention. In a recent report photoresponse of bimetallic Au–Ag nanoparticle embedded soda glass (Au–Ag@SG) substrate was reported [40] for surface plasmon assisted optical switching using 808 nm excitation, which was away from the characteristic SPR peaks of Ag (~400 nm) and Au (~550 nm) nanoparticles. The observation suggested the possible role of TPA owing to the presence of interacting electric dipole in these systems.

In order to understand the electrical transport process in noble metal-insulator nanocomposite or dispersed noble metal nanocluster in dielectric matrix, various models are discussed in the literature. Banking on high $\chi^{(3)}$ values, optical switching was reported in the percolation threshold of Au nanoclusters in SiO_2 matrix [9], and Cu nanoclusters on Si_3N_4 film [41]. A reversible electronic threshold switching was reported in percolative Ag nanoclusters embedded in polymer matrix [42]. The optical switching originating from the excitation of the surface plasmon was recorded for metal–oxide–metal tunnelling junctions [43]. The SPP in the form of drifting hot electrons across the oxide barrier and tunnelling to the counter electrode in the evanescent field of SPR was made responsible for electrical transport mechanism. In the specially designed experiment [34], optical switching in Au–GaN nanowires with a sub-band gap excitation of 532 nm suggested possible role of SPP assisted transport of electron in the system. The same mechanism was proposed for the propagation of electronic carrier belonging to the conduction electron of noble bimetals in the Au–Ag@SG system in understanding the observed photoresponse [40]. Contributions of interband and intraband transitions in noble metals and inter-particle separation were discussed in understanding the plasmonic coupling mechanism. These models, for the first time, are discussed in ambit of having a sub-band gap feature in the SPP assisted photoresponse where transport of electrical carriers may manifest either at the percolation threshold with enhanced electro-magnetic field, and in the form of tunnelling current through the potential barrier at the Fermi level or in the propagation of plasmon coupled electrons at SPR for metal-dielectric composites.

1.1.1 A Percolative Pathway for Electrical Transport

Optical absorption and electrical transport properties of noble metal-dielectric nanocomposites can be understood from the theoretical investigation in the calculation of the dielectric function for a heterogeneous composite medium [44–46]. The Maxwell-Garnett theory (MGT), dealing predictions related to the existence of the optical dielectric anomaly observed in granular metal films (presently understood as absorption peak due to SPR of the noble metal clusters), was modeled for the calculation of optical properties [44]. However, MGT was limited in predicting observed percolation threshold in granular metals for the volume fraction of the metallic phase comparable to that of the matrix phase. On the other hand, the effective medium theory (EMT) was used for the calculation of the dielectric constant for composites, as well as predicted a percolation threshold for electrical conductivity [45]. However, unlike the MGT, dielectric anomaly could not be inferred in the EMT. Moreover, the predicted value of the percolation threshold was lower than that compared with the experimental reports. Later, based on a phenomenological model considering distribution of conducting and insulating phases, both the optical dielectric anomaly and the percolation threshold was developed for the unified understanding of the optical absorption and percolation transport properties of granular composite media [46]. At a specific composition with the fraction of conducting phase $p > 0.35$, a percolative pathway for electrical conduction was correctly estimated for Au–SiO_2 composite (Fig. 1.2). At the same time, as observed in experiments for the SPR absorption of Au clusters embedded in dielectric matrices was well described under the same formalism in the limit of the percolative compositions $0.1 < p < 0.8$ (Fig. 1.3). The percolating threshold was not predictable under MGT with optical anomaly existing even for $p = 1$. Transition in infrared transmission for $p > 0.7$ was another success of the model where also MGT failed to estimate. Subsequently, a hopping model in a percolative pathway, based on the microstructure of composites, was developed where the effect of charging energy E_c as a function of the conductance of paths linking grains of separation '*s*'

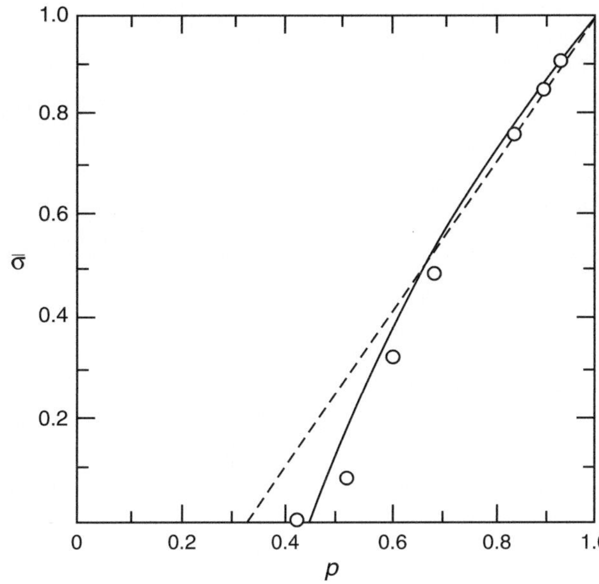

Fig. 1.2 Normalized conductivity $\overline{\sigma}$ as a function of p for Au–SiO_2 cermets. Data from Ref. [43]. *Dashed line* denotes EMT result (Ref. [46]; Copyright (1980) by the American Physical Society)

and diameter '*d*' were considered with the assumption of constant s/d throughout the specimen [47]. Deviating from the phenomenological description, a band model was proposed in the percolating clusters [48], independent of microstructure, for estimating the activation energy from the intra-grain energy level splitting due to the finite size of the grains, apart from the usual electrostatic charging energy, E_c. Later on drawing an analogy with Efros-Shklovskii model [49], the role of 'site energy' was conceptualized in introducing correlated Coulomb gap [50]. A detailed study on electrical conductivity in the Ag nanoparticle embedded in glass matrix reported inter-grain hopping of charge carrier [51]. The characteristic temperature of the conductivity, $T_0 = 4\chi s E_c/k_B$, where k_B is the Boltzmann's constant and $\chi = (2m^*\varphi)^{1/2}/\hbar$ is the effective inverse tunneling width. Here, m^* is the effective mass of the charge carrier, and φ is the energy barrier over which the charge must hop, $\hbar$ is the reduced Planck's constant. The expression for $E_c = 2q^2 s/[\kappa d^2(1/2 + s/d)]$, where κ is the permittivity of the insulating phase and q is the electronic charge. The values T_0 was obtained from the slope of the temperature-dependent resistivity plot (Fig. 1.4a). In the scope of the sub-band gap hopping model, the trend in inter-particle separation with increased density of metallic phase, as estimated from the SPR peak analysis (Fig. 1.4b), was correctly calculated from the expression of T_0 and E_c with $\varphi \sim 4$ eV for glass matrix and d, the average size of the Ag clusters ~10 nm. Thus the role of sub-band gap states in the percolating noble metal clusters embedded in the dielectric matrices can be understood for the electrical transport in achieving optical switching.

Electrical conductivity at a threshold noble metal composition, $p > 0.6$ of percolative Au–SiO_2 composite was reported to increase abruptly along with a

Fig. 1.3 Optical transmission as a function of light wavelength for a series of Au–SiO_2 composites. Data are from Ref. [43]. For clarity, the curves are displaced with respect to one another. The theoretical curves are normalized to the experimental values at 0.3 μm. The theoretical values of p are labelled to the right of pairs of curves, whereas the experimental values of p and the film thickness are given above each pair of curves (Ref. [46]; Copyright (1980) by the American Physical Society)

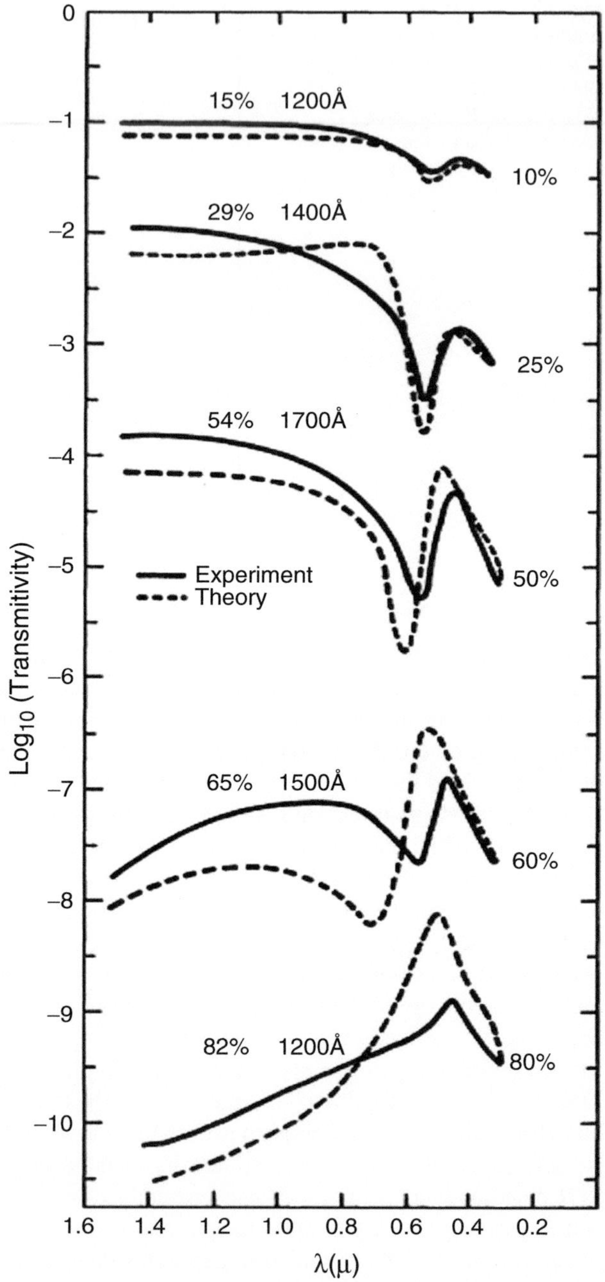

distinct change in the absorption pattern with infrared transmission increasing above the same threshold value. A collective effect of local field enhancement of the Au nanoclusters in combination with the increased light amplitude in resonant cavities formed between the surfaces of the optimally dense Au clusters embedded

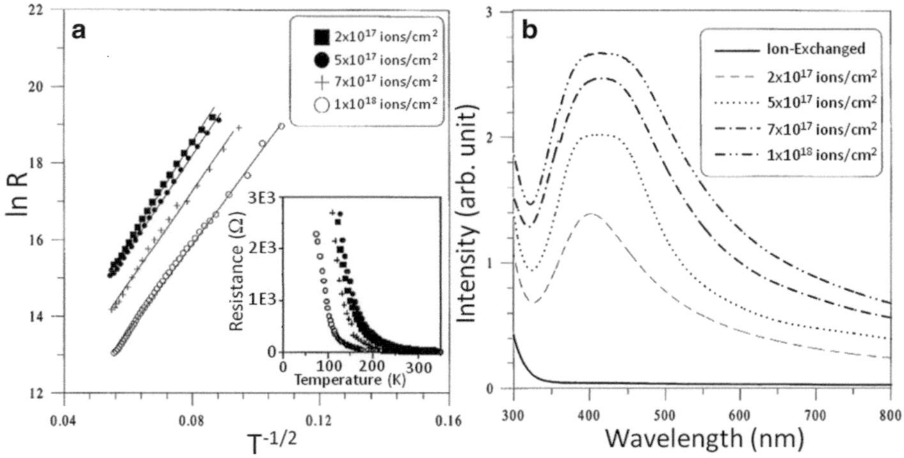

Fig. 1.4 (**a**) Plot of ln *R* versus $T^{-1/2}$ plot at various fluences. The inset shows resistance (*R*) versus temperature (*T*) plot. (**b**) Optical absorption spectra of ion exchanged and irradiated samples at various fluences (Ref. [51]; Copyright (2001) by the Elsevier Science B.V.)

in the SiO_2 matrix in the percolation limit was made responsible for the experimental observations [9]. In a unique study, light induced electrical conduction was reported in randomly-distributed Cu nanoparticles with varying surface coverage on an optically transparent Si_3N_4 [40]. At a percolation threshold of 64 % metal coverage, an optical switching was recorded corresponding to the peak wavelengths of the peak SPR (Fig. 1.5). Finite difference time domain (FDTD) simulation of the dielectric constants based on Drude-Lorentz-Sommerfeld model [52] in case of the Cu nanoparticles/Si_3N_4 system showed (Fig. 1.6) field enhancement at the percolation threshold. Thus, the report demonstrates that Cu nanoparticle-embedded device can detect the SPR by simply monitoring the current. The value of $\chi^{(3)}$ was found to depend on noble metal content in Ag:BiO_2 [53] and Cu:Al_2O_3 [54] composites with highest values achieved at the percolation threshold. The strong $\chi^{(3)}$ is responsible for optical switching in metal composite systems.

1.1.2 A Tunnelling Route to Electrical Transport

A reversible electronic switching effect was observed in Ag nanoparticles embedded in polymer films. A sharp change of up to six orders of magnitude in the current–voltage behavior are highly reversible for these nanocomposite materials, and are defined as threshold switching at a percolation threshold, 0.78 of the metallic coverage (Fig. 1.7) [42]. At the percolation threshold microstructure is characterized by near continuous chain of particles, with no conductive metallic path formed between the electrodes. At a specific gap of 2 nm or less between the particles, electric field of ~ 10^7 V/cm was estimated in between the Ag particles for

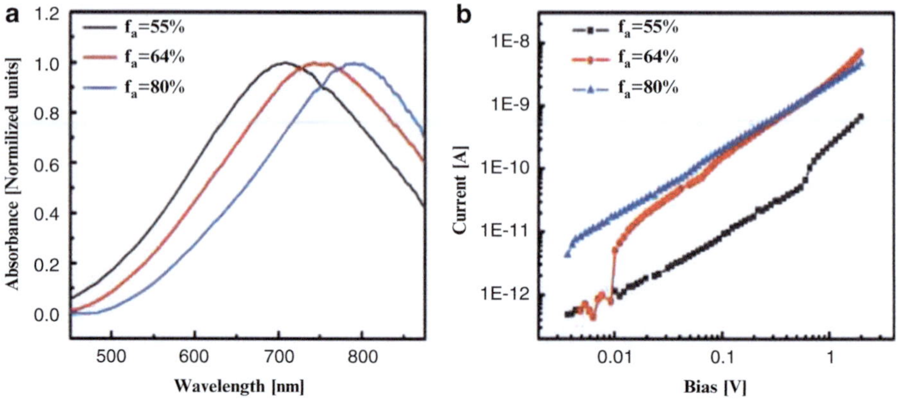

Fig. 1.5 (**a**) LSPR measurement results from samples with a surface coverage, f_a of 55 %, 64 %, and 80 %; each samples has a different resonance peak (55 %: 706 nm; 64 %: 743 nm; 80 %: 789 nm). (**b**) I–V characteristics of samples with a surface coverage of 55 %, 64 %, and 80 %; the plot of $f_a = 64$ % shows a jump of about one order of magnitude (for the percolation threshold) (Ref. [41]@2010, Copyright ©American Optical Society)

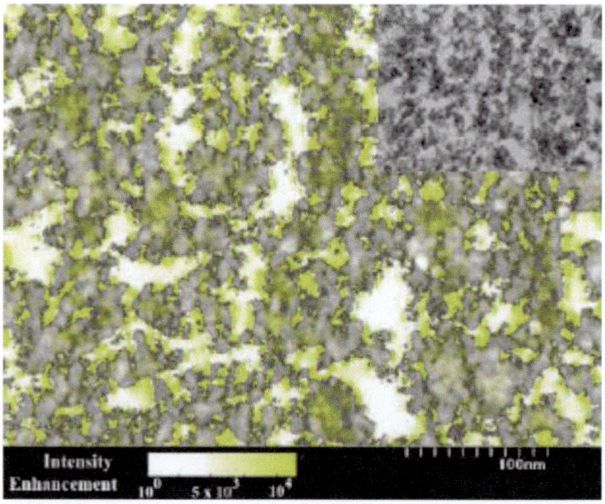

Fig. 1.6 Calculated electric field intensity enhancement in the plane of the deposited nanoparticles (64 % coverage sample). The local field intensity enhancement is depicted in the TEM image (inset) using the linear color bar (Ref. [41] @2010, Copyright©American Optical Society)

an applied voltage of 1 V between electrodes. A tunneling of electronic charges with relatively high current densities would occur through the potential barrier at the Fermi level in the field emission process at such a high field.

Narrow band photoresponsivities up to 60 mV /W into a 100 Ω impedance at 632.8 nm was observed in Ag–Al_2O_3–Al and Al–Al_2O_3–Al metal-insulator oxide-metal (MOM) devices [43]. The wavelength sensitivity of the detector could easily be varied over a wide range by coating the top electrode with different dielectrics, the upper limit being the SPP frequency of the top electrode. Optimum metal

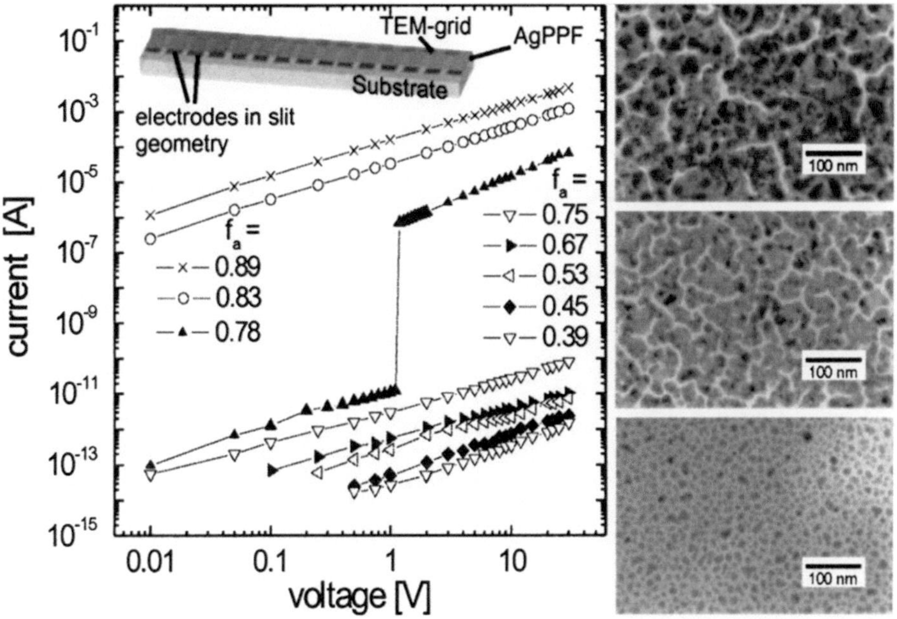

Fig. 1.7 *Left*: *I–V* characteristics of a plasma polymer film containing Ag nanoparticles (AgPPF) with different area filling factors f_a . *Top left*: Sample setup: Coplanar electrode arrangement (slit dimensions 500 μm × 6 nm); placement of TEM grids is matched to film positions (*slits*) for the electrical measurements. *Right*: TEM micrographs of different nanostructural types (*dark*: metal; *light*: plasma polymer) (Ref. [42]; Copyright (2003) by the American Institute of Physics)

grating parameters lead to high efficient SPP excitation. The data showed significantly improved photon assisted tunneling (SPP photo-signal to background ratio) with an applied bias. Inside the metals, the SPP was assumed to decay partly into single particle excitations of electron which could drift to the oxide barrier and tunnel to the counter electrode. The photoresponse was thus determined by the generation of the electrons in metal films at the evanescent field of SPR and by the tunneling rates through the barrier. A mean free path of the hot electrons was estimated to be ~25 nm for the MOM junction. In an interesting report, detection of visible light by two dimensional (2D) Ag–Al_2O_3–Al based MOM tunnel junctions on glass with periodic array of bumps was reported [55]. The SPP at the periodic gratings decayed into single particle excitations, which had sufficient energy to tunnel through the oxide barrier. They obtained a rectification of the photoresponse at 476.9 nm due to the nonlinear *I–V* characteristic of the tunnel junction. In a fresh approach, a strong wavelength-dependent and reversible photoresponse was reported in a two-terminal device using a self-assembled ensemble of Au nanopeapodded silica nanowires under light illumination, whereas no photoresponse was observed for the plain silica nanowires (Fig. 1.8). The switching wavelength was found to match the SPR absorption leading to the generation of electrons in the evanescent field where the photogenerated electrons tunnel through

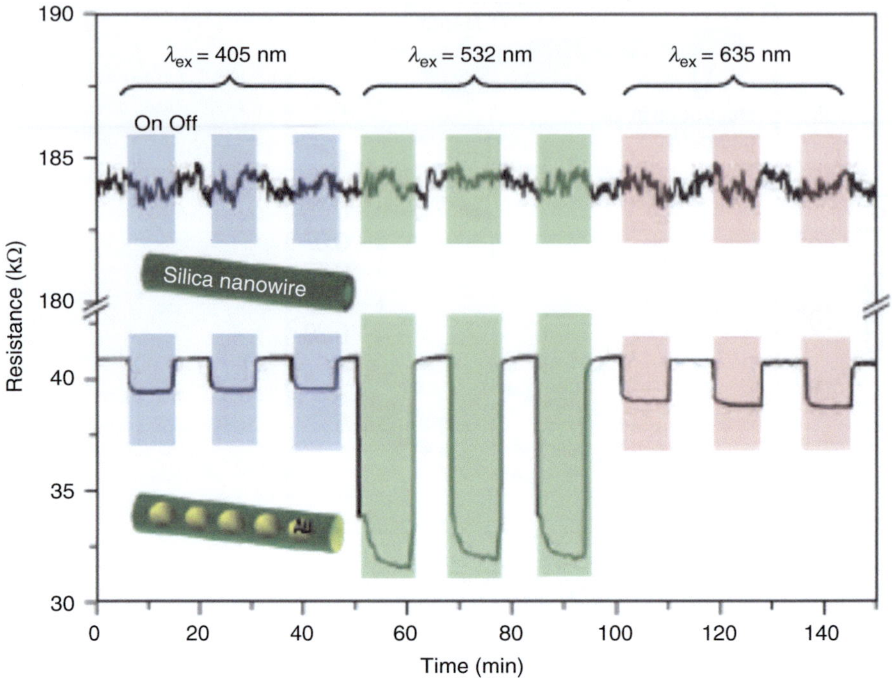

Fig. 1.8 Photoresponse measurements. The room-temperature resistance response as a function of time to light illumination for plain silica nanowires (*upper part*) and gold nanopeapodded silica nanowires (*lower part*). *Shaded* (*pink*, excitation wavelength $\lambda_{ex} = 635$ nm; *green*, $\lambda_{ex} = 532$ nm; *purple*, $\lambda_{ex} = 405$ nm) and *unshaded* regions mark the light-on and light-off periods, respectively (Ref. [33]; Copyright (2006) by the Nature Publication Group)

the oxide nanowires owing to the propagation of SPP. The propagating hot electron in the 1D nanoscale wave guide was observed to travel an inter-particle separation of 100 nm which was one order higher than that reported for MOM structures [43].

1.1.3 A Propagative Surface Plasmon for Electrical Transport

Albeit the origin of percolative pathway or tunneling of electrons at percolation threshold in explaining the observed photocurrent in noble metal-dielectric matrices, a plasmon coupling based propagation of SPP can always be envisaged in the understanding of optical switching. It is particularly important when pure quantum mechanical electron tunneling cannot be used for inter-particle separations above 10 nm. Thus the conception of SPP, where plasmonic coupling played a vital role in the formation of hot electron in evanescent field and its propagation, was the principal driving mechanism for the observed optical switching in noble

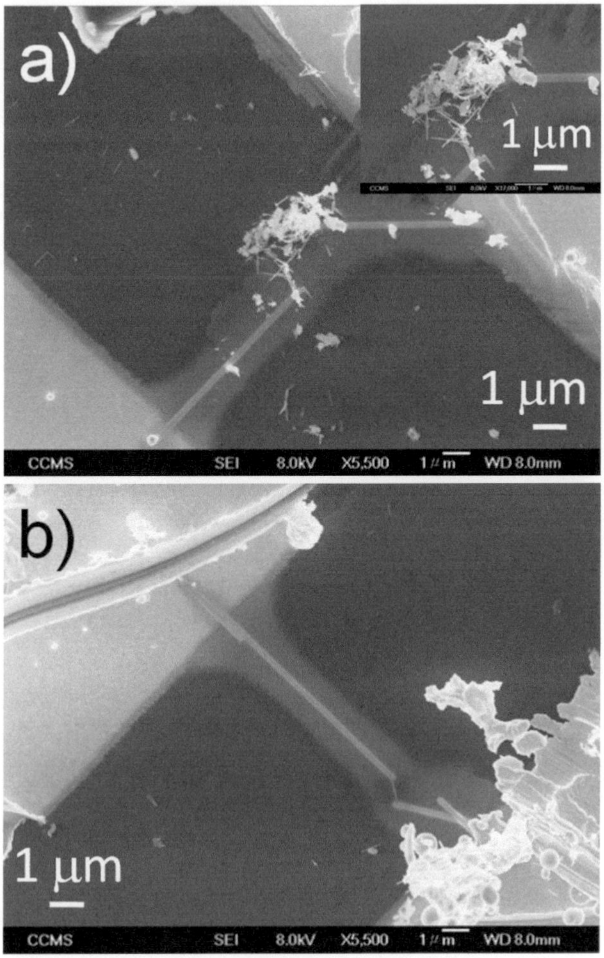

Fig. 1.9 Focused ion beam assisted Pt contact electrodes in (**a**) ensemble (**b**) single GaN nanowire system (Ref. [34]; Copyright (2014) by the Springer)

metal-dielectric matrices. The role of sub-band gap states, however was clearly understood for both percolative model considering 'site energy' correlated to Coulomb gap in the granular system [48, 50] and in tunneling model by definition with electron travelling through the potential barrier at the Fermi level.

In a recent study, optical switching for 1D Au–GaN nanowires and in a single nanowire (Fig. 1.9) was demonstrated [34] for sub-band gap (532 nm; GaN band gap at 365 nm) excitation with surface plasmon resonance peak of Au around 550 nm. Conduction process, responsible for the observed photoresponse (Fig. 1.10), was conceived as SPP assisted transport of electron with propagative resonating electromagnetic radiation coupled to Au nanoclusters generating the carriers (schematic in Fig. 1.11) [56]. Resistivity measurements in the single nanowire showed lowering of resistivity from the bulk value of 0.3 Ω-cm (measured at dark) to 0.05 Ω-cm for the 532 nm illumination as a measure of surface plasmon polariton assisted electrical conduction process [34].

Fig. 1.10 Photoresponse in (**a**) ensemble (**b**) single GaN nanowire samples with periodical dark and 532 nm illumination conditions (Ref. [34]; Copyright (2014) by the Springer)

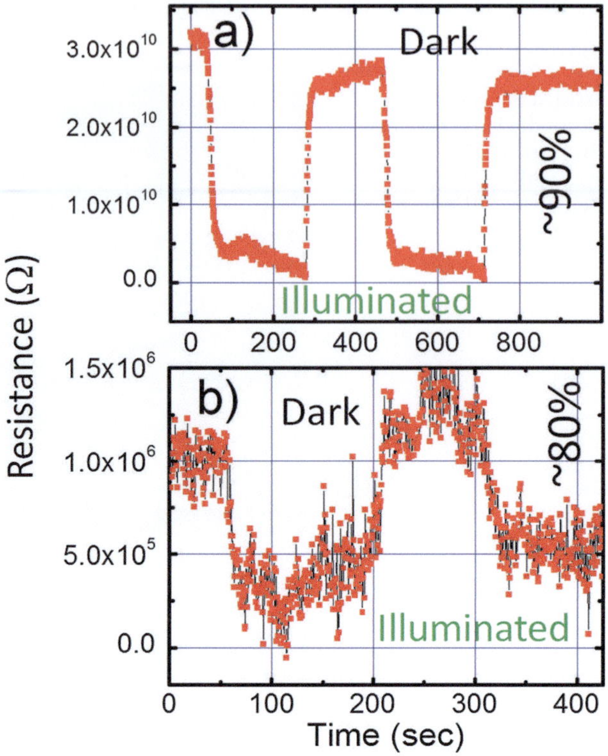

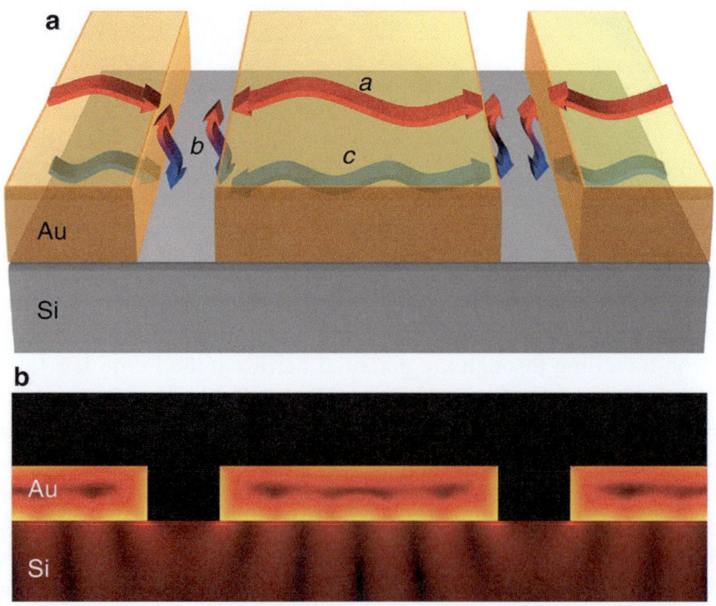

Fig. 1.11 Propagation of surface plasmons on gratings. (**a**) There are three forms of surface plasmon polaritons (SPPs) around each gold pitch: *a*, SPPs oscillating at the top surface of gratings; *b*, SPPs oscillating though the slits and; *c*, SPPs oscillating at the bottom surface of gratings at the Ti–Si Schottky interface. (**b**) Plasmonic heat absorption calculated by FDTD (shown with a logarithmic scale for clarity). Most of the hot electron generation occurs at the bottom surface of the *gold layer* (Ref. [56]; Copyright (2013) by the Nature Publication Group)

In a novel study, optical switching in the Au–Ag@SG system using 808 nm excitation was reported [40]. A change in current was almost twice for 808 nm excitation than that observed for the exposure with 532 nm (Fig. 1.12) indicating a possible role of TPA in the conduction process. The effect of TPA, unlike the conventional wisdom, was realized in its manifestation of doubling the photocurrent with the excitation wavelength of 808 nm as compared to 532 nm excitation where a single photon was involved in the SPR assisted photoresponse. In the sub-band gap picture, the SPP assisted generation of photocurrent does not occur across the band so a quadratic increase in the current cannot be expected. The photocurrent was because of the transport of conduction electrons belonging to Au–Ag system, so the role of TPA was limited to the plasmonic coupling the conduction electrons of Au–Ag system. In order to understand the possible role of competitive inter- and intraband transition in Au and Ag [57], FDTD calculations were analyzed showing very strong dipole coupling of Au–Ag@SG system for both 532 nm and 808 nm excitations for an inter-particle separation between Au and Ag nanoparticles of 2 nm (Fig. 1.13a). On the other hand, absence of electromagnetic coupling for 405 nm for the same system was noteworthy. The observation was correlated to the absence of interband contribution of Ag [56] to the electromagnetic light absorption process in the bimetallic system of Au and Ag. Dipole interaction between Au and Ag nanoparticles was reported active up to a gap of 4 nm for both 532 nm and 808 nm excitations (Fig. 1.13b). The role of larger layer being involved in the electrical conduction process at 808 nm (with low absorption

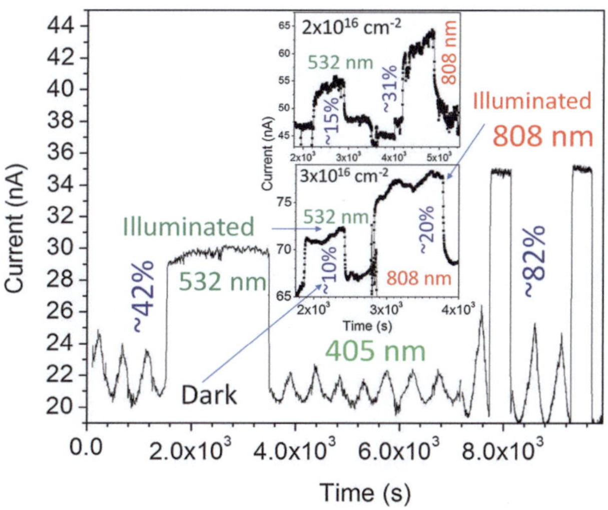

Fig. 1.12 Typical photoresponse of Au–Ag embedded in soda glass sample grown with Au^+ implantation at a fluence of 5×10^{16} ions.cm^{-2} with periodical dark and illumination to different laser wavelengths. Inset shows similar photoresponse of samples grown with Au^+ implantation at fluences of 2×10^{16} and 3×10^{16} ions.cm^{-2}. The studies show a double amount of change in current for 808 nm excitation than that observed for 532 nm exposure (Ref. [40]; Copyright (2015) by the American Institute of Physics)

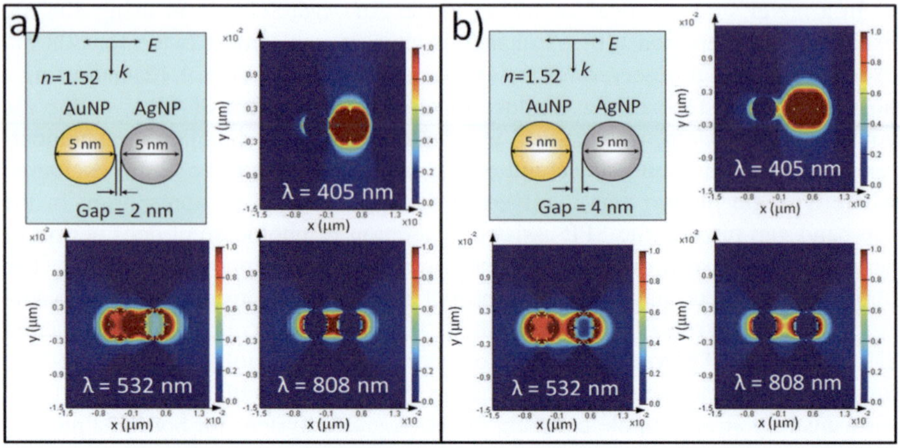

Fig. 1.13 Comparison of the electric field intensity enhancement contours for the interaction of electromagnetic radiation of different wavelengths with 5 nm Au and Ag nanoparticles separated by (**a**) 2 nm and (**b**) 4 nm in a medium with refractive index $n = 1.52$. Here k is the electromagnetic wave propagation vector, E is the electrical field vector and intensity bars indicate $|E|^2$ (Ref. [40]; Copyright (2015) by the American Institute of Physics)

of 808 nm in the bimetallic system) excitation and increasing the photocurrent to almost twice the value of that observed in for 532 nm exposure was ruled out in the presence of strong coupling of electromagnetic wave of both 532 and 808 nm in the Au–Ag system (Fig. 1.13). The observation suggested possibility of optical switching in noble bimetallic nanocluster system where long wavelength with higher skin depth can be used for the communication purpose [40].

References

1. MacDonald KF, Samson ZL, Stockman MI, Zheludev NI (2009) Ultrafast active plasmonics. Nat Photonics 3:55–58
2. Gramotnev DK, Bozhevolnyi SI (2010) Plasmonics beyond the diffraction limit. Nat Photonics 4:83–91
3. Bouhelier A, Bachelot R, Lerondel G, Kostcheev S, Royer P, Wiederrecht GP (2005) Surface plasmon characteristics of tunable photoluminescence in single gold nanorods. Phys Rev Lett 95:267405
4. Atwater HA, Polman A (2010) Plasmonics for improved photovoltaic devices. Nat Mater 9:205–213
5. Sonnichsen C, Reinhard BM, Liphardt J, Alivisatos AP (2005) A molecular ruler based on plasmon coupling of single gold and silver nanoparticles. Nat Biotechnol 23:741–745
6. Verhagen E, Spasenović M, Polman A, Kuipers L (2008) Nanowire plasmon excitation by adiabatic mode transformation. Phys Rev Lett 102:203904
7. Inouye H, Tanaka K, Tanahashi I, Hattori T, Nakatsuka H (2000) Ultrafast optical switching in silver nanoparticle system. Jpn J Appl Phys 39:5132–5133

8. Inouye H, Tanaka K, Kondo Y, Hirao K (1998) Femto second Optical Kerr effect in the gold nanoparticle system. Jpn J Appl Phys 37:L1520–L1522
9. Liao HB, Xiao RF, Fu JS, Yu P, Wong GKL, Sheng P (1997) Large third-order optical nonlinearity in Au:SiO_2 composite films near the percolation threshold. Appl Phys Lett 70:1–3
10. Haglund RF Jr, Yang L, Magruder RH III, Wittig JE, Becker K (1993) Picosecond nonlinear optical response of a Cu:silica nanocluster composite. Opt Lett 18:373–375
11. Maier SA, Kik PG, Atwater HA, Meltzer S, Harel E, Koel BE, Requicha AAG (2003) Local detection of electromagnetic energy transport below the diffraction limit in metal nanoparticle plasmon waveguides. Nat Mater 2:229–232
12. Andrew P, Barnes WL (2004) Energy transfer across a metal film mediated by surface plasmon polaritons. Science 306:1002–1005
13. Maier SA, Atwater HA (2005) Plasmonics: localization and guiding of electromagnetic energy in metal/dielectric structures. J Appl Phys 98:011101
14. Aihara T, Fukuhara M, Takeda A, Lim B, Futagawa M, Ishii Y, Sawada K, Fukuda M (2013) Monolithic integration of surface plasmon detector and metal-oxide-semiconductor field-effect transistors. IEEE Photonics J 5:6800609
15. Bozhevolni S, Volkov VS, Devaux E, Laluet JY, Ebbesen TW (2006) Channel plasmon subwavelength waveguide components including interferometers and ring resonators. Nature 440:508–511
16. Lezec HJ, Degiron A, Devaux E, Linke RA, Martin-Moreno L, Garcia-Vidal FJ, Ebbesen TW (2002) Beaming light from a subwavelength aperture. Science 297:820–822
17. Lewis A, Isaacson M, Harootunian A, Murray A (1984) Development of a 500 Å spatial resolution light microscope I. Light is efficiently transmitted through λ/16 diameter apertures. Ultramicroscopy 13:227–232
18. Pendry JB (2000) Negative refraction makes a perfect lens. Phys Rev Lett 85:3966–3969
19. Fang N, Lee H, Sun C, Zhang X (2005) Sub-diffraction-limited optical imaging with a silver superlens. Science 308:534–537
20. Hell SW (2007) Far-field optical nanoscopy. Science 316:1153–1158
21. Harootunian A, Betzig E, Isaacson MS, Lewis A (1986) Superresolution fluorescence near-field scanning optical microscopy (NSOM). Appl Phys Lett 49:674–676
22. Betzig E, Harootunian A, Lewis A, Isaacson M (1986) Near-field diffraction from a slit: implications for superresolution microscopy. Appl Optics 25:1890–1900
23. Kim S, Jin J, Kim Y-J, Park I-Y, Kim Y, Kim S-W (2008) High-harmonic generation by resonant plasmon field enhancement. Nature 453:757–760
24. McMahon MD, Lopez R, Haglund RF Jr, Ray EA, Bunton PH (2006) Second-harmonic generation from arrays of symmetric gold nanoparticles. Phys Rev B 73:041401
25. Bouhelier A, Beversluis M, Hartschuh A, Novotny L (2003) Near-field second-harmonic generation induced by local field enhancement. Phys Rev Lett 90:013903
26. Hubert C, Billot L, Adam P-M, Bachelot R, Royer P, Grand J, Gindre D, Dorkenoo KD, Fort A (2007) Surface plasmon spectralcharacteristics of second harmonic generation in gold nanorods. Appl Phys Lett 90:181105–181107
27. Nie SM, Emery SR (1997) Probing single molecules and single nanoparticles by surface-enhanced Raman scattering. Science 275:1102–1106
28. Bailoa E, Deckert V (2008) Tip-enhanced Raman scattering. Chem Soc Rev 37:921–930
29. Zhang R, Zhang Y, Dong ZC, Jiang S, Zhang C, Chen LG, Zhang L, Liao Y, Aizpurua J, Luo Y et al (2013) Chemical mapping of a single molecule by plasmon-enhanced Raman scattering. Nature 498:82–86
30. Brongersma ML, Hartman JW, Atwater HA (2000) Electromagnetic energy transfer and switching in nanoparticle chain arrays below the diffraction limit. Phys Rev B 62:R16356–R16359
31. Quinten M, Leitner A, Krenn JR, Aussenegg FR (1998) Electromagnetic energy transport via linear chains of silver nanoparticles. Opt Lett 23:1331–1333

32. Rycenga M, Cobley CM, Zeng J, Li W, Moran CH, Zhang Q, Qin D, Xia Y (2011) Controlling the synthesis and assembly of silver nanostructures for plasmonic applications. Chem Rev 111:3669–3712
33. Hu MS, Chen HL, Shen CH, Hong LS, Huang BR, Chen KH, Chen LC (2006) Photosensitive gold nanoparticle-embedded dielectric nanowires. Nat Mater 5:102–106
34. Dhara S, Lu C-Y, Chen K-H (2015) Plasmonic switching in Au functionalized GaN nanowires in the realm of surface plasmon polatriton propagation : a single nanowire switching device. Plasmonics 10:347–350
35. Kruss S, Srot V, van Aken PA, Spatz JP (2011) Au–Ag hybrid nanoparticle patterns of tunable size and density on glass and polymeric supports. Langmuir 28:1562–1568
36. Karthikeyan B (2012) Optical studies on thermally surface plasmon tuned Au, Ag and Au:Ag nanocomposite polymer films. Spectrochim Acta A Mol Biomol Spectrosc 96:456–460
37. Xu L, Tan LS, Hong MH (2011) Tuning of localized surface plasmon resonance of well-ordered Ag/Au bimetallic nanodot arrays by laser interference lithography and thermal annealing. Appl Optics 50:G74–G79
38. Tuan NA, Mizutani G (2009) Metal-interface second harmonic generation from pt/cu bimetallic nanowire arrays on NaCl(110) faceted templates. e-J Surf Sci Nanotech 7:831–835
39. Hoffbauer MA, McVeigh VJ (1990) Ultrafast optical probes of interface dynamics and structure. Proc SPIE Laser Photoionization Desorption Surf Anal Tech 1208:117
40. Dhara S, Lu C-Y, Magudapathy P, Huang Y-F, Tu W-S, Chen K-H (2014) Surface plasmon polariton assisted optical switching in noble bimetallic nanoparticle system. Appl Phys Lett 106:023101
41. Yang KY, Choi KC, Kang I-S, Ahn CW (2010) Surface plasmon resonance enhanced photoconductivity in Cu nanoparticle films. Opt Express 18:16379–16386
42. Kiesow A, Morris JE, Radehaus C, Heilmann A (2003) Switching behavior of plasma polymer films containing silver nanoparticles. J Appl Phys 94:6988–6990
43. Berthold K, Höpfel RA, Gornik E (1985) Surface plasmon polariton enhanced photoconductivity of tunnel junctions in the visible. Appl Phys Lett 46:626–628
44. Cohen RW, Cody GD, Coutts MD, Abeles B (1973) Optical properties of granular silver and gold films. Phys Rev B 8:3689–3701
45. Bruggeman DAG (1935) Berechnung verschiedener physikalischer Konstanten von heterogenen substanzen. Ann Phys (Leipzig) 24:636–679
46. Sheng P (1980) Theory for the dielectric function of granular composite media. Phys Rev Lett 45:60–63
47. Sheng P, Abeles B, Arie Y (1973) Hopping conductivity in granular metals. Phys Rev Lett 31:44–47
48. Simanek E (1980) The temperature dependence of the electrical resistivity of granular metals. Solid State Commun 40:1021–1023
49. Efros AL, Shklovskii BI (1975) Coulomb gap and low-temperature conductivity of disordered systems. J Phys C 8:L49–L51
50. Sheng P, Klafter J (1983) Hopping conductivity in granular disordered systems. Phys Rev B 27:2583–2586
51. Magudapathy P, Gangopadhyay P, Panigrahi BK, Nair KGM, Dhara S (2001) Electrical transport studies of Ag nano-clusters embedded in glass matrix. Phys B 299:142–146
52. Kreibig U, Vollmer M (1995) Optical properties of metal clusters. Springer, Berlin
53. Zhou P, You GJ, Li YG, Han T, Li J, Wang SY, Chen LY, Liu Y, Qian SX (2003) Linear and ultrafast nonlinear optical response of Ag:Bi_2O_3 composite films. Appl Phys Lett 83:3876–3878
54. del Coso R, Requejo-Isidro J, Solis J, Gonzalo J, Afonso CN (2005) Third order nonlinear optical susceptibility of Cu:Al_2O_3 nanocomposites: from spherical nanoparticles to the percolation threshold. J Appl Phys 95:2755–2762
55. Glass AM, Liao PF, Olson DH, Humphrey LM (1892) Optical metal-oxide tunnel detectors with microstructured electrodes. Opt Lett 7:575–577

56. Sobhani A, Knight MW, Wang Y, Zheng B, King NS, Brown LV, Fang Z, Nordlander P, Halas NJ (2013) Narrowband photodetection in the near-infrared with a plasmon-induced hot electron device. Nat Commun 4:1–6
57. Dhara S, Kesavamoorthy R, Magudapathy P, Premila M, Panigrahi BK, Nair KGM, Wu CT, Chen KH (2003) Quasiquenching size effects in gold nanoclusters embedded in silica matrix. Chem Phys Lett 370:254–260

Chapter 2
Modeling and Interpretation of Hybridization in Coupled Plasmonic Systems

Saïd Bakhti, Nathalie Destouches, and Alexandre V. Tishchenko

Abstract The coupled mode formalism is introduced to provide a phenomenological understanding of the coupling effects in finite systems of particles. Within this approach, a metal nanoparticle can be viewed as an optical resonator and the formation of hybrid modes, resulting from the coupling between particles, can be anticipated. An efficient numerical algorithm is proposed to extract the characteristics (complex poles and amplitudes) of each resonance of the system.

The spectral behavior of the eigen modes of a single metal sphere is analyzed. The redshift and broadening of the different modes with the increase of the particle size and the local refractive index are characterized. Optimal conditions can be found to maximize the particle absorption as well as the near-field enhancement. Sub-radiant and super-radiant hybrid modes of a dimer are identified from the extinction spectrum of each particle. These hybrid modes have different energetic behavior depending on the inter-particle distance, and can then be compared to bounding (attractive) and anti-bounding (repulsive) states. The near-field enhancement resulting from the hybrid mode excitation is maximized by optimizing the dimer geometry and the surrounding refractive index.

The hybrid modes in a quadrumer are identified. For small particles with a reduced coupling via scattering, the system exhibits an anti-crossing behavior of the hybrid modes typical for weakly coupled resonators. When the particles are sufficiently large to induce a strong coupling in the system, the extinction spectrum of the quadrumer present Fano-like resonances, i.e. resonances with an asymmetric line shape. The hybrid modes at the origin of these particular resonances are identified as sub- and super-radiant modes of the system. The sharp Fano-like resonance has a high figure of merit, making such system promising for sensing applications.

Keywords Plasmon resonances • Coupled mode approach • Metal nanoparticles • Hybrid modes • Fano resonances

---

S. Bakhti • N. Destouches • A.V. Tishchenko (✉)
CNRS, UMR5516, Hubert Curien Laboratory, Saint-Etienne F-42000, France
e-mail: tishchen@univ-st-etienne.fr

C.D. Geddes (ed.), *Reviews in Plasmonics 2015*,
DOI 10.1007/978-3-319-24606-2_2

2.1 Introduction

Plasmon resonances induced by the electromagnetic excitation of metal nanoparticles are generally associated with the presence of resonance bands in the spectral response of these nanoparticles [1], and lead to an increase in the local electromagnetic field at resonance wavelengths [2]. Each band corresponding to a particular mode of resonance is characterized by its spectral position and width. These spectral characteristics are strongly dependent on the geometry of the particles as well as on their environment. Changes in the resonance bands result in a red shift of the resonance position as well as in a broadening of the bands when the size of the particle or the local refractive index increases.

In general, the optical response of metal nanoparticles is largely dominated by bands of plasmon resonances. As part of a study of the spectral evolution of these different bands based on certain geometric and environmental parameters, it may be more convenient to deal only with resonance parameters (position and width) rather than considering the whole optical spectra. Different methods allow for retrieving these parameters; they are based on Mie theory for spheres [3], the eigenvalues of surface integrals [4], or the hybridization theory for more complex systems [5, 6]. We propose here to develop a method to extract precisely the resonance parameters from the total optical response of a system of particles based on the T-matrix method and its generalization to multi-particle systems [7, 8]. The main objective of this development is to provide an efficient and flexible numerical tool for the characterization of plasmon resonances together with a phenomenological approach to interpret their physical behavior [9–11].

We first introduce the coupled mode model applied to the phenomenological description of the plasmon resonance amplitudes resulting from an external excitation, allowing a treatment of metal nanoparticles as optical resonators. This model is applied to a single resonance mode and extended to a pair of coupled modes, providing a set of phenomenological parameters including losses, coupling with excitation and mutual coupling coefficients. This approach anticipates the formation of hybrid modes as resulting from the coupling between two plasmon modes. The coupled mode model is applied to the complex valued extinction coefficient of the individual particles, defined as an extension of the classical extinction cross-section. This coefficient, having the same phase than the dipolar moment of the particles, appears as a convenient parameter to identify the nature of the hybrid modes in a coupled system. The solutions of the coupled mode equations being in the form of a singular function of the pulsation, we propose an efficient numerical algorithm to extract the characteristics (complex poles and amplitudes) of each resonance of the system from the rigorous computation of complex valued extinction coefficients. All phenomenological parameters describing the coupling between particles can be deduced from these characteristics.

Within this theoretical framework, we first analyze the spectral behavior of the eigenmodes of a single silver sphere. The redshift and broadening of the different modes with the increase of the particle size and the local refractive index are characterized and optimal conditions are found to maximize the particle absorption as well as the near-field enhancement. The case of a silver dimer is then studied, where sub-radiant and super-radiant hybrid modes are identified from the extinction spectrum of each particle. These hybrid modes have different energetic behaviors depending on the inter-particle distance, and can then be compared to bounding (attractive) and anti-bounding (repulsive) states. Local field enhancement resulting from hybrid mode excitation can be maximized by optimizing the dimer geometry and the surrounding refractive index. Our method is finally applied to silver quadrumers, where hybrid modes exhibit an anti-crossing behavior in the case of weakly coupled small particles. For larger particles, their strong coupling induces the appearance of a Fano-like resonance in extinction spectra. The hybrid modes at the origin of this sharp asymmetric resonance line shape are identified and their behavior is analyzed, showing their potential for sensing applications.

2.2 Coupled Mode Model Applied to Interacting Plasmon Modes

We determine in this section an equation governing the coupled plasmon mode amplitudes, with a few phenomenological parameters characterizing the coupling behavior in simple plasmonic systems. The proposed formalism is well known in the classical coupled mode theory [12] and can be applied to more complex geometries.

A single plasmon mode can be described by a first order differential equation giving the time variations of the mode amplitude $a(t)$ when excited by an incident wave with electric field $f_0(t)$

$$\frac{da(t)}{dt} = -j\mathrm{Re}\{\omega_p\}a(t) - \frac{1}{\tau}a(t) + \kappa f_0(t) \tag{2.1}$$

where τ is the time decay of the plasmon (representing the total losses in the resonant system including absorption as well as re-radiation attributed to scattering), κ is the coupling coefficient and ω_p is the complex resonant angular frequency of the plasmon mode. The real part of ω_p corresponds to the resonance position and the imaginary part to its half width at half maximum (HWHM). When both the incident radiation and the plasmon amplitude oscillate at angular frequency ω, they can be expressed in terms of their modulation amplitudes $\widetilde{f}_0(t)$ and $\tilde{a}(t)$ respectively

$$\begin{cases} f_0(t) = \widetilde{f}_0(t)\exp\{-j\omega t\} \\ a(t) = \tilde{a}(t)\exp\{-j\omega t\} \end{cases} \tag{2.2}$$

Substituting these expressions into Eq. (2.1) and fixing condition $\widetilde{f}_0(t) = 1$ which corresponds to a plane wave with unit amplitude we get the following particular solution of Eq. (2.1) in steady state

$$\tilde{a}(\omega) = \frac{j\kappa}{\omega - \mathrm{Re}\{\omega_p\} + \frac{j}{\tau}} \tag{2.3}$$

Final modulation amplitude $\tilde{a}$ depends on angular frequency ω only and the obtained expression can be written in a very simple form as:

$$\tilde{a}(\omega) = \frac{a_p}{\omega - \omega_p} \tag{2.4}$$

where the coupling coefficient is related with amplitude as $a_p = j\kappa$, and the time decay corresponds to the imaginary part of the complex pulsation $\mathrm{Im}\{\omega_p\} = -1/\tau$. The coupling coefficient quantifies the coupling between the incident excitation and the plasmon mode. Its amplitude provides the coupling strength and its phase corresponds to the oscillator phase at resonance relative to the excitation.

In the case of close particles, plasmon resonances strongly interact, resulting in the formation of hybrid modes in the system. To generalize the phenomenological description to coupled systems, consider first two interacting plasmon modes. One can write two coupled-mode equations on the basis of Eq. (2.1). Two modes with amplitudes $a_1(t)$ and $a_2(t)$ and complex angular eigen frequencies ω_1 and ω_2 are coupled to the incident radiation with coupling coefficients κ_1 and κ_2. Phenomenological equations managing these modes can be written in the following form:

$$\begin{cases} \dfrac{da_1(t)}{dt} = -j\mathrm{Re}\{\omega_1\}a_1(t) + \mathrm{Im}\{\omega_1\}a_1(t) + \kappa_1 f_0(t) + \kappa_{12}a_2(t) \\ \dfrac{da_2(t)}{dt} = -j\mathrm{Re}\{\omega_2\}a_2(t) + \mathrm{Im}\{\omega_2\}a_2(t) + \kappa_2 f_0(t) + \kappa_{21}a_1(t) \end{cases} \tag{2.5}$$

where κ_{12} and κ_{21} are the coupling coefficients between the two modes. The temporal amplitudes of the modes can be expressed in terms of their temporal envelope similarly to the case of a single mode:

$$\begin{cases} a_1(t) = \tilde{a}_1(t)\exp(-j\omega t) \\ a_2(t) = \tilde{a}_2(t)\exp(-j\omega t) \end{cases} \tag{2.6}$$

Substituting expressions (2.6) into Eq. (2.5), we get rid of the fast oscillating term $\exp(-j\omega t)$. The particular solution of the coupled mode equations is found in steady state, when $\tilde{f}_0(t) = 1$:

$$\begin{cases} \tilde{a}_1(\omega) = \dfrac{a_1^+}{\omega - \omega^+} + \dfrac{a_1^-}{\omega - \omega^-} \\ \tilde{a}_2(\omega) = \dfrac{a_2^+}{\omega - \omega^+} + \dfrac{a_2^-}{\omega - \omega^-} \end{cases} \tag{2.7}$$

The solutions appear as a linear superposition of two singular functions, showing that the coupling between the plasmon modes results in the formation of two hybrid modes with complex resonance angular frequencies ω^+ and ω^- distinct from the original modal angular frequencies. The values of the phenomenological parameters a_1^-, a_1^+, a_2^-, a_2^+, ω^+ and ω^- of these hybrid modes can be found by fitting the optical response of the system with a meromorphic function of the pulsation that has two singular points. All phenomenological parameters introduced in Eqs. (2.5) are then related with the resonance parameters of hybrid modes. The coupling coefficients depend on the polar amplitudes as:

$$\begin{cases} \kappa_1 = -j\left(a_1^+ + a_1^-\right) \\ \kappa_2 = -j\left(a_2^+ + a_2^-\right) \end{cases} \tag{2.8}$$

Since these coefficients are intrinsic for the given modes, the sum of hybrid mode amplitudes $a_1^+ + a_1^-$ and $a_2^+ + a_2^-$ must remain constant whatever the strength of the coupling between the modes. The mutual coupling coefficients are

$$\begin{cases} \kappa_{21} = -\dfrac{a_2^+ a_2^-}{a_1^+ a_2^- - a_1^- a_2^+} \Delta\omega^{\pm} \\ \kappa_{12} = \dfrac{a_1^+ a_1^-}{a_1^+ a_2^- - a_1^- a_2^+} \Delta\omega^{\pm} \end{cases} \tag{2.9}$$

with $\Delta\omega^{\pm} = \omega^+ - \omega^-$. Considering that these coupling coefficients reflect the coupling strength in the system and regarding to their expressions, the difference between the complex angular frequencies of hybrid modes, $\Delta\omega^{\pm}$, as well as products of their amplitudes give a direct estimate of the coupling effects in particle aggregates. Similarly to the coupling coefficients, the mutual coupling coefficients are complex parameters whose amplitude provides a direct estimate of the coupling strength between the particles, and their phase corresponds to the phase detuning with which one mode acts on the other.

Finally, the initial and hybrid mode angular frequencies are linked as follows:

$$\begin{cases} \omega^+ = \omega_1 + j\dfrac{a_2^+}{a_1^+}\kappa_{12} \\ \omega^- = \omega_2 + j\dfrac{a_1^-}{a_2^-}\kappa_{21} \end{cases} \tag{2.10}$$

These relations show that with given initial modes, the shift in hybrid modes depends directly on the mutual coupling coefficients, as expected in strongly coupled oscillators [13].

In the limit of uncoupled modes, e.g. by sufficiently distancing two particles so that they no longer interact with each other, the mutual coupling coefficients nullify and the Eq. (2.5) become independent. In this case, vanishing mode amplitudes a_1^- and a_2^+ ensure that the mutual coupling coefficients become zero and that the hybrid mode angular frequencies tend to the initial values ($\omega^+ \rightarrow \omega_1$ and $\omega^- \rightarrow \omega_2$).

2.3 Definition of the Complex Valued Extinction of Nanoparticle Assemblies

In the general case of a system whith interacting particles our strategy is to consider a single parameter characterizing the response of each particle. A practical way to characterize the optical response of a system of particles is to introduce the optical cross-sections [14]. In particular, the extinction cross-section C_{ext}, which expresses the total optical losses in the system due to absorption and scattering processes. In the case of a single particle, this quantity is expressed using the optical theorem:

$$C_{ext} = \frac{4\pi}{k_s|\mathbf{E}_0|^2} \mathrm{Im}\left[\mathbf{E}_0^*.\mathbf{E}_{sca}^{\infty}(\mathbf{e}_k)\right] \tag{2.11}$$

where $\mathbf{E}_{sca}^{\infty}(\mathbf{e}_k)$ is the far-field scattered in the forward direction $\mathbf{e}_k$ of the incident field. The latter is assumed to be a monochromatic incident plane wave $\mathbf{E}_{inc} = \mathbf{E}_0$ exp $(j\mathbf{k}_s \cdot \mathbf{r})$ of angular frequency ω propagating in the homogeneous medium of refractive index n_s with wavevector k_s. For a multi-particle system, the total extinction cross-section can be defined as the sum of the individual particle extinction cross-sections.

In order to generalize the concept of extinction cross-section, we introduce the complex valued extinction $\widetilde{C}_{ext}^{i}$ for each particle and $\widetilde{C}_{ext}$ for the whole system on the basis of the conventional extinction cross-section defined in Eq. (2.11)

$$\widetilde{C}_{ext} = \sum_i \widetilde{C}_{ext}^{i} = -\sum_i \frac{4\pi}{k_s|\mathbf{E}_0|^2} \mathbf{E}_0.\mathbf{E}_{sca}^{i,\infty *}(\mathbf{e}_k) \tag{2.12}$$

where $\mathbf{E}_{sca}^{i,\infty}(\mathbf{e}_k)$ is the far-field scattered by the ith particle. The optical theorem expressed in Eq. (2.11) defines the extinction cross-section as an attenuation of the incident radiation in the forward direction. In contrast to its conventional analogue, the complex valued extinction defined in Eq. (2.12) gives access to the phase information resulting from the interaction of the scattered field with the incident one. Even if the individual contributions $\widetilde{C}_{ext}^{i}$ to the total extinction do not represent any measurable quantity, they allow for characterizing contribution of each particle to the optical response of the whole system. To interpret the resonance curves we study the phase as well as the real and imaginary parts of the individual complex valued extinctions, noting that their imaginary parts correspond to the conventional extinction cross-section. The phase of the complex valued extinction coincides with that of the dipolar moment oscillation.

Complex partial extinctions provide information about relative oscillations of the particle dipole moments. When particles are illuminated with a constant phase of the incident plane wave (Fig. 2.1), the phase difference between their scattering in the far field (and thus between their partial extinction according to Eq. (2.12)) results from a phase difference in their dipolar moments. The complex extinction

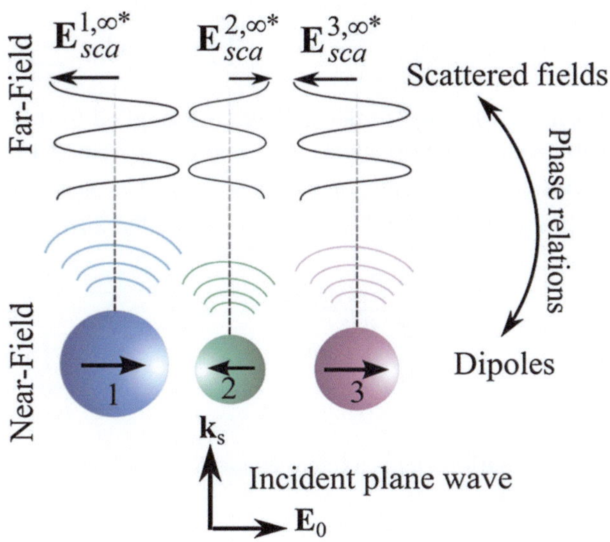

Fig. 2.1 Schematic representation of phase relations between partial scattered fields and oscillating dipolar moments of a system of particles (Reprinted with permission from Ref. [10]. Copyright 2015 American Chemical Society)

can therefore be conveniently used to characterize the radiation of each particle in a coupled system.

2.4 Numerical Extraction of Resonance Parameters

We describe the complex extinction coefficient (or each individual contribution to the total extinction) by a meromorphic function of angular frequency ω, with m poles corresponding to m resonances:

$$\widetilde{C}_{ext}(\omega) = \sum_{j=1}^{m} \frac{a_j}{\omega - \omega_j} + \sum_{k=0}^{\infty} b_k \omega^k \tag{2.13}$$

This function is composed of singular and regular parts. Each singular term corresponds to a particular resonance characterized by its complex amplitude a_j and a complex angular frequency ω_j. The real part of the latter corresponds to the spectral position of the resonance and its imaginary part to the HWHM. We propose a numerical algorithm for the determination of the characteristics of each resonance. The different steps of the approach are the following. First, in order to obtain an analytical function without singular function, $\widetilde{C}_{ext}(\omega)$ is multiplied by polynomial $P_m(\omega)$ which zeros are the poles of $\widetilde{C}_{ext}(\omega)$:

$$P_m(\omega) = \prod_{j=1}^{m} (\omega - \omega_j) = \sum_{k=0}^{m} p_k \omega^k \tag{2.14}$$

Then, an $(m+n)^{th}$ order derivation is applied to this product, with $m+n$ being sufficiently large to cancel all the regular terms in (2.13) for $k < n$. The derivation is performed numerically on product $\widetilde{C}_{ext}(\omega)P(\omega)$ at $m+n$ discrete values of ω, and by applying the $(m+n)^{th}$ order Newton divided difference formula:

$$\sum_{i=0}^{m+n} \frac{\widetilde{C}_{ext}(\omega_i)P_m(\omega_i)}{\prod\limits_{\substack{l=0\\ l\neq i}}^{m+n} (\omega_i - \omega_l)} = \sum_{k=0}^{m} p_k \sum_{i=0}^{m+n} \frac{\widetilde{C}_{ext}(\omega_i)\omega_i^k}{\prod\limits_{\substack{l=0\\ l\neq i}}^{m+n} (\omega_i - \omega_l)} \approx 0 \tag{2.15}$$

The relation yields a linear equation on polynomial coefficients p_k considered as unknowns

$$\sum_{k=0}^{m-1} A_k p_k = -A_m \tag{2.16}$$

where

$$A_k = \sum_{i=0}^{m+n} \left[\omega_i^k \widetilde{C}_{ext}(\omega_i) \prod_{\substack{l=0\\ l\neq i}}^{m+n} (\omega_i - \omega_l)^{-1} \right] \tag{2.17}$$

Taking m different sets of $m+n$ couples of points $\left(\omega_i\,;\, \widetilde{C}_{ext}(\omega_i)\right)$ leads to a system of m linear equations, whose resolution delivers m polynomial coefficients $p_k,\ k = 0, 1, 2, \ldots, m-1$. The last step consists in searching the roots of polynomial $P_m(\omega)$ defined by coefficients p_k. This can be efficiently done by factorizing the companion matrix $\mathbf{C}_m$ of polynomial $P_m(\omega)$ using the QR algorithm [15]:

$$\mathbf{C}_m = \begin{pmatrix} 0 & 0 & \cdots & 0 & -p_0 \\ 1 & 0 & \cdots & 0 & -p_1 \\ 0 & 1 & \cdots & 0 & -p_2 \\ \vdots & \vdots & \ddots & \vdots & \vdots \\ 0 & 0 & \cdots & 1 & -p_{m-1} \end{pmatrix} \tag{2.18}$$

Each eigenvalue of matrix $\mathbf{C}_m$ is equal to one of the polynomial roots and, consequently, to one of the poles of $\widetilde{C}_{ext}(\omega)$.

Once the poles of $\widetilde{C}_{ext}(\omega)$ known, the amplitude of each singular term is obtained from the Lagrangian form of the analytical product, using all $N = 2m + n$ discrete points

$$\widetilde{C}_{ext}(\omega)\prod_{k=1}^{m}(\omega-\omega_k)=\sum_{i=0}^{N}\left\{\left[\widetilde{C}_{ext}(\omega_i)\prod_{k=1}^{m}(\omega_i-\omega_k)\right]\prod_{\substack{l=0\\ l\neq i}}^{N}\frac{\omega_p-\omega_l}{\omega_i-\omega_l}\right\} \tag{2.19}$$

Resonant amplitude a_p is found in the limit $\omega \rightarrow \omega_p$:

$$a_p=\prod_{\substack{k=1\\ k\neq p}}^{m}(\omega_p-\omega_k)^{-1}\sum_{i=0}^{N}\left\{\left[\widetilde{C}_{ext}(\omega_i)\prod_{k=1}^{m}(\omega_i-\omega_k)\right]\prod_{\substack{j=0\\ j\neq i}}^{N}\frac{\omega_p-\omega_j}{\omega_i-\omega_j}\right\} \tag{2.20}$$

The presented algorithm determines efficiently the resonance characteristics, and only few discretization points are required for one searched pole. In practical application, two sets of points can differ by only one single point. Therefore, the algorithm can be implemented by calculating the values $\widetilde{C}_{ext}(\omega_i)$ at only $N = 2m + n$ different angular frequencies. The first linear equation is constructed using the $m + n$ first values of $\widetilde{C}_{ext}(\omega_i)$, the second equation uses values from the 2nd to the $(m + n + 1)^{th}$, and so on.

After truncation of the infinite series in Eq. (2.13) the equation becomes a rational approximation of the complex extinction function. Formally, it is comparable to the Pade approximation [16] which is designed to approximate analytic functions. We are looking for a decomposition which better fits a meromorphic function near its poles. Therefore, we aim at the most accurate search of complex poles ω_j and amplitudes a_j, and apply the algorithm based on the numerical derivation [17] instead of conventionally used Baker's algorithm designed for the best analytic function approximation in a fixed spectral interval.

In the simplest case of a single resonance present in the extinction spectrum, the complex pole ω_p related to this resonance can be directly expressed as function of N discrete values of the complex extinction. Then, taking Eq. (2.15) with $P_m(\omega) = \omega - \omega_p$, it follows immediately:

$$\omega_p=\left[\sum_{i=0}^{N}\frac{\widetilde{C}_{ext}(\omega_i)}{\prod_{\substack{j=0\\ j\neq i}}^{N}(\omega_i-\omega_j)}\right]^{-1}\sum_{i=0}^{N}\frac{\omega_i\widetilde{C}_{ext}(\omega_i)}{\prod_{\substack{j=0\\ j\neq i}}^{N}(\omega_i-\omega_j)} \tag{2.21}$$

Amplitude a_p of a single mode is found as

$$a_p=\sum_{i=0}^{N}\left[\widetilde{C}_{ext}(\omega_i)(\omega_i-\omega_p)\prod_{\substack{j=0\\ j\neq i}}^{N}\frac{\omega_p-\omega_j}{\omega_i-\omega_j}\right] \tag{2.22}$$

In the following examples, the extinction coefficients are computed at N equidistant wavelengths in a spectral range containing all searched resonances. Computations on various systems showed that accurate results (i.e stable resonance poles and amplitudes) are obtained within a number of discrete points only a little larger than the number of poles searched: generally from 3 for a single mode and 5 for two or three modes, to about 15 for about ten modes.

Examples demonstrating capabilities of the method are presented in Fig. 2.2. The extinction coefficient of a silver sphere in vacuum is calculated using the Mie theory and a modified Drude model to fit the metal dielectric permittivity (the used parameters are the same as in Ref. [10]). The plots of extinction denote the presence of a single dipolar resonance in the case of a 10 nm in radius particle (Fig. 2.2a) and both dipolar and quadripolar resonances for a 40 nm in radius particle (Fig. 2.2b). In these two cases, the parameters of equation (2.13) are found using only three discrete points around the single resonance and six points for the particle exhibiting two resonance modes. The singular part is also plotted to demonstrate that the optical response of this system is mainly resonant. The difference between the total extinction and its singular part correspond to the regular part of the meromorphic function (2.13). This regular part, which only adds a slow-varying real part of the extinction, corresponds to non-resonant contributions to the total extinction. Since our analysis is focused on the purely resonant response of metal particles, this regular part will be neglected. Another possibility contained in Eq. (2.13) is to separate the contribution of each mode to the total extinction, as shown in Fig. 2.2c where the extinction due to both dipolar and quadripolar resonance modes are given.

In the context of the physical interpretation of plasmon resonances and their coupling behavior, the proposed method not only allows to compute the resonant parameters of plasmons modes, but also to determine the phenomenological quantities introduced in a previous section. This extraction method combined to the coupled mode model provides a powerful tool to analyze and interpret the resonant effects in coupled systems.

2.5 Eigenmodes of Single Spheres

The polar decomposition of the complex extinction previously described is applied to analyze the resonant behavior of a single silver sphere when its radius R varies. We can note here that different strategies can be adopted for the determination of resonant parameters in the case of a single particle. Since the contribution of each mode to the total extinction can be treated separately by considering the different electric modes in the spherical wave decomposition of the scattered field, the analysis can be efficiently performed by searching a single resonance on each electric mode instead of dealing with several modes in the total extinction of the particle.

Position and width variations of the dipolar resonance are shown in Fig. 2.3a–b versus the particle radius for different refractive indexes of the surrounding medium. These results are in perfect agreement with those obtained in a previous

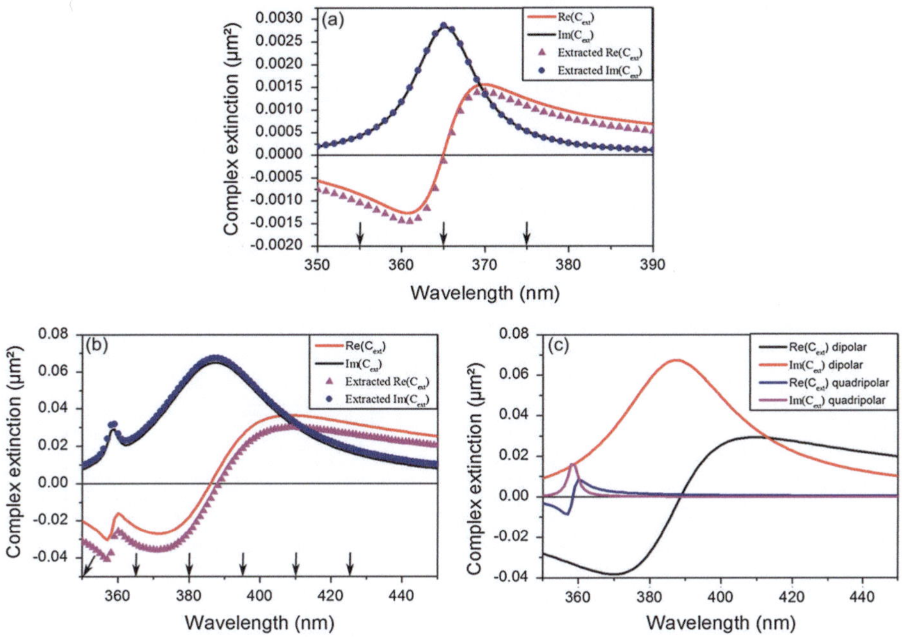

Fig. 2.2 Complex extinction coefficient of a single silver sphere of (**a**) 10 nm in radius and (**b**) 40 nm in radius, with the its singular part reconstructed after the extraction of resonant characteristics. *Arrows* indicate the wavelengths used for the extraction algorithm. The dipolar and quadripolar contributions to the total extinction in the case of a 40 nm in radius particle are given in (**c**) (Reprinted with permission from Ref. [11]. Copyright 2015 Springer)

numerical study [9]. The resonance position (Fig. 2.3a) is monotonically red-shifted by increasing the sphere radius or the surrounding refractive index n, with a nearly linear dependence of the wavelength position with n (at constant sphere radius). The dipolar resonance bandwidth also exhibits a nearly linear dependence with the ambient refractive index. Its variations versus the particle radius (Fig. 2.3b) meanwhile shows that the bandwidth passes by a minimum value at a radius between 10 and 15 nm depending on the surrounding refractive index. Since this resonance bandwidth can be physically related to the damping strength of the resonance mode due to the absorption and radiative losses, we can deduce that there is an optimal particle size for which the total losses of a given mode are minimal. This optimal particle size is reduced when increasing the local refractive index, from $R = 15$ nm for $n = 1$ to $R = 10$ nm for $n = 2$. The same resonant behavior is observed for higher order modes. As shown in Fig. 2.3c–d, the quadrupolar mode resonates at lower wavelength than the dipolar one, with a lower bandwidth and exhibits minimal losses for higher particle radii.

Once the resonance characteristics are found for a given plasmon mode, we propose to determine the contribution of this mode to the optical cross-sections. To do this within the Mie theory, we remind that the lth plasmon mode cross-sections are easily obtained by considering the lth Lorentz-Mie coefficient corresponding to

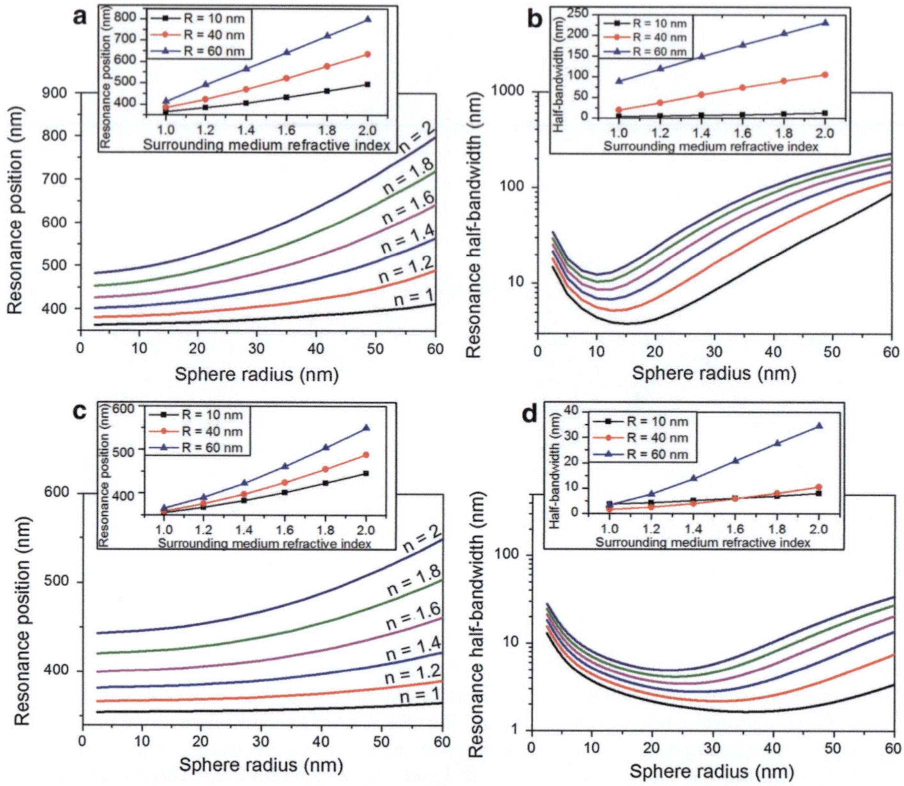

Fig. 2.3 (**a**) Position and (**b**) half-bandwidth of the dipolar resonance computed for a single silver sphere versus its radius, for various surrounding medium refractive indexes n. (**c**) Position and (**d**) half-bandwidth of the quadrupolar resonance computed for a single silver sphere versus its radius, for various surrounding medium refractive indexes n (Reprinted with permission from Ref. [11]. Copyright 2015 Springer)

electric modes instead of their infinite sum. The absorption, scattering and extinction cross-sections computed at the resonance wavelength of the dipolar and the quadrupolar modes are given in Fig. 2.4a and d in function of the particle radius. It appears that a sphere radius of $R = 22$ nm gives a maximum absorption cross-section and in the same time corresponds to the size from which the scattering cross-section becomes greater than the absorption. This particular behavior is observed whatever the mode and the value of the surrounding refractive index. It then appears that an optimal absorption coincides with equal absorption and scattering cross-sections. This result has been pointed out in the case of point dipoles [18] and for realistic particles [19]. In Ref. [19], the following condition has been established to obtain the lth plasmon resonance mode of a sphere as an ideal absorber:

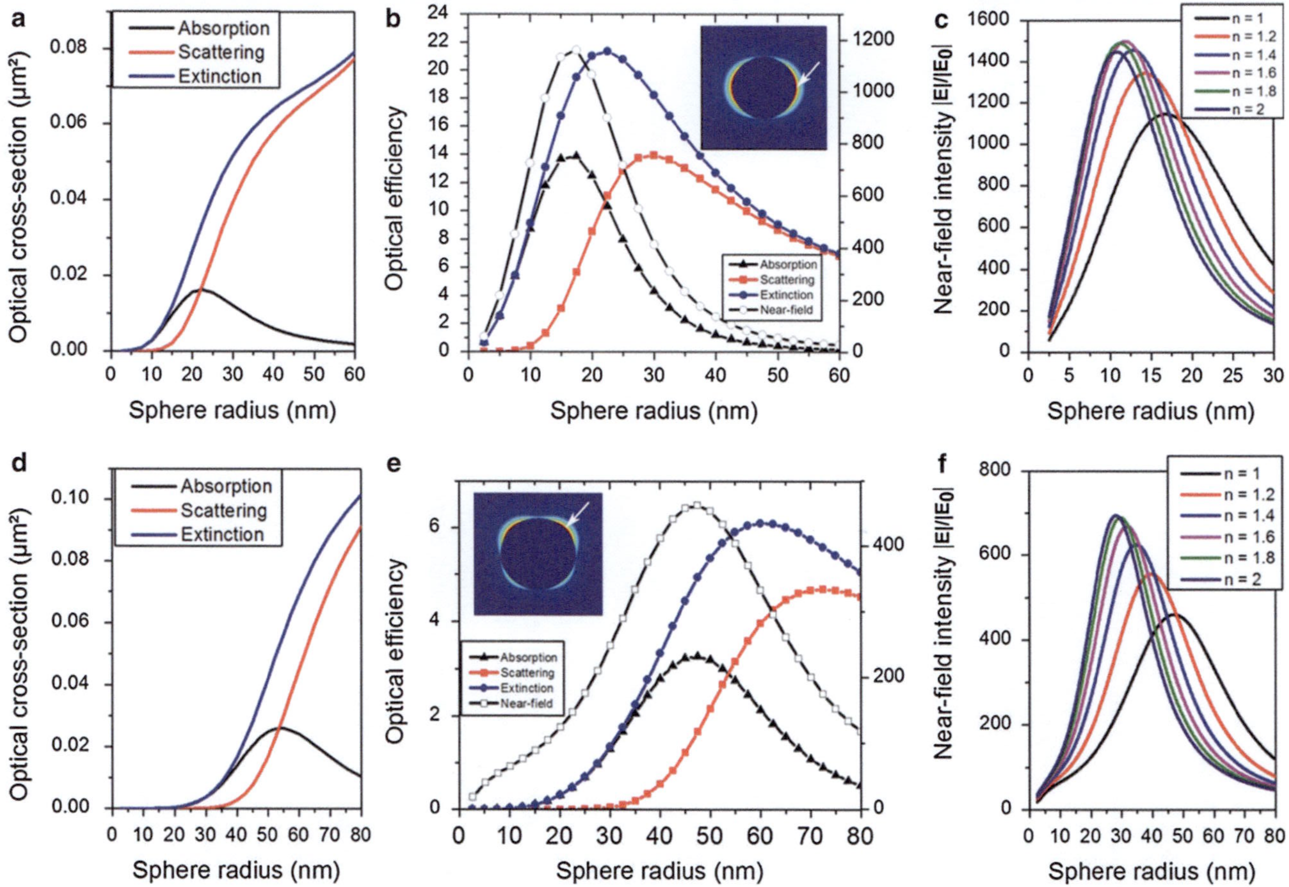

Fig. 2.4 Optical cross-sections of the (**a**) dipolar and (**d**) quadrupolar modes of a silver sphere in vacuum versus radius. Optical efficiencies and maximum near-field intensity of (**b**) the dipolar and (**e**) quadrupolar modes around the particle in vacuum (at the point indicated by the *white arrow* in the near field intensity mapping given in insets and obtained by illuminating a 17 (40) nm in radius sphere at the dipolar (quadrupolar) resonance position with an horizontally polarized and vertically travelling plane wave). Maximum near-field intensity versus radius for various surrounding refractive indexes, for (**c**) the dipolar mode and (**f**) the quadrupolar mode (Reprinted with permission from Ref. [11]. Copyright 2015 Springer)

$$\frac{\varepsilon_1 \tilde{j}_l(k_2R)}{\varepsilon_2 j_l(k_2R)} - \frac{\tilde{h}_l^{(1)}(k_1R)}{h_l^{(1)}(k_1R)} = 0 \tag{2.23}$$

where $j_l(x)$ is the spherical Bessel function, $h_l^{(1)}(x)$ is the spherical Hankel function of the first kind, $\tilde{j}_l$ and $\tilde{h}_l^{(1)}$ are the derivatives $\tilde{j}_l(x) = [xj_l(x)]'$ and $\tilde{h}_l^{(1)}(x) = \left[xh_l^{(1)}(x)\right]'$. ε_1 and k_1 are the dielectric permittivity and the wavenumber of the surrounding medium, ε_2 and k_2 are the dielectric permittivity and the wavenumber of the particle. For a fixed permittivity of the ambient medium and particle, and considering the wavenumbers at resonance position depending on the sphere radius, the condition (2.23) is never strictly satisfied but it has been verified that the optimal absorption of each plasmon mode observed in Fig. 2.4 corresponds to a particle size that minimizes the left-hand of Eq. (2.23).

Other interesting features of the radiative and non-radiative processes at resonance can be drawn from the analysis of the optical efficiencies, i.e. the optical cross-sections normalized by the physical cross-section of the particle (πR^2). Figure 2.4b shows that the absorption and scattering efficiencies for the dipolar mode are maximum for two different radii, R = 17 nm and with R = 30 nm, respectively. When calculating the maximum near-field intensity produced on the particle surface at dipolar resonance, we observe that its value passes by a maximum in coincidence with the absorption efficiency. This property is quite interesting since it provides an easy way to optimize the near-field enhancement. Complementary calculations prove that variations of the maximum near-field intensity are correlated to variations of the absorption efficiency whatever the considered mode and refractive index of the surrounding medium. Figure. 2.4e–f shows that the sizes for which the optical efficiencies and the maximum near-field intensity of the quadrupolar resonance are maximum, are larger than those calculated for the dipolar resonance and decrease when increasing the surrounding refractive index. Interestingly, the maximum near-field intensity can also be optimized by varying the surrounding refractive index. So, the maximal near-field intensity produced by the dipolar mode is obtained for n = 1.5 and R = 13 nm.

2.6 Hybrid Modes in Dimers

2.6.1 Hybrid Modes and Their Energetic Behavior

We start the analysis of plasmon modes in dimers by first considering two spheres of 20 nm (S1) and 10 nm (S2) in radius, with a gap of 5 nm between their surfaces and illuminated by a plane wave polarized parallel to the dimer axis. These particles are small enough to mainly exhibit a dipolar resonance when taken separately. We compute separately, from the rigorous resolution of the multiple scattering problem, the partial complex extinctions $\widetilde{C}^1_{ext}$ and $\widetilde{C}^2_{ext}$ of S1 and S2. Each of these complex quantities exhibits two resonances expected to correspond to hybrid modes resulting from the coupling between the dipolar modes of particles. Fitting their spectral variations with a meromorphic function (being the sum of two singular functions), as described in Eq. (2.7), gives the values of the hybrid pulsations ω^+ and ω^- and amplitude parameters a_1^-, a_1^+, a_2^- and a_2^+ (Fig. 2.5a). The reconstruction of each singular function of Eq. (2.7) from the extracted parameters is shown in Fig. 2.5c–d. These curves correspond to the contribution of each particle to each hybrid mode. The sum of all these functions, which corresponds to the total complex valued extinction of the system, is compared to the spectral variations calculated rigorously in Fig. 2.5b. The good agreement between the curves proves that the phenomenological approach is accurate for this system.

Looking at the separated contributions of S1 and S2 to each hybrid mode in Fig. 2.5c–d also inform on the nature of these modes. When the phase of the singular functions corresponding to S1 and S2 are the same their mode can be interpreted as resulting from dipoles oscillating in phase (Fig. 2.5c). We can note here that the same sign of the real and imaginary part of the singular functions can also be interpreted as in phase dipolar oscillations. This results in a large dipolar moment and a highly radiative system whose mode is qualified of super-radiant. When the real or the imaginary parts have opposite signs (Fig. 2.5d), the dipolar moments of both particles oscillate out-of-phase leading to a small resulting dipolar moment of the dimer and to a poorly radiative system. The mode is then said sub-radiant. The resonance bandwidth of a given mode is related to the total losses in the system. In the case of a sub-radiant mode, the reduced resulting dipolar moment results in lower radiative losses, leading to a sharper resonance bandwidth. For a super-radiant mode the opposite effect is observed, and the highly radiative behavior results in high radiative losses and hence to a broader resonance bandwidth. These behaviors appear in the two hybrid modes of the dimer.

Similar results are obtained when considering an incident polarization perpendicular to the dimer axis, where both sub-radiant and super-radiant modes are identified from the fitted resonance characteristics.

An important feature in coupled systems is the dependence of the coupling strength with the distance separating the particles. Following the phenomenological analysis in the previous section, the mutual coupling coefficients can serve to quantify the interaction between nanoparticles. Figure 2.6 depicts the modulus of both coupling and mutual coupling coefficients as a function of the gap between the particles, for an incident polarization parallel or perpendicular to the dimer axis. Coupling coefficients κ_1 and κ_2 are intrinsic characteristics of each sphere and then are expected to be independent of the dimer configuration. However, the plotted values show a slight decrease of these coefficients with increasing the gap. This can be interpreted as a consequence of coupling between dipolar modes and higher order modes. Indeed, the phenomenological analysis of the system only consider the ideal case of coupling between two (here dipolar) modes. Actually, coupling

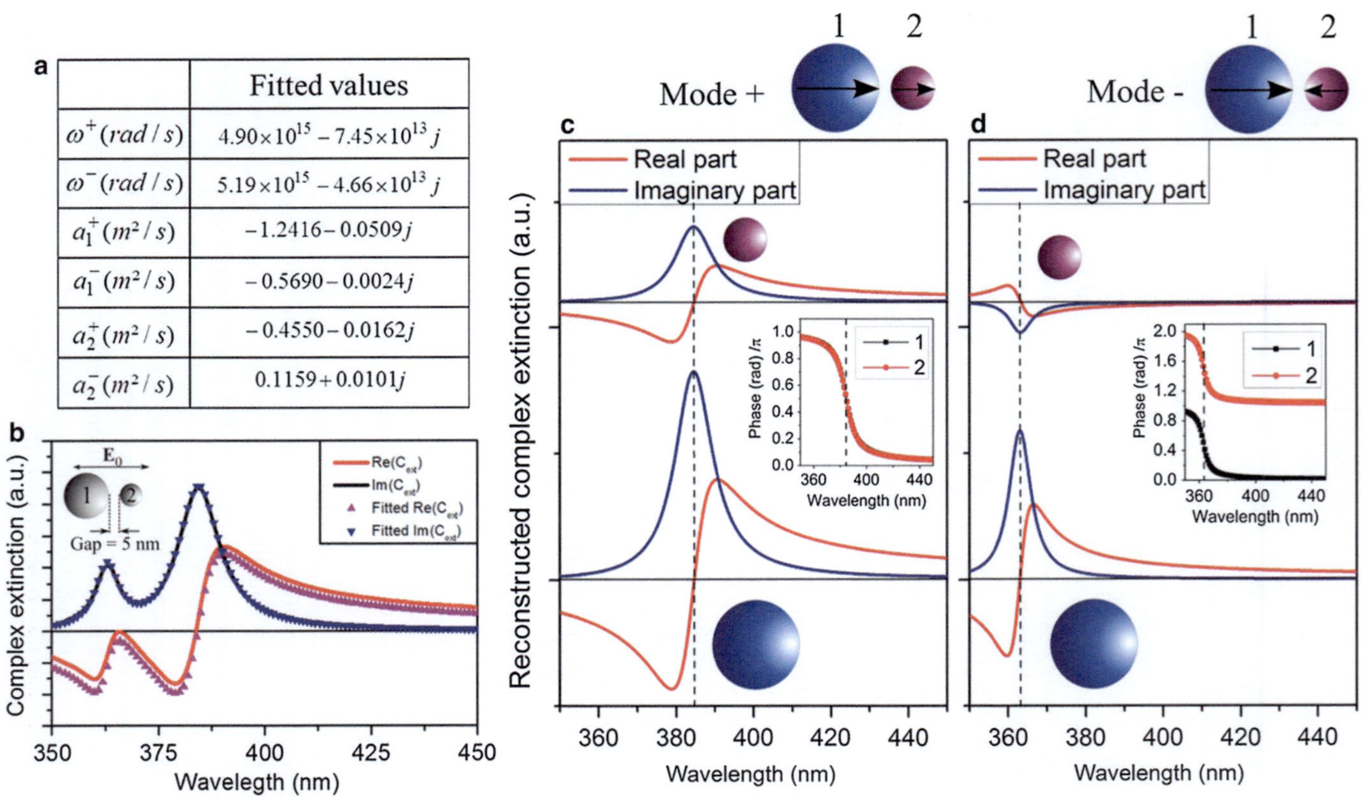

	Fitted values
$\omega^{+}(rad/s)$	$4.90\times10^{15}-7.45\times10^{13}j$
$\omega^{-}(rad/s)$	$5.19\times10^{15}-4.66\times10^{13}j$
$a_1^{+}(m^2/s)$	$-1.2416-0.0509j$
$a_1^{-}(m^2/s)$	$-0.5690-0.0024j$
$a_2^{+}(m^2/s)$	$-0.4550-0.0162j$
$a_2^{-}(m^2/s)$	$0.1159+0.0101j$

Fig. 2.5 (**a**) Resonance characteristics of hybrid modes for a dimer composed of 20 and 10 nm in radius spheres, with a gap of 5 nm (**b**) Comparison of the total complex extinction cross-sections calculated from a rigorous approach and from singular functions of **c–d**. (**c**) and (**d**) Contribution of each particle to the extinction of each hybrid mode (these two graphs have the same ordinate amplitude) (Reprinted with permission from Ref. [10]. Copyright 2015 American Chemical Society)

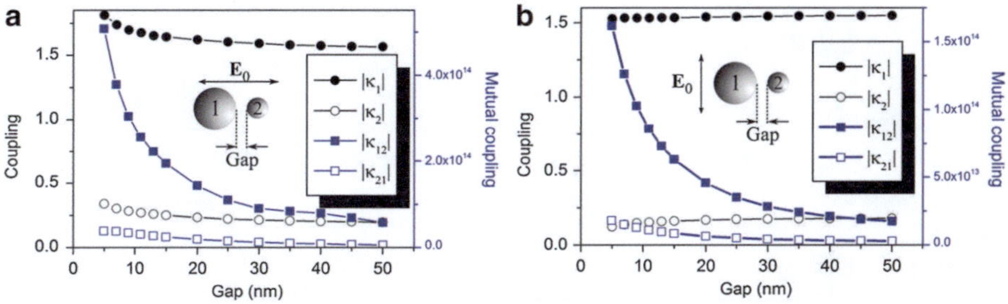

Fig. 2.6 Coupling (on the left y-axis) and mutual coupling (on the right y-axis) coefficients versus the gap between particles for an incident polarization (**a**) parallel or (**b**) perpendicular to the dimer axis (Reprinted with permission from Ref. [10]. Copyright 2015 American Chemical Society)

between dipolar and quadrupolar modes also occurs, inducing disparities between expected and computed phenomenological values.

The computed mutual coupling coefficients κ_{12} and κ_{21} show for each incident polarization a fast decreasing coupling strength when increasing the gap between particles. By comparing the amplitudes for each polarization, the coupling strength in the case of a parallel polarization appears to be larger (by a factor of about 3) than in the perpendicular case. When considering each sphere as oscillating dipoles, this result indicates a better coupling in the case of parallel dipolar moments. Regarding again the coupling coefficients, their values are more perturbed for a parallel incident polarization. In this case and for close particles, relatively strong coupling effects can be expected between dipolar and higher order modes, compared with a perpendicular polarization where the coupling coefficients are more stable.

We can note here the different orders of magnitude between the coupling and mutual coupling coefficients. This difference comes from the different physical inputs to which they apply: $\kappa_1 f_0(t)$, $\kappa_2 f_0(t)$, $\kappa_{12} a_2(t)$ and $\kappa_{21} a_1(t)$. These terms have the same dimension and their comparison could inform about the relative importance of the coupling and mutual coupling effects in the resonant behavior of the system.

The hybrid modes resulting from the coupling between particles resonate at complex pulsations ω^+ and ω^- different from the pulsation of initial modes ω_1 and ω_2. Their determination gives an energetic diagram of the system, as shown in Fig. 2.7a. The latter highlights different behaviors of hybrid modes. The two modes resonating at higher wavelengths (and lower energy) are red-shifted when increasing the inter-particle coupling. They correspond to an energetically favorable configuration of the dipolar moments of the particles, *i.e.* when the dipoles are attractive. These modes can be seen as bounding states of the system (Fig. 2.7c). The modes resonating at lower wavelengths correspond to a repulsive configuration of dipoles and can be seen in turn as anti-bounding states of the system. Another interesting feature is that the resonance position of bounding states tends to the dipolar mode position of the larger particle, whereas the resonance position of anti-bounding states tends to dipolar mode position of the smaller particle.

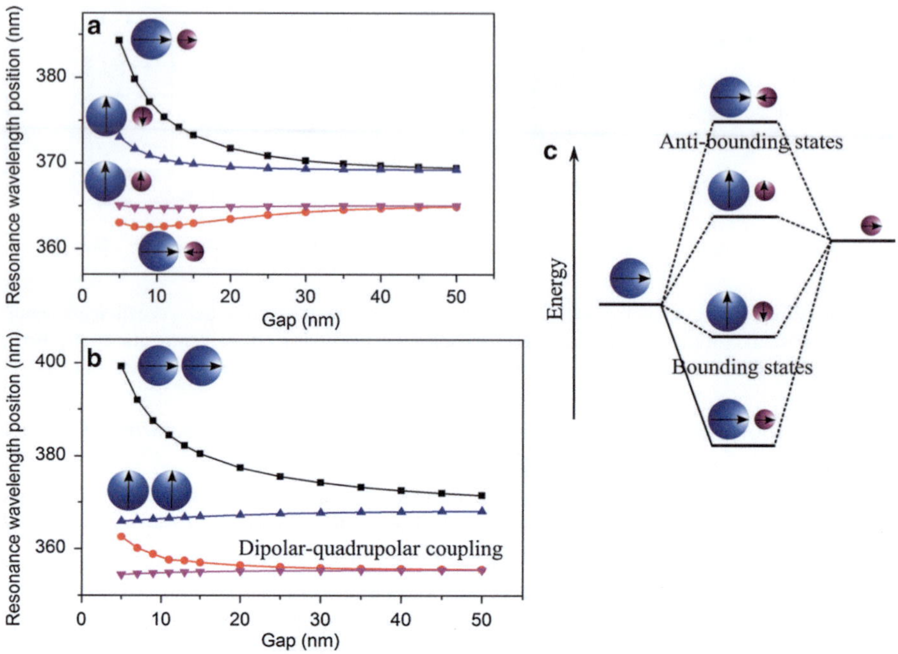

Fig. 2.7 Resonance position of hybrid modes versus the gap between particles for a dimer composed of (**a**) 10 and 20 nm in radius spheres and (**b**) identical spheres 20 nm in radius. (**c**) Schematic representation of the energetic repartition of hybrid modes (Reprinted with permission from Ref. [10]. Copyright 2015 American Chemical Society)

Consider now the special case of homo-dimers, composed of identical particles. In such a case, the initial modes have identical coupling coefficients κ_0 with the incident excitation and resonate at the same pulsation ω_p. Anticipating the mutual coupling coefficients to be equal ($\kappa_{12} = \kappa_{21} = \kappa$) as well as the mode amplitudes ($a_1(t) = a_2(t) = a(t)$) the coupled mode equations (2.5) can be simplified in a single equation:

$$\frac{da(t)}{dt} = -j\mathrm{Re}\{\omega_p\}a(t) + \mathrm{Im}\{\omega_p\}a(t) + \kappa a(t) + \kappa_0 f_0(t) \tag{2.24}$$

Including in Eq. (2.24) the expression of the mode amplitude given by Eq. (2.4), yields in steady-state with the condition $\widetilde{f}_0(t) = 1$

$$\widetilde{a}(\omega) = \frac{j\kappa_0}{\omega - \omega_p - j\kappa} = \frac{a^+}{\omega - \omega^+} \tag{2.25}$$

As a consequence, the coupling between two identical particles leads to a single excited hybrid mode with a complex pulsation $\omega^+ = \omega_p + j\kappa$ and a resonance amplitude $a^+ = j\kappa_0$. This hybrid mode has the same amplitude as the initial mode, and the resonance pulsation is only shifted from the initial position ω_p by the mutual coupling term $j\kappa$. Obviously, the resonant response of both particles is exactly the same and the hybrid mode corresponds systematically to a super-radiant mode (in-phase oscillations). Out-of-phase oscillations of dipolar moments appear as forbidden (*i.e.* dark) modes.

The energetic diagram of a homo-dimer composed of 20 nm in radius spheres is plotted in Fig. 2.7b. In addition to the identified hybrid modes, different modes resulting from a coupling between dipolar and quadrupolar initial modes are also represented. For a complete description of these interactions with higher modes, the phenomenological description used has to be generalized to the case of M cross-coupled modes.

2.6.2 Near-Field Enhancement

Let us characterize the spectral behavior of hybrid modes in a silver homo-dimer. Figure 2.8a provides the optical cross-sections of a silver dimer composed of 10 nm in radius spheres with a gap of 5 nm between the particles and illuminated with an incident plane wave polarized with an angle of 45° relative to the dimer axis. These spectra show the presence of three distinct resonance modes. The mode resonating at the lowest energy is identified as the longitudinal hybrid mode with in-phase oscillation of the dipolar moments oriented along the dimer axis. The second marked peak corresponds in turn to the transverse hybrid mode with in-phase oscillation of dipolar moments oriented perpendicularly to the dimer axis. The third mode, not studied in following, resonating at the highest energy originates from a quadrupolar mode. The resonance position and half-bandwidth of the transverse and longitudinal dipolar hybrid modes are reported in Fig. 2.8b–c in function of the gap between the spheres and for various surrounding medium refractive indexes. As expected from the previous analysis, the parallel hybrid mode is red-shifted and slightly broadened when decreasing the inter-particle distance, and therefore when increasing the coupling strength between the particles. By contrary, the parallel dipolar hybrid mode is blue-shifted and slightly narrowed when decreasing the particle gap between the particles. As in the case of a single particle, the resonance position of each hybrid mode has an almost linear dependence with the variation of the local refractive index.

A particular interest of hybrid modes in coupled particle systems relies on the large near-field enhancement that occurs at resonance compared to a single particle. Especially, the hot-spot resulting from the parallel hybrid mode excitation and located in the dimer gap may reach several thousand times the incident wave intensity. Still considering the same dimer with the same illumination condition, the variation of the near-field enhancement at the gap center, obtained at the longitudinal mode resonance position and for various local refractive indexes is reported in Fig. 2.9a. A maximum field enhancement occurs for $n = 1.5$, which corresponds to the surrounding medium refractive index maximizing the dipolar

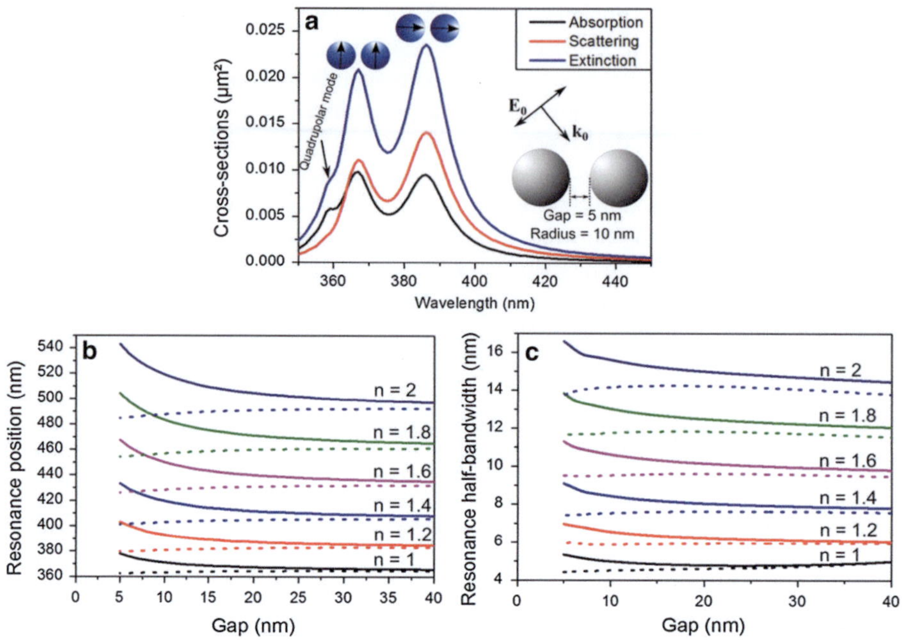

Fig. 2.8 (**a**) Optical cross-sections of a silver dimer composed of 10 nm in radius spheres with a gap of 5 nm between the particles and illuminated with an incident plane wave polarized with an angle of 45° relative to the dimer axis. The instantaneous dipolar moment orientations of the two dipolar hybrid modes are represented on their respective resonance peak. Resonance (**b**) position and (**c**) half-bandwidth of the dipolar hybrid modes versus the gap between the particles, for different local refractive indexes. *Lines* correspond to the longitudinal hybrid mode and *dotted lines* correspond to the transverse one (Reprinted with permission from Ref. [11]. Copyright 2015 Springer)

mode near-field intensity for a single particle. As in the same way as for uncoupled modes, the local field enhancement varies when considering different particle sizes. Figure 2.9b plots the near-field intensity for various particle sizes and local refractive indexes, keeping the ratio $R/gap = 2$. An optimal condition to optimize the hot-spot intensity is then found for $R = 11$ nm and $n = 1.5$.

2.7 Weak and Strong Coupling in Quadrumers

2.7.1 Weak Coupling in Small Size Systems

Here we consider a quadrumer composed of two pairs of homo-dimers. This system is schematically described in Fig. 2.10. The first dimer D1 is composed of two 20 nm in radius spheres disposed vertically with a gap of 20 nm. The second dimer

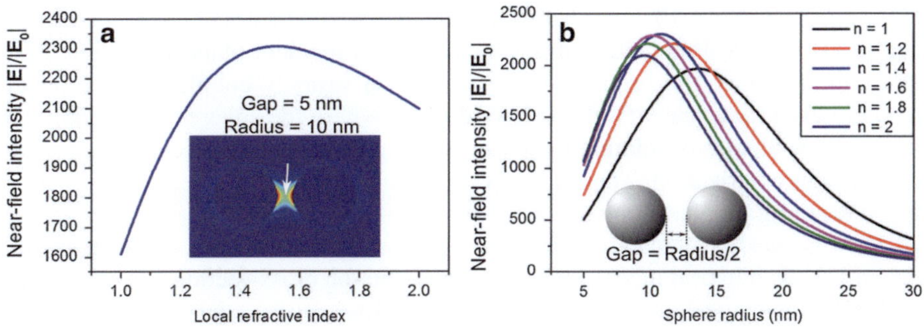

Fig. 2.9 (**a**) Near-field enhancement versus the local refractive index, at the gap center (shown in inset) of a dimer composed of 10 nm in radius spheres with a gap of 5 nm between the particles and obtained at the longitudinal hybrid mode resonance position and for various local refractive indexes. (**b**) Near-field intensity versus radius for different surrounding medium refractive indexes. The ratio between the particle radius and the gap is kept equal to 2. In all cases, the incident plane wave is polarized with an angle of 45° relative to the dimer axis (Reprinted with permission from Ref. [11]. Copyright 2015 Springer)

Fig. 2.10 Extinction cross-section of the system sketched on the right composed of two interacting dimers, versus the gap of the horizontal dimer (Reprinted with permission from Ref. [10]. Copyright 2015 American Chemical Society)

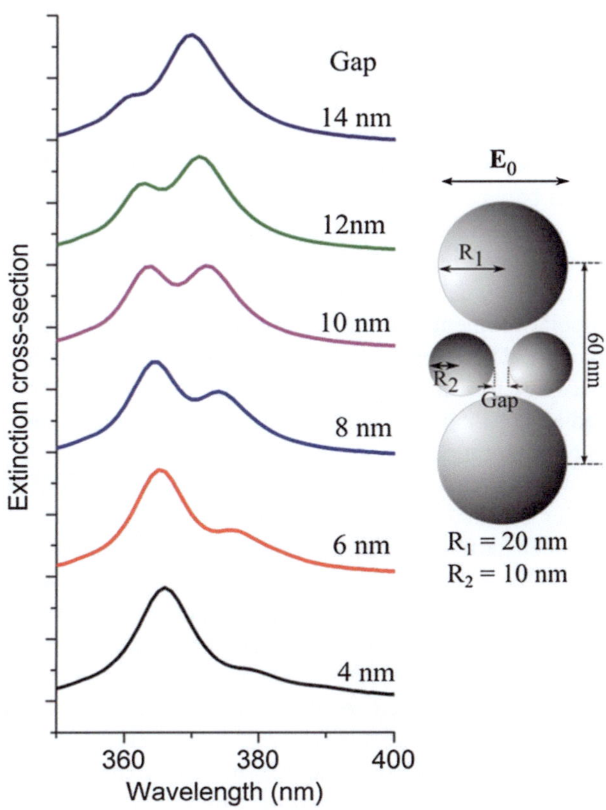

D2, composed of two 10 nm in radius spheres disposed horizontally, is placed in the gap of D1. When illuminating this system with an incident plane wave polarized along the D2 axis, the original dimer modes are strongly coupled. The original mode in each dimer results in one hybrid mode where the spheres composing the dimer act as dipoles oscillating in-phase, with dipolar moments oriented parallel to the incident polarization. Then each dimer can be viewed as a single dipole with a given resulting dipolar moment.

The rigorously calculated extinction spectra are plotted in Fig. 2.10 versus the gap of D2. Two hybrid modes appear in the extinction cross-section. As in the case of simple dimers, the coupling between the two dimer modes leads to the formation of hybrid modes with distinct resonance positions.

In order to determine the coupling characteristics of such a quadrumer, we compute the partial complex extinction cross-sections of each dimer $\widetilde{C}_{ext}^{D1}$ and $\widetilde{C}_{ext}^{D2}$. As expected, each partial cross-section exhibits two modes located at the same pulsations, which can be fitted with the meromorphic functions of Eq. 2.7. The resonance parameters of the hybrid modes (for a gap of 8 nm in D2) are gathered in Fig. 2.11a and the reconstructed singular functions corresponding to the contribution of each dimer to both hybrid modes are shown in Fig. 2.11c–d. The sum of all these contributions, deduced from the phenomenological approach, is in good agreement with the total complex valued extinction cross-section calculated rigorously (Fig. 2.11b). This proves the validity of the coupled-mode approach for this kind of system too.

As previously, the nature of hybrid modes can be identified through the phase shifts between the dipolar oscillations deduced from the comparison of the sign of the real part of singular functions plotted in Fig. 2.11c–d. For the hybrid mode resonating at pulsation ω^-, the dipolar modes associated to D1 and D2 oscillate in phase (Fig. 2.11c). This hybrid mode corresponds therefore to a high resultant dipolar moment making this mode highly radiative and can be qualified as super-radiant. The hybrid mode resonating at ω^+ is characterized by a $\pi/3$ phase shift between the dipolar oscillations of the two dimers. This configuration is less radiative than a super-radiant mode, and then will be qualified as a sub-radiant mode.

When decreasing the gap between particles in D2, the resonance position of the dimer original mode is red-shifted due to an increase in the coupling strength between particles. This shift changes the position of hybrid resonances. Figure 2.12a shows the energetic diagram of both coupled and isolated dimers in function of the gap in D2. The hybrid modes in the coupled system exhibit an anti-crossing behavior of the energy branches at a gap around 11 nm, where the difference between the two hybrid mode positions presents a minimum. For gaps below 11 nm, the extracted resonance parameters indicate that the super-radiant mode resonates at the lowest wavelengths and conversely for the higher frequency mode. The anti-crossing point induces an energy inversion of the hybrid modes and for higher gaps, the situation is reversed. This behavior is characterized by a sign inversion in the mutual coupling coefficients at crossing point. Since the hybrid

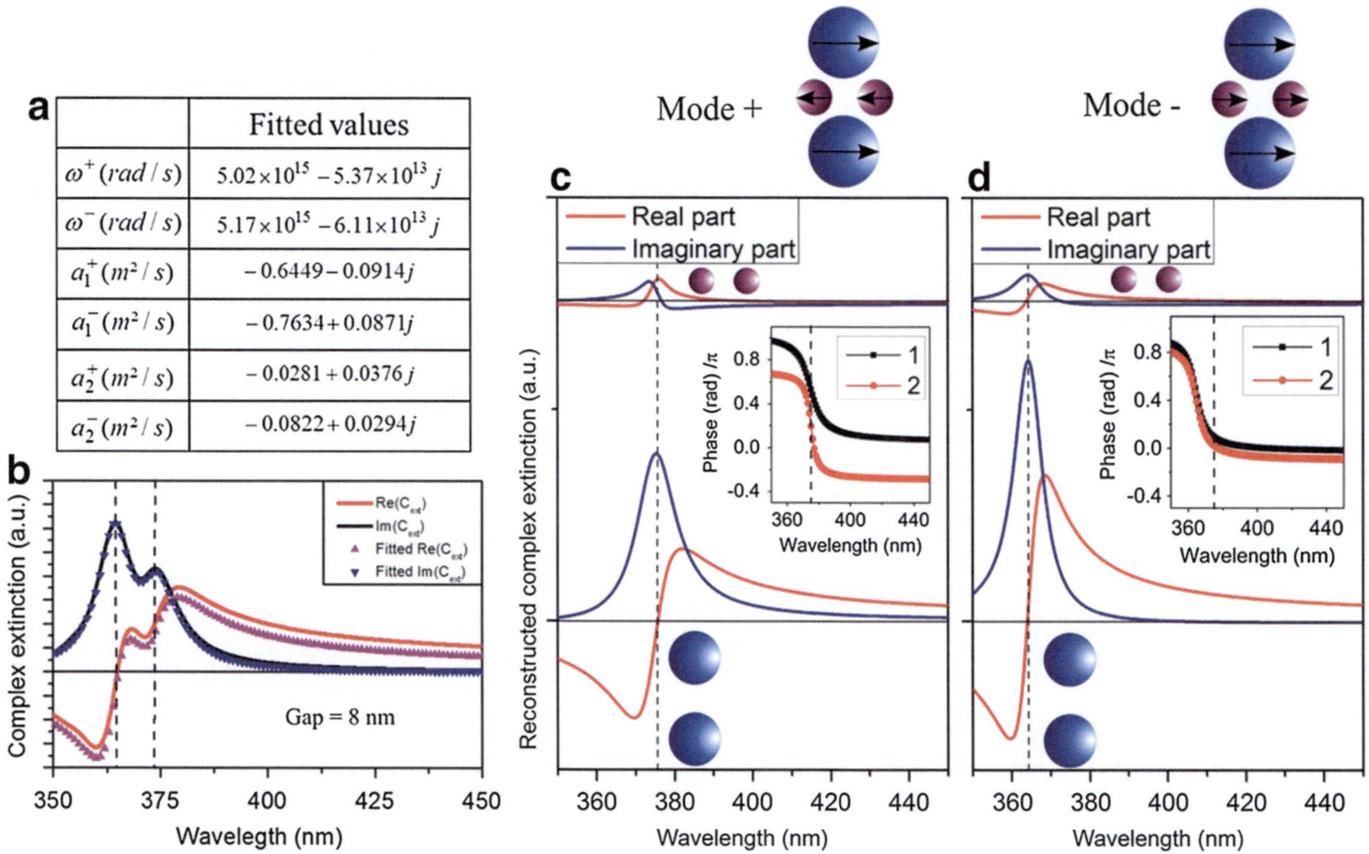

Fig. 2.11 (**a**) Resonance characteristics fitted from the partial complex valued extinction cross-section of each dimer. The gap of D2 is fixed to 8 nm. (**b**) Total complex extinction cross-section from rigorous calculation and the phenomenological modeling. (**c**) and (**d**) Singular functions giving the contribution of each dimer to each hybrid mode (Reprinted with permission from Ref. [10]. Copyright 2015 American Chemical Society)

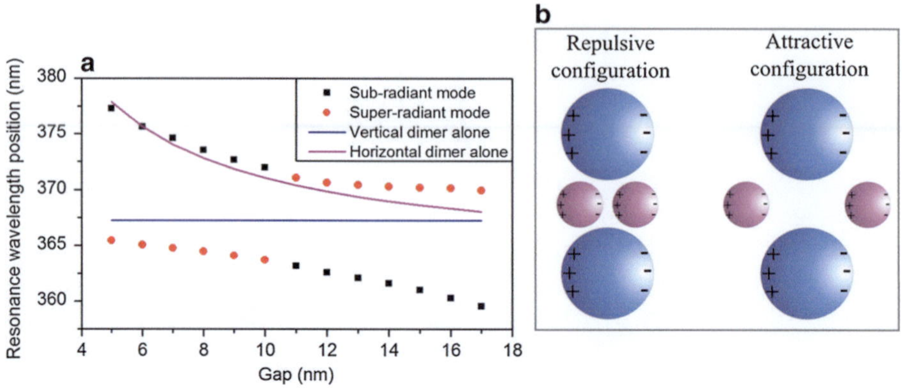

Fig. 2.12 (**a**) Energetic diagram of coupled and isolated dimers and (**b**) an illustration of Coulomb interactions between the particles for a small and a large gap of D2 (Reprinted with permission from Ref. [10]. Copyright 2015 American Chemical Society)

pulsations are directly related to these mutual coupling coefficients (see Eq. (2.10)), a sign inversion in the latter induces an energy inversion in the former. As illustrated in Fig. 2.12b, a qualitative interpretation of the energy inversion is given by considering changes in Coulomb interactions between the particles [20]. For short gaps between the particles in D2, repulsive forces act between D1 and D2 particles. When increasing the gap in D2, the electrostatic interactions are reversed, leading to attractive forces between spheres of D1 and D2.

2.7.2 *Hybridization and Fano-Like Resonances in Strongly Coupled Systems*

We deal now with a quadrumer composed of larger particles where stronger coupling between plasmon modes occurs. This system is schematically described in Fig. 2.13, and consists in a quadrumer composed of identical spherical particles 30 nm in radius. As previously, this system can be viewed as a pair of vertical (D1) and horizontal (D2) homo-dimers. The extinction spectra of the quadrumer are plotted in Fig. 2.13 versus the gap of D2, when illuminating with an plane wave polarized along the D2 axis. These spectra exhibit Fano resonances, *i.e.* asymmetric resonant line profiles, noting also the presence of a quadrupolar resonance at about 360 nm.

In order to analyze the coupling effects in the quadrumer, we consider Eqs. (2.5) extended to four coupled equations to describe the mutual coupling between the four particles. Due to the symmetry of this system and considering the given incident polarization, these equations can be reduced into two coupled equations describing the resonance behavior of the quadrumer

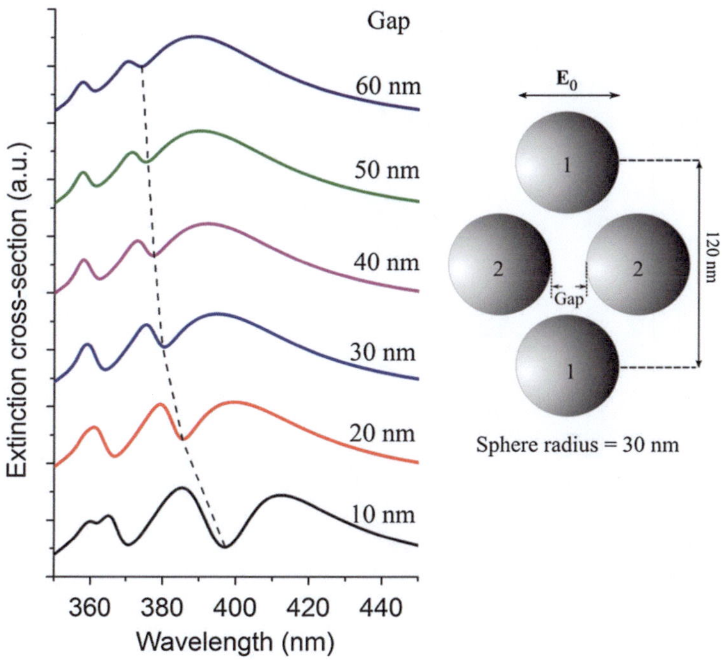

Fig. 2.13 Extinction cross-section of the system sketched on the right composed of two interacting dimers, versus the gap of the horizontal dimer. The *dashed line* indicates the position of the Fano dips in the extinction spectra (Reprinted with permission from Ref. [10]. Copyright 2015 American Chemical Society)

$$\begin{cases} \dfrac{da_1(t)}{dt} = -j\mathrm{Re}\{\omega_0\}a_1(t) + \mathrm{Im}\{\omega_0\}a_1(t) + \kappa_0 f_0(t) + \kappa_{11}a_1(t) + 2\kappa_{12}a_2(t) \\ \dfrac{da_2(t)}{dt} = -j\mathrm{Re}\{\omega_0\}a_2(t) + \mathrm{Im}\{\omega_0\}a_2(t) + \kappa_0 f_0(t) + \kappa_{22}a_2(t) + 2\kappa_{12}a_1(t) \end{cases} \tag{2.26}$$

where $a_1(t)$ and $a_2(t)$ are the resonance amplitudes in the dimers D1 and D2, respectively. ω_0 is the original complex pulsation of the particles' dipolar resonance and κ_0 is their coupling coefficient with the incident radiation. κ_{11} and κ_{22} are the mutual coupling coefficients between the particles in the dimers D1 and D2 respectively, κ_{12} is the mutual coupling coefficient between the D1 and D2 particles. The coupled equations (2.26) describing the system anticipates the formation of two different hybrid modes from the coupling of the initial dipolar resonances of the particles.

In order to determine the coupling characteristics of such a quadrumer, we compute the partial complex extinction cross-sections of each dimer $\widetilde{C}_{ext}^{D1}$ and $\widetilde{C}_{ext}^{D2}$. The resonance parameters of the hybrid modes (for a gap of 60 nm in D2) are gathered in Fig. 2.14a and the reconstructed singular functions corresponding to

a

	Fitted values
$\omega^{+}(rad/s)$	$4.90\times10^{15}-3.33\times10^{14}j$
$\omega^{-}(rad/s)$	$5.07\times10^{15}-3.47\times10^{13}j$
$a_1^{+}(m^2/s)$	$-6.4132-2.86558j$
$a_1^{-}(m^2/s)$	$-0.8037+0.7487j$
$a_2^{+}(m^2/s)$	$-8.8770-1.7735j$
$a_2^{-}(m^2/s)$	$0.6553-0.3418j$

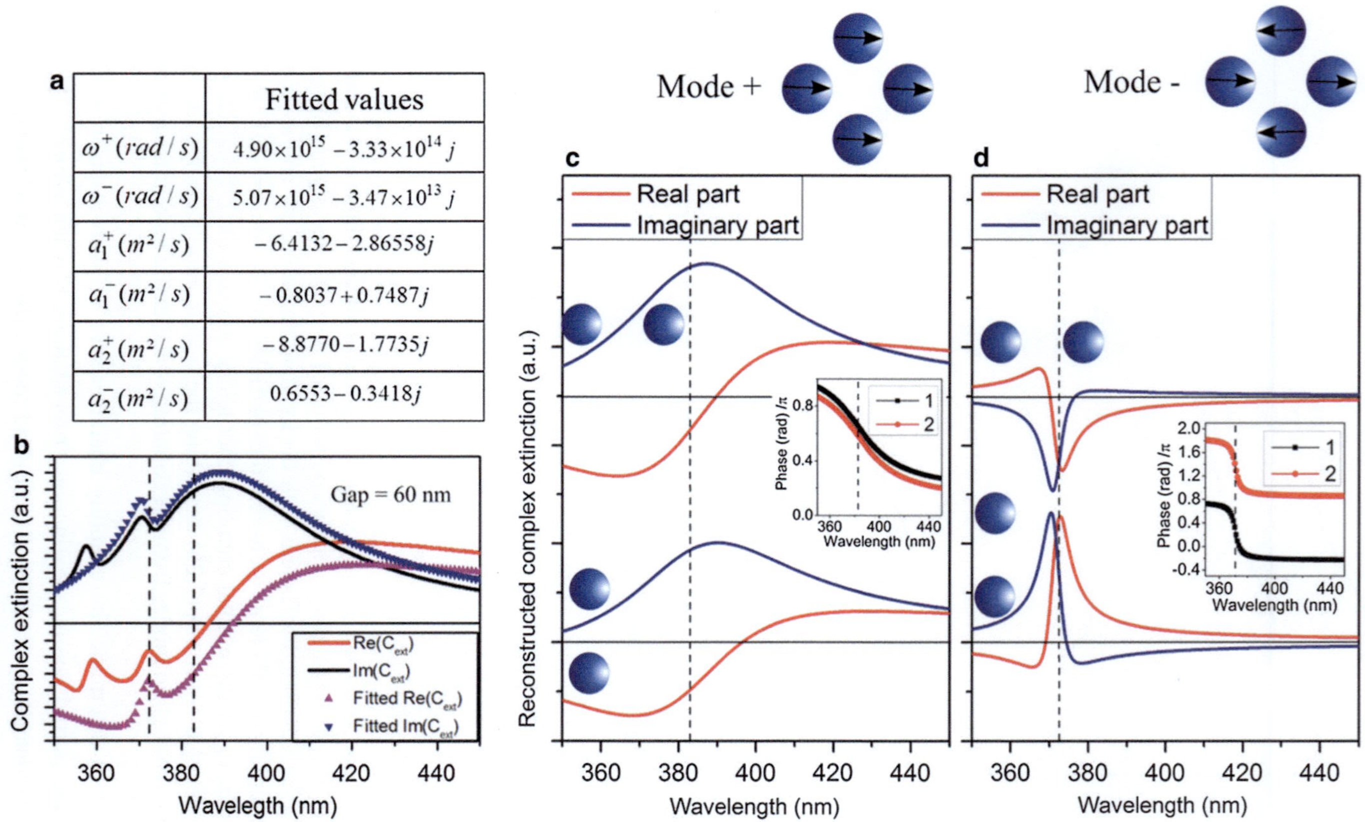

Fig. 2.14 (**a**) Resonance characteristics fitted from the partial complex valued extinction cross-section of each dimer. The gap of D2 is fixed to 60 nm. (**b**) Total complex extinction cross-section from rigorous calculation and the phenomenological modeling. (**c**) and (**d**) Singular functions giving the contribution of each dimer to each hybrid mode (these two graphs have the same ordinate amplitude) (Reprinted with permission from Ref. [10]. Copyright 2015 American Chemical Society)

the contribution of each dimer to both hybrid modes are shown in Fig. 2.14c–d. The sum of all these contributions, deduced from the phenomenological approach, is in good agreement with the total complex valued extinction cross-section calculated rigorously (Fig. 2.14b), except for the quadrupolar mode (at about 355 nm) not taken into account in our model.

As previously, the nature of hybrid modes can be identified through the phase shifts between the dipolar oscillations deduced from singular functions plotted in Fig. 2.14c–d. For the hybrid mode resonating at angular frequency ω^+, the dipolar modes associated to D1 and D2 oscillate nearly in phase (Fig. 2.14c). This hybrid mode corresponds therefore to a high resultant dipolar moment making this mode highly radiative and qualified of super-radiant. The hybrid mode resonating at ω^- is characterized by an out-of-phase oscillation of the dipolar moments, making this mode poorly radiative that can be qualified of sub-radiant. A particular feature of this system comes from highly asymmetric line profiles in the contribution of both dimers to the sub-radiant mode. The superposition of these asymmetric line profiles with the contribution of the super-radiant mode explains the Fano line-shape in the total extinction of the quadrumer.

Figure 2.15b shows the plot of relative dipolar moment phases for the two hybrid modes in function of the gap in D2, revealing two different coupling regimes. When increasing the gap between the particles in D2 (corresponding to weak coupling), the super-radiant mode tends to a zero phase shift between all the dipolar moments resulting in a large bandwidth (Fig. 2.15c), and the sub-radiant mode tends to a π phase shift with a reduced bandwidth. This regime results in a sharp Fano resonance in the extinction spectra (Fig. 2.13). In the strong coupling regime, *i.e.* when reducing the gap in D2, a reduction in the phase shifts of the super-radiant mode is observed while its bandwidth is reduced. Contrariwise, the phase difference between the dipolar moments of the sub-radiant mode tends to be reduced and its bandwidth increases. Then under this regime the super-radiant and sub-radiant modes become respectively less and more radiative. Moreover, the plot of the spectral position of these hybrid modes (Fig. 2.13b) shows that the difference between their spectral positions decreases with the gap in D2. The combination of these two modes forms a broader Fano resonance in the structure total extinction spectrum.

An actual interest of Fano resonances in plasmonic structures lies in their extreme sensitivity to the local environment, making such resonances well suited for sensing applications. The sensing capabilities of plasmon resonances can be evaluated through their Figure of Merit [21] (FoM) defined as the ratio of the plasmon energy shift per refractive index unit change in the surrounding medium, divided by the width of the resonance band. Since the FoM of a given mode mainly depends on its spectral width, the Fano resonance of the quadrumer in the weak coupling regime appears as the best configuration for sensing because of its sharp width compared to the Fano resonance in the strong coupling regime. The plot of the quadrumer extinction spectra in the weak coupling regime (Fig. 2.16a) shows the redshift of the asymmetric Fano line profile when increasing the local refractive

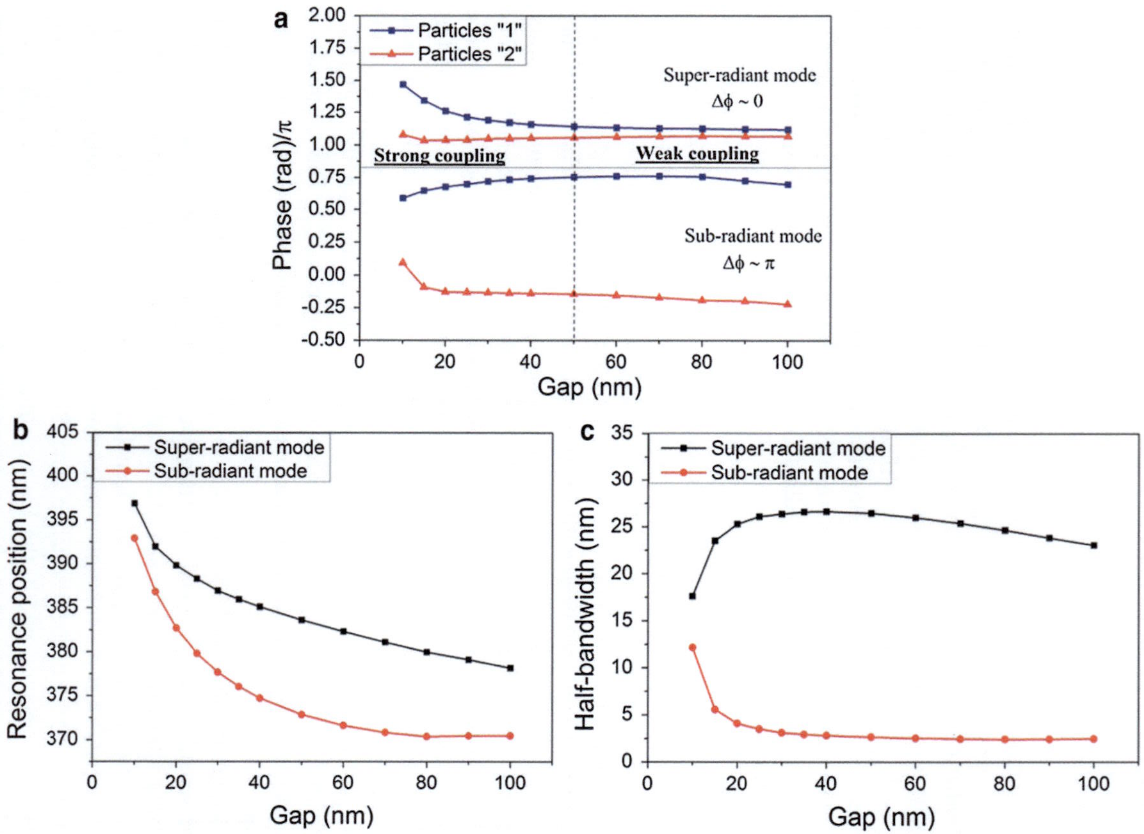

Fig. 2.15 (**a**) Relative phase between dipolar moment oscillations, (**b**) resonance position and (**c**) half-bandwidth of the hybrid mode in the quadrumer versus the gap of the dimer D2 (Reprinted with permission from Ref. [10]. Copyright 2015 American Chemical Society)

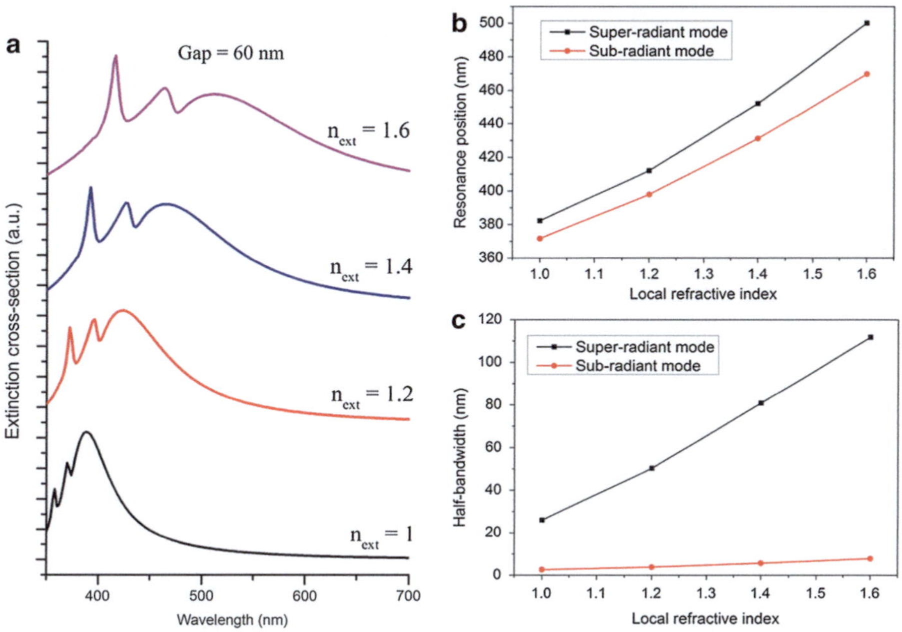

Fig. 2.16 (**a**) Extinction spectrum of the quadrumer for different surrounding medium refractive indexes. (**b**) Resonance position and (**c**) half-bandwidth of the hybrid modes versus the local refractive index. A gap of 60 nm in the dimer D2 is fixed (Reprinted with permission from Ref. [10]. Copyright 2015 American Chemical Society)

index. The plot of the two hybrid modes resonance positions in Fig. 2.16b exhibits a nearly identical energy shift when increasing the surrounding medium refractive index, while the spectral width of the super-radiant mode increases much more significantly than the sub-radiant one (Fig. 2.16c). The latter then retains its low radiative behavior. Using a linear regression of the energy shift versus the local refractive index and the sub-radiant mode FWHM in vacuum, the FoM of the Fano resonance in the weak coupling regime is estimated to 19.4, which is much larger than in previously studied finite structures [22, 23]. By comparison, the FoM in the strong coupling regime is estimated to 6.7, demonstrating the advantage of a weak coupling configuration for sensing applications.

2.8 Conclusion

In this chapter we developed an efficient method for analyzing natural resonant modes of unique particles and their coupling in simple structures. The principle of this method is based on an analytical representation of the optical response of nanostructures, as a unique function of the angular frequency. Thus, each mode of resonance is characterized by a complex pole in the optical response corresponding to its frequency and amplitude. An extraction algorithm is described first for the determination of the characteristics in the case of single resonances, then it is generalized to the presence of multiple resonances in the extinction spectrum of a multi-particle system.

The analytical representation of singular functions of resonance modes allows to establish phenomenological equations describing these resonances in the time domain. In the case of single particles, temporal modal amplitude can be described through a first-order differential equation with phenomenological parameters realizing losses and coupling with the incident excitation. This approach can also be extended to the interacting modes by introducing a system of coupled equations where mutual coupling terms are involved to account for interaction between modes. The formation of hybrid modes in coupled systems can be predicted according to this formalism.

The method for the extraction of resonance characteristics was first applied to single silver particles with the determination of the position, spectral width and resonance amplitude of eigenmodes of spheres as a function of the sphere radius. Hybrid modes of dimers were then studied. The initial dipole modes of two spheres form in the general case four hybrid modes of the dimer differentiated by the direction of the resulting dipole moment and phase relationships between the oscillations of the dipole moment of each particle. From energetic point of view, these hybrid modes can be seen as bounding and anti-bounding states, characterized by red and blue spectral shifts, respectively, and their resonance positions gives the measure of the power coupling between particles.

Finally, a more complex system consisting of two dimer in interaction was studied. The optical response of this system has asymmetric resonance profiles of Fano type very well taken into account by the singular representation of our model and perfectly characterized by the extraction method.

We conclude that the numerical extraction method of resonance characteristics combined with the phenomenological approach enables efficient characterization and physical interpretation of plasmon resonances and of coupling of their different modes.

References

1. Kreibig U, Vollmer M (1995) Optical properties of metal clusters. Springer Series in Materials Science. Springer, Berlin/New York. doi:10.1007/978-3-662-09109-8
2. Hao E, Schatz GC (2004) Electromagnetic fields around silver nanoparticles and dimers. J Chem Phys 120:357–366. doi:10.1063/1.1629280
3. Kolwas K, Derkachova A (2010) Plasmonic abilities of gold and silver spherical nanoantennas in terms of size dependent multipolar resonance frequencies and plasmon damping rates. Opto-Electron Rev 18:429–437. doi:10.2478/s11772-010-0043-6
4. Mayergoyz ID, Fredkin DR, Zhang Z (2005) Electrostatic (plasmon) resonances in nanoparticles. Phys Rev B 72:155412. doi:10.1103/PhysRevB.72.155412

5. Prodan E, Radloff C, Halas NJ, Nordlander P (2003) A hybridization model for the plasmon response of complex nanostructures. Science 302:419–422. doi:10.1126/science.1089171
6. Nordlander P, Oubre C, Prodan E et al (2004) Plasmon hybridization in nanoparticle dimers. Nano Lett 4:899–903. doi:10.1021/nl049681c
7. Mackowski DW (1991) Analysis of radiative scattering for multiple sphere configurations. Proc R Soc Lond Ser Math Phys Sci 433:599–614. doi:10.1098/rspa.1991.0066
8. Mackowski D (2012) The extension of Mie theory to multiple spheres. In: Hergert W, Wriedt T (eds) Mie theory. Springer, Berlin/Heidelberg, pp 223–256
9. Bakhti S, Destouches N, Tishchenko AV (2014) Analysis of plasmon resonances on a metal particle. J Quant Spectrosc Radiat Transf 146:113–122. doi:10.1016/j.jqsrt.2014.01.014
10. Bakhti S, Destouches N, Tishchenko AV (2015) Coupled mode modeling to interpret hybrid modes and Fano resonances in plasmonic systems. ACS Photonics 2:246–255. doi:10.1021/ph500356n
11. Bakhti S, Destouches N, Tishchenko AV (2015) Singular representation of plasmon resonance modes to optimize the near- and far-field properties of metal nanoparticles. Plasmonics, 10:1391–1399. doi:10.1007/s11468-015-9937-y
12. Haus HA (1984) Waves and fields in optoelectronics. Prentice-Hall series in solid state physical electronics. Prentice-Hall, Englewood Cliffs
13. Novotny L (2010) Strong coupling, energy splitting, and level crossings: a classical perspective. Am J Phys 78:1199–1202. doi:10.1119/1.3471177
14. Mackowski DW (1994) Calculation of total cross sections of multiple-sphere clusters. J Opt Soc Am 11:2851–2861. doi:10.1364/JOSAA.11.002851
15. Golub GH, Van Loan CF (1996) Matrix computations, 3rd edn. Johns Hopkins University Press, Baltimore
16. Baker GA (1974) Essentials of padé approximants. Academic Press, New York
17. Tishchenko AV, Hamdoun M, Parriaux O (2003) Two-dimensional coupled mode equation for grating waveguide excitation by a focused beam. Opt Quant Electron 35:475–491. doi:10.1023/A:1022921706176
18. Tretyakov S (2014) Maximizing absorption and scattering by dipole particles. Plasmonics 9:935–944. doi:10.1007/s11468-014-9699-y
19. Grigoriev V, Bonod N, Wenger J, Stout B (2015) Optimizing nanoparticle designs for ideal absorption of light. ACS Photonics. doi:10.1021/ph500456w
20. Lovera A, Gallinet B, Nordlander P, Martin OJF (2013) Mechanisms of Fano resonances in coupled plasmonic systems. ACS Nano 7:4527–4536. doi:10.1021/nn401175j
21. Sherry LJ, Chang S-H, Schatz GC et al (2005) Localized surface plasmon resonance spectroscopy of single silver nanocubes. Nano Lett 5:2034–2038. doi:10.1021/nl0515753
22. Mirin NA, Bao K, Nordlander P (2009) Fano resonances in plasmonic nanoparticle aggregates†. J Phys Chem A 113:4028–4034. doi:10.1021/jp810411q
23. Lassiter JB, Sobhani H, Fan JA et al (2010) Fano resonances in plasmonic nanoclusters: geometrical and chemical tunability. Nano Lett 10:3184–3189. doi:10.1021/nl102108u

Chapter 3
Radiolytically Synthesized Noble Metal Nanoparticles: Sensor Applications

Nilanjal Misra, Narender Kumar Goel, Lalit Varshney, and Virendra Kumar

Abstract Nanotechnology is a discipline that has grown exponentially over the last few decades. The unique properties of nanomaterials, including optical, thermal, electromagnetic etc., make them a highly attractive proposition for a vast array of applications, ranging from medical science to defense technologies. The singular optical properties of noble metal nanoparticles, in particular, have helped develop highly sophisticated sensor systems for estimating trace levels of various analytes. This chapter brings into focus some of the recent developments in the field of sensors based on the Localized Surface Plasmon Resonance (LSPR) properties of Gold (Au) and Silver (Ag) nanoparticles. The use of ionizing radiation, such as ^{60}Co gamma, for fabrication of polymer stabilized Au and Ag nanoparticles via radiolytic route has proved to be one of the most attractive and convenient techniques for nanosynthesis. Using gamma radiation, Au and Ag nanoparticles have been synthesized and successfully employed as LSPR based sensors for estimation of trace levels of analytes such as H_2O_2, Hg^{2+}, dopamine and Uric Acid. In a nutshell, the chapter highlights how the wedlock between radiation and nanotechnology is fast emerging as a powerful tool to design new materials that can cater to high end applications.

Keywords Nanomaterials • Sensors • Localized Surface Plasmon Resonance (LSPR) • Radiation

N. Misra • N.K. Goel
Radiation Technology Development Division, Bhabha Atomic Research Centre, Trombay, Mumbai 400085, India

L. Varshney • V. Kumar (✉)
Radiation Technology Development Division, Bhabha Atomic Research Centre, Trombay, Mumbai 400085, India

Homi Bhabha National Institute, Anushaktinagar, Mumbai 400094, India
e-mail: vkrawat75@gmail.com; vkumar@barc.gov.in

C.D. Geddes (ed.), *Reviews in Plasmonics 2015*,
DOI 10.1007/978-3-319-24606-2_3

3.1 Introduction

Metal nanoparticles (NPs) find application in a wide array of fields including catalysis [1] chemical and bio sensors [2, 3], antibacterial substances [4] and drug delivery systems [5]. Noble metal nanoparticles, in particular, such as those of Au and Ag show intense Localized Surface Plasmon Resonance (LSPR) absorption in the visible region [6] and have potential applications as optical biosensors [7, 8] or catalytic systems [9, 10] for initiating a variety of organic reactions. The high surface area provided by these nanoparticles make them extremely efficient as catalytic materials to carry out reactions that are otherwise difficult to initiate [11–16]

The use of metal nanoparticles as analytical and bioanalytical sensors has been receiving significant attention because of their unusual optical, electronic, and chemical properties [17–19]. A large number of methods have been developed for the fabrication of metal nanoparticles. However, the major disadvantages in most of the methods include use of toxic reducing chemicals, poor nanoparticle size distribution, and inhomogeneous dispersion of the nanoparticles in the polymer host [20]. Ionizing radiation (Gamma radiation, Electron Beam, etc.) induced radiolytic route is an efficient alternative technique for fast and one step synthesis of uniformly dispersed metal nanoparticles [21–24]. The use of ionizing radiation for the synthesis of noble metal nanoparticles is promising, as reactive water radiolytic species (e_{aq}^{-}, $H^{\cdot}$) with sufficiently high reduction potential, are generated in-situ on irradiation of aqueous precursor solutions. The easy handling of radiation parameters, namely, absorbed dose and dose rate, offers better control over the size and the size distribution of the metal nanoparticles. Moreover, radiation technology offers the added advantage of being a clean, ambient and room temperature process devoid of use of any external chemical reducing agents. The reduced metal atoms gradually coalesce to form metal nanoparticles whose size is regulated by suitable capping/stabilizing agents, such as PVP. The Localized Surface Plasmon Resonance (LSPR) band of these metal nanoparticles can be utilized as a highly sensitive tool for estimation of various analytes, thereby providing an appropriate, simplified, cheap and rapid alternative to more sophisticated detection techniques. The LSPR wavelength is extremely sensitive to the local environment around the nanoparticles, which facilitates their use as sensing devices [25].

The choice of capping/stabilizing agent during synthesis of metal nanoparticles is also crucial since it decides the overall stability and morphology of the particles generated, which in turn decides the long term utility of the system as a facile catalytic or a sensor system. Therefore, it is imperative to use capping agents that provide optimum stability to the nanoparticle systems and make their use economically viable. For instance, the use of biocompatible polymers like Poly (N-vinyl-2-pyrrolidone) (PVP) as the capping agent [26, 27] further enhances the viability of radiation synthesized metal nanoparticles to detect biologically relevant molecules without disturbing the natural environment of the biological samples in which the estimation is usually done.

In this chapter, we highlight some of the recent applications of radiolytically synthesized noble metal nanoparticles (Au and Ag) as LSPR based optical sensors for the estimation of trace levels of various important analytes.

3.2 Synthesis of Ag/Au Nanoparticles

^{60}Co - gamma and Electron beam irradiation are the two primary radiation sources conventionally used for radiolytic synthesis of noble metal nanoparticles. Prior to irradiation of the samples, radiation doses of the sources are determined by suitable dosimetry techniques, such as Fricke dosimetry [28] for ^{60}Co gamma sources and radiochromic film dosimeter for electron beam sources. When an aqueous solution containing Ag^+ or Au^{3+} precursor ion solution of known concentration, a stabilizer (PVP, Methacrylic acid etc.) and 2-propyl alcohol is purged with N_2 and subjected to radiation, radiolysis of water takes place. As a result, reactive transient species, namely, e_{aq}^-, $H^•$, $^•OH$, etc., are generated (Eq. 3.1). 2-propyl alcohol present in the reaction medium reacts with $H^•$ and •OH to give 2-propyl radical, a mild reducing agent (Eq. 3.2). Simultaneously, the oxidizing radical $^•OH$ gets eliminated in the process. The reducing species present in the system *viz.* e_{aq}^- and 2-propyl radical reduce the metal ion to metal in zero valent state.

$$H_2O \xrightarrow{\gamma\text{-radiation}} e_{aq}^-, H^\bullet, {}^\bullet OH, \text{ etc.} \tag{3.1}$$

$$^\bullet OH/H^\bullet + (CH_3)_2CH - OH \longrightarrow (CH_3)_2C^\bullet - OH + H_2O/H_2 \tag{3.2}$$

$$M^{n+} + ne_{aq}^- \longrightarrow nM^0 \tag{3.3}$$

$$M^{n+} + n(CH_3)_2C^\bullet - OH \longrightarrow nM^0 + n(CH_3)_2C = O + nH^+ \tag{3.4}$$

Subsequently, metal atoms in the zero valent state undergo coalescence to form metal colloids whose size and shape are controlled by the concentration and nature of capping/stabilizing agent used. The functional groups present in the capping agent, such as >C=O and >N- groups (in case of PVP), facilitate anchorage and subsequent stabilization of metal nanoparticles. The progress of the reduction reaction can be monitored spectrophotometrically since Ag/Au nanoparticles absorb strongly in the UV-visible region (LSPR band).

The extent of reduction of metal ions to metal atoms depends on the amount of absorbed radiation dose. The minimum dose required to achieve near complete reduction of a given concentration of precursor ions is termed as the saturation dose. Therefore, irradiations are generally carried out so as to impart saturation dose to the precursor ions in order to ensure near complete reduction of metal ions to metal atoms in the zero valent state.

3.3 Characterization of Metal Nanoparticles

The characterization of metal nanoparticles is important to determine the extent of reduction, the size, shape and morphology of the particles formed. The most commonly used techniques for nanoparticle characteriazation are UV-visible spectrophotometry, Transmission Electron Microscopy (TEM), X-ray Diffraction (XRD) Analysis, Atomic Force Microscopy (AFM) and Light Scattering based techniques, such as Particle Size Analyzer (PSA).

Noble metal nanoparticles, such as those of gold and silver, show characteristic intense absorption in the UV-visible region, which is attributed to the Localized Surface Plasmon Resonance (LSPR) based absorption. The intensity, wavelength of absorption maxima and the width of the LSPR band rely on several factors, such as the precursor ion concentration, morphology of the particles formed, as well as on the local environment of the nanoparticles (nature of solvent, capping agents, etc.). These parameters, therefore, provide us with a fair idea about the overall nature of the nanoparticles formed. For instance, spherical Ag nanoparticles in the size range of 8–10 nm show characteristic LSPR band at around 400 nm. The yield of Ag nanoparticles, manifested by the intensity of the LSPR band of Ag nanoparticles, increases with increase in absorbed radiation dose till all precursor Ag^+ ions are exhausted. However, an increase in particle size (when the interparticle spacings become less than the nanoparticle diameter) conventionally leads to a shift in the LSPR band toward higher wavelength (red shift), whereas a decrease in particle size can lead to a blue shift in the peak position. The FWHM of the LSPR band can also be used to predict the distribution in particle size, with a broad size distribution typically resulting in a correspondingly broad LSPR band. Anisotropic nanoparticles, if formed, can also result in the appearance of more than one band depending on the aspect ratio of the particles formed (transverse and longitudinal bands), instead of a single band as is the case with spherical particles.

Although UV-visible spectroscopy provides us with a rough idea about the nature of particles formed, the most reliable technique for accurately determining the size and shape of metal nanoparticles is Transmission Electron Microscopy (TEM). TEM analysis method can be used to capture images of particles with diameters in the nanometer regime. The particle size distribution and the shape of the particles formed can also be determined accurately using this technique. Besides TEM analysis, Atomic Force Microscopy (AFM), Scanning Electron Microscopy (SEM) and Light Scattering techniques, such as Particle Size Analyzer are also used to gather information about nanoparticle morphology. X-ray Diffraction (XRD) analysis can also provide valuable information about the composition of nanoparticles, their purity, size, shape, distribution, orientation, etc. XRD, in fact, offers unparalleled accuracy in the measurement of atomic spacing and is the technique of choice for determining strain states in thin films. The intensities measured with XRD can provide quantitative and accurate information on the atomic arrangements at interfaces. Zeta Potential analysis is another technique that is often used to determine the stability of nano colloids. The magnitude of the zeta potential gives an indication of the stability of the nanoparticle suspension system. If all the particles in suspension have a large negative or positive zeta potential then they will tend to repel each other and there will be less tendency for the particles to come together to form agglomerates.

All these techniques have been widely used, either individually or in combination, to characterize metal nanoparticle systems. These techniques have been immensely useful in providing new dimensions to the field of nanomaterials and have opened up new areas of research, which in turn, have yielded remarkable outputs.

3.4 Applications of Radiation Synthesized Noble Metal Nanoparticles: LSPR Based Sensor Applications

The LSPR band of noble metal nanoparticles is a highly sensitive and potent tool that can be employed for estimation of a vast array of biologically and environmentally relevant analytes. Radiolytically synthesized Au and Ag nanoparticles in particular offer an attractive option as LSPR based optical sensors for estimation of a number of important analytes. The highly intense, unique and tunable LSPR bands of Au and Ag NPs can be efficiently exploited owing to the distinct changes they exhibit in response to variations in their chemical environment. For instance, DNA-modified gold nanoparticles [29–31] and thiol-functionalized gold nanoparticles have been recently used for colorimetric estimation of Hg^{2+} ions [32–34]. Gold NPs functionalized with an antibody anti-CA15-3-HRP were also implemented in a traditional ELISA immunoassay to detect a breast cancer biomarker present in blood [35]. Pregnancy test kits, one of the most widely used devices in the world today, also make use of gold nanoparticles and their colorimetric properties. Carbohydrate stabilized nanoparticles have been successfully used to detect carbohydrate binding proteins, such as lectin concanavalin A [36]. Aggregation based chemical sensing has also been used to monitor a large number of enzymatic reactions. The list of such applications is endless as large scale, cutting edge research in the field of nanoscience is opening up new dimensions in the field with each passing day. Newer, more sophisticated materials and nano based techniques are being designed which have immensely contributed in areas previously unheard of. All these possibilities make nanomaterials one of the brightest prospects for the future. Some of the recent work done in the field of radiolytically synthesized metal nanoparticles based sensors is discussed in this chapter.

A sensor is a device that uses a particular reaction for detecting target analytes in the presence of interfering substances. Schematic of a typical sensor device with different essential components is shown in Fig. 3.1. In case of metal nanoparticles based sensors, the nanoparticle system acts as the recognition element that selectively interacts with the target analyte under study. The change in the LSPR band

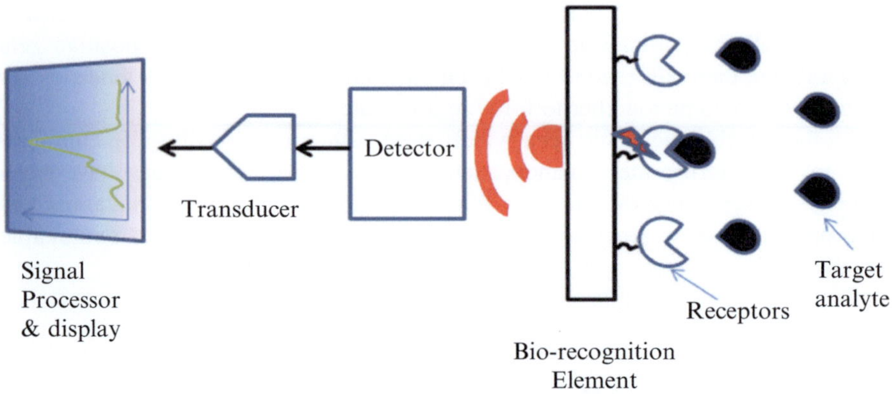

Fig. 3.1 Schematic representation of the principle of operation of a sensor

characteristics of metal nanoparticles induced by nanoparticle-analyte interaction serves as a measurable output, which can be correlated to the analyte concentration. The detector most commonly used to detect such outputs is a simple UV-visible spectrophotometer. Based on the spectrophotometer output, linear relations can be obtained between analyte concentrations and LSPR band intensities, which can subsequently be used to determine unknown concentrations of the given analyte.

3.4.1 PVP Stabilized-Au NPs for H_2O_2 Estimation

Hydrogen peroxide is widely used as an oxidant, a disinfectant and a bleaching agent in various industries, such as textile, paper and pulp, pharmaceutical industries [37]. Its presence in the environment is detrimental to human health since it causes irritation to eye, skin and mucous membrane. Hydrogen peroxide is also produced in stoichiometric amounts during the oxidation of biological analytes (e.g., glucose) by dissolved oxygen in the presence of corresponding oxidase. Hence micro and trace level determination of hydrogen peroxide is considerably important in clinical chemistry, analytical biochemistry and environmental science. Existing methods for the determination of hydrogen peroxide include titrimetry [38], spectrophotometry [39], kinetic flow-injection method [40], fluorescence [41], enzymatic method [42], chromatographic techniques [43] and electrochemical methods [44]. However, most of these conventional techniques are invariably complicated, time-consuming, expensive as well as inconvenient for point-of-use applications. Recently developed methods for determination of hydrogen peroxide include the use of Au NPs and a peroxidase-catalyzed reaction [45]. Radiolytically synthesized Au NPs have been proved to be highly efficient for such detection applications since they are green, clean systems devoid of any external chemical reagents (which are generally present in chemically synthesized nanoparticle systems) and therefore, provide minimum scope of interference during detection or estimation [8].

We have recently developed a detection method for H_2O_2 estimation, which is based on radiolytically synthesized PVP stabilized-Au NPs [8]. Substrate orthophenylenediammine (o-PDA) undergoes catalytic oxidation by H_2O_2 in presence of enzyme Horse radish peroxidise (HRP). The oxidation product of o-PDA has a weak absorption peak at 427 nm. Addition of PVP stabilized-Au NPs of pre determined concentration has been found to enhance the absorption intensity of the oxidation product of o-PDA. This may be due to the interaction of Au NPs with the functional groups of 2, 3- diaminophenazine, which is the final oxidation product of o-PDA [46, 47] (Fig. 3.2). With varying concentration of H_2O_2 there is a systematic increase in the intensity of the absorption peak of o-PDA ($\lambda_{max} = 427$ nm) in presence of PVP stabilized Au NPs, as shown in Fig. 3.3. The response is linear in the range of 2.5×10^{-6} mol dm^{-3} to 2.0×10^{-4} mol dm^{-3} H_2O_2 concentration (Fig. 3.3 inset). Lower concentrations of H_2O_2 can also be determined by simply reducing the concentration of the substrate o-PDA while keeping all other reagents and experimental procedures intact. The overall response of the system is linear in the H_2O_2 concentration range of 1.0×10^{-7} mol. dm^{-3} to 2.0×10^{-4} mol dm^{-3}, with a minimum detection limit of 1.0×10^{-7} mol. dm^{-3}

3.4.2 *PVP Stabilized-Au NPs for* Hg^{2+} *Estimation*

Au NPs have also been found to be effective in the detection of Hg^{2+} in aqueous solutions. Mercury is one of the most widespread pollutants found in nature, existing in a variety of different forms such as metallic, ionic and as part of organic and inorganic salts and complexes [29]. The highly toxic nature of mercury necessitates its quantification, particularly in aquatic ecosystems, up to extremely low concentration levels [48]. Colorimetric methods of Hg^{2+} detection using metal nanoparticles, such as those of gold and silver, have emerged as an attractive technique owing to their simplicity, accuracy and robustness. Techniques based on Au-Hg amalgam formation induced aggregation of Au NPs, catalytic reduction of Hg^{2+} by Ag NPs [30] have been reported.

Recently, radiolytically synthesized PVP stabilized spherical Au nanoparticles have been employed by us for the detection and estimation of trace levels of Hg^{2+} in aqueous solutions [49]. These sensors work on the principle of preferential interaction of Hg^{2+} with PVP, which gets manifested as a change in the LSPR band characteristics of Au NPs with varying concentration of Hg^{2+} ions (Fig. 3.4). The newly designed sensor system was found to have reasonably good specificity and showed linear response within a concentration range of 0–100 nM Hg^{2+} ion concentration (Fig. 3.4 inset). The simplicity and acuracy of this technique makes it a potential candidate for real life applications.

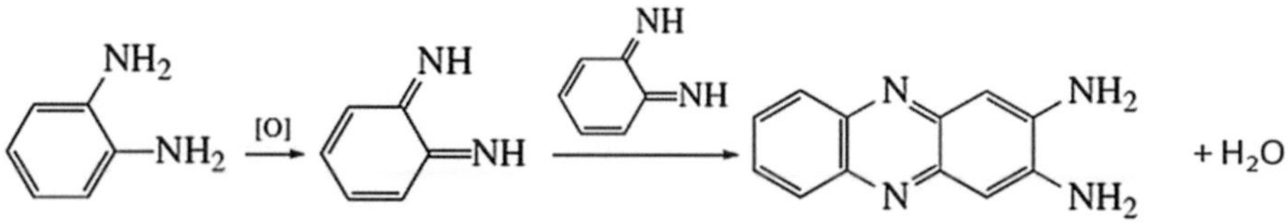

Fig. 3.2 Enzymatic oxidation of o-PDA

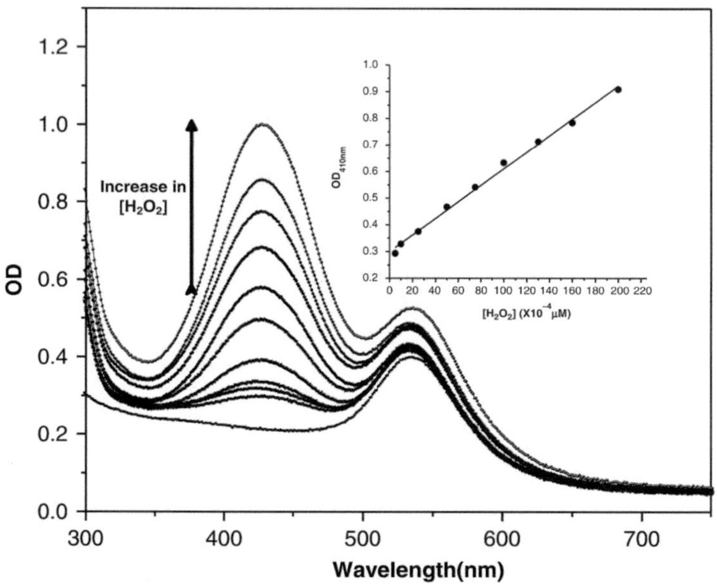

Fig. 3.3 UV–vis spectra of reaction medium containing o-PDA, HRP, H_2O_2 and Au NPs in citrate buffer with increasing H_2O_2 concentrations Inset: Linear plot of $OD_{410\ nm}$ vs H_2O_2 concentration (Revised from Ref. [8])

3.4.3 PVP Stabilized-Ag NPs for Uric Acid Estimation

Uric acid represents the major catabolite of purine breakdown in humans. The normal concentration of uric acid in blood samples is reported to be in the range 150–420 μM [50]. High levels of uric acid in the blood (hyperuricemia or Lesch-Nyhan syndrome) are linked with gout and other conditions including increased alcohol consumption, obesity, diabetes, high cholesterol, high blood pressure, kidney disease, and heart disease [51, 52]. On the other hand, abnormally low uric acid levels are symptoms of diseases, such as multiple sclerosis. Hence estimation of uric acid in blood can be used as a diagnostic tool for monitoring a large number of diseases. Furthermore, uric acid is an antioxidant in human adult plasma and is involved in various pathological changes [53]. In view of this, numerous techniques have been developed over the years for detection and estimation of uric acid levels.

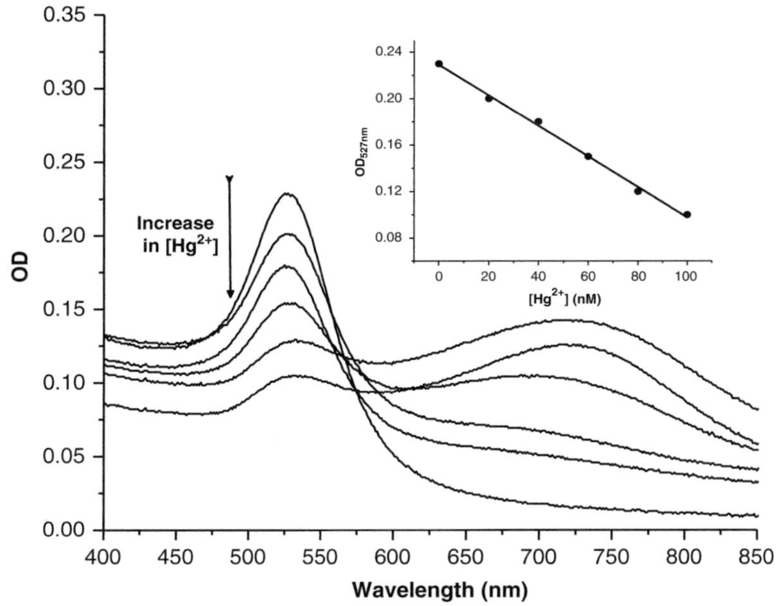

Fig. 3.4 UV–vis spectra of PVP-Au-NPs solution at varying concentrations of Hg^{2+} ion. Inset: Linear plot of OD vs Hg^{2+} ion concentration (Revised from Ref. [49])

Recently, a gamma radiolytically synthesized PVP stabilized-Ag NPs based optical sensor has been fabricated for use as a uric acid biosensor [7]. Uric acid undergoes enzymatic degradation in presence of enzyme Uricase under optimum assay conditions of 37 °C and pH 7.4 (Fig. 3.5). Hydrogen peroxide is generated as one of the reaction products, which is known to be a strong oxidizing agent. This, in turn, causes oxidation/degradation of silver nanoparticles, resulting in a decrease in intensity of the LSPR band. The schematic of working principle of PVP stabilized-Ag NPS based uric acid biosensor is presented in Fig. 3.5.

For the evaluation of PVP stabilized-Ag NPs solution as a uric acid biosensor, a predetermined volume of Uricase stock solution is added to different concentrations of uric acid in a phosphate buffer medium at room temperature. Addition of Ag NPs solution to these reaction mixtures results in a variation in the LSPR band of the silver nanoparticles with increasing concentration of uric acid, as evident from the UV-visible spectra recorded in the wavelength range 250–650 nm (Fig. 3.6). The LSPR band intensity decreases gradually with increase in uric acid concentration due to the degradation of Ag NPs. The decrease is accompanied by a slight red shift in the absorption maxima of the band, which is attributed to the partial oxidation of nano-Ag [54–57]. The shift in λ_{max} might also be due to the slight aggregation caused by the destruction of the PVP shell stabilizing the nanoparticles followed by decrease in the distance between the nanoparticles [58].

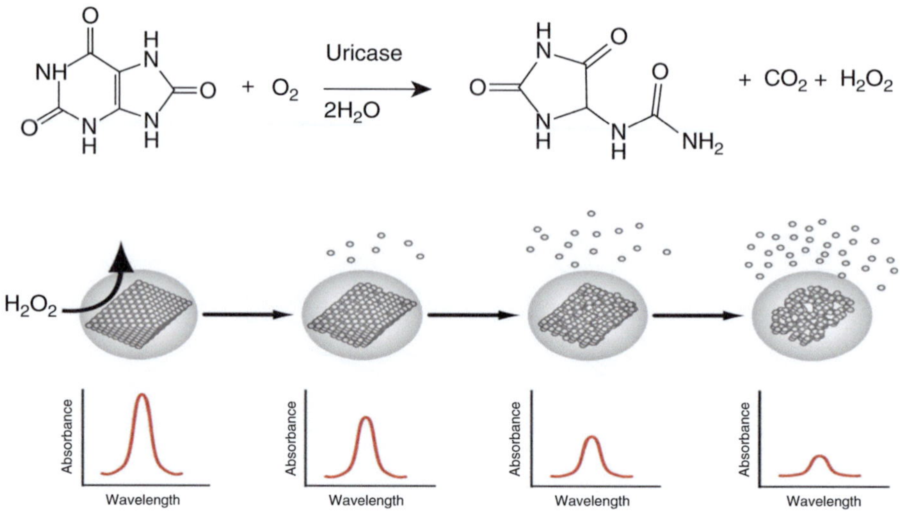

Fig. 3.5 Enzymatic degradation of uric acid in presence of Uricase and schematic of working principle of PVP stabilized-Ag NPS based uric acid biosensor

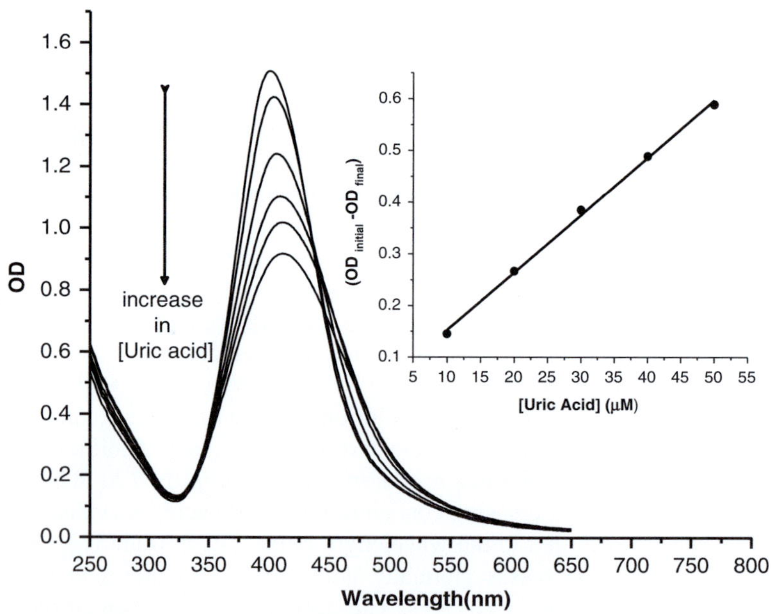

Fig. 3.6 UV–vis spectra of PVP stabilized-Ag NPs solution (PVP, Mol wt. = 40kD) with increasing concentration of uric acid. Inset: Linear plot of change in OD of Ag NPs vs uric acid concentration (Revised from Ref. [7])

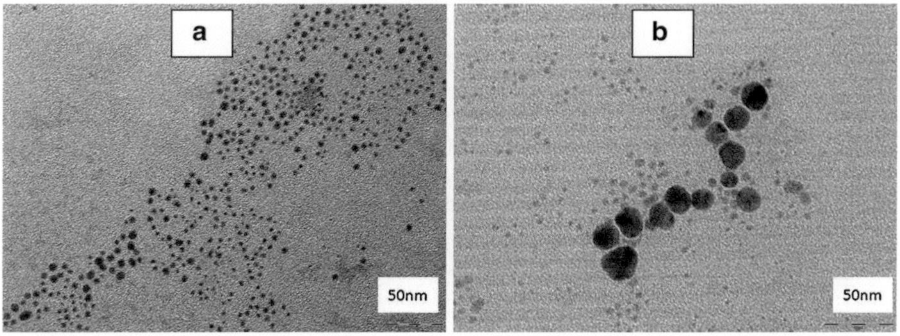

Fig. 3.7 TEM micrograph and particle size distribution of PVP stabilized-Ag NPs prepared using PVP of molecular weight 40 kD (**a**) before addition of uric acid (**b**) after addition of uric acid

The change in particle size is evident from the TEM images of PVP stabilized-Ag NPs recorded before and after addition of uric acid (Fig. 3.7a and b). While the control (Ag-NPs) exhibited an average particle size of 8–10 nm, those in presence of uric acid were found to have size in the range of 10–20 nm. These observations were also substantiated using AFM analysis.

The decrease in the magnitude of zeta potential indicates the tendency of the particles to form agglomerates [59]. In addition, zeta potential also indicates the presence of an oxidized surface layer. Zeta potential measurements can, therefore, be used in this case to further confirm the partial oxidation of the silver nanoparticles by in-situ generated hydrogen peroxide. With increase in uric acid concentration, the zeta potential values become less negative (Fig. 3.8), although the stabilities of the nanoparticle suspensions are not disturbed significantly. This is probably due to partial neutralization of the negative charge by Ag^+ ions generated via partial oxidation of nano Ag. It has been well established that electrostatic stabilization of nanoparticles would typically require a zeta potential above 30 mV or below −30 mV [60]. Therefore, in the present study, the values of zeta potentials clearly suggest that the stability of the PVP stabilized-Ag NPs suspension is predominantly based on steric stabilization by the PVP polymer. In contrast to electrostatic stabilization, steric stabilization with nonionic polymers is independent of pH and electrolyte concentration. Accordingly, steric stabilization is useful for prevention of agglomeration of nanoparticles in physiological media.

3.4.3.1 Estimation of Uric Acid in Bovine and Human Serum Samples

The concentration of uric acid in serum samples is normally found in the micromolar range. Therefore, the proposed method was effectively applied for determination of uric acid concentration in serum samples. To minimize interference from

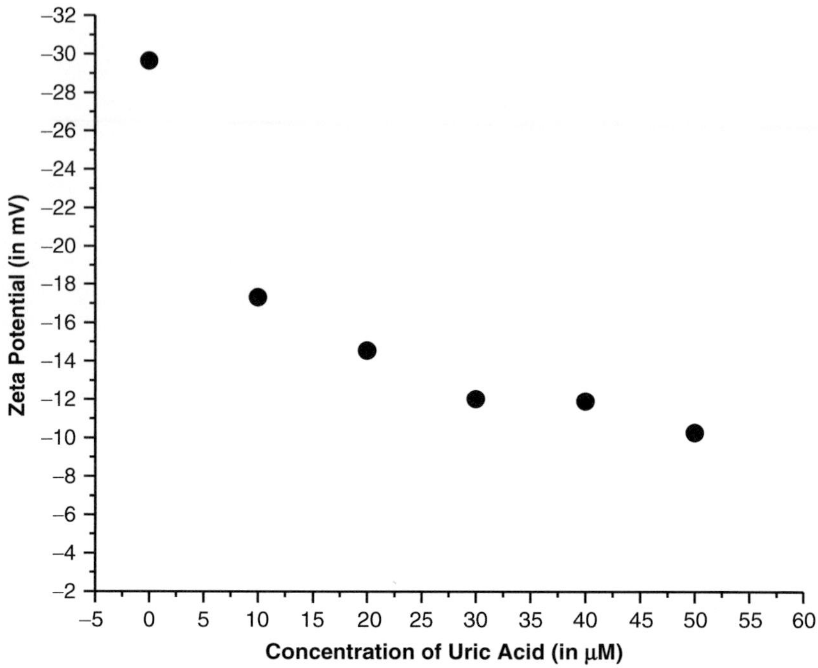

Fig. 3.8 Zeta potential of Ag NPs as a function of uric acid concentration (Revised from Ref. [7])

proteins present in the serum sample, samples were initially deproteinized and then subjected to the estimation protocol. The concentrations of uric acid in both bovine and human serum samples were estimated using the calibration plot generated using standard uric acid solutions of known concentrations.

3.4.4 PMA Stabilized-Ag NPs for Dopamine Estimation

Dopamine (DA) belongs to the class of catecholamines, which are compounds containing a dihydroxyphenyl group and an amine group [61]. These biomolecules act as neurotransmitters and aid in the functioning of the brain and nerve signal transductions. Dopamine deficiency can lead to diseases, such as parkinsons, whereas mental disorders, such as schizophrenia and bipolar disorder have been attributed to excess dopamine in the system. In recent years, many methods have been developed for the estimation of dopamine in pharmaceutical preparations and biological samples. Many of these techniques for neurotransmitter detection, however, require expensive and sophisticated instrumentation or complicated sample preparation and time consuming processes. Thus, the development of a new, sensitive, fast and practical method for neurotransmitter detection still remains a great challenge.

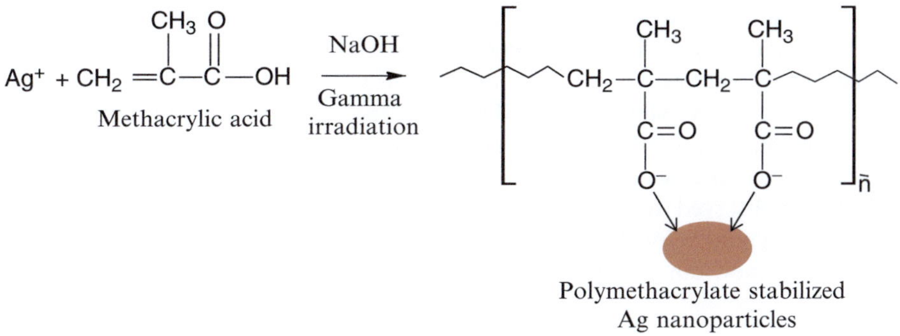

Fig. 3.9 Schematic of Ag NPs stabilization by polymethacrylate chains

Recently, we have developed a sensor for Dopamine, which was based on radiolytically synthesized polymethacrylate (PMA) stabilized-Ag NPs system. Dopamine, or 3,4 dihydroxy phenethylamine is a two electron donor and behaves as a mild reductant. The functional groups can interact with polymethacrylate stabilized-Ag NPs (Fig. 3.9) when present in solution, resulting in the alteration of the LSPR band characteristics of the nanoparticle system [13]. In fact, in this case, a pronounced colour change is observed, thereby making the interaction very conspicuous. This phenomenon can be quantified by recording the absorption spectra of PMA stabilized-Ag NPs in presence of varying concentrations of DA. The observed increase in LSPR band intensity with increase in DA concentration can be attributed to the reduction of residual Ag^+ ions present in the system, since the dose delivered in this case was less than the saturation dose required for complete reduction of Ag^+ ions. These unreduced Ag^+ ions are adsorbed onto the surface of the reduced Ag nanoclusters, resulting in an increase in their reduction potentials. These surface adsorbed Ag^+ ions are subsequently easily reduced by DA molecules, which act as mild reducing agents in presence of NaOH by losing an H^+ ion. Higher DA concentrations will lead to increased concentrations of freshly reduced Ag^+ ions, which are either deposited on the radiolytically formed Ag clusters or can form new Ag nanoclusters on their own. Thus, there is a possibility of increase in the Ag NP size as well as in the size distribution, which is manifested by the increase in intensity and broadening of the LSPR band. There is no shift in the peak position, because for spherical nanoparticles, a small change in particle size does not affect the plasmon band position.

3.4.4.1 Estimation of Dopamine in Presence of Ascorbic Acid

Conventionally, DA was estimated by electrochemical analysis. A common problem associated with electrochemical analyses at unmodified electrodes is the lack of selectivity due to the presence of interfering compounds. For instance, ascorbic acid (AA) oxidizes at nearly the same potential as DA. Also AA coexists with DA in the extracellular fluids of mammalian brain. Therefore, it is imperative to study the interference of AA in the estimation of DA.

The interference of AA in DA estimation was studied by adding 1.0×10^{-4} mol. dm^{-3} AA to PMA stabilized-Ag NPs in presence of different concentrations of DA. The intensity of the LSPR band of PMA stabilized-Ag NPs at ~415 nm gradually increases with increasing DA concentration even in the presence of AA. It was observed that with optimization of experimental parameters, the detection system could be made free of interference from AA concentrations as high as 1.0×10^{-4} mol dm^{-3} within the DA estimation range of 5.27×10^{-7} to 1.05×10^{-5} mol.dm^{-3}.

3.5 Conclusion

In the recent past, there has been rapid growth in the field of nanoscience and technology. The applications of nanomaterials have permeated into areas previously undreamt and unheard of. Countless sensors and analytical tools have been designed and developed which have immensely benefited mankind and shall continue to do so in the near future. It can only be envisaged that their popularity will grow manifold in the coming years with each new breakthrough and innovation. The future of mankind, we can sanguinely conclude, lies in switching over from the macro to the nano.

References

1. Crooks RM, Zhao M, Sun L, Chechik V, Yeung LK (2001) Dendrimer-encapsulated metal nanoparticles: synthesis, characterization, and applications to catalysis. Acc Chem Res 34:181–190
2. Liu J, Lu Y (2003) A colorimetric lead biosensor using DNAzyme-directed assembly of gold nanoparticles. J Am Chem Soc 125:6642–6643
3. Kerman K, Saito M, Yamamura S, Takamura Y, Tamiya E (2008) Nanomaterial-based electrochemical biosensors for medical applications. Trends Anal Chem 27:585–592
4. Zhao Y, Tian Y, Cui Y, Liu W, Ma W, Jiang X (2010) Small molecule-capped gold nanoparticles as potent antibacterial agents that target gram-negative bacteria. J Am Chem Soc 132:12349–12356
5. Brown SD, Nativo P, Smith J, Stirling D, Edwards PR, Venugopal B, Flint DJ, Plumb JA, Graham D, Wheate NJ (2010) Gold nanoparticles for the improved anticancer drug delivery of the active component of oxaliplatin. J Am Chem Soc 132:4678–4684
6. Creighton JA, Eadont DG (1991) Ultraviolet–visible absorption spectra of the colloidal metallic elements. J Chem Soc Faraday Trans 87:3881–3891
7. Misra N, Kumar V, Borde L, Varshney L (2013) Localized surface plasmon resonance-optical sensors based on radiolytically synthesized silver nanoparticles for estimation of uric acid. Sens Actuators B 178:371–378

8. Misra N, Biswal J, Gupta A, Sainis JK, Sabharwal S (2012) Gamma radiation induced synthesis of gold nanoparticles in aqueous polyvinyl pyrolidone solution and its application for H_2O_2 estimation. Radia Phys Chem 81:195–200
9. Fenger R, Fertitta E, Kirmse H, Thu¨nemann AF, Rademann K (2012) Size dependent catalysis with CTAB-stabilized gold nanoparticles. Phys Chem Chem Phys 14:9343–9349
10. Seoudi R, Said DA (2011) Studies on the effect of the capping materials on the spherical gold nanoparticles catalytic activity. World J Nano Sci Eng 1:51–61
11. Hvolbæk B, Janssens TVW, Clausen BS, Falsig H, Christensen CH, Nørskov JK (2007) Catalytic activity of Au nanoparticles. Nano Today 2:14–18
12. Gasaymeh SS, Radiman S, Heng LY, Saion E, Saeed GHM (2010) Synthesis and characterization of silver/polyvinilpirrolidone (Ag/PVP) nanoparticles using gamma irradiation techniques. Am J App Sci 7:892–901
13. Biswal J, Misra N, Borde L, Sabharwal S (2013) Synthesis of silver nanoparticles in methacrylic acid solution by gamma raiolysis and their applications for estimation of dopamine in low concentration. Radiat Phys Chem 83:67–73
14. Lin Y, Qiao Y, Wang Y, Yan Y, Huang J (2012) Self-assembled laminated nanoribbon-directed synthesis of noble metallic nanoparticle-decorated silica nanotubes and their catalytic applications. J Mater Chem 22:18314–18320
15. Samim M, Kaushik NK, Maitra A (2007) Effect of size of copper nanoparticles on its catalytic behaviour in Ullman reaction. Bull Mater Sci 30:535–540
16. Lee H, Kim C, Yang S, Han JW, Kim J (2012) Shape-controlled nanocrystals for catalytic applications. Catal Surv Asia 16:14–27
17. Schultz S, Smith D, Mock J, Schultz D (2000) Single-target molecule detection with nonbleaching multicolor optical immunolabels. Proc Natl Acad Sci U S A 97:996–1001
18. Taton T, Mirkin C, Letsinger R (2000) Scanometric DNA array detection with nanoparticle probes. Science 289:1757–1760
19. Yguerabide J, Yguerabide E (1998) Light-scattering submicroscopic particles as highly fluorescent analogs and their use as tracer labels in clinical and biological applications: II. Experimental characterization. Anal Biochem 262:157–176
20. Gasaymeh SS, Radiman S, Heng LY, Saion E, Saeed GHM (2010) Synthesis and characterization of silver/polyvinilpirrolidone (Ag/PVP) nanoparticles using gamma irradiation techniques. Afr Phys Rev 4:31–41
21. Choi S, Lee S, Hwang Y, Lee K, Kang H (2003) Interaction between the surface of the silver nanoparticles prepared by gamma irradiation and organic molecules containing thiol group. Rad Phys Chem 67:517–521
22. Choi S, Park HG (2005) Surface-enhanced Raman scattering (SERS) spectra of sodium benzoate and 4-picoline in Ag colloids prepared by gamma irradiation. Appl Surf Sci 243:76–81
23. Abedini A, Daud AR, Hamid MAA, Othman NK, Saion E (2013) A review on radiation-induced nucleation and growth of colloidal metallic nanoparticles. Nanoscale Res Lett 8:474
24. Misra N, Biswal J, Dhamgaye VP, Lodha GS, Sabharwal S (2013) A comparative study of gamma, electron beam, and synchrotron X-ray irradiation method for synthesis of silver nanoparticles in PVP. Adv Mater Lett 4:458–463
25. Filippo E, Serra A, Manno D (2009) Poly(vinyl alcohol) capped silver nanoparticles as localized surface plasmon resonance-based hydrogen peroxide sensor. Sensor Actuators B-Chem 138:625–630
26. Borodko Y, Habas SE, Koebel M, Yang P, Frei H, Somorjai GA (2006) Probing the interaction of poly(vinylpyrrolidone) with platinum nanocrystals by uv-raman and FTIR. J Phys Chem B 110:23052–23059
27. Wang H, Qiao X, Chen J, Wang X, Ding S (2005) Mechanisms of PVP in the preparation of silver nanoparticles. Mater Chem Phys 94:449–453
28. McLaughlin WL, Boyd AK, Chadwick KN, McDonald JC, Miller A (eds) (1980) Dosimetry for radiation processing. Taylor and Francis, London

29. Lee JS, Han MS, Mirkin CA (2007) Colorimetric detection of mercuric ion (Hg^{2+}) in aqueous media using DNA-functionalized gold nanoparticles. Angew Chem Int Ed 46:4093–4096
30. Yan Z, Yuen MF, Hu L, Suna P, Lee CS (2014) Advances for the colorimetric detection of Hg^{2+} in aqueous solution. RSC Adv 4:48373–48388
31. Du J, Jiang L, Shao Q, Liu X, Marks RS, Ma J, Chen X (2013) Colorimetric detection of mercury ions based on plasmonic nanoparticles. Small 9:1467–1481
32. Huang CC, Yang Z, Lee KH, Chang HT (2007) Synthesis of highly fluorescent gold nanoparticles for sensing mercury(II). Angew Chem Int Ed 46:6824–6828
33. Darbha GK, Singh AK, Rai US, Yu E, Yu H, Ray PC (2008) Highly selective detection of Hg^{2+} ion using NLO properties of gold nanomaterials. J Am Chem Soc 130:8038–8042
34. Liang GX, Wang L, Zhang H, Han Z, Wu X (2012) A colorimetric probe for the rapid and selective determination of mercury(II) based on the disassembly of gold nanorods. Microchim Acta 179:345–350
35. Ambrosi A, Airó F, Merkoçi A (2010) Enhanced gold nanoparticle based ELISA for a breast cancer biomarker. Anal Chem 82:1151–1156
36. Schofield CL, Haines AH, Field RA, Russell DA (2006) Silver and gold glyconanoparticles for colorimetric bioassays. Langmuir 22:6707–6711
37. Li Y (1996) Biological properties of peroxide-containing tooth whiteners. Food Chem Toxicol 34:887–904
38. Hurdis EC, Romeyn H (1954) Accuracy of determination of hydrogen peroxide by cerate oxidimetry. Anal Chem 26:320–325
39. Amin VM, Olson NF (1967) Spectrophotometric determination of hydrogen peroxide in milk. J Dairy Sci 50:461–464
40. Maeuchihara N, Nakano S, Kawashima T (2001) Kinetic flow-injection determination of hydrogen peroxide by use of iron(III)-catalyzed coloration and its application to the determination of biological substances. Anal Sci 17:255–258
41. Onoda M, Uchiyama T, Mawatari K, Kaneko K, Nakagomi K (2006) Simple and rapid determination of hydrogen peroxide using phosphine-based fluorescent reagents with sodium tungstate dihydrate. Anal Sci 22:815–817
42. Meyer J, Karst U (1999) Workplace monitoring of gas phase hydrogen peroxide by means of fluorescence spectroscopy. Anal Chim Acta 401:191–196
43. Takahashi A, Hashimoto K, Kumazawa S, Nakayama T (1999) Determination of hydrogen peroxide by high performance liquid chromatography with a cation exchange resin gel column and electrochemical detector. Anal Sci 15:481–483
44. Brestovisky A, KirowaEisner E (1983) Direct and titrimetric determination of hydrogen peroxide by reverse pulse polarography. Anal Chem 55:2063–2066
45. Wu ZS, Zhang SB, Guo MM, Chen CR, Shen GL, Yu RQ (2007) Homogeneous unmodified gold nanoparticle based colorimetric assay of hydrogen peroxide. Anal Chim Acta 584:122–128
46. Jiao K, Zhang SS, Wei L (1996) Determination of HRP with coupling reaction voltametric enzyme-linked immunoassay based on OPD-H_2O_2-HRP system. Sci China B 26:57–60
47. Peter JT, Victor PC, Dave W (1987) 2,3-diaminophenazin is the product from the horseradish peroxidase catalyzed oxidation of o-phenylenediamine. Anal Biochem 165:230–233
48. Yan L, Chen Z, Zhang Z, Qu C, Chen L, Shen D (2013) Fluorescent sensing of mercury (II) based on formation of catalytic gold nanoparticles. Analyst 138:4280–4283
49. Misra N, Kumar V, Goel NK, Varshney L (2015) Radiation synthesized Poly(n-vinyl-2-pyrrolidone)-stabilized-gold nanoparticles as LSPR based optical sensor for Mercury ions estimation. J Nanopar Res 17:279(1-9)
50. Miller E, Kędziora J (2011) Effect of whole body cryotherapy on uric acid concentration in plasma of multiple sclerosis patients. Int Rev Allergol Clin Immunol 17:20–23
51. Wurzner G, Gerster JC, Chiolero A, Maillard M, Fallab-Stubi CL, Brunner HR, Burnier M (2001) Comparative effects of losartan and irbesartan on serum uric acid in hypertensive patients with hyperuricemia and gout. J Hypertens 19:1855–1860

52. Tan L, Yang GM, Wang P, Xie ZY, Bai HP, Lu XX, Yang YH (2008) Direct determination of uric acid based on Pd nanoparticles electrodepositing onto anatase-type TiO2 nanoparticles/ chitsan film-modified electrode. Anal Lett 41:2860–2876
53. Goyal RN, Gupta VK, Sangal A, Bachheti N (2005) Voltammetric determination of uric acid at a fullerene-C60-modified glassy Carbone electrode. Electroanal 17:2217–2223
54. Lok C, Ho C, Chen R, He Q, Yu W, Sun H, Tam P, Chiu J, Che C (2007) Silver nanoparticles: partial oxidation and antibacterial activities. J Biol Inorg Chem 12:527–534
55. Henglein A (1998) Colloidal silver nanoparticles: photochemical preparation and interaction with O2, CCl4 and some metal ions. Chem Mater 10:444–450
56. Henglein A (1993) Physiochemical properties of small metal particles in solution: "microelectrode" reactions, chemisorptions, composite metal particles and the atom-to-metal transition. J Phys Chem 97:5457–5471
57. Li XJ, Lenhart J, Walker HW (2010) Dissolution-accompanied aggregation kinetics of silver nanoparticles. Langmuir 26:16690–16698
58. Vasileva P, Donkova B, Karadjova I, Dushkin C (2011) Synthesis of starch stabilized silver nanoparticles and their application as a surface plasmon resonance based sensor of hydrogen peroxide. Colloid Surf A 382:203–210
59. Dougherty GM, Rose KA, Tok JBH, Pannu SS, Chuang FYS, Sha MY, Chakarova G, Penn SG (2007) The zeta potential of surface functionalized metallic nanorod particles in aqueous solution. Electrophoresis 29:1131–1139
60. Bihari P, Vippola M, Schultes S, Praetner M, Khandoga AG, Reichel CA, Coester C, Tuomi T, Rehberg M, Krombach F (2008) Optimized dispersion of nanoparticles for biological in vitro and in vivo studies. Part Fibre Toxicol 5:1–14
61. Pratesi P (1963) Chemical structure and biological activity of catecholamines. Pure Appl Chem 6:435–449

Chapter 4
Construction, Modeling, and Analysis of Transformation-Based Metamaterial Invisibility Cloaks

Branislav M. Notaroš, Milan M. Ilić, Slobodan V. Savić, and Ana B. Manić

Abstract A summary review of construction, modeling, and analysis of transformation-based metamaterial invisibility cloaks is presented. In particular, we present a simplified and unified theory of vector, tensor, and operator changes under coordinate transformations, and relate them specifically to electromagnetic field vectors, medium permittivity and permeability tensors, and the curl operator in Maxwell's equations. The presented theory sets a basis for arbitrary manipulations of electromagnetic fields by coordinate transformations, known as the transformation electromagnetics or optics. We also present the examples of coordinate transformations which lead to construction of linear and nonlinear spherical metamaterial cloaks, as well as the resulting transformations of the permittivity/permeability tensors in both spherical and Cartesian systems. Similar principles are then used in construction of a cubical metamaterial cloak, where only Cartesian system is employed. The performance of all constructed spherical and cubical cloaks is verified by numerical simulations. Specifically, we perform full-wave rigorous modeling and analysis of the cloaking structures using a higher order finite element method for discretization of the cloaking region based on large continuously inhomogeneous anisotropic curved hexahedral finite elements with arbitrary material-representation orders and polynomial field expansions and a higher order method of moments for numerical termination of the computational domain.

B.M. Notaroš (✉) • A.B. Manić
Department of Electrical and Computer Engineering, Colorado State University, Fort Collins, CO 80523-1373, USA
e-mail: notaros@colostate.edu;
Url: http://www.engr.colostate.edu/~notaros

M.M. Ilić
Department of Electrical and Computer Engineering, Colorado State University, Fort Collins, CO 80523-1373, USA

School of Electrical Engineering, University of Belgrade, Belgrade 11120, Serbia

S.V. Savić
School of Electrical Engineering, University of Belgrade, Belgrade 11120, Serbia

C.D. Geddes (ed.), *Reviews in Plasmonics 2015*,
DOI 10.1007/978-3-319-24606-2_4

Keywords Invisibility cloaks • Transformation optics • Coordinate transformations • Metamaterials • Spherical cloaks • Cubical cloaks • Scattering • Numerical analysis • Finite element method • Anisotropic inhomogeneous media • Higher order modeling

4.1 Introduction

In recent years, there has been great and growing interest in metamaterials, which can broadly be defined as engineered artificial composite materials exhibiting some unusual (and desirable) properties beyond "standard" or natural media (the term comes from the Greek word "meta" meaning "beyond"). These emerging materials provide tremendous opportunities for designs of new components, devices, and systems at both optical wavelengths and RF/microwave frequencies [1–7]. One important research direction in this area is based on the concept of controlling the electromagnetic (EM) fields, namely, conveniently guiding and redirecting optical paths (or lines of the Poynting vector), in metamaterials designed using transformation-optics techniques. Indeed, transformation optics (or transformation electromagnetics) has recently become one of the mainstream areas of theoretical and applied optics (electromagnetics) [8–17], where of particular and emerging interest are transformation-based metamaterials. Such techniques, enabling designs of devices that possess novel wave-material interaction properties, are founded on the invariance property of Maxwell's equations under coordinate transformations. In the design process, the transformations are first employed to reshape the original coordinate system (and thus redirect the optical paths) in a desired fashion, and then the newly formed spatial curvatures are transformed into the corresponding scaling of medium parameters.

Perhaps the best-known application of the transformation-optics techniques is the design of EM/optical invisibility cloaks. Essentially, reshaping of the original coordinate system and guiding the optical paths using the transformations is done in order to create a void region (spatial region that is not spanned by any curved coordinate lines) in the new coordinate system, which is going to be invisible for observation locations outside the cloak. Upon transforming of the new spatial curvatures into the scaling of medium parameters, the resulting materials for the cloaks turn out to be continuously inhomogeneous and anisotropic, with permittivity and permeability tensors being (continuously varying) functions of spatial coordinates.

There are two basic types of transformation-based metamaterial cloaks: (i) line-transformed (cylindrical) cloaks and (ii) point-transformed (spherical) cloaks. In the case of line-transformed or two-dimensional (2-D) cloaks, an infinite straight line is transformed into an infinitely long spatial domain with a finite cross section of an arbitrary (designed) shape. To obtain point-transformed or three-dimensional (3-D) cloaks, on the other side, a point is transformed into a spatial domain with a finite volume of an arbitrary shape. In both cases, a goal is to have a void region in the new coordinate system, invisible from outside. The first announcements of the cloaking paradigm, given in [8] and [9], rely on transformation-optics and analytical (asymptotic in the geometrical optics limit) ray-tracing derivations to describe and quantify the behavior of both 2-D and 3-D cloaks. These works predict that considerable reductions in scattering cross sections of objects can be achieved, imperfect to the degree of satisfying the required derived material parameters, and they demonstrate a connection between cloaks and previously studied reflectionless interfaces such as perfectly matched layers (PMLs). They also raise questions regarding the influence of the singularity in the ray tracing, when a ray headed directly toward the center of a cloaked sphere (or cylinder) is considered, and whether the cloaking effect is broadband or specific to a single frequency.

A great majority of more detailed studies in the field are those of 2-D cloaks [10–15]. Full-wave numerical analyses of cylindrical cloaks in [10–12] show that low reflection of a cloaking structure is not too sensitive to modest permittivity and permeability variations, that the cloaking performance degrades smoothly with increasing losses, that the effective shielding can be achieved with a cylindrical shell composed of eight piecewise homogeneous layers as an approximation of the ideal continuous medium, and that mirage effect can be observed in the structure (i.e., the source appears to radiate from a shifted location, which is in accordance with the involved geometrical transformation). A numerical study of a 2-D structured cloak, shown to work for different wavelengths provided that they are ten times larger than the size of outermost sectors of the cloaking layers, is given in [13]. Another study of cylindrical cloaks, with homogeneous and isotropic layers, can be found in [14]. An attempt to physically realize a cylindrical cloak is described in [15], where a metallic cylinder was "hidden" inside a cloak constructed from artificially structured metamaterials, designed for operation over a band of microwave frequencies. Some recent studies, based on analytical methods, investigate in more detail problems of material singularities and performances of simplified (non-exact) cloaks [16], general cloaking designs of noncircular annular geometries and the application of cloaking to RF/microwave antenna shielding [17], methodologies for cloak designs based on a linear inhomogeneous field transformation (that does not include space compression as in transformation-optics), which result in a new type of bianisotropic metamaterials that do not scatter arbitrarily incident fields [18], and cloak designs that could be implemented by silver nanowires [19]. Note that a very useful full-wave Mie scattering analytical model of spherical 3-D cloaks is given in [20]. More recently, nonlinear ("high-order") coordinate transformations have been investigated. They are introduced in order to obtain more degrees of freedom in designing the material parameters, which is essential for improving the performance of the cloak when it is approximated (segmented) into a sequence of piecewise-homogeneous layers. Detailed investigations of the optimal discretization (e.g., thickness control of each layer, nonlinear factor, etc.) for both linear and nonlinear transformation-based spherical cloaks and their effects on invisibility performance are presented in [21].

However, while most of the reported theoretical works and investigations of transformation-based metamaterial cloaking concepts and designs rely to greater or lesser extent on analytical derivations and analyses (based on geometrical optics or other methods), an alternative approach, a full-wave rigorous numerical analysis of 2-D, and especially 3-D, cloaking structures, based on concepts and techniques of computational electromagnetics (CEM) and guided by analytical formulas for cloaking materials, seems to have not been fully exploited. Effective numerical modeling of 3-D cloaks, in particular, not only provides an alternative solution for spherical and other canonical shapes that can as well be analyzed analytically (e.g., using Mie's series) but is welcome and actually unavoidable in analysis and design of more complex real-world geometries and material compositions that cannot be treated purely analytically. Of course, one should always have in mind that CEM solutions inevitably introduce some error, due to the involved modeling and numerical approximations and imperfections.

Excellent examples of 2-D numerical studies of cloaking phenomena are those in [10–12], which present full-wave simulations of cylindrical (infinitely long) cloaks using commercial COMSOL Multiphysics software, based on a 2-D finite element method (FEM) for electromagnetic field computations. In addition, the used FEM technique relies on a piecewise homogeneous graded approximation of continuous electromagnetic parameters (permittivity and permeability) of the material, which, in turn, results in the necessity to introduce as many as 32 homogeneous material layers in the model to obtain a smooth enough approximation of the cloaking medium [10]. Finally, as electrically very small low-order finite elements are employed in the technique, it is reported that the simulation requires as many as 340,000 unknowns. A notable example of 3-D full-wave numerical analysis of spherical transformation-based cloaks appears in a study of optical nanotrapping [22], which as well is based on COMSOL Multiphysics (3-D FEM software) and reports utilization of 23,285 small cubic finite elements, domain truncation using PMLs, and 422,233 unknowns in the numerical analysis. Note that comprehensive full-wave CEM simulations of 3-D spherical plasmonic cloaks, which do not involve any anisotropy or inhomogeneity of the material, using commercial CST Microwave Studio software [based on the finite integration technique (FIT)] are presented in [23].

Most recently, highly efficient and versatile full-wave rigorous 3-D numerical modeling of linear and nonlinear transformation-based spherical cloaking structures has been proposed [24], and is further discussed in this summary. This modeling is based on a hybridization of a higher order FEM for discretization of the cloaking region and a higher order method of moments (MoM) for numerical termination of the computational domain. The cloaking region is modeled by Lagrange-type generalized curved parametric hexahedral finite elements of arbitrary geometrical orders and curl-conforming hierarchical polynomial vector basis functions of arbitrary field-expansion orders for the approximation of the electric field vector within the elements [25]. The external surface of the cloak is modeled by generalized quadrilateral MoM patches in conjunction with divergence-conforming polynomial bases of arbitrary orders for the expansion of electric and magnetic equivalent surface currents over the patches [26, 27]. The FEM elements can be large and filled with anisotropic inhomogeneous materials with continuous spatial variations of complex permittivity and permeability tensors described by Lagrange interpolation polynomials of arbitrary material-representation orders [28, 29], so there is no need for a discretization of the permittivity and permeability profiles of the cloak, namely, for piecewise homogeneous (layered) approximate models, with material tensors replaced by appropriate piecewise constant approximations; such a discretization is typical for all other existing approaches to FEM analysis of transformation-based metamaterial cloaks. Furthermore, because curved elements are used, a small number of elements suffice for the geometrical precision of the model, for curvature modeling. Finally, because higher order basis functions are implemented, there is no need for a discretization of the computational domain into electrically very small elements, with low-order basis functions for field/current modeling.

Approaches for construction and analysis of transformation-based metamaterial invisibility cloaks with other, non-spherical and non-circular, shapes, have been investigated as well. Most theoretical considerations of such cloaks are based on a straight-line compression of space, where there exists at least one point inside the cloak's void region positioned so that all of the cloak's outer surface points could be reached from it along straight lines inside the cloak [17, 30]. A more general approach is considered in [31], where the coordinate transformation is represented in the form of a homogeneous topological mapping where possible, while in the case of cloaks with arbitrary (potentially complex) shapes, for which homogeneous topological mapping functions cannot be (easily) obtained in a closed form, an approximation approach is proposed. Coordinate transformations for irregular particles in the form of ellipsoids, rounded cuboids, and rounded cylinders, respectively, are studied in [32], where the scattered field distribution is simulated using the discrete dipole approximation (DDA). As far as transformation-based metamaterial invisibility cloaks with sharp edges and corners are considered, on the other hand, [33] provides a design and demonstration of a square 2-D (cylinder with a square-shaped cross section) cloak. However, a generalization of this approach to the corresponding 3-D case, namely, to a construction, analysis, and evaluation of a cubical 3-D cloak, is certainly of great interest, from both theoretical and practical aspects.

Most recently, a conformal cubical transformation-based metamaterial invisibility cloak has been proposed [34], and is further discussed in this summary. The cloak is constructed invoking a classical coordinate transformation, similar to that in [33]. The numerical characterization is carried out employing the full-wave rigorous higher order anisotropic continuously inhomogeneous FEM modeling approach and the FEM-MoM hybridization. Additionally, we emphasize that for all situations where conformal cloaks for cloaking of objects with sharp edges and corners are employed, a full-wave numerical analysis is unavoidable, as these problems cannot be treated by the Mie scattering theory or any other analytical method.

4.2 Theory of Controlling EM Fields by Coordinate Transformation

We next present a simplified and unified theory of vector, tensor, and operator changes under coordinate transformations, and relate them specifically to electric and magnetic field intensity vectors, tensor permittivity and permeability of the medium, and the curl operator in Maxwell's equations. The presented theory sets a basis for arbitrary manipulations of EM fields by coordinate transformations, also known as the transformation electromagnetics or optics.

4.2.1 Derivation of Medium Parameter Tensors under Coordinate Transformation

Consider a time-harmonic EM field of radian frequency ω in a source-free domain. The field is governed by Maxwell's equations [35],

$$\nabla \times \mathbf{E} = -\mathrm{j}\omega \mathbf{B}, \tag{4.1a}$$

$$\nabla \times \mathbf{H} = \mathrm{j}\omega \mathbf{D}, \tag{4.1b}$$

$$\nabla \cdot \mathbf{D} = 0, \tag{4.1c}$$

$$\nabla \cdot \mathbf{B} = 0, \tag{4.1d}$$

where **E** and **H** are electric and magnetic field intensity vectors, **D** and **B** are electric and magnetic flux density vectors, $\mathrm{j} = \sqrt{-1}$ is the imaginary unit, and time-harmonic convention $\mathrm{e}^{\mathrm{j}\omega t}$ is used, with constitutive relations

$$\mathbf{D} = \overline{\overline{\varepsilon}} \cdot \mathbf{E}, \tag{4.2a}$$

$$\mathbf{B} = \overline{\overline{\mu}} \cdot \mathbf{H}, \tag{4.2b}$$

where $\overline{\overline{\varepsilon}}$ and $\overline{\overline{\mu}}$ stand for the tensor permittivity and permeability of the medium, respectively. From the vector identity $\nabla \cdot (\nabla \times \mathbf{F}) = 0$, where **F** is an arbitrary nonsingular vector function, we conclude that Eqs. (4.1c) and (4.1d) will be automatically satisfied provided that (4.1a) and (4.1b) are satisfied and $\omega \neq 0$; in other words, the EM field is completely described by Maxwell's first two equations, (4.1a) and (4.1b) and constitutive relations, (4.2a) and (4.2b), which, combined, give

$$\nabla \times \mathbf{E} = -j\omega\,\bar{\bar{\mu}} \cdot \mathbf{H}, \tag{4.3a}$$

$$\nabla \times \mathbf{H} = j\omega\,\bar{\bar{\varepsilon}} \cdot \mathbf{E}. \tag{4.3b}$$

We next consider two arbitrary (curved) coordinate systems. The systems will be designated as the g-coordinate system with coordinates g^1, g^2, and g^3 and the h-coordinate system with coordinates h^1, h^2, and h^3, where the two sets of coordinates are related by nonsingular single-valued analytical functions. Equation (4.3a) can equally be expressed in any of the coordinate systems; interpreted in the g-coordinate system, it becomes

$$(\nabla_{(g)} \times \mathbf{E})^i = -j\omega \sum_{j=1}^{3} \mu_{(g)}^{ij} H_j^{(g)}, i = 1,2,3, \tag{4.4a}$$

where $\mu_{(g)}^{ij}$ $(i,j = 1,2,3)$ represent the contravariant components of the medium permeability tensor in the g-coordinate system and $H_j^{(g)}$ $(j = 1,2,3)$ are the covariant components of the magnetic field intensity vector in the g-coordinate system. In addition, $(\nabla_{(g)}\times)^i$ $(i = 1,2,3)$ stands for the curl operator expressed by contravariant components in the g-coordinate system and defined as [36]

$$(\nabla_{(g)} \times \mathbf{E})^i = \frac{1}{J_{(g)}} \sum_{j,k=1}^{3} \frac{\partial E_k^{(g)}}{\partial g^j} e_{ijk}, i = 1,2,3, \tag{4.4b}$$

where $J_{(g)} = \mathbf{a}_1^{(g)} \cdot \left(\mathbf{a}_2^{(g)} \times \mathbf{a}_3^{(g)}\right)$ is the Jacobian of the g-coordinate system, $E_k^{(g)}$ $(k = 1,2,3)$ are the covariant components of the electric field intensity vector in this coordinate system, e_{ijk} $(i,j,k = 1,2,3)$ is the Levi-Civita symbol, and vectors $\mathbf{a}_i^{(g)}$ $(i = 1,2,3)$ are the primary vectors of the g-coordinate system [36]. Equation (4.3a) can then be interpreted in the h-coordinate system as follows:

$$(\nabla_{(h)} \times \mathbf{E})^i = -j\omega \sum_{j=1}^{3} \mu_{(h)}^{ij} H_j^{(h)}, i = 1,2,3, \tag{4.5a}$$

with an appropriate definition of the curl operator in that coordinate system,

$$(\nabla_{(h)} \times \mathbf{E})^i = \frac{1}{J_{(h)}} \sum_{j,k=1}^{3} \frac{\partial E_k^{(h)}}{\partial h^j} e_{ijk}, i = 1,2,3. \tag{4.5b}$$

Covariant components of the electric field intensity vector [$E_k^{(g)}$ and $E_k^{(h)}$ $(k = 1,2,3)$] represent the same vector in two coordinate systems; hence they have to be related in accordance with the rule of transformation of the vector covariant components between the two coordinate systems [37]. Using the index notation, this rule is given by

$$E_k^{(h)} = \sum_{m=1}^{3} \frac{\partial g^m}{\partial h^k} E_m^{(g)}, k = 1, 2, 3. \tag{4.6a}$$

By analogy, covariant components of the magnetic field intensity vector are transformed as

$$H_j^{(h)} = \sum_{n=1}^{3} \frac{\partial g^n}{\partial h^j} H_n^{(g)}, j = 1, 2, 3. \tag{4.6b}$$

Equating the corresponding terms in Eqs. (4.4a)-(4.5b) and applying the rules of covariant vector component transformations in Eqs. (4.6a) and (4.6b), we obtain that contravariant components of the medium permeability tensor are transformed between the two coordinate systems in the following way:

$$\mu_{(h)}^{ij} = \sum_{m,n=1}^{3} \frac{\partial h^i}{\partial g^n} \mu_{(g)}^{nm} \frac{\partial h^j}{\partial g^m}, i, j = 1, 2, 3, \tag{4.7a}$$

which can be recast in the matrix (tensor) notation

$$\bar{\bar{\mu}}_{(h)} = \mathbf{J} \bar{\bar{\mu}}_{(g)} \mathbf{J}^{\mathrm{T}}, \tag{4.7b}$$

with $\mathbf{J}$ being the 3×3 Jacobian transformation matrix between the g- and h-coordinate systems.

Starting from Maxwell's second equation (4.3b) and applying the relations analogous to Eqs. (4.4a)-(4.5b) and the covariant component vector transformation rule, Eqs. (4.6a) and (4.6b), it can be shown that the contravariant components of the permittivity tensor are transformed between the two coordinate systems as

$$\varepsilon_{(h)}^{ij} = \sum_{m,n=1}^{3} \frac{\partial h^i}{\partial g^n} \varepsilon_{(g)}^{nm} \frac{\partial h^j}{\partial g^m}, i, j = 1, 2, 3, \tag{4.8a}$$

of which the matrix (tensor) form is given by

$$\bar{\bar{\varepsilon}}_{(h)} = \mathbf{J} \bar{\bar{\varepsilon}}_{(g)} \mathbf{J}^{\mathrm{T}}. \tag{4.8b}$$

Maxwell's first equation written as in Eqs. (4.4a)-(4.5b) is referred to as the form-variant equation between the two coordinate systems; in general, $J_{(g)} \neq J_{(h)}$ in

Eqs. (4.4b) and (4.5b), and thus the curl operator form is different in the two coordinate systems. In the general case, the medium parameter tensors are transformed between the two coordinate systems as tensors of rank 2 with tensor density of weight 0 [37].

We next consider a different form of Eqs. (4.5a) and (4.5b) in the h-coordinate system. The main idea is to make it identical to the form of the corresponding Eqs. (4.4a) and (4.4b) in the g-coordinate system. Eqs. (4.5a) and (4.5b) thus become

$$(\hat{\nabla}_{(h)} \times \mathbf{E})^i = -\mathrm{j}\omega \sum_{j=1}^{3} \hat{\mu}^{ij}_{(h)} H_j^{(h)}, i = 1, 2, 3, \tag{4.9a}$$

$$(\hat{\nabla}_{(h)} \times \mathbf{E})^i = \frac{1}{J_{(g)}} \sum_{j,k=1}^{3} \frac{\partial E_k^{(h)}}{\partial h^j} e_{ijk}, i = 1, 2, 3, \tag{4.9b}$$

where $\hat{\nabla}_{(h)}\times$ represents the new (modified) curl operator in the h-coordinate system, obtained by replacing the Jacobian $J_{(h)}$ with $J_{(g)}$ in Eq. (4.5b). Contravariant components $\hat{\mu}^{ij}_{(h)}$ $(i, j = 1, 2, 3)$ of the new (modified) permeability tensor must change (with respect to $\mu^{ij}_{(h)}$) in order to compensate for the change of the operator form, so that vectors **E** and **H** in Eqs. (4.9a) and (4.9b) ultimately remain unchanged with respect to corresponding vectors in Eqs. (4.4a)-(4.4b). Maxwell's first equation written as in (4.9a) and (4.9b) is referred to as the form-invariant equation.

Similarly to the form-variant Maxwell's equations, it can be shown that contravariant tensors of the medium parameters for the form-invariant Maxwell's equations are transformed between the two coordinate systems as

$$\bar{\bar{\hat{\varepsilon}}}_{(h)} = \frac{\mathbf{J}\bar{\bar{\varepsilon}}_{(g)}\mathbf{J}^{\mathrm{T}}}{\det(\mathbf{J})}, \tag{4.10a}$$

$$\bar{\bar{\hat{\mu}}}_{(h)} = \frac{\mathbf{J}\bar{\bar{\mu}}_{(g)}\mathbf{J}^{\mathrm{T}}}{\det(\mathbf{J})}, \tag{4.10b}$$

namely, as tensors of rank 2 with tensor density of weight 1 [37–39].

Since the form of Eqs. (4.9a) and (4.9b) is now the same as that of Eqs. (4.4a) and (4.4b), covariant components of vectors **E** and **H** in the h-coordinate system [Eqs. (4.9a) and (4.9b)] can now be considered as the covariant components of **E** and **H** in the g-coordinate system [Eqs. (4.4a) and (4.4b)]. In addition, contravariant components of the media parameter tensors in the h-coordinate system [Eqs. (4.10a) and (4.10b)] can now be considered as contravariant components of the media parameter tensors in the g-coordinate system. Hence, a basis for arbitrary manipulations of EM fields constituting the main idea behind the transformation electromagnetics and optics is set: by choosing the media parameter

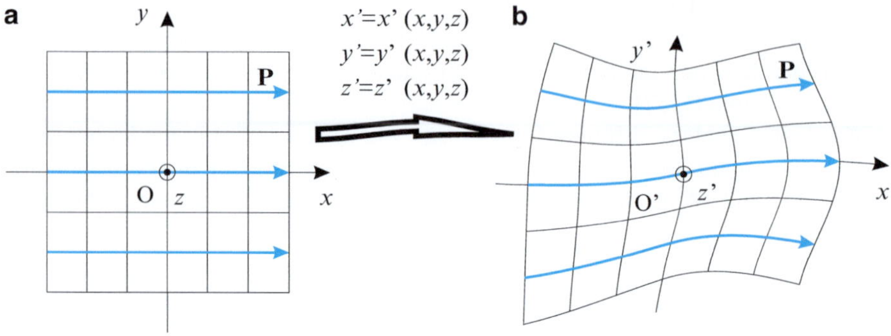

Fig. 4.1 Coordinate transformation causing deformation of light ray trajectories: (**a**) original rectangular coordinate system with parallel rays and (**b**) transformed curvilinear coordinate system with curved ray trajectories [24]

tensors as in Eqs. (4.10a) and (4.10b) and interpreting them in the original coordinate system, the EM field can be bent in a desired fashion.

4.2.2 Redirection of Optical Paths by Coordinate and Medium Transformation

Let us consider a homogeneous isotropic medium (e.g., air) with a traveling uniform plane wave, as shown in Fig. 4.1a. The light rays (or lines of the Poynting vector, $\mathbf{P} = \mathbf{E} \times \mathbf{H}^*$ [35], of the wave) are parallel to each other. Distorting the original space, with a spatial transformation applied to change the coordinate system (e.g., by employing appropriate mapping functions), will cause the rays to bend in the same manner, as depicted in Fig. 4.1b, where the original coordinate system (g-coordinate system) is denoted as (x, y, z) and the new system (h-coordinate system) is denoted using primed coordinates (x', y', z'). This is enabled by material relations in Eqs. (4.10a) and (4.10b), i.e., by

$$\bar{\bar{\varepsilon}}' = \frac{\mathbf{J}\bar{\bar{\varepsilon}}\mathbf{J}^{\mathrm{T}}}{\det(\mathbf{J})}, \quad \bar{\bar{\mu}}' = \frac{\mathbf{J}\bar{\bar{\mu}}\mathbf{J}^{\mathrm{T}}}{\det(\mathbf{J})}, \tag{4.11}$$

where $\bar{\bar{\varepsilon}}'$ and $\bar{\bar{\mu}}'$ are material property tensors in the transformed space, $\bar{\bar{\varepsilon}} = \varepsilon_0\mathbf{I}$ and $\bar{\bar{\mu}} = \mu_0\mathbf{I}$ ($\mathbf{I}$ is the 3×3 identity matrix) are the respective tensors in the original space (air), and $\mathbf{J}$ stands for the 3×3 Jacobi matrix of the transformation from $\mathbf{r} = x\mathbf{i}_x + y\mathbf{i}_y + z\mathbf{i}_z$ to $\mathbf{r}' = x'\mathbf{i}'_x + y'\mathbf{i}'_y + z'\mathbf{i}'_z$, with $\mathbf{i}$ and $\mathbf{i}'$ being the corresponding coordinate unit vectors.

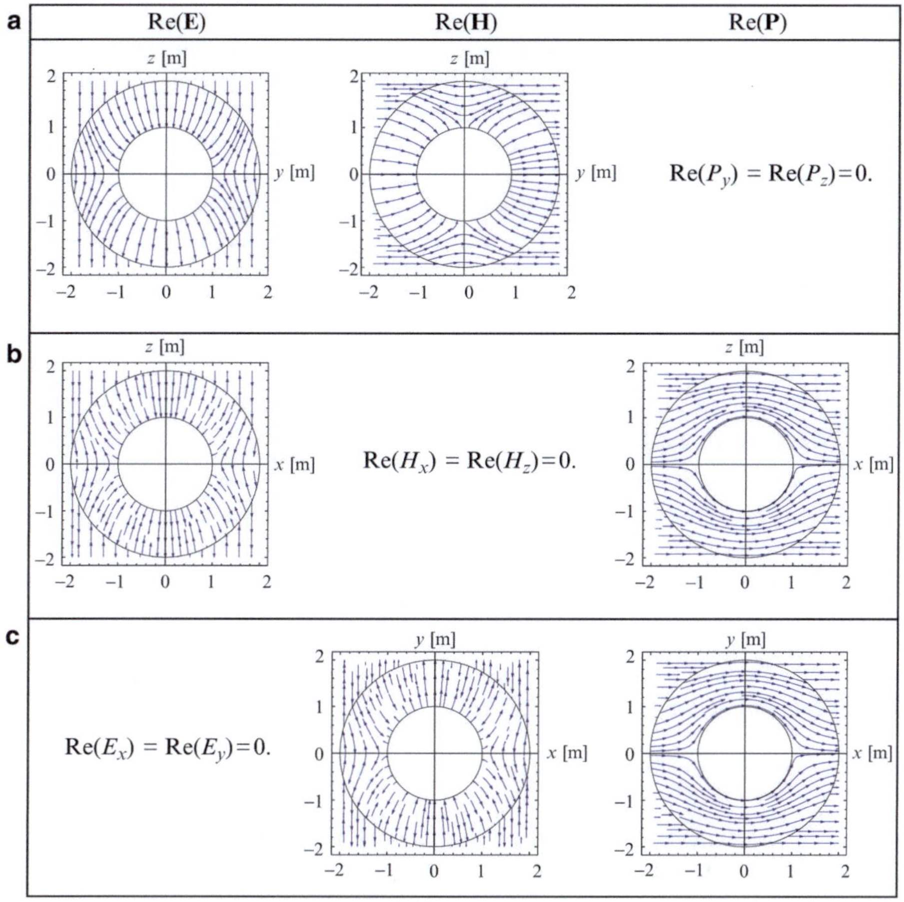

Fig. 4.2 Example of field deformation of Fig. 4.1 in a linear spherical cloak: lines of the real parts of the electric field (*left panel*), magnetic field (*central panel*), and Poynting vector (*right panel*) in three Cartesian coordinate planes containing the center of the cloak in Fig. 4.3 (for $R_2/R_1 = 2$) illuminated by a uniform plane wave: (**a**) $x = 0$ plane, (**b**) $y = 0$ plane, and (**c**) $z = 0$ plane

We next illustrate the complete process described above using the example of a uniform linear spherical metamaterial cloak that will be introduced in Sect. 4.3.1., in Fig. 4.3, and the details of which will be given in examples in Sect. 4.6.1. We assume that the outer to inner radius ratio of the cloak is $R_2/R_1 = 2$ and that the cloak is excited by a uniform plane wave with the free-space wavelength $\lambda_0 = 1$ m. Shown in Fig. 4.2 are the real parts of vectors **E**, **H**, and **P** in three characteristic planes (Cartesian coordinate planes $x = 0$, $y = 0$, and $z = 0$), when covariant components of vectors in the primed system (h-coordinate system), for the form-invariant Maxwell's equations, are interpreted as the corresponding covariant components of the vectors in the original (non-primed g-coordinate system). The field lines in Fig. 4.2 are obtained starting from a z-polarized uniform plane wave of wavelength $\lambda_0 = 1$ m traveling in free space in the x-direction. We then transform the electric and magnetic field vector components according to Eqs. (4.6a) and (4.6b), for the coordinate transformation given in Eq. (4.12). Thus obtained field components in the new coordinate system are interpreted as the components in the original coordinate system and plotted in Fig. 4.2.

From Fig. 4.2 it can be clearly seen that vectors **E** and **H** are always perpendicular to the surface $r = R_1$, while the Poynting vector is always tangential to that surface. This, in turn, directly implies that the space bounded by the surface $r = R_1$ is completely electromagnetically isolated from the surrounding space; the field neither penetrates it, nor is in any way perturbed by its presence.

4.3 Theory of Spherical Transformation-Based Metamaterial Cloaks

We present here the examples of coordinate transformations which lead to construction of linear and nonlinear spherical metamaterial cloaks, as well as the resulting transformations of the permittivity and permeability tensors in both spherical and Cartesian systems. Similar principles are then used in Sect. 4.4 in construction of a cubical metamaterial cloak, where only Cartesian system is employed. The performance of all presented cloak examples is finally verified by numerical simulations in Sect. 4.6.

4.3.1 Construction of Linear Spherical Metamaterial Cloaks

A linear spherical 3-D cloak is obtained using the following linear transformation described in the spherical coordinate system with the transformation domain $0 \leq r \leq R_2$, $0 \leq \phi < 2\pi$, and $-\pi/2 \leq \theta \leq \pi/2$ [8, 9]

$$r' = R_1 + \frac{R_2 - R_1}{R_2} r, \quad \theta' = \theta, \quad \phi' = \phi, \tag{4.12}$$

where R_1 and R_2 are the inner and outer radii, respectively, of the spherical cloak. A cross section of the visual representation of this transformation (in $z = 0$ plane) is shown in Fig. 4.3, for the original (**a**) and transformed (**b**) coordinate systems. The region $0 \leq r \leq R_2$ is mapped into the spherical shell (transformation range) $R_1 \leq r'$ $\leq R_2$ in the transformed coordinate system, with the Jacobi matrix of the transformation given by

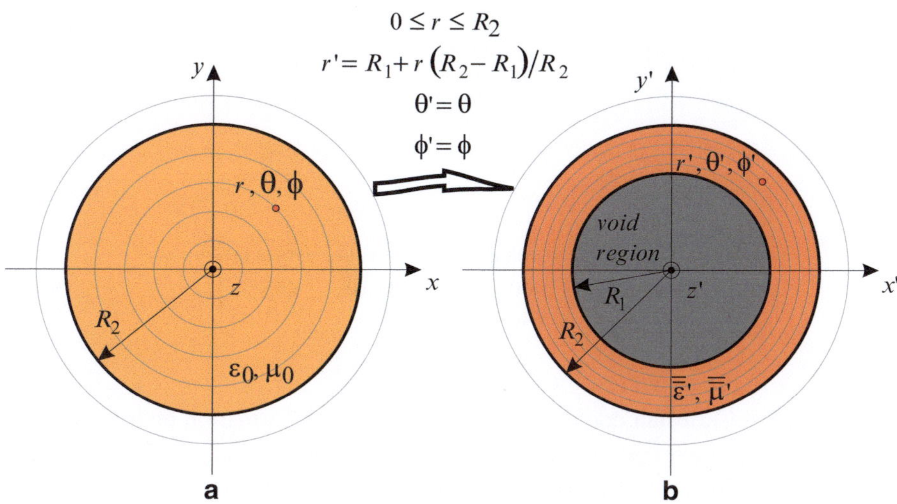

Fig. 4.3 Coordinate transformation yielding a spherical 3-D cloak: cross sections of (**a**) the original space with an air-filled homogeneous sphere and (**b**) the transformed space with a void region and an anisotropic continuously inhomogeneous spherical shell [24]

$$\mathbf{J} = \begin{bmatrix} \dfrac{R_2 - R_1}{R_2} & 0 & 0 \\ 0 & \dfrac{r'}{r} & 0 \\ 0 & 0 & \dfrac{r'}{r} \end{bmatrix}. \tag{4.13}$$

Substituting Eq. (4.13) into Eq. (4.11) results in the following expressions for permittivity and permeability tensors in the mapped space in the spherical coordinate system:

$$\frac{\overline{\overline{\varepsilon}}'(r',\theta',\phi')}{\varepsilon_0} = \frac{1}{\varepsilon_0}\begin{bmatrix} \varepsilon'_{rr}(r') & 0 & 0 \\ 0 & \varepsilon'_{\theta\theta} & 0 \\ 0 & 0 & \varepsilon'_{\phi\phi} \end{bmatrix} = \begin{bmatrix} \dfrac{R_2(R_1 - r')^2}{(R_2 - R_1)r'^2} & 0 & 0 \\ 0 & \dfrac{R_2}{R_2 - R_1} & 0 \\ 0 & 0 & \dfrac{R_2}{R_2 - R_1} \end{bmatrix} = \frac{\overline{\overline{\mu}}'}{\mu_0}. \tag{4.14}$$

Finally, $\overline{\overline{\varepsilon}}'(r',\theta',\phi')$ is analytically transformed to its Cartesian equivalent,

$$\bar{\bar{\varepsilon}}'(x',y',z') = [Q]\bar{\bar{\varepsilon}}'(r',\theta',\phi')[Q]^{\mathrm{T}} = \begin{bmatrix} \bar{\bar{\varepsilon}}'_{xx}(x',y',z') & \bar{\bar{\varepsilon}}'_{xy}(x',y',z') & \bar{\bar{\varepsilon}}'_{xz}(x',y',z') \\ \bar{\bar{\varepsilon}}'_{yx}(x',y',z') & \bar{\bar{\varepsilon}}'_{yy}(x',y',z') & \bar{\bar{\varepsilon}}'_{yz}(x',y',z') \\ \bar{\bar{\varepsilon}}'_{zx}(x',y',z') & \bar{\bar{\varepsilon}}'_{zy}(x',y',z') & \bar{\bar{\varepsilon}}'_{zz}(x',y',z') \end{bmatrix}, \tag{4.15}$$

where

$$[Q] = \begin{bmatrix} \sin\theta\cos\phi & \cos\theta\cos\phi & -\sin\phi \\ \sin\theta\sin\phi & \cos\theta\sin\phi & \cos\phi \\ \cos\theta & -\sin\theta & 0 \end{bmatrix} \tag{4.16}$$

is Spherical to Cartesian coordinate system vector transformation matrix and $[Q]^{\mathrm{T}} = [Q]^{-1}$ is its inverse (since $[Q]$ is orthogonal), and analogously for $\bar{\bar{\mu}}'$. Hence, the elements of the permittivity tensor in Eq. (4.15) are obtained to be

$$\begin{aligned} &\varepsilon'_{xx} = \frac{x'^2\varepsilon'_{rr} + (y'^2+z'^2)\varepsilon'_{\theta\theta}}{x'^2+y'^2+z'^2}, && \varepsilon'_{xy} = \frac{x'y'(\varepsilon'_{rr}-\varepsilon'_{\theta\theta})}{x'^2+y'^2+z'^2}, && \varepsilon'_{xz} = \frac{x'z'(\varepsilon'_{rr}-\varepsilon'_{\theta\theta})}{x'^2+y'^2+z'^2}, \\ &\varepsilon'_{yx} = \varepsilon'_{xy}, && \varepsilon'_{yy} = \frac{y'^2\varepsilon'_{rr} + (x'^2+z'^2)\varepsilon'_{\theta\theta}}{x'^2+y'^2+z'^2}, && \varepsilon'_{yz} = \frac{y'z'(\varepsilon'_{rr}-\varepsilon'_{\theta\theta})}{x'^2+y'^2+z'^2}, \\ &\varepsilon'_{zx} = \varepsilon'_{xz}, && \varepsilon'_{zy} = \varepsilon'_{yz}, && \varepsilon'_{zz} = \frac{z'^2\varepsilon'_{rr} + (x'^2+y'^2)\varepsilon'_{\theta\theta}}{x'^2+y'^2+z'^2}, \end{aligned} \tag{4.17}$$

where $\varepsilon'_{\theta\theta} = \varepsilon'_{\phi\phi}$, and analogous expressions hold for the components of $\bar{\bar{\mu}}'$.

4.3.2 Construction of Nonlinear Spherical Metamaterial Cloaks

In addition to linear-transformation 3-D spherical cloaks, two classes of nonlinear-transformation 3-D spherical cloaks are introduced in [21] and classified in terms of the positive (concave-up) or negative (concave-down) sign of the second derivative of the transformation function. The nonlinear transformation used to construct a spherical cloak can be analytically described in the transformation domain $0 \le r \le R_2$, $0 \le \phi < 2\pi$, and $-\pi/2 \le \theta \le \pi/2$ as

$$r' = \sqrt[l]{\frac{R_2^l - R_1^l}{R_2}r + R_1^l}, \quad \theta' = \theta, \quad \phi' = \phi, \tag{4.18}$$

for the concave-up case, and as

$$r' = R_1 R_2 \sqrt[l]{\frac{R_2}{R_2^{l+1} - (R_2^l - R_1^l) r}}, \quad \theta' = \theta, \quad \phi' = \phi, \tag{4.19}$$

for the concave-down case. In both cases, l represents the degree of the nonlinearity in the transformation (l does not need to be an integer). Cartesian equivalent tensors for the permittivity and permeability of a nonlinear spherical cloak are derived performing algebraic manipulations analogous to those in the linear spherical cloak case, i.e., as in Eq. (4.17). The resulting models are then used as additional examples in the evaluation of the higher order full-wave FEM-MoM technique, to demonstrate the validity of the method when more complex tensor profiles are involved.

4.4 Conformal Cubical Transformation-Based Metamaterial Cloaks

To construct a linear cubical 3-D cloak, we start with a rectangular version of the linear transformation in Eq. (4.12), namely, by introducing a simple coordinate transformation that maps the original Cartesian coordinate system (transformation domain), with x, y, z coordinates [Fig. 4.4a], into a transformed one (transformation range), with x', y', z' coordinates [Fig. 4.4b]:

$$x' = R_1 + \frac{R_2 - R_1}{R_2} x, y' = y\left(\frac{R_2 - R_1}{R_2} + \frac{R_1}{x}\right), z' = z\left(\frac{R_2 - R_1}{R_2} + \frac{R_1}{x}\right), \tag{4.20}$$

where R_1 and R_2 represent half-side lengths of the inner and the outer cubical shells, respectively. Note that the transformation given in Eq. (4.20) can be considered as a generalization of the transformation for the square 2-D cloak construction [33]. It maps one sixth of the original homogeneous isotropic (air filled in our case) cube ($0 < x \leq R_2$, $-R_2 \leq y \leq R_2$, $-R_2 \leq z \leq R_2$, and $|y|, |z| < |x|$) to one sixth of the transformed cubical shell ($R_1 < x' \leq R_2$, $-R_2 \leq y' \leq R_2$, $-R_2 \leq z' \leq R_2$, and $|y'|, |z'| < |x'|$), as shown by the shaded regions in Fig. 4.4, and similarly for the mappings of the remaining five parts of the cubical shell. This way, a hole [free of any exterior field and marked as the void region in Fig. 4.4b] is opened in the transformed space. Hence, the foundation is set for the cubical shell, comprised of six pyramidal frusta (which symmetrically surround the void region), to be designed as a cloak.

We next derive the material interpretation [9] of the space transformation given in Eq. (4.20) to obtain the (inhomogeneous and anisotropic) parameters of the cubical cloak in the form of permittivity and permeability tensors in the transformed space, $\overline{\overline{\varepsilon}}'$ and $\overline{\overline{\mu}}'$. To this end, we first calculate the Jacobi matrix in Eq. (10) for the transformation in Eq. (4.20) as

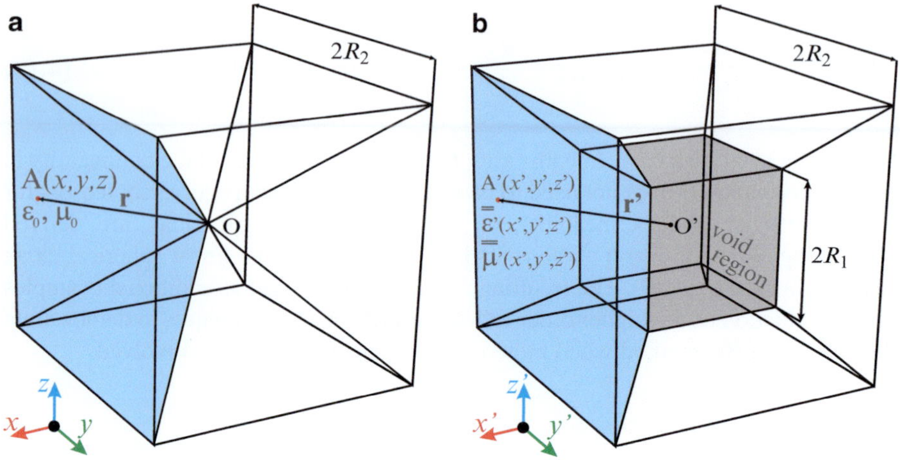

Fig. 4.4 Illustration of the effects of the coordinate transformation given in Eq. (4.20), which leads to the construction of the cubical cloak: (**a**) The original space with an air-filled homogeneous isotropic cube and (**b**) the transformed space with a void region and an anisotropic continuously inhomogeneous cubical-shell cloak comprised of six pyramidal frusta. One sixth of the cloak constructed by transformation in Eq. (4.20) is highlighted in both figures [34]

$$\mathbf{J} = \begin{bmatrix} \dfrac{(R_2 - R_1)}{R_2} & 0 & 0 \\ -\dfrac{R_1 y}{x^2} & \dfrac{(R_2 - R_1)}{R_2} + \dfrac{R_1}{x} & 0 \\ -\dfrac{R_1 z}{x^2} & 0 & \dfrac{(R_2 - R_1)}{R_2} + \dfrac{R_1}{x} \end{bmatrix}. \quad (4.21)$$

After simple algebra, from Eq. (4.20), we obtain the permittivity tensor components for one sixth of the cloak highlighted in Fig. 4.4b:

$$\begin{aligned}
&\varepsilon'_{xx}/\varepsilon_0 = a(R_1 - x')^2/x'^2, \varepsilon'_{yy}/\varepsilon_0 = a + aR_1{}^2 y'^2/x'^4, \\
&\varepsilon'_{zz}/\varepsilon_0 = a + aR_1{}^2 z'^2/x'^4, \varepsilon'_{xy}/\varepsilon_0 = \varepsilon'_{yx}/\varepsilon_0 = aR_1(R_1 - x')y'/x'^3, \\
&\varepsilon'_{xz}/\varepsilon_0 = \varepsilon'_{zx}/\varepsilon_0 = aR_1(R_1 - x')z'/x'^3, \varepsilon'_{yz}/\varepsilon_0 = \varepsilon'_{zy}/\varepsilon_0 = aR_1{}^2 y'z'/x'^4,
\end{aligned} \quad (4.22)$$

where $a = (R_2 - R_1)/R_2$ is introduced to simplify the final expressions, and similarly for the permeability tensor. For constructing the remaining five parts of the cloak, an analogous procedure can be used. Alternatively, the material-property tensor for one side of the cloak in Eq. (4.22) could be rotated around the Cartesian axes to obtain the material-property tensors for the remaining five parts of the cloak.

Nonlinear cubical cloaks, based on concave-up and concave-down nonlinear transformations similar to those in Eqs. (4.18) and (4.19), and for different nonlinearity degrees, are also possible.

As explained in Sect. 4.2.1, medium parameter tensors in the primed coordinate system will be interpreted as those in the non-primed (original) coordinate system. We can now omit the primes from the notation of coordinates and use (r, θ, ϕ) in place of (r', θ', ϕ') and (x, y, z) in place of (x', y', z') for medium tensor descriptions.

4.5 Higher Order EM Modeling and Analysis of Metamaterial Cloaks

In our full-wave 3-D computational EM model, the cloak is tessellated using Lagrange-type generalized curved parametric hexahedral finite elements of arbitrary geometrical orders K_u, K_v, and K_w (K_u, K_v, $K_w \geq 1$), shown in Fig. 4.5 and analytically described as [25]

$$\mathbf{r}(u,v,w) = \sum_{i=0}^{K_u}\sum_{j=0}^{K_v}\sum_{k=0}^{K_w} \mathbf{r}_{ijk}\, L_i^{K_u}(u)\, L_j^{K_v}(v)\, L_k^{K_w}(w), \quad -1 \leq u,v,w \leq 1, \tag{4.23}$$

where $\mathbf{r}_{ijk} = \mathbf{r}(u_i, v_j, w_k)$ are position vectors of interpolation nodes and $L_i^{K_u}(u)$ represent Lagrange interpolation polynomials in the u coordinate, with u_i being the uniformly spaced interpolating nodes defined as $u_i = (2i - K_u)/K_u$, $i = 0, 1, \ldots, K_u$, and similarly for $L_j^{K_v}(v)$ and $L_k^{K_w}(w)$.

The electric field vector in each element (Fig. 4.5), $\mathbf{E}(u, v, w)$, is approximated by means of curl-conforming (automatically satisfying the boundary condition for the tangential component of $\mathbf{E}$ at interfaces between elements) hierarchical (each lower-order set of functions is a subset of all higher-order sets) polynomial vector basis functions $\mathbf{f}$ as follows [25]:

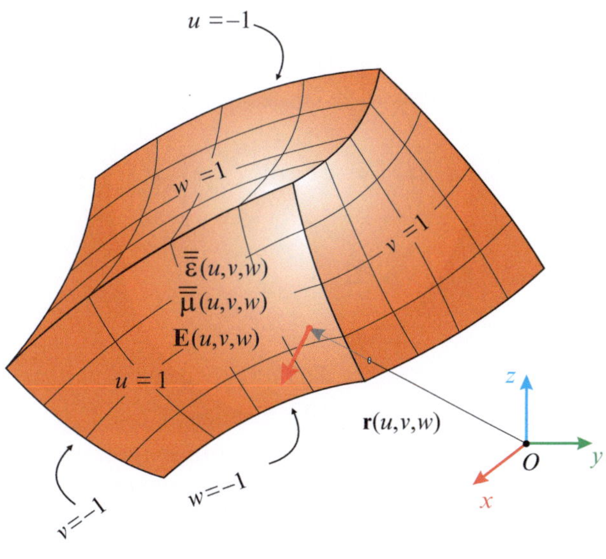

Fig. 4.5 Generalized curved Lagrange-type parametric hexahedral finite element defined by Eq. (4.23), with continuous spatial variations of complex permittivity and permeability tensors of the material defined by Eq. (4.29) [29]

$$\mathbf{E} = \sum_{m=0}^{N_u-1}\sum_{n=0}^{N_v}\sum_{p=0}^{N_w} \alpha_{mnp}^{(u)}\,\mathbf{f}_{mnp}^{(u)} + \sum_{m=0}^{N_u}\sum_{n=0}^{N_v-1}\sum_{p=0}^{N_w} \alpha_{mnp}^{(v)}\,\mathbf{f}_{mnp}^{(v)} + \sum_{m=0}^{N_u}\sum_{n=0}^{N_v}\sum_{p=0}^{N_w-1} \alpha_{mnp}^{(w)}\,\mathbf{f}_{mnp}^{(w)}, \tag{4.24}$$

with N_u, N_v, and N_w (N_u, N_v, $N_w \geq 1$) being arbitrarily high field-approximation orders within the element. The basis functions for different field components are defined by

$$\begin{aligned} \mathbf{f}_{mnp}^{(u)} &= u^m P_n(v) P_p(w)\,\mathbf{a}^u \\ \mathbf{f}_{mnp}^{(v)} &= P_m(u)\,v^n P_p(w)\,\mathbf{a}^v, \\ \mathbf{f}_{mnp}^{(w)} &= P_m(u) P_n(v)\,w^p \mathbf{a}^w \end{aligned} \quad P_m(u) = \begin{cases} 1-u, & m=0 \\ u+1, & m=1 \\ u^m-1, & m\geq 2,\ \text{even} \\ u^m-u, & m\geq 3,\ \text{odd} \end{cases}, \tag{4.25}$$

where the reciprocal unitary vectors $\mathbf{a}^u$, $\mathbf{a}^v$, and $\mathbf{a}^w$ are obtained as

$$\mathbf{a}^u = \frac{\mathbf{a}_v \times \mathbf{a}_w}{J}, \qquad \mathbf{a}^v = \frac{\mathbf{a}_w \times \mathbf{a}_u}{J}, \qquad \mathbf{a}^w = \frac{\mathbf{a}_u \times \mathbf{a}_v}{J}, \tag{4.26}$$

from the unitary vectors $\mathbf{a}_u$, $\mathbf{a}_v$, $\mathbf{a}_w$ and the Jacobian (J) of the covariant transformation,

$$\mathbf{a}_u = \frac{\mathrm{d}\mathbf{r}}{\mathrm{d}u}, \qquad \mathbf{a}_v = \frac{\mathrm{d}\mathbf{r}}{\mathrm{d}v}, \qquad \mathbf{a}_w = \frac{\mathrm{d}\mathbf{r}}{\mathrm{d}w}, \qquad J = (\mathbf{a}_u \times \mathbf{a}_v)\cdot\mathbf{a}_w, \tag{4.27}$$

with $\mathbf{r}$ given in Eq. (4.23). In Eq. (4.24), the basis functions are multiplied by unknown complex field-distribution coefficients $\{\alpha\}$, which are determined in a numerical solution of the standard Galerkin weak-form discretization of the electric-field vector wave equation

$$\int_V \overline{\overline{\mu}}^{-1}(\nabla \times \mathbf{f}_{\hat{m}\hat{n}\hat{p}})\cdot(\nabla \times \mathbf{E})\,\mathrm{d}V - k_0^2 \int_V \overline{\overline{\varepsilon}}\mathbf{f}_{\hat{m}\hat{n}\hat{p}}\cdot\mathbf{E}\,\mathrm{d}V = \mathrm{j}k_0 Z_0 \oint_S \mathbf{f}_{\hat{m}\hat{n}\hat{p}}\cdot\mathbf{n}\times\mathbf{H}\,\mathrm{d}S, \tag{4.28}$$

where V is the volume of the FEM computational region (bounded by the surface S), $\mathbf{f}_{\hat{m}\hat{n}\hat{p}}$ (standing for any of the functions $\mathbf{f}_{\hat{m}\hat{n}\hat{p}}^{(u)}$, $\mathbf{f}_{\hat{m}\hat{n}\hat{p}}^{(v)}$ or $\mathbf{f}_{\hat{m}\hat{n}\hat{p}}^{(w)}$) are testing functions [the same as basis functions in Eqs. (4.24)-(4.27)], k_0 and Z_0 are the free-space wave number and intrinsic impedance, respectively, and $\mathbf{n}$ is the outward-looking normal on S. The tangential component of $\mathbf{H}$ over S appearing on the right-hand side of Eq. (4.28) is determined by the appropriate boundary conditions imposed at the surface, providing a foundation for an FEM mesh termination, i.e., for a numerical interface between the FEM domain and the remaining, unbounded, space.

The accuracy of an existing higher order solution can in general be improved by refining the mesh (*h*-refinement), increasing the field-approximation orders (N_u, N_v, N_w) over elements without subdividing the elements (*p*-refinement), or combining the two approaches (*hp*-refinement) [40].

Continuous spatial variations of Cartesian components of the medium tensors $\overline{\overline{\varepsilon}}$ and $\overline{\overline{\mu}}$ in the FEM generalized hexahedra, in Fig. 4.5, are implemented using the same Lagrange interpolating scheme for defining element spatial coordinates in Eq. (4.23) [28, 29], as follows:

$$\overline{\overline{\varepsilon}}(u,v,w) = \begin{bmatrix} \overline{\overline{\varepsilon}}_{xx}(u,v,w) & \overline{\overline{\varepsilon}}_{xy}(u,v,w) & \overline{\overline{\varepsilon}}_{xz}(u,v,w) \\ \overline{\overline{\varepsilon}}_{yx}(u,v,w) & \overline{\overline{\varepsilon}}_{yy}(u,v,w) & \overline{\overline{\varepsilon}}_{yz}(u,v,w) \\ \overline{\overline{\varepsilon}}_{zx}(u,v,w) & \overline{\overline{\varepsilon}}_{zy}(u,v,w) & \overline{\overline{\varepsilon}}_{zz}(u,v,w) \end{bmatrix}$$
$$= \sum_{q=0}^{M_u}\sum_{s=0}^{M_v}\sum_{t=0}^{M_w} \overline{\overline{\varepsilon}}_{qst}\, L_q^{M_u}(u)\, L_s^{M_v}(v)\, L_t^{M_w}(w), \; -1 \le u,v,w \le 1, \tag{4.29}$$

with M_u, M_v, and M_w (M_u, M_v, M_w $\geq$ 1) standing for arbitrary material-representation polynomial orders within the element and $\overline{\overline{\varepsilon}}_{qst} = \overline{\overline{\varepsilon}}(u_q, v_s, w_t)$ for the permittivity values at the points defined by position vectors of spatial interpolation nodes, $\mathbf{r}_{qst}$, corresponding to orders M_u, M_v, and M_w. Geometrical-mapping orders (K_u, K_v, K_w) in Eq. (4.23), field-expansion orders (N_u, N_v, N_w) in Eq. (4.24), and material-representation orders (M_u, M_v, M_w) in Eq. (4.29) are entirely independent from each other, and the three sets of parameters of a higher order model can be combined independently for the best overall performance of the method. Furthermore, all of the parameters can be adopted anisotropically in different directions within an element and nonuniformly from element to element in a model. This way, the generally adopted (and widely used by many researchers [10, 14, 17, 33]) strategy to treat the continuously inhomogeneous media by approximating them by layers with constant permittivity and permeability (which leads to unnecessarily finer finite element meshes and larger number of unknowns) is avoided. The strengths of the higher order modeling can thus be fully exploited, i.e., the elements can be kept large, regardless of the material-parameter variations, which ultimately results in highly efficient electromagnetic models [28]. Applied to the linear and nonlinear cloak modeling, the permittivity values are computed as $\overline{\overline{\varepsilon}}_{qst} = \overline{\overline{\varepsilon}}(x_q, y_s, z_t)$ using Eqs. (4.15) and (4.16) or (4.17), where (x_q, y_s, z_t) are Cartesian components of the position vector $\mathbf{r}_{qst} = \mathbf{r}(u_q, v_s, w_t)$, from Eq. (4.23), and similarly for $\overline{\overline{\mu}}(u,v,w)$.

For the FEM mesh truncation, we introduce, in accordance to the surface equivalence principle (generalized Huygens' principle), a set of unknown electric and magnetic equivalent surface currents, $\mathbf{J}_\mathrm{s}$ and $\mathbf{M}_\mathrm{s}$, at the outer surface S of the cloak, which are defined on curved quadrilateral patches representing external faces of the FEM hexahedra (in Fig. 4.5). The currents are expanded using a divergence-conforming (automatically satisfying continuity conditions for normal components of $\mathbf{J}_s$ and $\mathbf{M}_s$ at element joints) 2-D (in parametric coordinates u and v) version of FEM basis functions in Eq. (4.25), with unknown current-distribution coefficients being evaluated by the method of moments [26]. Scattered electric and magnetic fields due to $\mathbf{J}_s$ and $\mathbf{M}_s$, $\mathbf{E}_{\text{MoM}}$ and $\mathbf{H}_{\text{MoM}}$, and incident fields (due to an incident plane wave), $\mathbf{E}_{\text{inc}}$ and $\mathbf{H}_{\text{inc}}$, are coupled to the electric field in the FEM domain, in the cloak region, $\mathbf{E}_{\text{FEM}}$, given by Eq. (4.24) in individual finite elements, through boundary conditions for the tangential field components on the cloak external surface,

$$\begin{aligned} \mathbf{n} \times \mathbf{E}_{\text{MoM}}(\mathbf{J}_S, \mathbf{M}_S) + \mathbf{n} \times \mathbf{E}_{\text{inc}} &= \mathbf{n} \times \mathbf{E}_{\text{FEM}}, \\ \mathbf{n} \times \mathbf{H}_{\text{MoM}}(\mathbf{J}_S, \mathbf{M}_S) + \mathbf{n} \times \mathbf{H}_{\text{inc}} &= \mathbf{J}_S, \end{aligned} \tag{4.30}$$

thus providing the computational interface between the FEM and MoM regions, with $\mathbf{E}_{\text{FEM}}$, $\mathbf{J}_S$, and $\mathbf{M}_S$ as unknowns, and giving rise to a hybrid higher order FEM-MoM solution [27].

4.6 Results and Discussion

We next consider several examples of spherical and cubical transformation-based metamaterial invisibility cloaks in Figs. 4.3 and 4.4, respectively, with void regions filled with a perfect electric conductor (PEC). We perform full-wave rigorous modeling and analysis of the cloaking structures using the higher order FEM-MoM technique based on large continuously inhomogeneous anisotropic curved hexahedral finite elements in Fig. 4.5, with high-order polynomial field expansions. All simulations are performed on a modest Windows-7 PC with Intel® Core™ i5-760 CPU @ 2.8 GHz and 8 GB RAM.

4.6.1 Examples of Spherical Transformation-Based Metamaterial Cloaks

Consider a spherical scatterer of diameter $d = 2R_1$, situated in air, as shown in Fig. 4.6. The sphere is made of a PEC and its surface is geometrically modeled by six fourth-order ($K_u = K_v = 4$) curved quadrilateral patches. The sphere is coated by a spherical layer (a cloak) with outer radius R_2 and permittivity and permeability tensors $\overline{\overline{\varepsilon}}$ and $\overline{\overline{\mu}}$ given by Eq. (4.14) for a linear cloak and by Eqs. (4.18) and (4.19) for nonlinear ones. We analyze the far-field scattering properties of the cloaked sphere excited by a θ-polarized plane wave, incident from the direction defined by $\theta_{\text{inc}} = 90°$ and $\phi_{\text{inc}} = 0°$, as indicated in Fig. 4.6. The cloak is modeled using only six large cushion-like fourth-order ($K_u = K_v = K_w = 4$) curved hexahedral finite elements (Fig. 4.6), terminated by a PEC boundary condition on the inner side

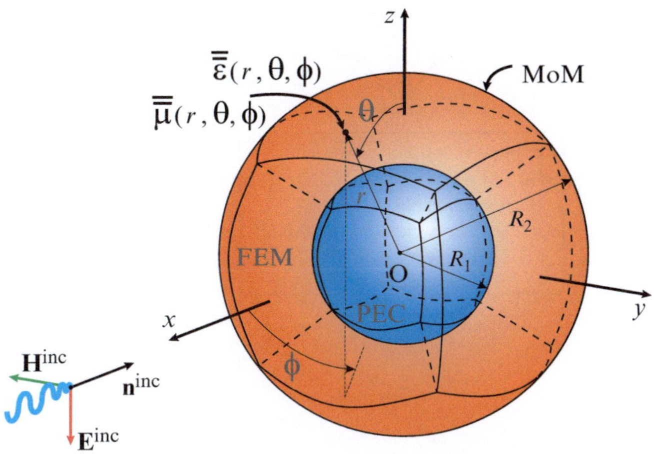

Fig. 4.6 FEM-MoM model of a PEC spherical scatterer with a spherical-shell cloak: the cloak is modeled using only six large continuously inhomogeneous anisotropic curved finite elements with high-order polynomial field expansions in parametric coordinates [24]

$(r = R_1)$ and by six large conformal fourth-order MoM quadrilateral elements, with equivalent surface electric and magnetic currents, on the outer side $(r = R_2)$. Continuous variations of the permittivity and permeability tensor components of the cloak are modeled within the FEM volume elements using fourth-order $(M_u = M_v = M_w = 4)$ Lagrange interpolating polynomials (where the inhomogeneity control nodes coincide with the nodes guiding the FEM element geometry). The adopted field approximation orders are $N_u = N_v = N_w = 6$ for all FEM hexahedra, which results in only 3,900 FEM unknowns (unknown field-distribution coefficients). The adopted current approximation orders are $N_u = N_v = 5$ for all MoM patches, yielding a total of as few as 600 MoM unknowns (current-distribution coefficients). The overall simulation takes 150 s of computational time per frequency (wavelength) point for analysis on the modest PC used. Note that the reduction in the number of unknowns is by two orders of magnitude when compared to FEM analysis in [22], where computational times are not given.

First, we assume that $R_2/R_1 = 1.1$ in Eq. (4.12) (a thin linear cloak) and show in Fig. 4.7 the normalized backscattering cross section, $\sigma/(R_1^2\pi)$, of the cloaked sphere obtained by the rigorous full-wave numerical FEM-MoM technique, against the normalized diameter d/λ_0 (λ_0 being the free space wavelength) of the PEC sphere. The results include those for the lossless cloak, as well as for lossy ones with the loss tangent ranging from 0.0001 to 0.01, obtained with the original 6-element model (shown in Fig. 4.6) and with an h-refined 24-element model, where each of the 6 original elements is symmetrically subdivided into 2×2 elements along the circumferential directions. For the purpose of validation of the numerical solution, the computed scattering cross section of the uncloaked sphere, with the continuously inhomogeneous anisotropic FEM elements constituting the cloaking layer

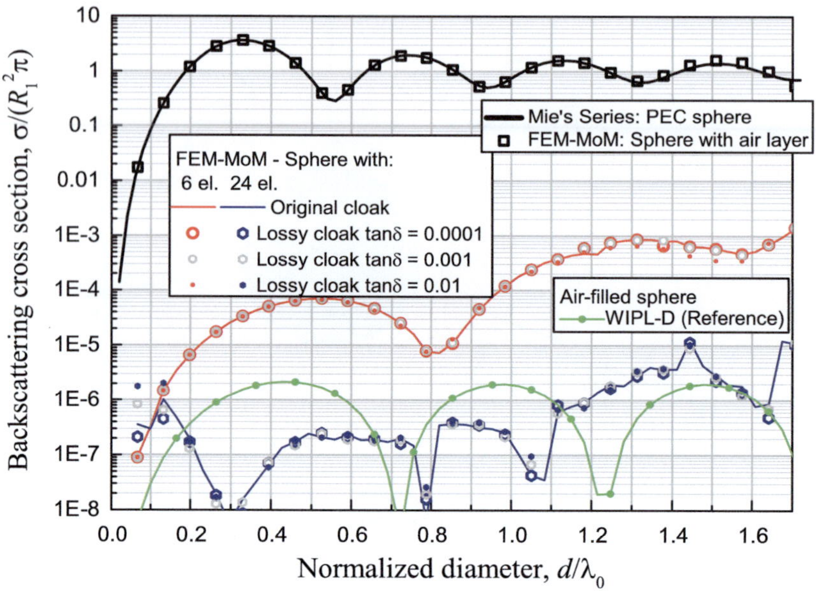

Fig. 4.7 Normalized backscattering cross sections of the PEC sphere with lossless and lossy linear cloaks and $R_2/R_1 = 1.1$ in Fig. 4.6, and of the uncloaked sphere, with the cloak replaced by a homogeneous air layer, obtained by the full-wave rigorous FEM-MoM numerical analysis vs. the normalized sphere diameter (λ_0 is the free-space wavelength) [24]. The uncloaked sphere is analyzed using the same 6-element numerical model as the cloaked ones, and the results are compared with the exact solution in the form of Mie's series. Also shown is the backscatter of a homogeneous air-filled sphere, obtained using WIPL-D (pure MoM commercial code), as a reference for verification of the best numerical approximation of the zero backscatter from an empty spherical region of the same size as the original scatterer

being replaced by homogeneous air-filled elements having all field and current expansions and other parameters in FEM and MoM analyses the same as in the 6-element cloak model, is also shown in Fig. 4.7, where it is compared with the exact Mie's series solution, and an excellent agreement of the two sets of results is observed. In addition, a fully converged *hp*-refined pure MoM solution for a homogeneous air-filled sphere, obtained by the commercial software tool for full wave EM analysis WIPL-D [41], is shown as a reference, giving a clear insight into what the best numerical solution for the given geometry and an ideal invisibility material (scattering from free-space) would be. Based on the cloaking numerical results in Fig. 4.7, excellent convergence properties of the method with significant reductions (three to five orders of magnitude for the 6-element model and five to seven orders of magnitude for the 24-element model, the two models being the two roughest possible) in the scattering cross section of the cloaked sphere is observed in the entire analyzed range of wavelengths, even though this is a rather thin cloak (its thickness is only one tenth of the PEC sphere radius). While having in mind that the cloak is theoretically ideal (backscatter theoretically vanishes), we note here that the (still rather rough) 24-element model yields a backscatter so low that it is on

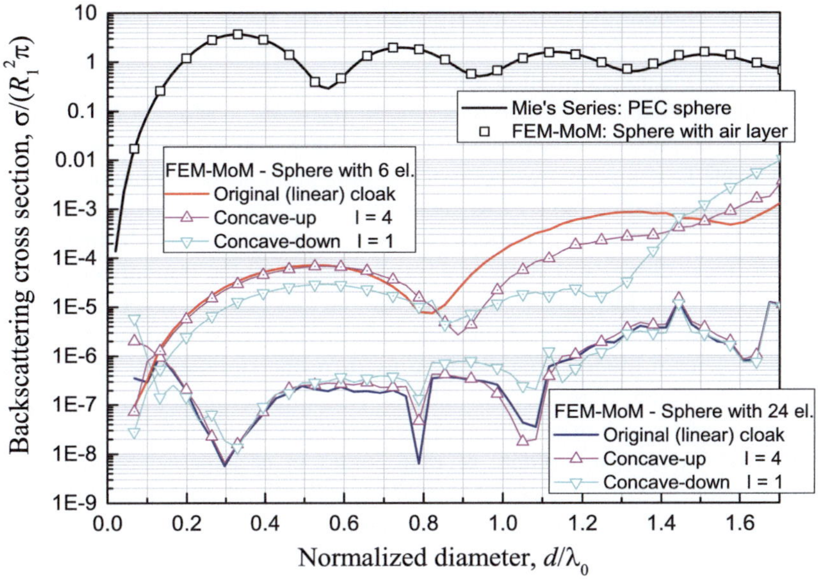

Fig. 4.8 Normalized backscattering cross sections of the PEC sphere with lossless nonlinear cloaks and $R_2/R_1 = 1.1$ in Fig. 4.6 computed by the higher order FEM-MoM technique vs. the normalized sphere diameter [24]. The nonlinear cloaks include both the concave-up and the concave-down cases with two nonlinearity degrees in the transformations [Eqs. (4.18) and (4.19) with $l = 4$ and $l = 1$, respectively]. The results for the uncloaked PEC sphere and for the sphere with linear cloaks (repeated from Fig. 4.7) are also included for easier comparison

par with the best numerical approximation of the zero backscatter from an empty spherical region of the same size as the original scatterer, as verified by WIPL-D and a pure surface (MoM) model. Note also that, given that some possible future applications of cloaking devices would likely include thin (for practical reasons) conformal cloaking layers on complex 3-D geometries, thin cloaks, as well as their analysis and design based on numerical simulations, rather than purely analytical methods, should be of great interest in the near future. A notable example of thin cloaks based on plasmonic cloaking and scattering cancellation techniques is given in [42]. Additionally, we see in Fig. 4.7 that adding loss to the cloaking material practically does not influence the numerical solution for backscattering.

We next analyze a thin nonlinear cloak, based on concave-up and concave-down nonlinear transformations, in Eqs. (4.18) and (4.19), and for two different nonlinearity degrees, namely, for $l = 4$ and $l = 1$, respectively, using, however, the same geometrical models, the same field and current expansions, and the same material-representation orders as in the previous example (results in Fig. 4.7). From Fig. 4.8, showing the normalized backscattering cross section of the cloaked PEC sphere, we observe that the higher order FEM-MoM simulation results for the nonlinear cloaks match the results for the linear cloaks very well and, hence, that the proposed analysis method can handle nonlinear coordinate transformations equally well, without the necessity to subdivide the mesh or introduce piecewise constant (e.g., layered) material regions. We also remark that in cases where the nonlinearity of the transformation is even more pronounced, material-representation polynomials of orders higher than four (orders $M_u = M_v = M_w = 4$ are found to be sufficient in the presented examples) can be employed in Eq. (4.29) to allow for a more adequate Lagrange interpolation of the more pronounced continuous change of media parameters within large FEM elements.

As the last example of spherical cloaks, we adopt $R_2 = 1.5R_1$ in Eq. (4.12) (a thicker linear cloak) and show in Figs. 4.9 and 4.10 the normalized scattering cross sections of the PEC sphere with and without lossy cloaks, respectively, for the normalized sphere diameter of $d/\lambda_0 = 0.3$, in two characteristic planes. The simplest 6-element models (Fig. 4.6) are used for the simulations. For the cloaks with the loss tangent ranging from 0.05 to 1, numerical results are compared to exact Mie's series solutions obtained as presented in [20]. A very substantial reduction (three to seven orders of magnitude) in σ achieved by the cloak is observed, with an excellent agreement of numerical and analytical solutions for all scattering angles except for those approaching the backscattering angle (appearing in the middle of each of the graphs), at which the scattering cross section theoretically vanishes. However, we realize that all numerical backscatter solutions are practically the same, regardless of loss, and again identify the lowest cross section limit that the code and the adopted model and single machine precision can numerically achieve in this example. Note, however, that the predicted backscatter is approximately two orders of magnitude lower than that for the thin cloak for the same particle electrical size and the same simple FEM-MoM model, namely, $\sigma/\left(R_1^2\pi\right) \approx 2 \times 10^{-7}$ for the thick cloak vs. $\sigma/\left(R_1^2\pi\right) \approx 3.5 \times 10^{-5}$ for the thin one, implying that thicker cloaks are numerically easier to solve, which, in turn, may be attributed to longer paths for the fields to relax.

4.6.2 Examples of Cubical Transformation-Based Metamaterial Cloaks

Consider next a PEC cubical scatterer of side length $d = 2R_1$ situated in air. The scatterer is illuminated by a θ-polarized plane wave, incident from the $\theta_{inc} = 90°$ and $\phi_{inc} = 0$ direction, as shown in Fig. 4.11. The scatterer is coated with a cubical conformal layer (a cloak) of half-side length R_2 with material tensors $\bar{\bar{\varepsilon}}$ and $\bar{\bar{\mu}}$ obtained from Eq. (4.22). The cloaking region is modeled by means of only 24 geometrically first-order ($K_u = K_v = K_w = 1$) large hexahedral finite elements (4 finite elements for each pyramidal frustum), terminated by a PEC boundary condition on the inner side and by 24 large conformal geometrically first-order MoM quadrilateral patches (with equivalent surface electric and magnetic currents) on the outer side, which provide a rigorous numerical FEM domain truncation. Spatial variation of material parameters within the FEM hexahedra is approximated

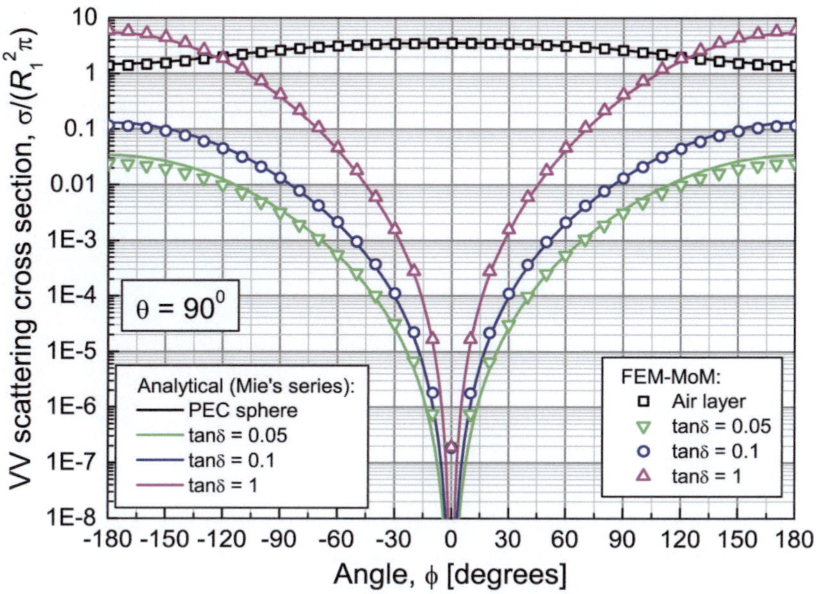

Fig. 4.9 Normalized VV scattering cross section of the cloaked and uncloaked PEC sphere in Fig. 4.6 for lossy linear cloaks, with $R_2/R_1 = 1.5$ and the normalized sphere diameter of $d/\lambda_0 = 0.3$ [24]

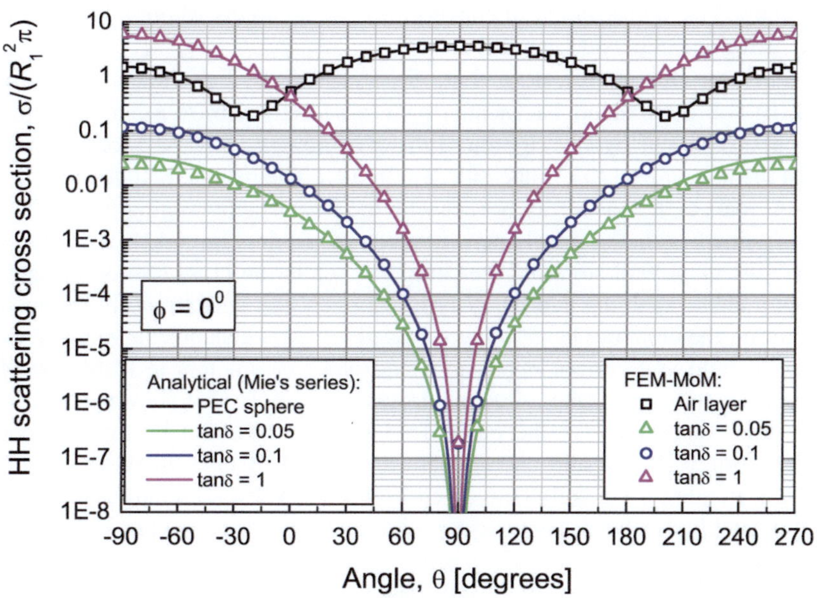

Fig. 4.10 Normalized HH scattering cross section of the structure in Fig. 4.6 (lossy linear cloaks, $R_2/R_1 = 1.5$, and $d/\lambda_0 = 0.3$) [24]

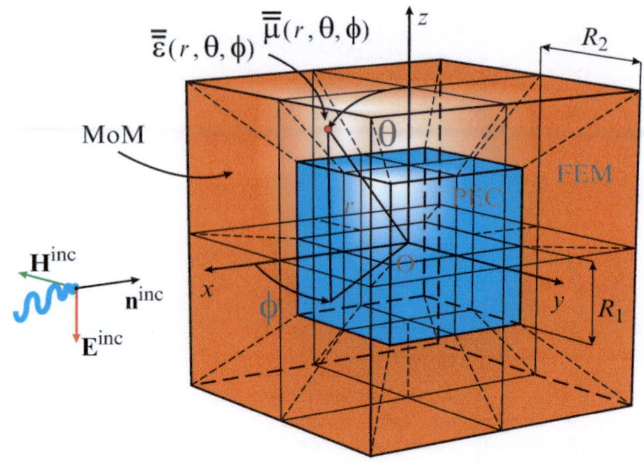

Fig. 4.11 FEM-MoM model of a PEC cubical scatterer with a cubical-shell cloak: the cloak is modeled using 24 large continuously inhomogeneous anisotropic finite elements with high-order polynomial field expansions in parametric coordinates [34]

by the fourth-order ($M_u = M_v = M_w = 4$) Lagrange polynomials. The adopted field approximation orders are $N_u = N_v = N_w = 6$ for all FEM hexahedra, which results in only 15,564 FEM unknowns (unknown field-distribution coefficients). The adopted current approximation orders are $N_u = N_v = 5$ for all MoM patches, yielding a total of as few as 2,400 MoM unknowns (unknown current-distribution coefficients). The total simulation time on the modest PC used is around 3 min per frequency (wavelength) point. Note that the analysis of the same structure using COMSOL Multiphysics employs as many as 82,549 small finite elements, 536,968 FEM unknowns, and it failed to converge for wavelengths where $d/\lambda_0 > 0.66$ ($d = 2R_1$ being the cube-side length).

We assume that the cloak is thin, $R_2/R_1 = 1.1$ in Eq. (4.20), and show in Fig. 4.12 the normalized backscattering cross section, $\sigma/(R_1^2\pi)$, of the cloaked PEC cube obtained by the rigorous full-wave numerical FEM-MoM technique, against the normalized PEC cube-side length d/λ_0. The results include those for the lossless cloak, as well as for lossy ones with the loss tangent ranging from 0.0001 to 0.01, obtained with the 24-element model (shown in Fig. 4.11). To rigorously validate the numerical solution, the computed scattering cross section of the cube without a cloak, with the continuously inhomogeneous anisotropic FEM elements constituting the cloaking layer being replaced by homogeneous air-filled elements having all field and current expansions and other parameters in FEM and MoM analyses the same as in the 24-element cloak model, is also shown in Fig. 4.12, where it is compared with the pure MoM solution obtained by the commercial software tool WIPL-D, and an excellent agreement of the two sets of results is observed. In addition, a fully converged *hp*-refined WIPL-D solution for a homogeneous air-filled cube is shown as an additional reference, giving a clear insight into what the best numerical solution for the given geometry and an ideal invisibility material (scattering from free-space) would be. Based on the cloaking

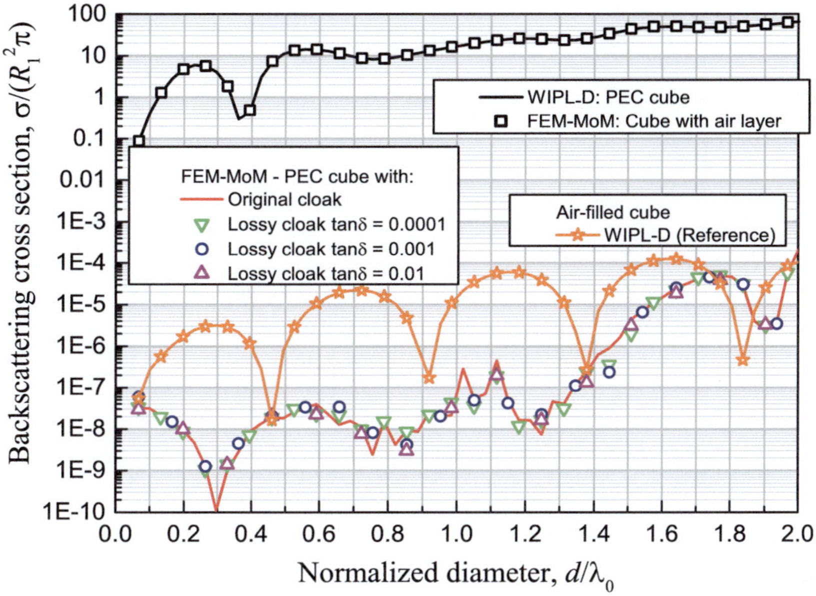

Fig. 4.12 Normalized backscattering cross section of the cloaked PEC cube with $R_2/R_1 = 1.1$ in Fig. 4.11, including lossless (original) and lossy cloaks and an uncloaked PEC cube (the cloak shell replaced by an air layer), obtained by the higher order full-wave rigorous FEM-MoM numerical analysis vs. the normalized PEC cube side length (λ_0 is the free-space wavelength) [34]. FEM-MoM results for the uncloaked PEC cubical scatterer (PEC cube with the air layer, analyzed using the same 24-element numerical model as the cloaked cube) are compared with WIPL-D results for a PEC scatterer. WIPL-D results for a homogeneous air-filled cube are also shown, as a reference for verification of the best numerical approximation of the zero backscatter from an empty cubical region of the same size as the original scatterer

numerical results in Fig. 4.12, significant reductions (five to ten orders of magnitude) in the scattering cross section of the cloaked PEC cube is observed in the entire analyzed range of wavelengths. While having in mind that the cloak is theoretically ideal (backscatter theoretically vanishes), we note here that the presented 24-element model yields a backscatter so low that it is on par with the best numerical approximation of the zero backscatter from an empty cubical region of the same size as the original scatterer, as verified by WIPL-D and a pure surface (MoM) model. Note that the far-field numerical results for the cubical cloak are similar or better than the results obtained for the linear spherical cloak in Sect. 4.6.1, which can be attributed to the fact that in the case of the cubical cloak, the geometrical models are exact, whereas in the spherical cloak example, the spherical geometry is approximated by the fourth-order Lagrange interpolatory functions. Finally, we can conclude that the incorporation of loss does not degrade the backscattering performance, which is consistent with the results in [10].

For the same model and incident wave shown in Fig. 4.11, Figs. 4.13 and 4.14 present, in two characteristic planes, the normalized scattering cross sections of the uncloaked and cloaked PEC cube with lossless and lossy cloaks, together with the

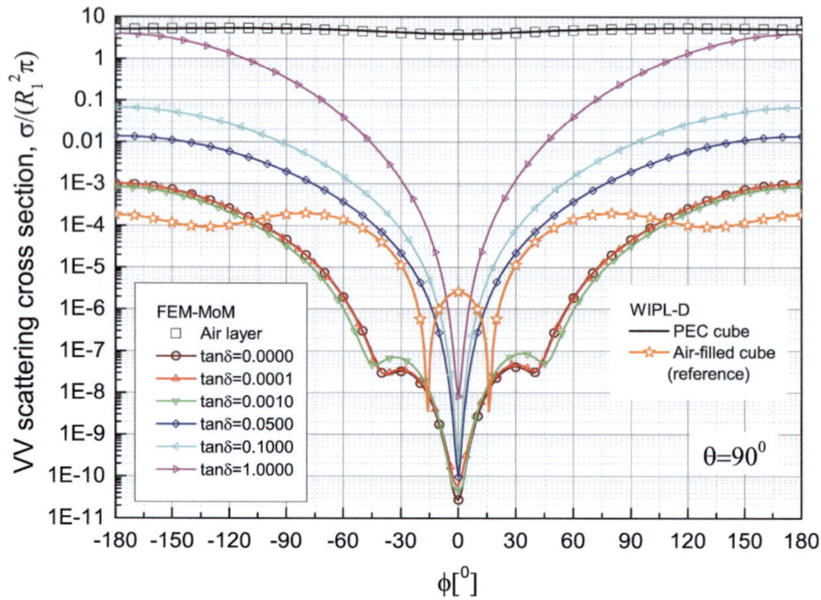

Fig. 4.13 Normalized VV scattering cross section of the cloaked (lossless and lossy, for different loss tangents) and uncloaked PEC cube in Fig. 4.11 ($R_2/R_1 = 1.1, d/\lambda_0 = 0.3, d = 2R_1$), obtained by the higher order FEM-MoM technique [34]. The results include WIPL-D solutions for the PEC cubical scatterer and for the air-filled cube, as specified in the caption to Fig. 4.12

reference solutions of the uncloaked PEC cube and an air-filled cubical region, all for the normalized PEC cube side length of $d/\lambda_0 = 0.3$. We observe a substantial reduction (four to ten orders of magnitude) in σ achieved by the cloak, as well as a smooth degradation of the cloak's forward scattering performance with the increase of loss, as in [10]. The results are also consistent with those for the spherical cloak in Sect. 4.6.1.

As the final example, we present the near field results for the same PEC cube in Fig. 4.11 with a slightly thicker cloak ($R_2/R_1 = 1.5$, so that the field in the cloak can be better observed), with the remaining modeling parameters kept the same as for the thin cloak. The near field results obtained by the higher order FEM-MoM and COMSOL Multiphysics (shown in Fig. 4.15) are practically the same for the wavelength of the incident plane wave given by $d/\lambda_0 = 0.66$ (which is the shortest wavelength for which the COMSOL solution converged), and hence only one set of the results is presented. The near field solutions plotted in Fig. 4.15 (for the two characteristic directions of the plane wave incidence) clearly demonstrate the wave transformation in the cloaking region and validate the design goal that the cloaked structure cannot be observed from outside.

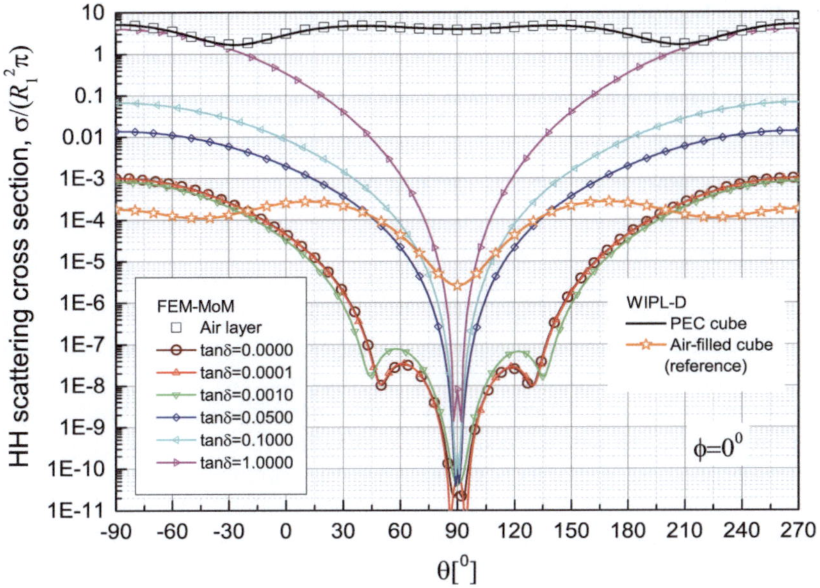

Fig. 4.14 Normalized HH scattering cross section of the cloaked (lossless and lossy) and uncloaked PEC cube in Fig. 4.11 ($R_2/R_1 = 1.1$, $d/\lambda_0 = 0.3$), simulated by the FEM-MoM. WIPL-D results are for structures specified in the caption to Fig. 4.12 [34]

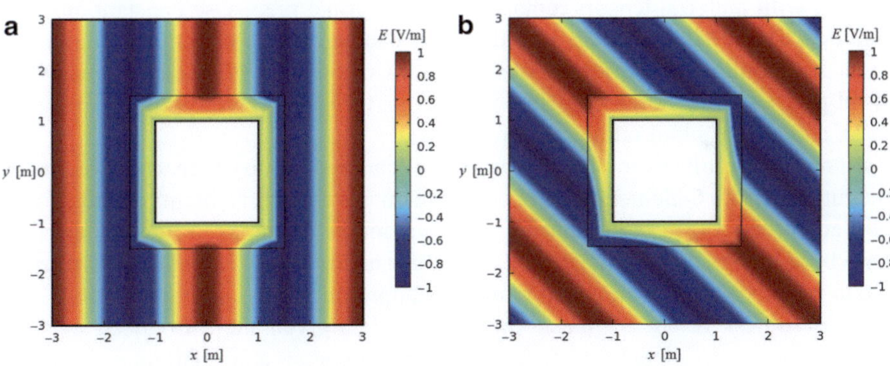

Fig. 4.15 Near field (in $z = 0$ plane) of the cloaked PEC cube in Fig. 4.11 ($R_2/R_1 = 1.5$, $d/\lambda_0 = 0.66$) excited by a uniform plane wave of magnitude $E_0 = 1$ V/m, incident from the direction defined by (**a**) $\theta_{inc} = 90°$, $\phi_{inc} = 0°$ and (**b**) $\theta_{inc} = 90°$, $\phi_{inc} = 45°$ [34]

4.7 Conclusions

Most of the reported theoretical works and investigations of 3-D transformation-based metamaterial invisibility cloaking concepts and designs rely to greater or lesser extent on analytical derivations and analyses. An alternative approach, a full-wave rigorous numerical analysis of cloaking structures, based on concepts and techniques of computational electromagnetics, provides an alternative solution and additional insight for spherical and other canonical shapes that can as well be analyzed analytically (e.g., using the Mie scattering theory). However, it is unavoidable in analysis and design of more complex geometries and material compositions that cannot be treated purely analytically, e.g., for a cubical cloak.

CEM analysis of 3-D transformation-based metamaterial cloaking structures can be performed in a highly efficient and versatile manner using a hybridization of a higher order finite element method for discretization of the cloaking region and a higher order method of moments for numerical termination of the computational domain. The technique allows for an effective modeling of the continuously inhomogeneous anisotropic cloaking region, for cloaks based on both linear and nonlinear coordinate transformations, using a very small number of large curved finite elements with continuous spatial variations of permittivity and permeability tensors and high-order p-refined field approximations throughout their volumes. In analysis, there is no need for a discretization of the permittivity and permeability profiles of the cloak, namely, for piecewise homogeneous (layered) approximate models, with material tensors replaced by appropriate piecewise constant approximations. The flexibility of the technique has enabled a very effective modeling of linear and nonlinear spherical cloaks by means of only six FEM elements and six MoM patches over the volume and external surface, respectively, of the cloaking layer, and a very small number of unknowns.

Numerical results have shown a very significant reduction (three to five orders of magnitude for a 6-element model and five to seven orders of magnitude for a 24-element model) in the scattering cross section of the cloaked PEC sphere in a quite broad range of wavelengths. CEM solutions inevitably introduce some error, due to the involved modeling and numerical approximations and imperfections, but their usefulness is not in question. In particular, given the introduced explicit approximations in modeling of the spherical geometry and continuous material tensor profiles (both by fourth-order Lagrange interpolating functions), and inherent numerical approximations involved in the FEM and MoM techniques and codes, a conclusion is that the cloaking effects can be predicted rather accurately by the presented full-wave higher order FEM-MoM numerical analysis method. The method and numerical model can be readily adapted for analysis and design of electrically larger and/or more complex 3-D cloaking devices (which can be arbitrarily inhomogeneous and can include sharp edges and reentrant corners) with proper h-, p-, and hp-refinements [40] of simple initial models.

Transformation-based metamaterial invisibility cloaks with other, non-spherical and non-circular, shapes are also of great theoretical and practical interest, and so are the approaches and tools for their construction, analysis, and design. The presented cubical cloak design and the full-wave CEM characterization of such a structure with sharp edges and corners should be indicative of other possibilities and generalizations to cloaks of arbitrary shapes.

The analysis of lossless and lossy cubical cloaks has required about 30 times fewer unknowns to obtain the results of a similar accuracy when compared to the COMSOL Multiphysics solution. Numerical results have shown a very substantial reduction, of five to ten orders of magnitude, in the backscattering cross section of the cloaked cube, with both lossless and lossy cloaks, in the entire analyzed range of wavelengths. They have also demonstrated the accuracy and efficiency of a simple 24-element model of the cubical cloak yielding a backscatter so low that it is on par with the best numerical approximation of the zero backscatter from an empty cubical region of the same size as the original scatterer, as verified by WIPL-D and a pure surface (MoM) model. The fact that the far-field numerical FEM-MoM results for the cubical cloak are similar or better than the respective results for the linear spherical cloak has been attributed to an exact geometrical representation of the cubical cloak vs. an approximate modeling of the spherical geometry using fourth-order Lagrange quadrilateral patches. The results have also shown the bistatic behavior of the lossless and lossy (with several characteristic loss tangents) cubical cloaks consistent with the corresponding results for the spherical and cylindrical cloaks – the incorporation of loss does not degrade the backscattering performance of the cloak, while a smooth degradation of the cloak's forward scattering performance occurs with the increase of loss in the cloak material.

The presented cubical cloak and its rather unconventional validation and evaluation can be of a significant interest to researchers seeking coordinate transformations needed for the conformal cloaking of cubical structures or for similar (more complex) structures with sharp edges and corners. Moreover, this study may be of a great importance for the development of conformal transformation-based perfectly matched layers, as mathematical concepts, used for local-type truncations of finite element meshes in FEM-based codes for numerical analysis of electromagnetics/optics phenomena. It may also serve as a starting point for developments of the respective reduced-parameter invisibility cloaks, constructed from isotropic homogeneous material layers and thus potentially enabling a simpler experimental demonstration [14, 43], which, in turn, may be treated (based on the effective medium approximation) as an effective anisotropic medium (e.g., [14]).

Acknowledgements This work was supported by the National Science Foundation under grants ECCS-1002385 and ECCS-1307863 and by the Serbian Ministry of Science and Technological Development under grant TR-32005.

References

1. Alu A, Engheta N (2011) Optical metamaterials based on optical nanocircuits. Proc IEEE 99(10):1669–1681
2. Engheta N, Ziolkowski RWE (2006) Metamaterials: physics and engineering explorations, 1st edn. Wiley-IEEE Press, New York, p 440
3. Christophe C, Tatsuo I (2005) Electromagnetic metamaterials: transmission line theory and microwave applications, 1st edn. Wiley-IEEE Press, New York, p 376
4. Capolino F (2009) Metamaterials handbook, 1st edn. CRC Press, New York, p 1736
5. Cai W, Shalaev VM (2010) Optical metamaterials: fundamentals and applications, 1st edn. Springer, New York, p 212
6. Pendry JB (2004) Negative refraction. Contemp Phys 45(3):191–202
7. Scherer A, Painter O, Vuckovic J, Loncar M, Yoshie T (2002) Photonic crystals for confining, guiding, and emitting light. IEEE Trans Nanotechnol 1(1):4–11
8. Pendry JB, Schurig D, Smith DR (2006) Controlling electromagnetic fields. Science 312(5781):1780–1782
9. Schurig D, Pendry JB, Smith DR (2006) Calculation of material properties and ray tracing in transformation media. Opt Express 14(21):9794–9804
10. Cummer SA, Popa B-I, Schurig D, Smith DR, Pendry J (2006) Full-wave simulations of electromagnetic cloaking structures. Phys Rev E 74(3):036621
11. Zolla F, Guenneau S, Nicolet A, Pendry JB (2007) Electromagnetic analysis of cylindrical invisibility cloaks and the mirage effect. Opt Lett 32(9):1069–1071
12. Ni Y, Gao L, Qiu C-W (2010) Achieving invisibility of homogeneous cylindrically anisotropic cylinders. Plasmonics 5(3):251–258
13. Farhat M, Guenneau S, Movchan AB, Enoch S (2008) Achieving invisibility over a finite range of frequencies. Opt Express 16(8):5656–5661
14. Huang Y, Feng Y, Jiang T (2007) Electromagnetic cloaking by layered structure of homogeneous isotropic materials. Opt Express 15(18):11133–11141
15. Schurig D, Mock JJ, Justice BJ, Cummer SA, Pendry JB, Starr AF, Smith DR (2006) Metamaterial electromagnetic cloak at microwave frequencies. Science 314(5801):977–980
16. Yan M, Yan W, Qiu M (2009) Invisibility cloaking by coordinate Transformation. In: Emil W (ed) Progress in optics, vol 52. Elsevier, Amsterdam, pp 261–304
17. DoHoon K, Werner DH (2010) Transformation electromagnetics: an overview of the theory and applications. IEEE Antennas Propag Mag 52(1):24–46
18. Maci S (2010) A cloaking metamaterial based on an inhomogeneous linear field transformation. IEEE Trans Antennas Propag 58(4):1136–1143
19. Xie Y, Chen H, Xu Y, Zhu L, Ma H, Dong JW (2011) An invisibility cloak using silver nanowires. Plasmonics 6(3):477–481
20. Chen H, Wu B-I, Zhang B, Kong JA (2007) Electromagnetic wave interactions with a metamaterial cloak. Phys Rev Lett 99(6):063903–4
21. Qiu C, Hu L, Zhang B, Wu B-I, Johnson SG, Joannopoulos JD (2009) Spherical cloaking using nonlinear transformations for improved segmentation into concentric isotropic coatings. Opt Express 17(16):13467–13478
22. Furlani EP, Baev A (2009) Optical nanotrapping using cloaking metamaterial. Phys Rev E 79(2):026607
23. Alù A, Engheta N (2007) Plasmonic materials in transparency and cloaking problems: mechanism, robustness, and physical insights. Opt Express 15(6):3318–3332
24. Savić SV, Manić AB, Ilić MM, Notaroš BM (2013) Efficient higher order full-wave numerical analysis of 3-D cloaking structures. Plasmonics 8(2):455–463
25. Ilić MM, Notaroš BM (2003) Higher order hierarchical curved hexahedral vector finite elements for electromagnetic modeling. IEEE Trans Microw Theory Technol 51(3):1026–1033

26. Djordjević M, Notaroš BM (2004) Double higher order method of moments for surface integral equation modeling of metallic and dielectric antennas and scatterers. IEEE Trans Antennas Propag 52(8):2118–2129
27. Ilić MM, Djordjević M, Ilić AŽ, Notaroš BM (2009) Higher order hybrid FEM-MOM technique for analysis of antennas and scatterers. IEEE Trans Antennas Propag 57(5):1452–1460
28. Ilić MM, Ilić AŽ, Notaroš BM (2009) Continuously inhomogeneous higher order finite elements for 3-D electromagnetic analysis. IEEE Trans Antennas Propag 57(9):2798–2803
29. Manić AB, Manić SB, Ilić MM, Notaroš BM (2012) Large anisotropic inhomogeneous higher order hierarchical generalized hexahedral finite elements for 3-D electromagnetic modeling of scattering and waveguide structures. Microw Opt Technol Lett 54(7):1644–1649
30. Pendry J (2009) Taking the wraps off cloaking. Physics 2:95
31. Ma H, Qu S, Xu Z, Wang J (2008) Approximation approach of designing practical cloaks with arbitrary shapes. Opt Express 16(20):15449–15454
32. You Y, Kattawar GW, Zhai P-W, Yang P (2008) Invisibility cloaks for irregular particles using coordinate transformations. Opt Express 16(9):6134–6145
33. Rahm M, Schurig D, Roberts DA, Cummer SA, Smith DR, Pendry JB (2008) Design of electromagnetic cloaks and concentrators using form-invariant coordinate transformations of Maxwell's equations. Photonics Nanostruct Fundam Appl 6(1):87–95
34. Savić SV, Notaroš BM, Ilić MM (2013) Conformal cubical 3D transformation-based metamaterial invisibility cloak. J Opt Soc Am A 30(1):7–12
35. Notaros BM (2010) Electromagnetics, 1st edn. Pearson Prentice Hall, Upper Saddle River, p 840
36. Tai C-T (1997) Generalized vector and dyadic analysis; applied mathematics in field theory, 2nd edn. Wiley-IEEE Press, New Jersey, p 208
37. Post EJ (1962) Formal structure of electromagnetics; general covariance and electromagnetics. North-Holland Publishing Company, Netherlands, Amsterdam, p 204
38. Ward AJ, Pendry JB (1996) Refraction and geometry in Maxwell's equations. J Mod Opt 43(4):773–793
39. Milton GW, Briane M, Willis JR (2006) On cloaking for elasticity and physical equations with a transformation invariant form. New J Phys 8(10):1–20
40. Notaroš BM (2008) Higher order frequency-domain computational electromagnetics. IEEE Trans Antennas Propag 56(8):2251–2276
41. *Wipl-D Pro*, 11.0; WIPL-D d.o.o. (2013) http://www.wipld.com
42. Guild MD, Haberman MR, Alù A (2011) Plasmonic cloaking and scattering cancelation for electromagnetic and acoustic waves. Wave Motion 48(6):468–482
43. Zhang B, Chen H, Wu B-I (2008) Limitations of high-order transformation and incident angle on simplified invisibility cloaks. Opt Express 16(19):14655–14660

Chapter 5
Interaction of Surface Plasmon Polaritons with Nanomaterials

Gagan Kumar and Prashant K. Sarswat

Abstract Surface plasmon polaritons are the surface electromagnetic excitations that exist at the metal-air or metal-dielectric interface. Because of their enhanced field and strong confinement near the metal surface, they offer variety of applications ranging from high sensitivity sensors to miniaturized photonic components. In this chapter, we present analytical formalisms of several optical phenomenon that occur in their interaction with nanomaterials and analyze their significance in absorption, enhanced Raman scattering, electron acceleration, optical fiber sensors etc. First, we discuss dispersion properties of surface plasmon polaritons in single and double metal film configurations. The analysis is extended to the examination of dispersion properties of multilayer thin film configurations. We also present dispersion properties of surface plasmon polaritons in thin film metal coated optical fibers and develop an analytical formalism for the calculation of amplitude of the laser mode converted surface plasmon wave. The chapter also presents a theoretical model for surface plasmon polariton assisted electron acceleration in thin film metal configuration. Furthermore, surface plasmon polariton interactions with metallic nanoparticles are examined and analytical formalism of their anomalous absorption by nanoparticles is presented. In this context, we present surface plasmon assisted surface enhanced Raman scattering by the molecules when they are adsorbed on nanoparticles.

Keywords Surface plasmons • Nanomaterials • Electron acceleration • Optical fibers • Surface enhanced Raman scattering • Anomalous absorption

---

G. Kumar (✉)
Department of Physics, Indian Institute of Technology Guwahati, Guwahati 781039, Assam, India
e-mail: gk@iitg.ernet.in

P.K. Sarswat
Department of Metallurgical Engineering, University of Utah, Salt Lake City, UT 84112, USA
e-mail: prashant.sarswat@utah.edu

C.D. Geddes (ed.), *Reviews in Plasmonics 2015*,
DOI 10.1007/978-3-319-24606-2_5

5.1 Introduction

Surface plasmon polaritons (SPPs) are the guided electromagnetic waves that propagates along the interface between a metal and a dielectric, often a vacuum [1]. The amplitude of the SPPs decays exponentially away from the interface in both the media. The magnetic field of the SPPs is parallel to the surface and perpendicular to its direction of propagation while electric field has components parallel to the direction of propagation and perpendicular to the surface. Surface waves were first observed by Wood in 1902 [2]. He noted anomalous behavior in the diffraction intensities for small angular and spectral variations from the grating of large and rapid changes. In 1907, Rayleigh gave the first theoretical explanation of these anomalies suggesting that such behavior was due to the cutoff or the appearance of a new spectral order [3]. However, more precisely Fano in 1941, theoretically suggested the excitation of surface plasmon polaritons over the grating surface [4]. He pointed out that such wave can be excited only by transverse magnetic (TM) incident wave and exists only at the interface of a dielectric and a medium with negative real part of effective relative permittivity viz. plasma and conductor. Major developments in the excitation of surface plasmon waves over solid state plasmas took place in 1968 when Otto proposed the SPPs excitation in attenuated total reflection (ATR) configuration [5]. Kretschmann and Raether [6] later proposed an alternative configuration of Otto's method.

The SPPs offers a variety of applications, ranging from high sensitivity sensors to supplementary heating in a fusion reactor, coherent radiation generation, enhanced material ablation, electron emission, and plasma diagnostics. Surface plasmons are confined near the metal surface and its amplitude at the surface is significantly larger than that of the laser. The enhanced field of the SPPs could increase the emission rate of luminescent dyes placed near the metal surface. This enhancement can increase the efficiency and the brightness of solid-state LEDs [7]. In last two decades, surface plasmon based sensors have emerged as a powerful tool for characterizing and quantifying bimolecular interactions. Surface plasmon polaritons are being explored for their potential in sub-wavelength optics, magneto-optic data storage, and microscopy [8]. The properties of surface plasmon polaritons can lead to miniaturized photonic circuits by creatively designing the structure of metal's surface. In this chapter, we will discuss the dispersion properties of the surface plasmon polaritons over the metal-air and metal-dielectric interfaces, and analyze their significance in absorption, electron acceleration, optical fiber sensors, and surface enhanced Raman scattering.

5.2 Dispersion Properties of Surface Plasmon Polaritons

5.2.1 Dispersion Relation Over a Single Metal Surface

The fields of the SPPs at metal-dielectric interface are governed by Maxwell's equations in each medium and the associated boundary conditions. Boundary conditions involve the continuity of the tangential component of the electric field

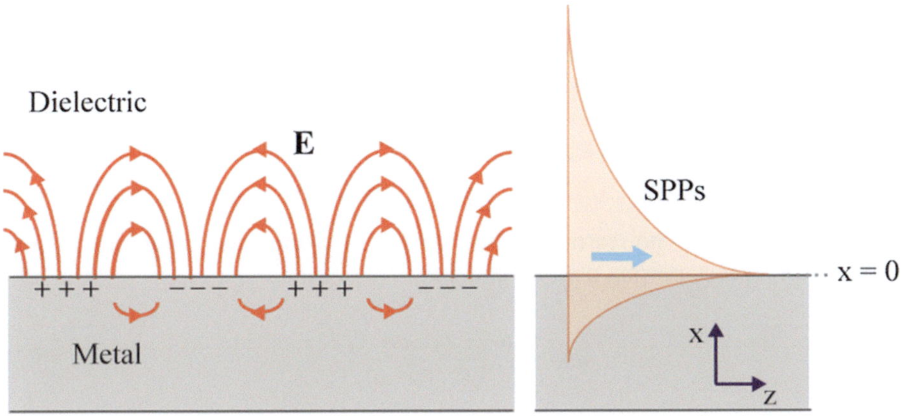

Fig. 5.1 Geometry of surface plasmon polaritons propagation at metal-air interface

and magnetic fields across the interface and vanishing of these fields away from the interface.

Consider a dielectric of relative permittivity ε_1 for $x > 0$ and a metal with effective relative permittivity $\varepsilon(\omega)$ in the half space $x < 0$ (Fig. 5.1). When a SPP propagates at the dielectric-metal interface with (t, z) variation as $e^{-i(\omega t - k_z z)}$, its behavior is governed by Maxwell's equations.

$$\nabla \times \vec{E} = -\frac{1}{c}\frac{\partial \vec{H}}{\partial t}, \tag{5.1}$$

$$\nabla \times \vec{\mathbf{H}} = \frac{\varepsilon'}{\mathrm{c}}\frac{\partial \vec{\mathbf{E}}}{\partial \mathrm{t}}, \tag{5.2}$$

which on taking curl of Eq. (5.1) and using Eq. (5.2) give

$$\nabla^2 \vec{E} - \frac{\varepsilon'}{c^2}\frac{\partial^2 \vec{E}}{\partial t^2} = 0, \tag{5.3}$$

where $\varepsilon' = \varepsilon(\omega) = \varepsilon_L - \omega_P^2/\omega^2$ for $x < 0$ and $\varepsilon' = \varepsilon_1$ for $x > 0$. ε_L is the lattice permittivity, ω_P is the plasma frequency. The continuity conditions on electromagnetic fields at the interface demand that the t and z variation of fields be the same in both media. Thus replacing $\partial/\partial z$ and $\partial/\partial t$ in Eq. (5.3) by ik_z and $-i\omega$ respectively, we get

$$\frac{\partial^2 \vec{E}}{\partial x^2} - \left(k_z^2 - \frac{\omega^2}{c^2}\varepsilon'\right)\vec{E} = 0. \tag{5.4}$$

The well behaved solutions of Eq. (5.4) are

$$\begin{aligned} \vec{E} &= \left(\hat{z} + \hat{x}\frac{ik_z}{\alpha_1}\right) A e^{-\alpha_1 x} e^{-i(\omega t - k_z z)}, \quad \text{for} \quad x > 0, \\ \vec{E} &= \left(\hat{z} - \hat{x}\frac{ik_z}{\alpha_2}\right) A e^{\alpha_2 x} e^{-i(\omega t - k_z z)}, \quad \text{for} \quad x < 0. \end{aligned} \tag{5.5}$$

where $\alpha_1^2 = k_z^2 - \varepsilon_1\omega^2/c^2$, $\alpha_2^2 = k_z^2 - \omega^2\varepsilon(\omega)/c^2$. The real part of α_1, α_2 must be positive for an electromagnetic wave localized to the dielectric-metal interface at x = 0. The continuity of E_z and $\varepsilon' E_x$ at the interface yields $A = B$,

$$\frac{\alpha_2}{\alpha_1} = -\frac{\varepsilon(\omega)}{\varepsilon_1}. \tag{5.6}$$

Squaring both sides and rearranging the terms, one obtains the explicit dispersion relation for surface plasma polaritons

$$k_{SPP} = \frac{\omega}{c}\left(\frac{\varepsilon_1\varepsilon(\omega)}{\varepsilon_1 + \varepsilon(\omega)}\right)^{1/2}. \tag{5.7}$$

Further, $\alpha_1^2 = -\frac{\omega^2}{c^2}\frac{\varepsilon_1^2}{\varepsilon_1+\varepsilon}$, $\alpha_2^2 = -\frac{\omega^2}{c^2}\frac{\varepsilon^2}{\varepsilon_1+\varepsilon}$. For α_1 and α_2 to be positive one needs $\varepsilon_1 + \varepsilon < 0$ or $\omega < \frac{\omega_P}{\sqrt{\varepsilon_L+\varepsilon_1}}$. In case of metal-free space interface, the surface plasmons dispersion relation reduces to $k_z = \frac{\omega}{c}\left(\frac{\varepsilon}{1+\varepsilon}\right)^{1/2}$. We have plotted the dispersion relation at dielectric-metal and air-metal interfaces in Fig. 5.2 for the parameters: $\varepsilon_1 = 2.25, \varepsilon_L = 4$. One may note that as the wavenumber increases, dispersion curve shifted away from the light line and finally saturates at high wavenumber value. It may be noted that the surface plasmon polaritons wavenumber increases with ε_1.

5.2.2 *Dispersion Relation of SPPs in Double Metal Surface Configuration*

The excitation of symmetric surface plasmon polartons in a double metal surface configuration is important in a variety of applications such as biochemical sensing, line narrowing of spectral features, and electron acceleration etc. Therefore it is

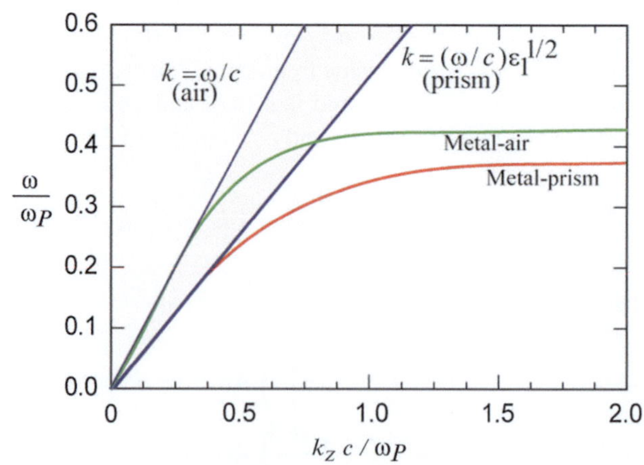

Fig. 5.2 Dispersion relations for surface plasmon polaritons at metal-air and metal prism interfaces

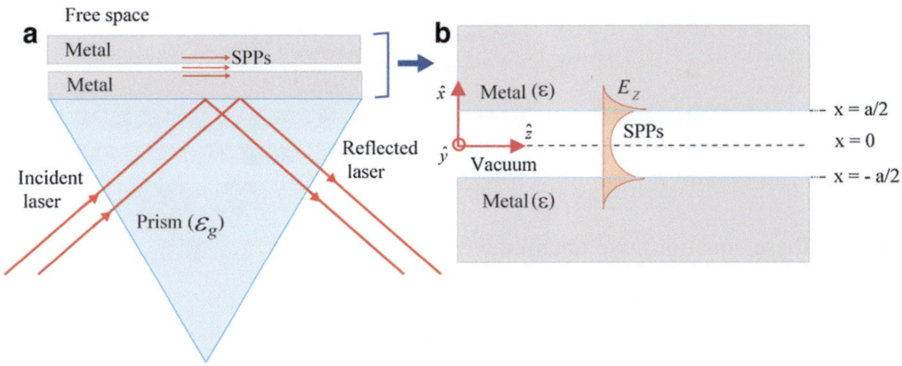

Fig. 5.3 Schematic of two metal surface plasmon structure. The metal surfaces at x = −a/2 and x = a/2 are separated by free space. The surface plasmon polaritons are excited via attenuated total reflection (*ATR*) configuration

crucial to understand the propagation properties of the surface plasmon polariton excitations in this configuration. Consider two parallel metal half spaces x < −a/2 and x > a/2, separated by a thin vacuum region –a/2 < x < a/2 as shown in Fig. 5.3. The effective permittivity of metal at frequency ω is $\varepsilon = \varepsilon_L - \omega_P^2/\omega^2$, where ε_L is the lattice permittivity and ω_P is the plasma frequency.

Surface plasmons propagate through two metal configuration with t, z variation as $\exp[-i(\omega t - k_z z)]$. The field variations are governed by Maxwell's third and fourth equations given by Eqs. (5.1) and (5.2) with

$$\begin{aligned} \varepsilon' &= 1 \quad \text{for} \quad -a/2 < x < a/2 \\ &= \varepsilon \quad \text{for} \quad x < -a/2 \quad \text{and} \quad x > a/2 \end{aligned}$$

As discussed earlier, taking curl of Maxwell's third equation and using fourth, we obtain wave equation. The wave equation governing E_z in three media is given by Eq. (5.4). The well behaved solution of Eq. (5.4), satisfying $\nabla.\vec{E} = 0$ in each region is

$$\begin{aligned} \vec{E} &= A\left(\hat{z} + \hat{x}\frac{ik_z}{\alpha_1}\right)e^{-\alpha_1 x}, && x > a/2, \\ &= \left[A_1\left(\hat{z} - \hat{x}\frac{ik_z}{\alpha_2}\right)e^{\alpha_2 x} + A_2\left(\hat{z} + \hat{x}\frac{ik_z}{\alpha_2}\right)e^{-\alpha_2 x}\right], && -a/2 < x < a/2, \\ &= A_3\left(\hat{z} - \hat{x}\frac{ik_z}{\alpha_1}\right)e^{\alpha_1 x}, && x < -a/2, \end{aligned} \tag{5.8}$$

Where $\alpha_1 = \left(k_z^2 - \omega^2\varepsilon/c^2\right)^{1/2}$, $\alpha_2 = \left(k_z^2 - \omega^2/c^2\right)^{1/2}$. Applying conditions of continuity of E_z and $\varepsilon' E_x$ at $x = a/2$ and $-a/2$, we get

$$
\begin{aligned}
A_1\, e^{\alpha_2 a/2} + A_2\, e^{-\alpha_2 a/2} &= A\, e^{-\alpha_1 a/2}, \\
A_1\, e^{\alpha_2 a/2} - A_2\, e^{-\alpha_2 a/2} &= \; - A \left(\frac{\varepsilon \alpha_2}{\alpha_1} \right) e^{-\alpha_1 a/2}, \\
A_1\, e^{-\alpha_2 a/2} + A_2\, e^{\alpha_2 a/2} &= A_3\, e^{-\alpha_1 a/2}, \\
A_1\, e^{-\alpha_2 a/2} - A_2\, e^{\alpha_2 a/2} &= A_3 \left(\frac{\varepsilon \alpha_2}{\alpha_1} \right) e^{-\alpha_1 a/2}.
\end{aligned}
\tag{5.9}
$$

Solving these equations, we obtain the dispersion relation

$$
\frac{\alpha_2^2}{\alpha_1^2} = \left| \frac{1 - e^{\alpha_2 a}}{1 + e^{\alpha_2 a}} \right|^2 \frac{1}{|\varepsilon|^2}. \tag{5.10}
$$

Using dimensionless quantities: $q = k_z c\ /\omega_P, \Omega = \omega/\omega_P, a' = a\omega_P/c$, Eq. (5.10) is normalized and plotted normalized frequency Ω versus normalized wavenumber q in Fig. 5.4 for the mode that has E_x symmetric about the $x = 0 \quad (A_1 = -A_2)$. The parameters are: $\varepsilon_L = 1$, $a\omega_P/c = 100$. From the figure, one may note that as the wave number increases, frequency increases and then saturates at higher wave numbers. For higher value of wave number q, the behavior of the curve is same as that of single metal surface structure. However in double metal surface, at lower value of wave number phase velocity is smaller. The phenomenon of reducing phase velocity can be useful in achieving wave electron synchronization at lower electron energies and hence accelerating electrons.

5.2.3 *Surface Plasmons in Multilayer Thin Films Configuration*

In case of metal-insulator-metal (MIM) configuration, surface plasmons are correlated with natural electromagnetic mode of the structures that can be obtained from Maxwell's equation solution using proper boundary conditions. The modes propagation behavior relies on permittivity of each layer. In case of multilayer structures, various analytical models and numerical methods have been utilized for the analysis of the splitting modes [9]. Other methods such as Eigen function [10] treatment and semi-analytical mode matching technique were also utilized for the analysis of the one dimensional photonic crystals and scattering of surface plasmons at abrupt surface (metal-dielectric interface), respectively [11]. Such analysis becomes more and more difficult when number of layers is increased. In order to find dispersion relation for these intricate problems, matrix method have been used, where one diagonal term is set as zero [12].

In case of multilayered structures, dielectric layers are aligned in XY plane. As each layer has a constant electric permittivity, it will result in a piecewise continuous solution. Knowing the fact that surface plasmons are excited due to presence

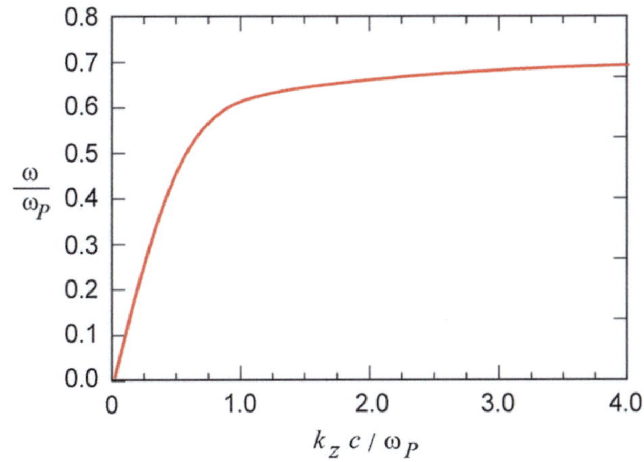

Fig. 5.4 Variation of normalized frequency $\Omega = \omega/\omega_P$ versus normalized wave number $q = k_z c / \omega_P$ of the surface plasmon polaritons in double metal surface configuration. The parameters are: $\varepsilon_L = 1$, $a\omega_P/c = 100$.

of perpendicular electric field or its component with respect to surface, electric field components are considered in X-Z plane and magnetic field is considered in Y direction. Let us consider Eqs. (5.1) and (5.2) again, if their 'x' dependence is considered as $\exp(i\alpha_n x)$ for the representation of a wave, traveling in positive 'x' direction, its magnetic field variation along z direction can be represented as [12]

$$\alpha_n^2 u_n(z) = \varepsilon(z) k^2 u_n(z) + \varepsilon(z)\left[\frac{1}{\varepsilon(z)}\frac{d^2 u_n(z)}{dz^2} + \frac{du_n(z)}{dz}\frac{d}{dz}\left(\frac{1}{\varepsilon(z)}\right)\right], \tag{5.11}$$

where, $u_n(z)$ is the nth eigenfunction and α_n [2] is corresponding n^{th} eigenvalue. As mentioned earlier that it is a piecewise continuous, the solution of Eq. (5.11) can be written as

$$u_n(z) = \sum \left(H_{n,l}^{+} e^{\left(i\gamma_{n,l}\left(z - z_l^0\right)\right)} + H_{n,l}^{-} e^{-\left(i\gamma_{n,l}\left(z - z_l^0\right)\right)}\right) \quad \text{(Summed over } l\text{)}, \tag{5.12}$$

where, z_l^0 is the location of the lower boundary for layer 'l'. $H_{n,l}^{+}$ and $H_{n,l}^{-}$ are the amplitudes of upward and downward propagating waves, respectively. The corresponding wavenumber for each layer, $\gamma_{n,l}$ can be evaluated as:

$$\gamma_{n,l} = \pm\sqrt{\varepsilon_l k^2 - \alpha_n^2}. \tag{5.13}$$

One should be careful while choosing the sign of $\gamma_{n,l}$. This sign should be chosen such that the imaginary part will always be positive, in order to ensure that there will be a decay in wave amplitude when it propagate. It is important to note that coefficients have different value for different layer but for a particular layer these coefficients will be constant. The magnetic field can be written as [12]

$$H(r,\omega) = \vec{Y}\sum \left(H^{+}_{n,l}e^{\left(i\gamma_{n,l}\left(z-z_l^0\right)\right)} + H^{-}_{n,l}e^{-\left(i\gamma_{n,l}\left(z-z_l^0\right)\right)}\right)e^{i\alpha_n x} \quad \text{(Summed over } l\text{)}, \tag{5.14}$$

Similarly, for electric field:

$$E(r,\omega) = \\ e^{i\alpha_n x}\sum\left(\frac{1}{k\varepsilon_l}{}^{l}H^{+}_{n,l}\left(\gamma_{n,l}\hat{x} - \alpha_n\hat{z}\right)e^{\left(i\gamma_{n,l}\left(z-z_l^0\right)\right)} - \frac{1}{k\varepsilon_l}H^{-}_{n,l}\left(\gamma_{n,l}\hat{x} + \alpha_n\hat{z}\right)e^{-\left(i\gamma_{n,l}\left(z-z_l^0\right)\right)}\right), \tag{5.15}$$

Across the layer, tangential components of field are continuous, α_n will be treated as a constant for entire structure. Also, $H^{+}_{n,l}$ and $H^{-}_{n,l}$ in particular layer 'l' are linked to other coefficients in adjacent layers. Considering the boundary condition, x-component of electric field and magnetic field between two adjacent layers (l and $l+1$) can be matched using following matrix equation [12]

$$\begin{bmatrix} H^{+}_{n,l+1} \\ H^{-}_{n,l+1} \end{bmatrix} = \frac{1}{2}\begin{bmatrix} 1 & \dfrac{\varepsilon_{l+1}}{\gamma_{n,l+1}} \\ 1 & -\dfrac{\varepsilon_{l+1}}{\gamma_{n,l+1}} \end{bmatrix}\begin{bmatrix} 1 & 1 \\ \dfrac{\gamma_{n,l}}{\varepsilon_l} & -\dfrac{\gamma_{n,l}}{\varepsilon_l} \end{bmatrix} \times \begin{bmatrix} e^{i\gamma_{n,l}t_l} & 0 \\ 0 & e^{-i\gamma_{n,l}t_l} \end{bmatrix}\begin{bmatrix} H^{+}_{n,l} \\ H^{-}_{n,l} \end{bmatrix}, \tag{5.16}$$

where, the thickness of layer 'l' is t_l. Using Eq. (5.16), correlation between adjacent layers can be established and magnetic field can be obtained in entire structure by multiplying matrix for individual layer. Hence, magnetic field at lowest layer ($l=0$) can be linked with magnetic field at a generalized ($l=L$) layer

$$\begin{bmatrix} H^{+}_{n,L} \\ H^{-}_{n,L} \end{bmatrix} = \begin{bmatrix} A_{11} & A_{12} \\ A_{21} & A_{22} \end{bmatrix}\begin{bmatrix} H^{+}_{n,0} \\ H^{-}_{n,0} \end{bmatrix}, \tag{5.17}$$

where, A_{xx} ($x=1$ or 2) are matrix elements. Now, for the top most layer and in absence of incident field: $H^{+}_{n,0} = H^{-}_{n,L} = 0$. For outward directed field: $H^{+}_{n,L} = A_{12}$ $H^{-}_{n,0}$; $0 = A_{22}$ $H^{-}_{n,0}$, that gives $A_{22}=0$; This is a constrain that is responsible for discrete set of surface plasmon modes in the structure. It is important to note that solution of these equations can be found using numerical methods. However, dispersion relation for single metal surface and thin film can be verified as follows

For a single metal dielectric combination, $L=1$, it can be written [12]

$$\begin{bmatrix} H^{+}_{n,1} \\ H^{-}_{n,1} \end{bmatrix} = \frac{1}{2}\begin{bmatrix} 1 & \dfrac{\varepsilon_1}{\gamma_{n,1}} \\ 1 & -\dfrac{\varepsilon_1}{\gamma_{n,1}} \end{bmatrix}\begin{bmatrix} 1 & 1 \\ \dfrac{\gamma_{n,0}}{\varepsilon_l} & -\dfrac{\gamma_{n,0}}{\varepsilon_l} \end{bmatrix} \times \begin{bmatrix} 1 & 0 \\ 0 & 1 \end{bmatrix}\begin{bmatrix} H^{+}_{n,0} \\ H^{-}_{n,0} \end{bmatrix}. \tag{5.18}$$

Further simplification gives

$$\begin{bmatrix} H_{n,1}^{+} \\ H_{n,1}^{-} \end{bmatrix} = \frac{1}{2} \begin{bmatrix} 1 + \dfrac{\varepsilon_1\ \gamma_{n,0}}{\gamma_{n,1}\ \varepsilon_0} & 1 - \dfrac{\varepsilon_1\ \gamma_{n,0}}{\gamma_{n,1}\ \varepsilon_0} \\ 1 - \dfrac{\varepsilon_1\ \gamma_{n,0}}{\gamma_{n,1}\ \varepsilon_0} & 1 + \dfrac{\varepsilon_1\ \gamma_{n,0}}{\gamma_{n,1}\ \varepsilon_0} \end{bmatrix} \times \begin{bmatrix} 1 & 0 \\ 0 & 1 \end{bmatrix} \begin{bmatrix} H_{n,0}^{+} \\ H_{n,0}^{-} \end{bmatrix}. \tag{5.19}$$

$$\text{It indicates } \mathrm{A}_{22} = \frac{1}{2}\left[1 + \frac{\varepsilon_1\ \gamma_{n,0}}{\gamma_{n,1}\ \varepsilon_0}\right]. \tag{5.20}$$

Using relation $\mathrm{A}_{22} = 0$, Eq. (5.20) can be written as

$$\left[1 + \frac{\varepsilon_1}{\sqrt{\varepsilon_1 k^2 - \alpha_n^2}} \frac{\sqrt{\varepsilon_0 k^2 - \alpha_n^2}}{\varepsilon_0}\right] = 0. \tag{5.21}$$

Solving for α_n, we obtain

$$\alpha = \pm k \left[\frac{\varepsilon_1 \varepsilon_0}{\varepsilon_1 + \varepsilon_0}\right]^{\frac{1}{2}}. \tag{5.22}$$

This is an established dispersion relation already discussed in Sect. 5.1 through Eq. (5.7).

5.3 Absorption of Surface Plasmons Polaritons by Metallic Nanoparticles

The absorption of laser radiation at material surfaces is significant to material ablation [13] and other applications. The optical absorption is very poor for smooth metal surfaces because the high free electron density of metals renders effective plasma permittivity to be negative and surface re-radiates light energy in the surrounding medium. However, for the metallic particles, absorption of the electromagnetic waves could be significantly high when the wave frequency is close to the natural frequency of oscillations of the electron cloud. The total absorption of electromagnetic wave can be achieved through the excitation of surface plasmon polaritons [1, 14]. Experiments have revealed total absorption of femtosecond laser due to surface modifications via the generation of surface electromagnetic waves [15]. In this section, we will understand the mechanism by which the total absorption of the surface electromagnetic waves can be achieved through the excitation of surface plasmon polaritons.

Consider a metal free-space interface ($x = 0$) with metal occupying half space ($x < 0$) and free space is $x > 0$ (cf. Fig. 5.5). The metal is having spherical particles

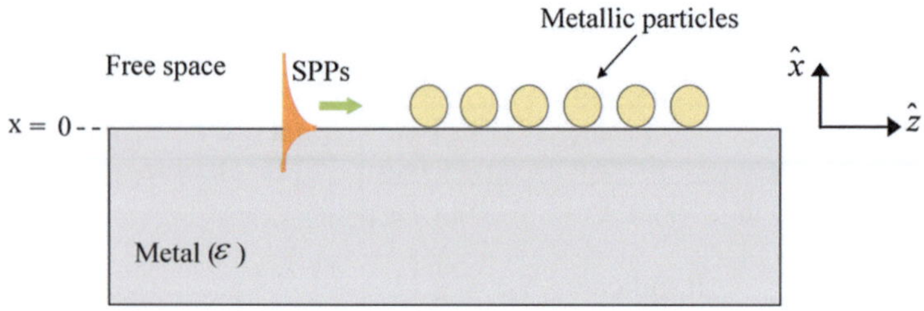

Fig. 5.5 Schematic for the absorption of surface plasmon polaritons by metallic particles adsorbed on metal surface

of size r_c and aerial density N per unit surface area, are placed over it. The effective relative permittivity ε of the metal is given as

$$\varepsilon = \left(\varepsilon_L - \frac{\omega_P^2}{\omega^2}\right) + i\frac{\gamma}{\omega}\frac{\omega_P^2}{\omega^2}, \tag{5.23}$$

where ω is the incident laser frequency. ε_L is the lattice permittivity, ω_P is the plasma frequency and γ is the electron phonon collision frequency in the metal film. Suppose a surface plasmon propagates through the configuration with field variation given by $\boldsymbol{E}_S = \boldsymbol{A}_S\, exp[-i(\omega t - k_z z)]$. The amplitude $\boldsymbol{A}_S$ in metal and free space regions are given by Eq. (5.5). The dispersion relation of surface plasmon polaritons as discussed in the earlier section is given by

$$k_{sp} = \frac{\omega}{c}\left(\frac{\varepsilon(\omega)}{1+\varepsilon(\omega)}\right)^{1/2}.$$

The surface plasmon polaritons decay with z over a propagation length (inverse of imaginary part of k_{sp})

$$L = {k_{si}}^{-1} = \frac{2\omega}{\gamma}\frac{\omega_P^2}{\omega^2}\left(\frac{\omega_P^2}{\omega^2} - \varepsilon_L - 1\right)^{1/2}\left(\frac{\omega_P^2}{\omega^2} - \varepsilon_L\right)^{1/2}. \tag{5.24}$$

When the surface plasmons field in the free space region interacts with the metal particle having internal electron density n_e, then the response of electrons of a particle is governed by equation of motion

$$m\frac{d^2\boldsymbol{s}}{dt^2} = -e\boldsymbol{E}_S - m\gamma\,\boldsymbol{v} - \frac{m\omega_{Pe}^2\,\boldsymbol{s}}{3}, \tag{5.25}$$

where '***s***' is the displacement of electrons of the particle from equilibrium, $\boldsymbol{v} = ds/dt$ is their velocity, m and $-e$ are the electronic mass and charge, ω_{Pe} is the plasma frequency of the metal particle which is same as ω_P when the metal particle and metal film are of same material. Taking x-component of Eq. (5.25), the velocity of the electron in x-direction turns out to be

$$v_x = \frac{e\,A\,\omega\,(k/\alpha_1)e^{-i\omega t}}{m\left(\omega^2 - \omega_{Pe}^2/3 + i\gamma\omega\right)}. \tag{5.26}$$

Similarly taking z-component one obtains

$$v_z = \frac{-ie\,A\,\omega\,e^{-i\omega t}}{m\left(\omega^2 - \omega_{Pe}^2/3 + i\gamma\omega\right)}. \tag{5.27}$$

Under the influence of the SPPs field, the energy absorbed by an electron per second is

$$\Gamma_{abs} = \frac{1}{2}\,Re\left[-e\boldsymbol{E}_S^*.\,\boldsymbol{v}\right], \tag{5.28}$$

where $\boldsymbol{E}_S$ is the electric field and * denotes the complex conjugate. Using Eqs. (5.5), (5.26) and (5.27) in Eq. (5.28), along with boundary conditions yielding A = B, we obtain

$$\Gamma_{abs} = \frac{e^2\omega^2\,A^2\,\gamma\,\left(1 + k^2/\alpha_1^2\right)}{2m\left(\left(\omega^2 - \omega_{Pe}^2/3\right)^2 + \gamma^2\omega^2\right)}. \tag{5.29}$$

If n_0 is the electron density in the metal particle, then the power absorbed per second per particle is

$$\Gamma_{abs} = \frac{\omega_{Pe}^2\,A^2\,\omega^2\,\gamma\,\left(1 + k^2/\alpha_1^2\right)\,r_c^3}{6\left(\left(\omega^2 - \omega_{Pe}^2/3\right)^2 + \gamma^2\omega^2\right)}, \tag{5.30}$$

When N is the number of particles per unit area on the metal surface with inter particle separation $d = N^{-1/2} \geq r_c$, such that

$$N = \int n_P\,\delta(x - a)dx. \tag{5.31}$$

Then the energy absorbed in distance dz is

$$dP = -\Gamma_{abs}\,n_P\,dz, \tag{5.32}$$

$$= \frac{\omega_{Pe}^2\,A^2\,\omega^2\,\gamma\,\left(1 + k^2/\alpha_1^2\right)\,r_c}{6\left(\left(\omega^2 - \omega_{Pe}^2/3\right)^2 + \gamma^2\omega^2\right)}\left(\frac{r_c^2}{d^2}\right)dz. \tag{5.33}$$

Here P is the SPPs power flow per unit y width in the vacuum region

$$P = \frac{c}{4\pi}\int_0^{\infty} Re\left(\frac{\boldsymbol{E_S} \times \boldsymbol{H_S}}{2}\right)_z dx, \tag{5.34}$$

where $\boldsymbol{E_S}$, $\boldsymbol{H_S}$ are the electric and magnetic field intensity of the surface plasmon polaritons in the free space region. Using the third Maxwell's equation, we have $\boldsymbol{H_S} = (c/i\omega)(\nabla \times \boldsymbol{E_S})$. Submitting $\boldsymbol{E_S}$ and $\boldsymbol{H_S}$, we get

$$P = \frac{A^2\, c^2\, k\, \left(k^2/\alpha_1 - \alpha_1\right)}{16\, \pi\, \alpha_1^2\, \omega}, \tag{5.35}$$

From Eqs. (5.33) and (5.35), we obtain

$$\int_{P_0}^{P} \frac{dP}{P} = -\int_0^z k_{ip} dz + C, \tag{5.36}$$

giving $P = P_0\, e^{-k_{ip}\, z}$, where P_0 is the power at $z = 0$. The absorption constant k_{ip} is given as

$$k_{ip} = \frac{8\, \pi\, \omega_{Pe}^2\, \alpha_1^2\, \omega^3\, \gamma\, \left(1 + k^2/\alpha_1^2\right) r_c}{3\, k\, c^2 \left(\left(\omega^2 - \omega_{Pe}^2/3\right)^2 + \gamma^2\omega^2\right)\left(k^2/\alpha_1 - \alpha_1\right)} \frac{r_c^2}{d^2}, \tag{5.37}$$

One may note that the resonant enhancement in k_{ip} occurs at $\omega^2 = \omega_{Pe}^2/3$, corresponding to strong absorption of the wave. The resonant absorption depends upon the number density and the size of particles. In order to have numerical appreciation, normalized absorption constant k_{ip} versus $\delta/\omega_{Pe} = \left(\omega - \omega_{Pe}/\sqrt{3}\right)$ is plotted in Fig. 5.6 for two different particle sizes of $r_c = 6.5\ nm$ and $r_c = 8\ nm$ for the parameters: $\varepsilon = -5.5$, $\omega_{Pe} = 4.1 \times 10^{15}\ rad/sec$, $d = 112\ nm$, $\gamma = 7 \times 10^{12}\ sec^{-1}$.

5.4 Laser Mode Conversion into SPPs in a Metal Coated Optical Fiber

Fiber optic based sensors are important in a variety of applications including environmental monitoring, detection of bio-molecules and biochemical monitoring [16] etc. An optical fiber coated with silver or gold film instead of cladding have been employed for SPR (surface plasmon resonance) [17] and SERS (surface enhanced Raman scattering) sensor applications [18]. In this section, we examine how does the laser mode in a metal coated optical fiber converts to the surface plasmon polaritons. In our analytical formalism of mode conversion, we presume a

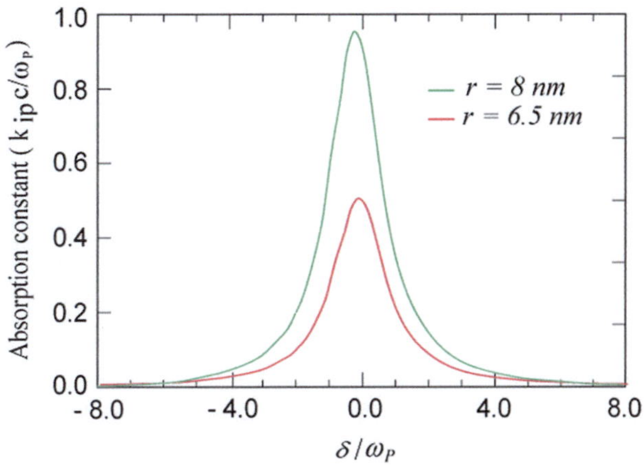

Fig. 5.6 Variation of normalized absorption constant, $k_{ip}c/\omega_{Pe}$ versus normalized frequency $(\omega - \omega_{Pe}/\sqrt{3})/\omega_{Pe}$ for two particles of different size (i) $r_c = 6.5\ nm$ and (ii) $r_c = 8\ nm$ for the parameters: $\varepsilon = -5.5$, $\omega_{Pe} = 4.1 \times 10^{15}\ rad/sec$, $d = 112\ nm$, $\gamma = 7 \times 10^{12}\ sec^{-1}$.

ripple at the fiber metal interface to assist k-matching for the excitation of surface plasmon polaritons. When the body wave i.e. laser propagates through the fiber it induces oscillatory velocity to the electrons of the metal. The oscillatory velocity beats with the space modulated density to produce a current driving the surface plasmon polaritons on the metal free space interface. In our analysis we will first analyze the dispersion properties of surface plasmon polaritons and body waves followed by the mode conversion.

5.4.1 *Dispersion Relations of Body Waves and Surface Plasmon Polaritons*

Consider an optical fiber of radius ‘a’ and permittivity ε_g, coated with metal ($a < r < b$) of effective permittivity ε_m (cf. Fig. 5.7). The figure shows ripple at the glass-metal interface. Here we will confine our discussion to the dispersion relations of the body waves and surface plasmon waves. The role of ripple will be discussed in the next section of mode conversion.

A laser propagates through this structure in azimuthally symmetric TM mode with $t - z$ variation as $\exp[-i(\omega t - k_z z)]$. The propagation of field is governed by wave Eq. (5.3). The r-variation of fields in is given as

$$\frac{\partial^2 E_z}{\partial r^2} + \frac{1}{r}\frac{\partial E_z}{\partial r} + \left(\frac{\omega^2}{c^2}\varepsilon' - {k_z}^2\right)E_z = 0, \tag{5.38}$$

where $\varepsilon' = \varepsilon_g$ for $r < a$, $\varepsilon' = \varepsilon_m = \varepsilon_L - \omega_P^2/\omega(\omega + i\nu)$ for $a < r < b$, and $\varepsilon' = 1$ for $r > b$. ε_L is the lattice permittivity. ω_P is the plasma frequency of free electrons inside the metal and γ is the electron collision frequency. The well behaved solutions of Eq. (5.38) in different regions are

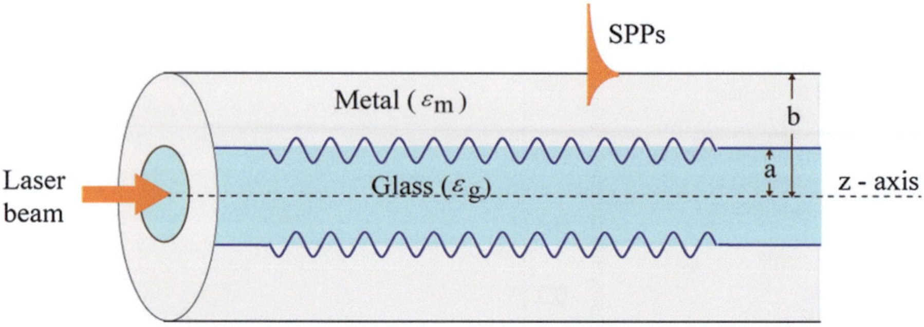

Fig. 5.7 Schematic of a metal coated optical fiber. The rippled interface between fiber and metal coating is used to assist k-matching for the conversion of a TM mode body wave to the surface plasmon polaritons

$$
\begin{aligned}
\vec{E} &= A_1\left[J_0(k_\perp r)\hat{z} + \frac{ik_z}{k_\perp}J_0'(k_\perp r)\hat{r}\right]e^{-i(\omega t - k_z z)}, && 0<r<a,\\
\vec{E} &= \left[(A_{21}I_0(\alpha_2 r) + A_{22}K_0(\alpha_2 r))\hat{z} - \left(\frac{ik_z}{\alpha_2}\right)(A_{21}I_0'(\alpha_2 r) + A_{22}K_0'(\alpha_2 r))\hat{r}\right]e^{-i(\omega t - k_z z)}, && a<r<b,\\
\vec{E} &= A_3\left(K_0(\alpha_3 r)\hat{z} - \left(\frac{ik_z}{\alpha_3}\right)K_0'(\alpha_3 r)\hat{r}\right)e^{-i(\omega t - k_z z)}, && r>b,
\end{aligned}
\tag{5.39}
$$

where $k_\perp = \left(\frac{\omega^2\varepsilon_g}{c^2} - k_z^{\,2}\right)^{1/2}$, $\alpha_2 = \left(k_z^{\,2} - \frac{\omega^2\varepsilon_m}{c^2}\right)^{1/2}$, $\alpha_3 = \left(k_z^{\,2} - \frac{\omega^2}{c^2}\right)^{1/2}$ and the prime denotes differentiation w. r. t. argument. Continuity of E_z and $\varepsilon' E_r$ at $r = a, b$ demands

$$
\begin{aligned}
A_{21}I_0(\alpha_2 a) + A_{22}K_0(\alpha_2 a) &= A_1 J_0(k_\perp a),\\
A_{21}I_0'(\alpha_2 a) + A_{22}K_0'(\alpha_2 a) &= -\left(\frac{\varepsilon_g\alpha_2}{\varepsilon_m k_\perp}\right)A_1 J_0'(k_\perp a),\\
A_{21}I_0(\alpha_2 b) + A_{22}K_0(\alpha_2 b) &= A_3 K_0(\alpha_3 b),\\
A_{21}I_0'(\alpha_2 b) + A_{22}K_0'(\alpha_2 b) &= \left(\frac{\alpha_2 A_3}{\varepsilon_m\alpha_3}\right)K_0'(\alpha_3 b).
\end{aligned}
\tag{5.40}
$$

leading to the dispersion relation

$$
\left[J_0(k_\perp a) + \left(\frac{\alpha_2\varepsilon_g}{k_\perp\varepsilon_m}\right)\left(\frac{J_0'(k_\perp a)}{K_0'(\alpha_2 a)}\right)K_0(\alpha_2 a)\right]\left[I_0(\alpha_2 b) - \left(\frac{\alpha_3\varepsilon_m}{\alpha_2}\right)\left(\frac{I_0'(\alpha_2 b)}{K_0'(\alpha_3 b)}\right)K_0(\alpha_3 b)\right] = Q,
\tag{5.41}
$$

where,

$$
Q = \frac{I_0'(\alpha_2 a)}{K_0'(\alpha_2 a)}\left[J_0(k_\perp a) + \frac{\alpha_2\varepsilon_g}{k_\perp\varepsilon_m}\frac{J_0'(k_\perp a)}{I_0'(\alpha_2 a)}I_0(\alpha_2 a)\right]\left[K_0(\alpha_2 b) - \frac{\alpha_3\varepsilon_m}{\alpha_2}\frac{K_0'(\alpha_2 b)}{K_0'(\alpha_3 b)}K_0(\alpha_3 b)\right]
$$

In the limit $b \to \infty$ or $a \to 0$, $Q = 0$. In the former, the first factor in Eq. (5.41) equated to zero gives the body wave (TM mode)

$$J_0(k_\perp a) = -\frac{\alpha_2 \varepsilon_g}{k_\perp \varepsilon_m} \frac{J_0'(k_\perp a)}{K_0'(\alpha_2 a)} K_0(\alpha_2 a). \tag{5.42}$$

For $a \to 0$ the second factor in Eq. (5.41) equated to zero gives the surface plasmon polaritons propagation

$$I_0(\alpha_2 b) = \frac{\alpha_3 \varepsilon_m}{\alpha_2} \frac{I_0'(\alpha_2 b)}{K_0'(\alpha_3 b)} K_0(\alpha_3 b). \tag{5.43}$$

In the limit $b \to \infty$, Eq. (5.43) reduces to $\varepsilon_m = -\alpha_2/\alpha_3$, giving the usual dispersion relation for SPPs over a planar surface

$$k_z^2 = \frac{\omega^2}{c^2} \frac{\varepsilon_m}{\varepsilon_m + 1}. \tag{5.44}$$

In the general case when a and b are finite the modes are significantly modified by finite Q. In the case of much interest, $\alpha_2 b >> 1$, Q is small and the coupling between the two factors on the left of Eq. (5.41) is weak. Equation (5.41) admits another excitation of surface plasmon polaritons that propagates along glass-metal interface with dispersion relation

$$K_0(\alpha_2 a) = -\frac{\varepsilon_m k_\perp}{\varepsilon_g \alpha_2} \frac{K_0'(\alpha_2 a)}{J_0'(k_\perp a)} J_0(k_\perp a), \tag{5.45}$$

In the limit $a \to \infty$, it takes the form $k_z^2 = (\omega^2/c^2)\left(\varepsilon_g \varepsilon_m / (\varepsilon_m + \varepsilon_g)\right)$. In Fig. 5.8, we have plotted normalized frequency $\Omega = \omega/\omega_P$ versus normalized wave number $q_z = k_z c/\omega_P$ for the Eqs. (5.42), (5.43) and (5.45) for the parameters: $\varepsilon_g = 2.13$, $\varepsilon_L = 10$, $b/a = 1.01$.

One may note that the frequency for TM mode begins with a cutoff and rises as q_z increases. The surface plasmon polartons have linear variation of Ω with q_z initially, however at large value of q_z, Ω tends to a saturation value. At a given Ω, the difference in q_z value for the TM mode and surface plasmon wave represents the wave number mismatch or the wave number required for resonant mode conversion. At $a\omega_P/c = 20$, one may calculate that a minimum of $q_z = 0.0184$ is required at $\Omega = 0.2453$ for the excitation of surface plasmon polaritons at the metal-free space interface. It is only in a narrow range of Ω, we can convert body wave into SPPs with a ripple of small q_z. Outside this range one would require a ripple of very high q_z, which is quite difficult to obtain.

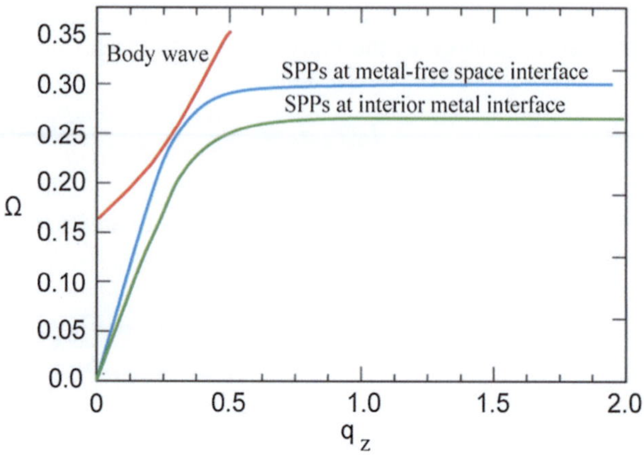

Fig. 5.8 Variation of normalized frequency, Ω with normalized wave number, q_z for TM mode body wave and SPPs. The parameters are: $\varepsilon_g = 2.13$, $\varepsilon_L = 10$, $a\omega_P/c = 20$, $b/a = 1.01$

5.4.2 Mode Conversion

Consider the fiber metal interface to be rippled (cf., Fig. 5.7), $r = a + h\cos(k_W z)$. On the outer side of the ripple, the electron density in the metal is n_0. In the ripple region $(a - h < r < a + h)$, it is periodic function of z, $n(z) = n(z + \lambda_W)$, where $\lambda_W = 2\pi/k_W$, h is the ripple wave amplitude and k_W is the ripple wave number.

Following Liu and Tripathi [19] we model the ripple by an electron density modulation,

$$n = \frac{n_0}{2}(1 + \sin(k_W z)), \quad \text{for} \quad a - h < r < a + h. \tag{5.46}$$

A laser of frequency ω_L and parallel wave number $k_z = k_L$ propagates through the fiber in TM mode. Its field in different regions is given by Eq. (5.39) with $A_S^{'}, k_\perp$, α_2, α_3 having a superfix L, designating laser. From Eq. (5.40), A_{21}^L, A_{22}^L, A_3^L are expressed in terms of A_1^L. Thus we may write the laser field as $\vec{E} = A_1^L\,\overrightarrow{\psi_L}\,(r)\exp(-i(\omega_L t - k_L z)$. This field imparts oscillatory velocity to electrons, $\vec{v} = e\,\vec{E}/mi\omega_L$, where $-e$ and m are the electronic charge and mass. Within the ripple region, $\vec{v}$ beats with n to produce a nonlinear current density at frequency ω_L and wave number $k_S = k_L + k_W$,

$$\vec{J}_S^{NL} = -\frac{n_0 e^2 A_1^L}{4mi\omega}\overrightarrow{\psi_L}\exp(-i(\omega_L t - k_S z)). \tag{5.47}$$

$\vec{J}_S^{NL}$ is localized in the ripple region and can be taken to be a delta function of r with $\overrightarrow{\psi_L}\,(r)$ replaced by $h\overrightarrow{\psi_L}\,(a)$. ω_L, k_S satisfy the dispersion relation for SPW, hence $\vec{J}_S^{NL}$ derive the surface plasmon polariton propagation. The relevant Maxwell's equations for the SPPs are

$$\nabla \times \vec{E}_S = \left(\frac{i\omega_L}{c}\right)\vec{H}_S,$$
$$\nabla \times \vec{H}_S = -\left(\frac{i\omega_L}{c}\right)\varepsilon'\vec{E}_S + \left(\frac{4\pi}{c}\right)\vec{J}_S^{NL}. \tag{5.48}$$

Let the solution of these equations when $\vec{J}_S^{NL} = 0$ be $\vec{E}_{SO}$ and $\vec{H}_{SO}$, i. e.

$$\nabla \times \vec{E}_{SO} = \left(\frac{i\omega_L}{c}\right)\vec{H}_{SO},$$
$$\nabla \times \vec{H}_{SO} = -\left(\frac{i\omega_L}{c}\right)\varepsilon'\vec{E}_{SO}. \tag{5.49}$$

$\vec{E}_{SO}$ is given by Eq. (5.39) with all $A'\,s$, $\alpha_{2,}, \alpha_3$ having superscript s, denoting SPPs. Using Eq. (5.40), $A_{21}^s, A_{22}^s, A_3^s$ are expressible in terms of A_1^s. When $\vec{J}_S^{NL}$ is retained, let the fields, following Liu and Tripathi [19] be

$$\vec{E}_S = A(z)\,\vec{E}_{SO}(r,z,t), \quad \vec{H}_S = B(z)\,\vec{H}_{SO}(r,z,t). \tag{5.50}$$

Using these in Eq. (5.48) and employing Eq. (5.49), we obtain

$$\frac{\partial A}{\partial z}\hat{z} \times \vec{E}_{SO} = \frac{i\omega_L}{c}(B-A)\,\vec{H}_{SO}, \tag{5.51}$$

$$\frac{\partial B}{\partial z}\hat{z} \times \vec{H}_{SO} = \frac{i\omega_L}{c}\varepsilon'(B-A)\vec{E}_{SO} + \frac{4\pi}{c}\vec{J}_S^{NL}. \tag{5.52}$$

Multiplying Eq. (5.51) by $\vec{H}_{SO}{}^{*}\,rdr$, Eq. (5.52) by $\vec{E}_{SO}{}^{*}\,rdr$ and integrating over r from 0 to ∞, we get

$$\frac{\partial A}{\partial z} = +\frac{i\omega_L}{c}(B-A)\frac{P_2}{P_3{}^*}, \tag{5.53}$$

$$\frac{\partial B}{\partial z} = -\frac{i\omega_L}{c}(B-A)\frac{P_1}{P_3} + RA_1^L, \tag{5.54}$$

where $R = \dfrac{\pi n_0 e^2 h a}{c P_3 m i \omega_L}\,\vec{\psi}_L(a).\vec{\psi}_S{}^*(a)$, $\qquad \vec{E}_{so} = \vec{\psi}_s \exp[-i(\omega_L t - k_S z)]$,

$\vec{\psi}_L(a) = J_0(k_\perp a)\,\hat{z} + \frac{ik_z}{k_\perp}J_0'(k_\perp a)\,\hat{r}$, $\qquad \vec{\psi}_S(a) = I_0(\alpha_2 a)\,\hat{z} - \frac{ik_z}{\alpha_2}I_0'(\alpha_2 a)\,\hat{r}$,

$P_1 = \varepsilon'\displaystyle\int_0^\infty \vec{E}_{SO}.\vec{E}_{SO}{}^*\,rdr$, $P_2 = \displaystyle\int_0^\infty \vec{H}so.\vec{H}so^*\,rdr$, $P_3 = \displaystyle\int_0^\infty \left(\vec{E}_{SO}{}^* \times \vec{H}_{SO}\right)_z rdr$.

* denotes the complex conjugate. Equations (5.53) and (5.54), with initial conditions at $z = 0$, A = B = 0 give

$$\frac{A}{A_1^L} = \frac{R}{(1 + P_1P_3{*}/P_2P_3)}\left(z + \left(e^{-i\gamma z} - 1\right)/i\gamma\right), \tag{5.55}$$

where $\gamma = (\omega_L/c)\,(P_1P_3{*}/P_2P_3)$. This treatment is valid as long as $z < k_{si}^{-1}$, where k_{si}^{-1} is the absorption coefficient of the SPPs given by

$$k_{si} = \mathrm{Im}\left[\frac{\omega_L}{c}\left(\frac{\varepsilon_m}{1 + \varepsilon_m}\right)^{1/2}\right]. \tag{5.56}$$

Putting $\varepsilon_m = \varepsilon_L - \omega_P^2/\omega_L\,(\omega_L + i\nu)$ in Eq. (5.56) we get

$$k_{si} = k_{SPW}\left(\frac{\omega_P^2\,\nu}{2\,\omega_L{}^3}\right)\frac{1}{\left[\omega_P^2/\omega_L^2 - (\varepsilon_L + 1)\right]\left[\omega_P^2/\omega_L^2 - \varepsilon_L\right]}, \tag{5.57}$$

where $k_{SPW} = (\omega_L/c)\left[\left(\omega_P^2/\omega_L^2 - \varepsilon_L\right)/\left(\omega_P^2/\omega_L^2 - (\varepsilon_L + 1)\right)\right]^{1/2}$, For $\gamma \approx 10^{12}\,s^{-1}$, $\varepsilon_L = 10$, $\omega_L/\omega_P = 0.2453$, then we obtain $z < 0.12\,cm$ for $\omega_L = 3.1970 \times 10^{15}\,rad/\sec$.

In order to have an appreciation of $|A|$ with z, we consider the case of frequency of the body wave, for a given $a\,\omega_P/c$, for which minimum ripple wave number is required. For $a\,\omega_P/c = 20$, this frequency is $\omega_L/\omega_P = \Omega = 0.2453$ with wave number $k_z\,c/\omega_P = q_z = 0.2727$ and corresponding ripple wave number $k_w\,c/\omega_P = q_w = 0.0184$. Also we use $\omega_L h/c \approx 0.2$. For these values, we get $P_1 = 3.5792\,cm^2$, $P_2 = 3.47176\,cm^2$, $P_3 = 2.34354\,cm^2$, $|R| = 47.214\,cm^{-1}$. This give $P_1P_3{*}/P_2P_3 = 1.03 \cong 1$. Putting this value in Eq. (5.55) and taking modulus gives

$$\left|\frac{A}{A_L^1}\right| = \left[\left(\frac{|R|^2}{(\omega_L/c)^2}\right)\left(\frac{1}{4}\left(\frac{\omega_L z}{c} - \sin\left(\frac{\omega_L z}{c}\right)\right)^2 + \sin^4\left(\frac{\omega_L z}{2c}\right)\right)\right]^{1/2}. \tag{5.58}$$

Eq. (5.58) gives the ratio of amplitude of the surface plasma wave to the amplitude of the incident laser.

5.5 Electron Acceleration by Surface Plasmon Polaritons

The excitations of symmetric surface plasmon polaritons in double metal surface configuration as well as over a single metal surface have been discussed in the Sect. 5.1. In this section, an analytical formalism resulting to the acceleration of electrons by surface plasmon polaritons and their trajectories in both single and double metals configurations will be discussed.

5.5.1 Double Metal Configuration

Let an electron be injected into the middle of the vacuum region bounded by two metal surfaces (cf. Fig. 5.3), in the presence of large amplitude surface plasmon wave. The electron response is governed by the equation of motion [20]

$$\frac{d\vec{p}}{dt} = -e(\vec{E} + \vec{v} \times \vec{B}), \tag{5.59}$$

where $-e$, m are the electronic charge and mass and $\vec{B} = (\nabla \times \vec{E})/i\omega$. Expressing $d/dt = v_z d/dz$, the x and z components of Eq. (5.59) can be written as

$$\frac{dp_x}{dz} = \left[\frac{em\gamma}{p_z}\left(\frac{k_z}{\alpha_2}\right) + \frac{e}{\omega}\left(\alpha_2 - \frac{k_z{}^2}{\alpha_2}\right)\right](e^{\alpha_2 x} + e^{-\alpha_2 x})A_1^{'} \sin(\omega t - kz + \phi), \tag{5.60}$$

$$\frac{dp_z}{dz} = \begin{bmatrix} -\dfrac{em\gamma}{p_z}(e^{\alpha_2 x} - e^{-\alpha_2 x})\cos(\omega t - kz + \phi) \\ +\dfrac{e}{\omega}\dfrac{p_x}{p_z}\left(-\alpha_2 + \dfrac{k_z{}^2}{\alpha_2}\right)(e^{\alpha_2 x} + e^{-\alpha_2 x})\sin(\omega t - kz + \phi) \end{bmatrix} A_1^{'}, \tag{5.61}$$

where $A_1^{'} = A_1 e^{-(t-z/v_g)^2/2\tau_L{}^2}$, $\gamma = (1 + p^2/m^2c^2)^{1/2}$, v_g is the group velocity of the surface plasmon wave, ϕ is the initial phase of the wave and we have considered a Gaussian temporal profile of the SPPs amplitude with τ_L pulse width. These equations are supplemented with

$$\frac{dx}{dz} = \frac{p_x}{p_z}, \tag{5.62}$$

$$\frac{dt}{dz} = \frac{\gamma m}{p_z}. \tag{5.63}$$

We introduce dimensionless quantities: $A_1^{''} \to eA_1^{'}/m\omega_p c$, $X \to \omega_p x/c$, $Z \to \omega_p z/c$, $P_x \to p_x/(mc)$, $P_z \to p_z/(mc)$, $T \to \omega_p t$, $\Omega = \omega/\omega_P$, $q = k_z c/\omega_P$, $v_g^{'} = v_g/c$. In terms of these, Eqs. (5.60), (5.61), (5.62), and (5.63) can be written as follows

$$\frac{\partial P_x}{\partial Z} = \left[\frac{\gamma}{P_z}\left(\frac{q}{\alpha_2^{'}}\right) + \frac{1}{\Omega}\left(\alpha_2^{'} - \frac{q^2}{\alpha_2^{'}}\right)\right]\left(e^{\alpha_2^{'} X} + e^{-\alpha_2^{'} X}\right)A_1^{''} \sin(\Omega T - qZ + \phi), \tag{5.64}$$

$$\frac{\partial P_z}{\partial Z} = \begin{bmatrix} -\dfrac{\gamma}{P_z}\left(e^{\alpha_2^{'} X} - e^{-\alpha_2^{'} X}\right)\cos(\Omega T - qZ + \phi) \\ +\dfrac{1}{\Omega}\dfrac{P_x}{P_z}\left(-\alpha_2^{'} + \dfrac{q^2}{\alpha_2^{'}}\right)\left(e^{\alpha_2^{'} X} + e^{-\alpha_2^{'} X}\right)\sin(\Omega T - qZ + \phi) \end{bmatrix} A_1^{''}, \tag{5.65}$$

$$\frac{dX}{dZ} = \frac{P_X}{P_Z}, \tag{5.66}$$

$$\frac{dT}{dZ} = \frac{\gamma}{P_Z}. \tag{5.67}$$

We solve Eqs. (5.64), (5.65), (5.66), and (5.67) numerically for electron energy and electron trajectory. In Fig. 5.9a, b, we plotted kinetic energy (in keV) gained by the electrons versus normalized distance $z\,\omega_p/c$ for different values of laser frequency, for the parameters: $P_x(0) = 0.0$, $P_z(0) = 0.1$, $x(0) = 0.0$, $t(0) = 0.0$, $\tau_L\omega_P = 200$, $\varepsilon_L = 1, \phi = \pi/2, E_{SP} = 1.2 \times 10^{11}\,V/m$, $\omega_P = 1.3 \times 10^{16}\,rad/\sec$. Here we have taken width of vacuum gap $a\omega_P/c = 100$ i.e. $a = 231\,\mu m$ In Fig. 5.9a, we obtain electron acceleration of 12.7 keV for laser frequency $\omega/\omega_P = 0.06$ while in Fig. 5.9b, we obtained electron acceleration of 8.8 keV for $\omega/\omega_P = 0.087$. We note that with the increase in the frequency of the incident laser, electron acceleration decreases. This appears due to the increase in the phase velocity of the SPPs with increasing $|\varepsilon|$. The SPPs can accelerate electrons to the velocities of the order of phase velocity. The trajectory of the accelerated electron, launched in the centre of vacuum region turns out to be a straight line at $x = -a/2$, i.e., the electron moves in z-direction without deviation from its path. The trajectory of the accelerated electron in double metal surface is shown in Fig. 5.10a alongwith the trajectory of the accelerated electron over a single metal surface.

5.5.2 *Single Metal Configuration*

Consider an interface separating free space $(x > 0)$ and a metal $(x < 0)$. As discussed in Sect. 5.1, one can obtain the dispersion relation of the SPPs as

$$k_z^2 = \frac{\omega^2}{c^2}\frac{\varepsilon}{1+\varepsilon}, \tag{5.68}$$

where ε is the dielectric constant of the metal.

An electron beam is launched parallel to the surface with initial velocity $\vec{v}_0$. Its response to the surface plasmon polaritons is governed by the wave Eq. (5.59). The x and z components alongwith supplementary equations can be written as

$$\frac{\partial P_x}{\partial z} = A\left[-\frac{\gamma me}{P_z}\left(\frac{k_z}{\alpha_1}\right)\sin(\omega t - k_z z + \phi) + \frac{e}{\omega}\left(\frac{k_z^2}{\alpha_1} - \alpha_1\right)\sin(\omega t - k_z z + \phi)\right]e^{-\alpha_1 x}, \tag{5.69}$$

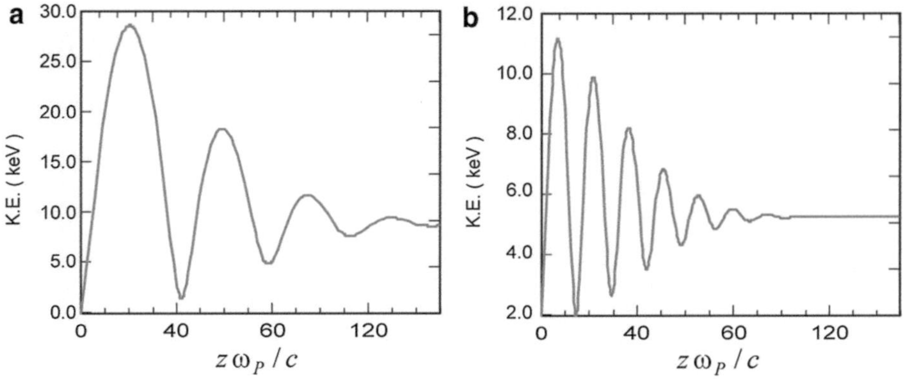

Fig. 5.9 Variation of kinetic energy of accelerated electrons, $(\gamma - 1)\, mc^2$ versus normalized distance $z\, \omega_P/c$ in double metal surface for two different laser frequencies (**a**) $\omega/\omega_P = 0.06$, and (**b**) $\omega/\omega_P = 0.087$. The other parameters are: $Px\ (0) = 0.0$, $Pz\ (0) = 0.009$, $x(0) = 0.1$, $t(0) = 0.0$, $\tau_L \omega_P = 200$, $\varepsilon_L = 1$, $\varphi = \pi/2$, $E_{SP} = 1.2 \times 10^{11}\, V/m$, $\omega_P = 1.3 \times 10^{16}\, rad/\sec$, $a\omega_P/c = 100$

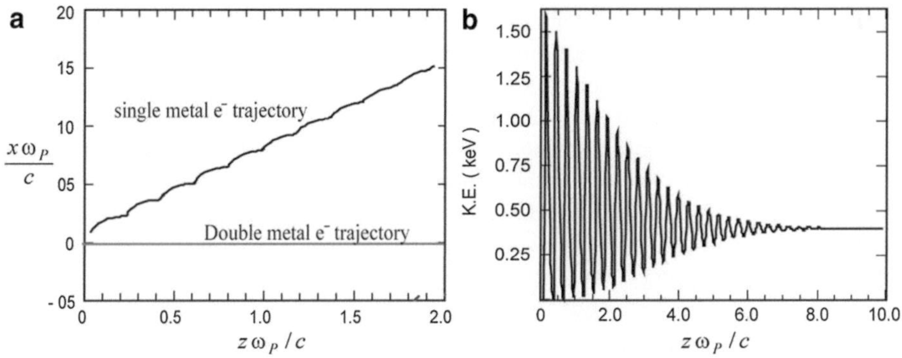

Fig. 5.10 (**a**) Trajectory of electron in single and double metal surface. (**b**) Variation of kinetic energy versus normalized distance $z\, \omega_P/c$ over the single metal surface. The parameters are: $Px\ (0) = 0.0$, $Pz\ (0) = 0.007$, $x(0) = 0.1$, $t(0) = 0.0$. $L_\tau = 27$ fs, $E = 10^6$ V/cm, $\varepsilon_L = 5$, $\omega = 375 THz$. $\varphi = \pi$

$$\frac{\partial P_z}{\partial z} = A\left[-\frac{\gamma m e}{P_z}\cos(\omega t - k_z z + \phi) - \frac{e}{\omega}\frac{p_x}{p_z}\left(\frac{k_z^2}{\alpha_1} - \alpha_1\right)\sin(\omega t - k_z z + \phi)\right] e^{-\alpha_1 x}, \tag{5.70}$$

$$\frac{dx}{dz} = \frac{p_x}{p_z}, \tag{5.71}$$

$$\frac{dt}{dz} = \frac{\gamma m}{p_z}. \tag{5.72}$$

where ϕ is the initial phase of the SPPs. Equations (5.69), (5.70), (5.71), and (5.72) in dimensionless form can be written as

$$\frac{\partial P_x}{\partial Z} = A''\left[-\frac{\gamma}{P_z}\left(\frac{q}{\alpha_1'}\right)\sin(\Omega T - qZ + \phi) + \frac{1}{\Omega}\left(\frac{q^2}{\alpha_1'} - \alpha_1'\right)\sin(\Omega T - qZ + \phi)\right], \tag{5.73}$$

$$\frac{\partial P_z}{\partial Z} = A''\left[-\frac{\gamma}{P_z}\cos(\Omega T - qZ + \phi) - \frac{1P_x}{\Omega P_z}\left(\frac{q^2}{\alpha_1'} - \alpha_1'\right)\sin(\Omega T - qZ + \phi)\right], \tag{5.74}$$

$$\frac{dX}{dZ} = \frac{P_X}{P_Z}, \tag{5.75}$$

$$\frac{dT}{dZ} = \frac{\gamma}{P_Z}. \tag{5.76}$$

where, $A'' = A'e^{-\alpha_1' X}e^{-\left(T - Z/v_g'\right)^2/2\tau_L'2}$, and X, Y and $\vec{P}$ are same as defined above.

We have solved these equations numerically for following parameters: $P_X(0) = 0.0,\ P_Z(0) = 0.007,\ X(0) = 1.0,\ T(0) = 0.0,\ \phi = \pi$. We express the surface plasmon wave amplitude as $|E_S| = \eta|E_L|$, where $|E_L|$ is the amplitude of the laser used to excite it in the attenuated total reflection (ATR) configuration and η is the enhancement factor. Presuming laser energy conversion to the SPPs as 50 %, enhancement factor comes out to be $\eta \cong 3.2$ corresponds to the laser intensity of 10^{13} W/cm^2. We choose $E_L = 2.9 \times 10^5$ V/cm. The results are displayed in Fig. 5.10a, b. The maximum kinetic energy gain by the electrons comes out to be ~ 0.39 KeV, which is close to the experimentally observed value of 0.4 KeV by Zawadzka et al. [21]. In Fig. 5.10a, we have plotted electron trajectory in x-z plane for the same parameters. One may note that the electron move away from the metal surface as it gains energy and hence not confined.

5.6 Surface Plasmon Excitations in Surface Enhanced Raman Spectroscopy

Surface enhanced Raman spectroscopy or surface enhanced Raman scattering (SERS) is an analytical technique that provides signals with enhanced intensity for Raman active molecule that have been adsorbed onto metal surface, patterns or nanostructures [22]. There are two main theories that explain the mechanism behind enhanced Raman scattering. One theory relies on the formation of charge complex and charge transfer [23] between chemisorbed species and the metal surface. The second theory believes that Raman enhancement occurs for the adsorbed material on a specific substrate (mainly metals or their nanocrystals) due to the enhancement of electric field provided by the substrate. There is generation of localized surface plasmons when a light beam is incident on substrate. In this case electric field enhancement is more at plasmon frequency [24].

Raman scattering is inelastic scattering of photon that often occurs when time harmonic field interacts with the molecular vibration. In this process, scattered radiation are also produced [24]. The frequency of this scattered radiation is little shifted from the incident radiation. Such a shift is equivalent to the vibrational frequency of the molecule. The origin of this vibrating frequency is oscillation between constituent atoms of the molecules. This oscillation and corresponding shift in frequency depends on molecular structure. The Raman scattering is weak effect and its scattering cross section order of magnitude is ~14–15 times smaller than the fluorescence cross section of an efficient dye [25]. Hence, field enhancement in many cases becomes very important. Such an enhancement factor is typically on the order of ~10^6. However, in some cases higher order enhancements were also achieved. In this chapter Raman signal enhancement from the enhanced electric fields at rough metal surface is discussed. Such an enhancement can be observed at places such as junction between metal crystals, grove, and cracks on the surfaces. The most common assumption is that Raman scattering enhancement is related to fourth power of electric field enhancement. Fig. 5.11 is shown to understand the phenomenon of SERS. Here a molecule is located at d_o, that is also in close proximity to metallic structure (local field enhancing device). There will be an interaction between incident field 'E_o' with the molecule, that result in a dipole moment associated with Raman scattering [25].

$$\mu(\omega_R) = \alpha(\omega_R + \omega)(E_o(d_o, \omega) + E_s(d, \omega)), \tag{5.77}$$

where, ω = frequency of the incident radiation, ω_R = A particular vibrationally shifted frequency ($\omega_R = \omega \pm \omega_{vib}$), α = polarizability (modulated at the vibrational frequency, ω_{vib}). Note that, there is an interaction of molecule with local field '$E_o + E_s$', here 'E_o' is the parent electric field, 'E_s' is the enhanced electric field due to interaction with nanostructures or pattern. E_s linearly depends on E_o, and can be represented as:

$$E_s = f_1(\omega).E_o, \; (f_1(\omega) = \text{field enhancement factor}) \tag{5.78}$$

The induced dipole (μ) radiated electric field can be written using Green's Function relation, after considering the presence of nano-structures [25]:

$$E(d_\infty, \omega_R) = \frac{\omega_R^2}{\varepsilon_o c^2} G(d_\infty, d_o)\mu(\omega_R). \tag{5.79}$$

Greens function can be split into a free space Green function 'G_o' (without any metal) and a scattered one 'G_s' (that originates from metallic nanostructure) [25]:

$$G(d_\infty, d_o) = G_o(d_\infty, d_o) + G_s(d_\infty, d_o), \tag{5.80}$$

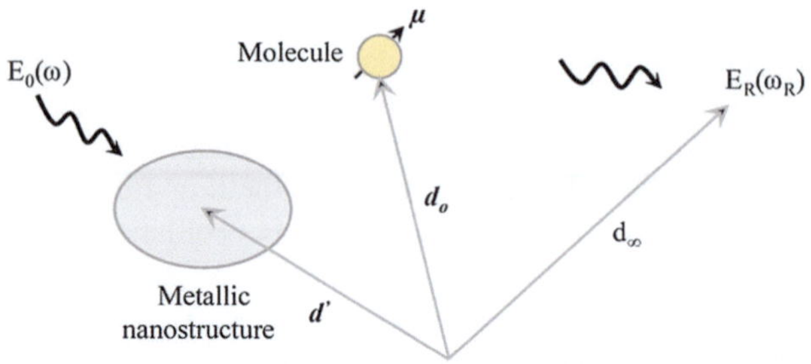

Fig. 5.11 An schematic representation of surface enhanced spectroscopy phenomenon. Here an interaction between a molecule (polarizability α) and the exciting electric field 'E_o' causes scattered Raman radiation with electric field E_R . The presence of metallic nanostructure in close proximity of molecule or compound of interest, enhances the exciting field as well as radiated field

$$G_s = f_2(\omega_R)G_o, \tag{5.81}$$

where $f_2(\omega_R) =$ second field enhancement factor. Now using above equations, intensity can be calculated, that is squared proportional to intensity [25]:

$$I(d_\infty, \omega_R) = \frac{\omega_R^4}{\varepsilon_o^2 c^4} |(1 + f_1(\omega_R))(1 + f_2(\omega_R))G(d_\infty, d_o)\alpha(\omega_R, \omega)|^2 I_o(d_o, \omega). \tag{5.82}$$

Hence it can be said that Raman scattered intensity varies linearly with intensity I_o, with factor:

$$f_{overall} = |(1 + f_1(\omega_R))(1 + f_2(\omega_R))|^2. \tag{5.83}$$

Now, when there is no metal structures, the enhancement factors f_1 and f_2 are zero; whereas in the presence of metallic nanostructures these factors are very high compared to 1 and hence, overall scattering enhancement can be written as [25]:

$$f_{overall} = |f_1(\omega_R)f_2(\omega_R)|^2, \tag{5.84}$$

(Under the assumption that $|\omega_R \pm \omega|$ is smaller compared to the spectral response of the metal nanostructures.) Fig. 5.12 shows Raman spectra of rutile TiO_2 nanocrystalline film coated on glass as well as gold substrate. These examinations were carried using an R 3000 QE Raman spectrophotometer at backscattering mode. The excitation wavelength was ~785 nm and laser power was ~100 mW. The peaks were observed at ~448 cm^{-1} and 611 cm^{-1}. It can be observed that peak intensity for TiO_2 coated on Au back substrate was several times higher compared to that coated on glass substrate.

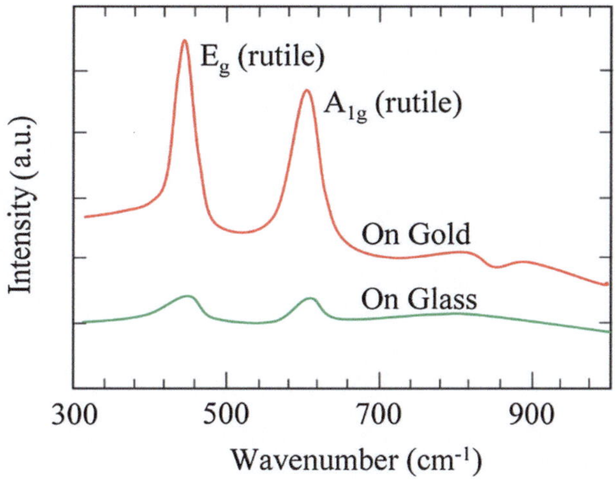

Fig. 5.12 Raman spectra of rutile TiO_2 nanocrystalline film coated on glass and gold substrate. The height of spectral features is several times higher when Au is used as a back contact. This is caused by the surface enhanced Raman scattering phenomenon

5.7 Surface Plasmon Plasmon Applications in Sensing and Solar Cell Technology

Surface plasmons and their associated resonance condition have been widely utilized in various sensing devices [26]. These groups of surface plasmons based refractometric sensing devices are one of the main sensing technique, used for biological species and chemical detections. In most of the cases these are SPR based refractometer that measure any variation in refractive index at metal film surface. Any change in refractive index of dielectric medium causes a change in surface plasmon propagation constant [26]. The change in coupling condition (coupling intensity, phase, wavelength, coupling angle) can be utilized for probing. For example, in case of angular modulation based SPR sensors, a monochromatic light wave is generally utilized for surface plasmons excitation [26]. Coupling strength between surface plasmon and incident light wave can be observed at multiple angles of incidence by using convergent light beam. For evaluating sensor output, the angle of incidence giving strongest coupling, is measured. In contrast, SPR sensor working on wavelength modulation, surface plasmons are generally excited using collimated light wave with multiple wavelengths. These SPR based sensors have been extensively used in detection of pathogens, toxins, hormones, proteins, vitamins, and diagnostic antibodies [26].

Surface plasmons have been widely utilized for effective light management in order to achieve high efficiency of photovoltaic devices with thin absorber layer [27]. Use of thin film is advantageous because it reduces material cost. Additionally, for such thin film solar cells, performance can be enhanced and in certain cases, efficiency can be improved if adequate light can be scattered inside the absorber layer film using metallic nanoparticles [27]. In many cases, surface texturing has been used to improve a short circuit current. However, surface recombination and difficulties in fabrication of good quality semiconducting layers in such cases have been a challenge. Enhanced light trapping in absorber layer can be achieved by tuning localized surface plasmons in arrays of Au or Ag nanocrystals [28]. Both thin film and wafer based Si solar cells have shown increased performance due to the presence of Ag nanoparticles. The main mechanisms that explain photocurrent enhancement by metallic nanoparticles are light scattering and near field concentration of light [27, 28]. The size of particle and electrical/optical properties of semiconductor absorber layer are factors that decide the contribution from each mechanism. The power conversion efficiency of ~15.3 % using metallic nanoparticles but only using ~18 μm thick Si layer, was reported [29]. It saves ~97 % of the absorber materials. Apart from Si, nanoparticles have also been utilized in organic bulk hetrojunction solar cells, hybrid solar cells, and perovskite materials, for efficiency improvement [30].

Acknowledgements Authors would like to greatly acknowledge Prof. V.K. Tripathi for his valuable guidance and inspiring discussions during the course of work reported in this chapter.

References

1. (a) Zayats AV, Smolyaninov II, Maradudin AA (2005) Nano-optics of surface plasmon polaritons. Phys Rep 408(3):131–314; (b) Raether H (1988) Surface plasmons on smooth surfaces. Springer, Berlin Heidelberg; (c) Agranovich VM, Mills DL (1982) Surface Polaritons. Elsevier, North-Holland, Amsterdam.
2. Wood R (1902) XLII. On a remarkable case of uneven distribution of light in a diffraction grating spectrum. Lond Edinb Dublin Philos Mag J Sci 4(21):396–402
3. Wood RW (1902) On a remarkable case of uneven distribution of light in a diffraction grating spectrum. Phil Mag 4:396–402
4. Fano U (1941) The theory of anomalous diffraction gratings and of quasi-stationary waves on metallic surfaces (Sommerfeld's waves). JOSA 31(3):213–222
5. Otto A (1968) Excitation of nonradiative surface plasma waves in silver by the method of frustrated total reflection. Z Phys 216(4):398–410
6. Kretschmann E, Raether H (1968) Radiative decay of non-radiative surface plasmons excited by light. Z Naturforsch A 23(12):2135–2136
7. Okamoto K, Niki I, Shvartser A, Narukawa Y, Mukai T, Scherer A (2004) Surface-plasmon-enhanced light emitters based on InGaN quantum wells. Nat Mater 3(9):601–605
8. (a) Barnes WL, Dereux A, Ebbesen TW (2003) Surface plasmon subwavelength optics. Nature 424(6950):824–830; (b) Genet C, Ebbesen T (2007) Light in tiny holes. Nature 445 (7123):39–46; (c) Klimov V, Ducloy M, Letokhov V (2002) A model of an apertureless scanning microscope with a prolate nanospheroid as a tip and an excited molecule as an object. Chem Phys Lett 358(3):192–198; (d) Krupin O, Asiri H, Wang C, Tait RN, Berini P (2013) Biosensing using straight long-range surface plasmon waveguides. Opt Express 21(1): 698–709
9. Saito H, Namura K, Suzuki M, Kurata H (2014) Dispersion relations for coupled surface plasmon-polariton modes excited in multilayer structures. Microscopy 63(1):85–93
10. Sheng P, Stepleman RS, Sanda PN (1982) Exact eigenfunctions for square-wave gratings: application to diffraction and surface-plasmon calculations. Phys Rev B 26(6):2907–2916
11. Stockman MI, Faleev SV, Bergman DJ (2001) Localization versus delocalization of surface plasmons in nanosystems: can one state have both characteristics? Phys Rev Lett 87(16):167401

12. Davis TJ (2009) Surface plasmon modes in multi-layer thin-films. Opt Commun 282(1):135–140
13. Stoian R, Rosenfeld A, Ashkenasi D, Hertel I, Bulgakova N, Campbell E (2002) Surface charging and impulsive ion ejection during ultrashort pulsed laser ablation. Phys Rev Lett 88 (9):097603
14. (a) Kumar G, Tripathi V (2007) Anomalous absorption of surface plasma wave by particles adsorbed on metal surface. Appl Phys Lett 91(16):161503; (b) Mishra Y, Adelung R, Kumar G, Elbahri M, Mohapatra S, Singhal R, Tripathi A, Avasthi D (2013) Formation of self-organized silver nanocup-type structures and their plasmonic absorption. Plasmonics 8 (2):811–815
15. Vorobyev A, Guo C (2005) Enhanced absorptance of gold following multipulse femtosecond laser ablation. Phys Rev Ser B 72(19):195422
16. (a) Wolfbeis OS (2006) Fiber-optic chemical sensors and biosensors. Anal Chem 78(12):3859–3874; (b) Chau L-K, Lin Y-F, Cheng S-F, Lin T-J (2006) Fiber-optic chemical and biochemical probes based on localized surface plasmon resonance. Sensors Actuators B Chem 113(1):100–105
17. Homola J, Yee SS, Gauglitz G (1999) Surface plasmon resonance sensors: review. Sens Actuators B 54(1):3–15
18. (a) Lucotti A, Zerbi G (2007) Fiber-optic SERS sensor with optimized geometry. Sensors Actuators B Chem 121(2):356–364; (b) Liu C, Kumar G, Tripathi V (2006) Laser mode conversion into a surface plasma wave in a metal coated optical fiber. J Appl Phys 100 (1):013304
19. Liu CS, Tripathi VK (1998) Diffraction-limited laser excitation of a surface plasma wave and its scattering on a rippled metallic surface. Quantum Electron IEEE J 34(8):1503–1507
20. Liu C, Kumar G, Singh D, Tripathi V (2007) Electron acceleration by surface plasma waves in double metal surface structure. J Appl Phys 102(11):113301
21. Zawadzka J, Jaroszynski DA, Carey JJ, Wynne K (2001) Evanescent-wave acceleration of ultrashort electron pulses. Appl Phys Lett 79(14):2130–2132
22. (a) Schlücker S (2014) Surface-enhanced Raman spectroscopy: concepts and chemical applications. Angew Chem Int Ed 53(19):4756–4795; (b) Kumar G, Singh D, Tripathi V (2006) Surface enhanced Raman scattering of a surface plasma wave. J Phys D Appl Phys 39(20):4436
23. Lombardi JR, Birke RL, Lu T, Xu J (1986) Charge-transfer theory of surface enhanced Raman spectroscopy: Herzberg–Teller contributions. J Chem Phys 84(8):4174–4180
24. Moskovits M (2005) Surface-enhanced Raman spectroscopy: a brief retrospective. J Raman Spectrosc 36(6–7):485–496
25. Novotny L, Hecht B (2012) Principles of nano-optics. Cambridge University Press, Cambridge
26. Homola J (2008) Surface plasmon resonance sensors for detection of chemical and biological species. Chem Rev 108(2):462–493
27. Catchpole KR, Polman A (2008) Design principles for particle plasmon enhanced solar cells. Appl Phys Lett 93(19):191113
28. Beck FJ, Polman A, Catchpole KR (2009) Tunable light trapping for solar cells using localized surface plasmons. J Appl Phys 105(11):114310
29. Zhang Y, Stokes N, Jia B, Fan S, Gu M (2014) Towards ultra-thin plasmonic silicon wafer solar cells with minimized efficiency loss. Sci Rep 4:4939
30. (a) Lee JH, Park JH, Kim JS, Lee DY, Cho K (2009) High efficiency polymer solar cells with wet deposited plasmonic gold nanodots. Org Electron 10(3):416–420; (b) Zhou F, Zhu J, Lai Z, Liu Y, Zhao X (2014) Surface plasmon resonances behavior in visible light of non-metal perovskite oxides AgNbO3. Appl Phys Lett 105(23): 231121

Chapter 6
Ultrafast Response of Plasmonic Nanostructures

Sunil Kumar and A.K. Sood

Abstract Ultrafast photoresponse of plasmonic nanostrucutres, specifically, nanoparticles has been discussed here. Femtosecond laser pulses are useful not only for the time-resolved investigations but also to look at the optical nonlinearities in materials which are primarily electronic at such time-scales. Like the linear photoresponse such as absorption and scattering cross-sections, ultrafast nonlinear optical response also gets emensely enhanced at wavelenths near the surface plasmon resonace of the nanosystem under study. In a time-resolved measurement, typically the electronic scattering processes are studied, however, the confined accoustic phonons due to the finite size effects of the nanostrucutres can also modify the ultrafast time-resolved response. Although, Raman spectroscopy and infrared absorption spectroscopy are the popular techniques for studying phononic properties of the nanostructures, terahertz time-domain spectroscopy using ultrashort terahertz pulses also has shown potential to become another important characterization technique for nanoparticles specially at very low frequencies where other techniques are difficult to reach. We have attempted to present a comprehensive account of the above topics by including essential background given in the beginning of the chapter and subsequently discussing some of the important experimental results from the recent literature along with our results on metal nanoparticles.

Keywords Surface plasmon • Time-resolved spectroscopy • Optical nonlinearities • Coherent phonons • Terahertz spectroscopy • Confined acoustic phonons

S. Kumar (✉)
Department of Physics, Indian Institute of Technology Delhi, New Delhi 110016, India

Department of Physics, Indian Institute of Science, Bangalore 560012, India
e-mail: kumarsunil@physics.iitd.ac.in; sunilvdabral@gmail.com

A.K. Sood
Department of Physics, Indian Institute of Science, Bangalore 560012, India

C.D. Geddes (ed.), *Reviews in Plasmonics 2015*,
DOI 10.1007/978-3-319-24606-2_6

6.1 Introduction

Plasmonic metal nanostructures are an integral part of nanophotonic device applications owing to their ability to generate strong localized electromagnetic fields when illuminated from the far-field. These nanostructures can be prepared in the laboratory with precise shape and size by various chemical and physical synthesis techniques as demonstrated in recent years. Typically, the size of these structures is in the range of few nm to a few hundred nm, i.e., much below the wavelength of light of observation or excitation. The focus of this chapter is to exploit the ultrafast response of metal nanostructures following femtosecond laser pulse excitation. Femtosecond laser pulses are very important experimental probes to explore the material response at an ultrafast time scale ($\sim$10–10,000 fs) on the order of the electronic and vibrational wave packet dynamics in solids and molecules. Broadly, after ultrafast optical excitation the system of study undergoes various microscopic dynamical changes which can be monitored in real time so as to understand the intermediate stages before the system returns to the equilibrium. This is depicted pictorially in Fig. 6.1 where after excitation by a femtosecond laser pulse (Fig. 6.1a), the system's response R(t) is monitored by an ultrafast probe pulse I (t) through the convolution between the two at each time t (Fig. 6.1b). Material excitations which have life-time broadening larger than the temporal width of the laser pulse can be coherently populated under proper experimental conditions. These coherent states are observed by collecting the time-resolved scattered signal. In Fig. 6.1c we have also shown a situation where due to the ultrahigh peak electric field associated with the femtosecond laser pulses, one can easily find various

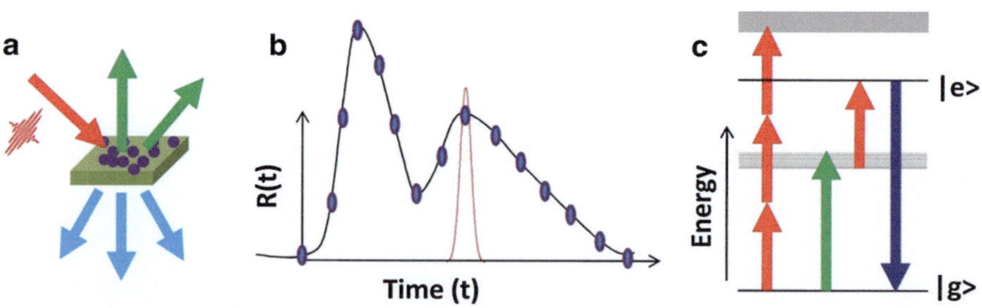

Fig. 6.1 (**a**) Photo-excitation of a material system by an ultrashort light pulse. The scattered light can be collected in either time-resolved or spectrally resolved manner using optical schemes which allow detection of smallest possible changes in the system's optical response created by the ultrafast laser pulse and bearing the state character of the dynamical system at a point of observation in the phase space. (**b**) Following ultrashort photo-excitation, the system evolves through various intermediate dynamical states which can be monitored by using an ultrashort optical probe pulse that measures the convolution between the system's response function and the probe itself. (**c**) The ultrahigh peak electric fields associated with the ultrashort laser pulses induce various possible nonlinear processes in the system such as three-photon absorption, cascaded two-photon absorption and two-photon excited fluorescence, to name a few which can be easily detected

possibilities for parametric and nonparametric nonlinear optical processes such as, multi-photon absorption, cascaded two-photon absorption, two-photon excited fluorescence, second or higher harmonic generation, Kerr effect and coherent anti-Stokes scattering.

Plasmons are coupled modes of electromagnetic field of the light and the collective oscillations of conduction electrons in metals. Also, the coupled electromagnetic fields are confined only within a thin skin layer on the surface of the metal leading to a large enhancement of the local electromagnetic field. Surface plasmons can be localized, for example, in nanoscale particles with large surface to volume ratio, or propagating, such as in interfaces between metals and dielectrics. Propagating surface plasmons which need proper optical arrangements for excitation by external light [1–3], have shown potential as a good candidate for future high-speed opto-electro-plasmonic chip scale integrated devices [4, 5]. Similarly, localized surface plasmons in nanoparticles have led to many interesting effects one of which is surface enhanced Raman scattering [6]. Strikingly, the electromagnetic resonance due to the localized surface plasmons in the metal nanoparticles is easily tunable by shape, size and choice of the surrounding dielectric medium, and occurs in the visible to near infrared region of the electromagnetic spectrum. The plasmons in nanoparticles can be directly excited by incident light and lead to strong light scattering and absorption due to large local optical field enhancements inside and in the neighbourhood of the particles. These appealing characters of metal nanoparticles make them very attractive from the point of view of fundamental studies as well as usability in various technological applications [7–11]. The surface confinement effects in metal nanoparticles cause additional intrinsic dynamical processes which get themselves involved in determining the overall response to an external stimuli such as a femtosecond laser pulse.

A very good understanding of the linear optical properties of plasmonic nanostructures can be achieved simply by applying Maxwell's classical theory of electromagnetism with appropriate boundary conditions and continuity equations at the interfaces between the metal and the surrounding medium and the other intrinsic and extrinsic size effects. The readers can find very good literature available on the synthesis, theoretical and experimental investigations of the linear optical properties of metal nanostructures [10–15]. In this chapter, we focus on the ultrafast response of plasmonic nanostructures, specifically the noble metal nanoparticles. We will discuss the optical nonlinearities and electron dynamics after ultrashort photo-excitation. The emphasis will be on these effects near the plasmonic resonance and away from it. The consequences of reduced dimensions in nanoparticles are not only related to the electronic confinement but also in other attributes such as the confined surface acoustic phonons. These excitations have unique signatures of the thermal and elastic properties of the material and have been studied for quite a long time using Raman scattering, infrared absorption and terahertz time-domain spectroscopy techniques. We will discuss some of these observations as well. Metal nanoparticles with only a few atoms termed as nanoclusters make another interesting area of active research due to very interesting optical phenomena observed in them [16–18]. Such metallic systems of sub-nanometer size lack surface plasmon resonance, the typical feature of bigger nanoparticles in the size range of ~2–100 nm. Instead they show characteristic peaks in the linear absorption spectrum which are signatures of their molecular nature. In this chapter, we have also discussed experimentally observed ultrafast nonlinearities in 15-atom gold clusters where it was seen that the optical limiting performance of the nanoclusters is many folds enhanced when placed in contact with a thin metallic film of indium tin oxide.

The chapter has been organized as follows. We begin with a brief introduction in Sect. 6.2 about the linear optical properties of metal nanoparticles which are essential to understand the ultrafast responses presented in subsequent sections. The absorption and scattering by nanoparticles is discussed in the light of Mie's theory [19] and Gans approximation [20]. We will consider mainly spherical and cylindrical nanoparticles to bring the notions of shape-dependence of the localized surface plasmon resonance. Since in this chapter our focus is on metal nanoparticles, the surface plasmon resonance means the localized surface plasmon resonance. In Sect. 6.3, we have discussed the electronic relaxation dynamics in metal nanoparticles following ultrafast optical-excitation at the surface plasmon resonance and away from it. Theoretical models for understanding the underlying relaxation mechanisms have been discussed. We should note that most of the experimental studies on metal nanoparticles have been performed and discussed for inhomogeneous distributions of the particles around a mean value of their size and hence ensemble averaging should be assumed unless mentioned otherwise. Single particle optical spectroscopy measurements are generally more sophisticated in terms of experimental realizations, some of which will be mentioned while discussing the ultrafast time-resolved response in Sect. 6.3. In Sect. 6.4, we will describe the ultrafast nonlinear optical response of metal nanoparticles. For a flavour of nonlinear optical response from subnanometric and nonplasmonic particles, we will discuss an interesting example of 15-atom gold clusters which show an improved optical limiting performance at wavelengths near the surface plasmon resonance of bigger nanoparticles which otherwise would have shown saturable absorption. Section 6.5 discusses effects of phonon confinement in nanoparticles measured by the ultrafast optical response from them. Experimental studies reporting the characterization of the confined acoustic phonons in spherical and rod-shaped nanoparticles will be discussed. Finally we will conclude by providing an assessment of the potential opportunities and challenges for metal nanoparticle plasmonics in Sect. 6.6.

6.2 Surface Plasmons in Metal Nanoparticles

The optical properties of materials can be completely described by the complex dielectric function ($\varepsilon = \varepsilon_1 + i\varepsilon_2$) or complex index of refraction $\left(\tilde{n} = \varepsilon^{1/2}\right)$. The dielectric function of metals consists of contributions from the free electrons or the conduction band electrons (intraband transitions) through the Drude term ε^D and the bound electrons from inner electronic bands (interband transitions such as by d-band electrons in noble metals) through a Lorentzian term ε^{ib},

$$\varepsilon(\omega) = \varepsilon^D + \varepsilon^{ib} = 1 - \frac{\omega_p^2}{\omega^2 + i\Gamma_0\omega} + \frac{\widetilde{\omega}_p^2}{\omega_0^2 - \omega^2 - i\gamma\omega} \tag{6.1}$$

Here ω_p is the lowest cut-off frequency of the collective motion of the free electrons inside metals termed as the bulk plasma frequency, given by $\omega_p = \left(n_e e^2/\varepsilon_0 m_e\right)^{1/2}$ for electronic density n_e, charge e, mass m_e and vacuum permeability ε_0. The parameter Γ_0 is the free-electron scattering rate, $\widetilde{\omega}_p$ and γ are analogous plasma frequency and scattering rate for the bound electrons with interband transition energy at $\hbar\omega_0$. Neglecting the interband contribution and at high frequencies (ω) such that $\Gamma_0/\omega << 1$, Eq. (6.1) provides the more commonly used expression, $\varepsilon(\omega) = 1 - \omega_p^2/\omega^2$. The plasma frequency ranges between ~1 to 6 eV in various metals below which they do not absorb light and this is the reason why metals are high reflectors in the visible and near infrared range. As the physical dimension of metal structures is reduced to be comparable to the de Broglie wavelength, i.e., of the order of the electron scattering length in metals, confinement effects become dominant leading to significantly different optical properties of metal nanostructure than the bulk metals. For large surface to volume ratios, the plasmons become localized due to the surface effects. The shape, size and composition of the nanoparticles as well as the surrounding medium in which they are embedded dictate the interaction of light with the nanoparticles. Below, we discuss the cases of spherical and cylindrical nanoparticles which are simplest in terms of obtaining analytical solutions for the extinction cross-section.

Gustav Mie [19] for the first time analytically solved the problem of scattering and absorption by a spherical metal nanoparticle. This allows theoretically predict the linear response of a metal sphere to an external electromagnetic field by incorporating the material dielectric properties and solving the Maxwell's equations under appropriate boundary conditions. Consider a spherical particle of radius/volume R/V and metal dielectric function ε_m that is embedded in a dielectric medium with dielectric function ε_d. For R much smaller than the wavelength of light such that the particle can be assumed to be an ideal dipole in the applied field, Mie's theory predicts the absorption (abs) and scattering (sca) cross-sections [12, 13] to be:

$$\sigma_{abs}(\omega) = \frac{k_d}{\varepsilon_0\varepsilon_d}\mathrm{Im}(\alpha) \tag{6.2a}$$

$$\sigma_{sca}(\omega) = \frac{k_d^4}{6\pi(\varepsilon_0\varepsilon_d)^2}(\alpha)^2 \tag{6.2b}$$

Here, k_d is the light wave-vector in the dielectric medium and α is the linear polarizability given by the Clasusius-Mosotti relation:

$$\alpha(\omega) = 3V\varepsilon_0\varepsilon_d \frac{\varepsilon_m - \varepsilon_d}{\varepsilon_m + 2\varepsilon_d} \tag{6.3}$$

It is clear that surface plasmon resonance in spherical nanoparticles is observed when the condition $\varepsilon_m = -2\varepsilon_d$ is satisfied.

In the second example, we consider rod-like nanostructures with dimensions still much smaller than the wavelength of light. For small ellipsoidal particles, scattering problem was solved by Gans in 1912 [20], according to which, the linear polarizability along one of the axes ($k = 1,2,3$) is given by [12, 13, 15]

$$\alpha_k(\omega) = V\varepsilon_0\varepsilon_d \frac{\varepsilon_m - \varepsilon_d}{g_k(\varepsilon_m - \varepsilon_d) + \varepsilon_d} \tag{6.4}$$

Here g_k are geometric factors along the three axes of the ellipsoid. For a particular case of spheroidal particles for which the two axes of the ellipsoid are equal and considering g_1 along the long axis (a > b = c), the geometrical factors can be expressed as:

$$g_1 = (1/e^2 - 1)\left[(1/2e)\ln\left(\frac{1+e}{1-e}\right) - 1\right]; \quad g_2 = g_3 = \frac{(1-g_1)}{2} \tag{6.5}$$

where $e = \sqrt{1 - b^2/a^2}$ is the eccentricity of the particle related to the particle aspect ratio ζ (length/width = a/b). Equation (6.4) in combination with Eq. (6.2a, b) is used to calculate the scattering and absorption cross-sections for nanorod structures. Clearly there are two resonances in the spectrum for the nanorods, one due to the electron oscillations along the diameter (short axis) called the transverse surface plasmon (TSP) resonance and the other along the length (long axis) of the particles called the longitudinal surface plasmon (LSP) resonance. The LSP resonance is tunable by varying the aspect ratio ζ but it has significantly small effect on the TSP resonance. Also, the cross-sections are larger for bigger aspect ratios simply because of increase in the volume of the particle. For randomly oriented ensemble of nanorods, the cross-sections are independent of the light polarization and one can use the average quantities, i.e., $\langle\alpha\rangle = (\alpha_1 + \alpha_2 + \alpha_3)/3$ and $\langle\alpha^2\rangle = (\alpha_1^2 + \alpha_2^2 + \alpha_3^2)/3$.

One may take note of few things. Firstly, for small particles (size << wavelength), the scattering is negligible and the experimentally measured extinction cross-section is dominated by the absorption cross-section. Secondly, not only the size of the particles but also the type of the surrounding dielectric medium can be varied for continuously tuning the surface plasmon resonance. We further note that the bulk dielectric functions ε_m and ε_d are used to calculate the nanoparticle scattering and absorption cross-sections. In the literature, experimental data by

Johnson and Christy [21] on the spectral dielectric function of noble metals are considered as bench marks.

The tunability of the surface plasmon resonance due to the finite size-effects in spherical gold (Au) nanoparticles and cylindrical gold nanorods [22] is summarized in Figs. 6.2 and 6.3, respectively. The red-shift in the surface plasmon absorption maximum as a function of the size of the gold nanospheres is shown in Fig. 6.2a while the dependence of the plasmon bandwidth ($\Delta\lambda$) on the particle radius is shown in Fig. 6.2b. Is can be seen that the bandwidth increases with decreasing nanoparticle radius in the intrinsic size region (<10 nm) and also with increasing radius in the extrinsic size region as predicted by the Mie theory. In Fig. 6.2c, the extinction coefficients of these gold nanoparticles at their respective plasmon absorption maxima are plotted against their volume on log-log scale where the solid line is a linear fit to the data illustrating an agreement with the Mie theory. Calculated absorption spectra of gold nanorods of different aspect ratio ζ using the theory developed by Gans are presented in Fig. 6.3a. The linear-dependence of the LSP wavelength λ_{LSP} on the gold nanorod aspect ratio ζ for a fixed $\varepsilon_d = 4$ and as a

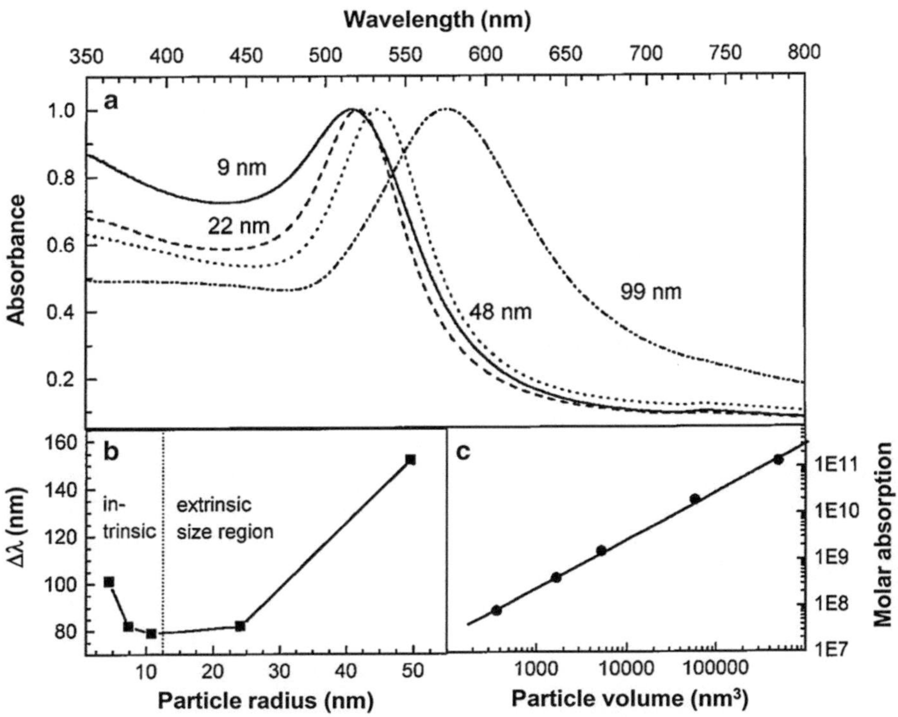

Fig. 6.2 (**a**) UV-Visible absorption spectra of colloidal solutions of spherical gold nanoparticles with diameters varying between 9 and 99 nm, (**b**) plasmon bandwidth as a function of the particle radius, and (**c**) the extinction coefficient at the respective plasmon absorption maxima plotted against their volume on a log-log scale. The *solid line* in (**c**) is a linear fit illustrating the agreement with the Mie theory (Adapted from Ref. [22] with permission)

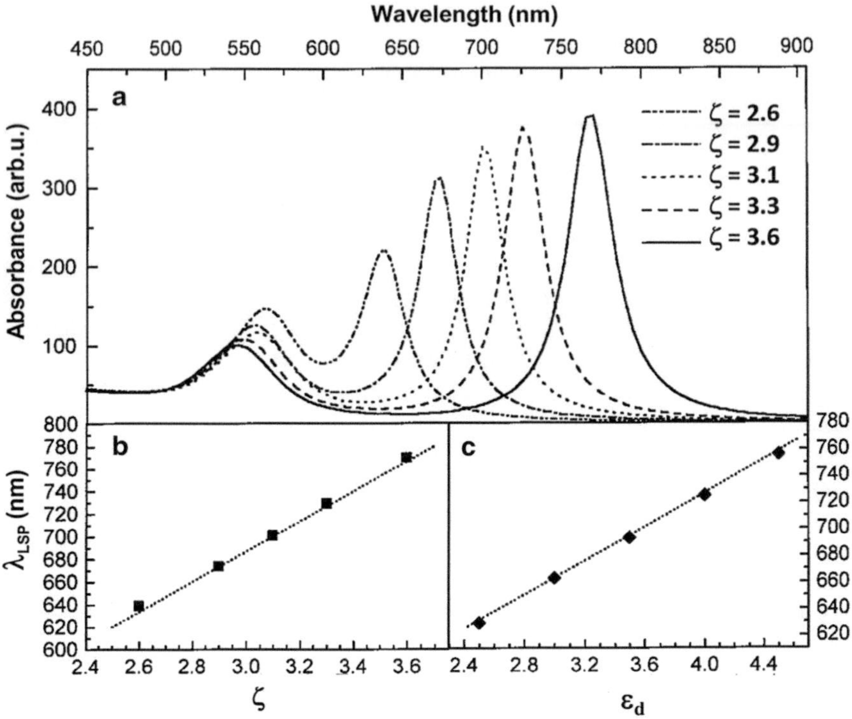

Fig. 6.3 (**a**) Tunability of the surface plasmon resonance in gold nanorods simulated as a function of the nanorod aspect ratio ζ. Linear-dependence of the longitudinal surface plasmon resonance wavelength λ_{LSP} on, (**b**) the nanorod aspect ratio for a fixed $\varepsilon_d = 4$, and (**c**) the dielectric constant ε_d of the surrounding material for a fixed $\zeta = 3.3$ (Adapted from Ref. [22] with permission)

function of the medium dielectric constant ε_d for fixed $\zeta = 3.3$ can be seen from Fig. 6.3b, c, respectively.

In the Drude model of free electron metals, the only scattering that an electron undergoes is with the other electrons inside the metal. However, for small nanoparticles with large surface to volume ratio, the electron surface scattering and size-dependent radiation damping [23] have to be taken into account. In order to satisfy the energy momentum conservation in the process of a photon being absorbed by a free electron, participation of an auxiliary electron or phonon or a defect is implicit. Therefore, the scattering rate Γ_0 in Eq. (6.1) has to be replaced by an effective scattering rate parameter Γ_{eff} which is dependent on the incident photon with angular frequency ω, electronic and lattice temperature (T_e and T_L, respectively) and contains atleast the additional surface scattering term. Ignoring the scattering from defects, the effective scattering rate [24] can be written as:

$$\Gamma_{\text{eff}}(\omega, \mathrm{T_e}, \mathrm{T_L}) = \Gamma_{\text{e-ph}}(\omega, \mathrm{T_e}, \mathrm{T_L}) + \Gamma_{\text{e-e}}(\omega, \mathrm{T_e}) + \Gamma_{\text{e-S}}(\omega, \mathrm{T_e}) \qquad (6.6)$$

The electron–phonon damping term ($\Gamma_{\text{e-ph}}$) is the dominant one which is computed by modelling the electron–phonon interaction via deformational potential connecting the bottom of the conduction band to the periodic deformation induced by lattice vibrations. Since phonons are much lower energy excitations than electrons, electron–phonon interaction modifies only the momentum of the excited electrons. The second term in Eq. (6.6), $\Gamma_{\text{e-e}}$ describes the electron–electron scattering which is possible only by exchange of a reciprocal lattice vector via Umklapp processes and hence its contribution in the photon absorption by the electron is small.

The last term, $\Gamma_{\text{e-S}}$ describes the electron scattering by surface, i.e., optical absorption assisted by electron-surface scattering. It is very significant for confined systems such as nanoparticles of size in the range of $\leq$30 nm, the mean free path of conduction electrons in metals. An empirical relation for the surface scattering term [24] is given as

$$\Gamma_{\text{e-S}}(\omega, \text{T}_\text{e}) = g(\omega, \text{T}_\text{e})\upsilon_\text{F}/\text{D} \tag{6.7}$$

where υ_F is the Fermi velocity of electrons and $\text{D} = (\text{S}_\text{p}/\pi)^{1/2}$ is the equivalent diameter of the particle with surface area S_p. g is another material parameter which depends on the electron level occupation number defined for quasi-continuous conduction band states of not too small particles.

6.3 Ultrafast Optical Response of Photoexcited Metal Nanoparticles

The surface confinement at the nanometric scale affects the internal dynamical processes in metal nanoparticles which modify the dielectric function and hence their optical response. As a consequence, after optical excitation by an ultrafast laser pulse, the electron thermalization, energy loss by interactions with mechanical degrees of freedom in the system and energy transfer by heat dissipation to the surrounding; all are modified as compared to those in the corresponding bulk metal. Time-resolved pump-probe spectroscopy offers a mean to investigate the internal dynamical processes by measuring the time evolution of the ultrafast optical response. More specifically, a strong pump pulse creates a dynamical state whose evolution in time is monitored by another weak probe pulse by collecting the scattered signal from the sample as a function of the time-delay between the pump and the probe pulses. As depicted in Fig. 6.4a, the routine pump-probe measurements are performed on the ensemble of nanoparticles [22, 25–27], however, experiments on single nanoparticles have also become possible recently [28–32]. Following the ultrafast photo-excitation by a femtosecond laser pulse, the laser deposited energy in the system under investigation relaxes via various possible pathways [24], for example, initial relaxation by electron gas thermalization

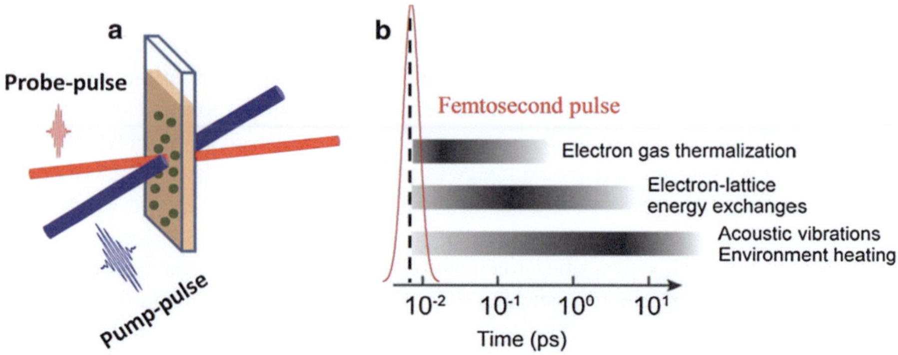

Fig. 6.4 (**a**) Pump-probe spectroscopy of an ensemble of nanoparticles arranged for measurement of the transient differential transmission spectra of the probe. (**b**) Typical time-scales of dynamical processes undergoing in photo-excited metal nanoparticles

followed by electron-lattice interaction and finally emission of acoustic vibrations and heat loss to the surroundings, as depicted in Fig. 6.4b.

An understanding of the underlying internal system kinetics from the experimentally measured time-resolved optical response can be made based on models which take into account the instantaneous time-dependent modifications of the material dielectric function in the desired spectral range. Detailed quantitative modelling have been performed by Fatti and Vallee [24] which include the impact of the out-of-equilibrium conditions (electron excitation and subsequent lattice heating) on the electron energy distribution, the induced modifications of the metal dielectric function, and their influence on the nanoparticle optical response. A brief overview of the same is presented here as the following.

For nanoparticles in size range of ~2–100 nm we can neglect the scattering losses, i.e., the extinction cross-section to be simply the absorption cross-section σ_{abs}. Ultrafast optical excitation of a nanoparticle leads to time-dependent changes in $\Delta\sigma_{sbs}$ which is directly related to experimentally measured change in the transmission (ΔT) of the probe at wavelength λ by the relation:

$$\frac{\Delta T}{T}(\lambda, t) = -\frac{\Delta\sigma_{abs}(\lambda, t)}{S} \tag{6.8}$$

S being the surface area of the probe laser spot on the sample. For weak pump-induced changes in the nanoparticles [33], the $\Delta\sigma_{abs}$ variation at the probe wavelength λ and pump-probe time-delay t, can be expressed as:

$$\begin{aligned}\Delta\sigma_{abs}(\lambda, t) &= \frac{\partial\sigma_{abs}}{\partial\varepsilon_1}(\lambda)\Delta\varepsilon_1(\lambda, t) + \frac{\partial\sigma_{abs}}{\partial\varepsilon_2}(\lambda)\Delta\varepsilon_2(\lambda, t)\\ &= a_1(\lambda)\Delta\varepsilon_1(\lambda, t) + a_2(\lambda)\Delta\varepsilon_2(\lambda, t)\end{aligned} \tag{6.9}$$

$a_{1,2}$ being the stationary spectral-dependent derivatives and $\Delta\varepsilon_{1,2}$ the dynamical spectral- pump-induced variations of the nanoparticle real and imaginary dielectric functions. Therefore, time-resolved signals result from a combination of the dynamical physical effects, which induce variations $\Delta\varepsilon_{1,2}$ and their optical detection, with a spectral sensitivity given by $a_{1,2}$. For all shapes of metal nanoparticles, the localized surface plasmon leads to a strong enhancement of both the linear absorption spectrum and its derivatives $a_{1,2}$ governing the amplitude of the out-of-equilibrium transient response. For spherical gold and silver (Ag) nanoparticles and cylindrical Au nanorods, the calculated spectra [24] are shown in Fig. 6.5. The dispersion like a_1 profile always crosses the horizontal axis near the surface plasmon resonance. This has consequence in determining the nature or polarity of the transient probe signal from pump-probe spectroscopy near the zero time-delay.

The time- and energy-dependent distribution function *f*(E,t) of the electronic states determines the transient variations of the dielectric function $\Delta\varepsilon_{1,2}$ and ultimately the optical response that is measured experimentally. The modification of the electron distribution induces a change in the interband term $\Delta\varepsilon^{ib}_{1,2}$ (dominant on short timescales) and smaller modifications in the Drude term $\Delta\varepsilon^{D}_{1,2}$, and the concomitant heating of the lattice at much longer time-scales. After the pump-pulse excitation of the nanoparticles, the evolution of the electron occupation number, electron–electron scattering and electron energy transfer to the lattice through electron–phonon scattering and heat dissipation to the surrounding medium, all can be computed by solving the Boltzmann equation provided a size-dependent correction to the electron–electron and electron–phonon coupling rates has been taken into account [24]. For pump-photon energy lower than the interband transition energy, electron–hole pairs are generated only in the conduction band. Electrons are thus described by an athermal distribution with a small fraction of electrons having absorbed a photon promoted to a state above the Fermi energy while most of the electrons still occupying initial unperturbed states. Immediately after excitation by a short femtosecond laser pulse, the amplitude of the excitation can be quantified by an elevated electron temperature $T_{exc} = T_0 + \Delta T_{exc}$, T_0 being the initial temperature of thermalized electrons before the optical excitation, corresponding to the equilibrium temperature of an electron gas with the same total energy as the excited system. The electron gas thermalizes through internal electron–electron interactions which is described by a screened Coulomb interaction potential containing a sum over all possible two-electron scattering processes satisfying energy and momentum conservation. In low perturbation regime, typical internal thermalization timescales are of the order of $\sim$500 fs. A deformation potential coupling is assumed to compute the electron–phonon coupling matrix element which is then integrated over all available electronic and phonon states satisfying energy and momentum conservation to obtain the contribution from electron–phonon scattering. The time-scale for electron–phonon scattering processes is typically in the picosecond range and depends on the excitation conditions.

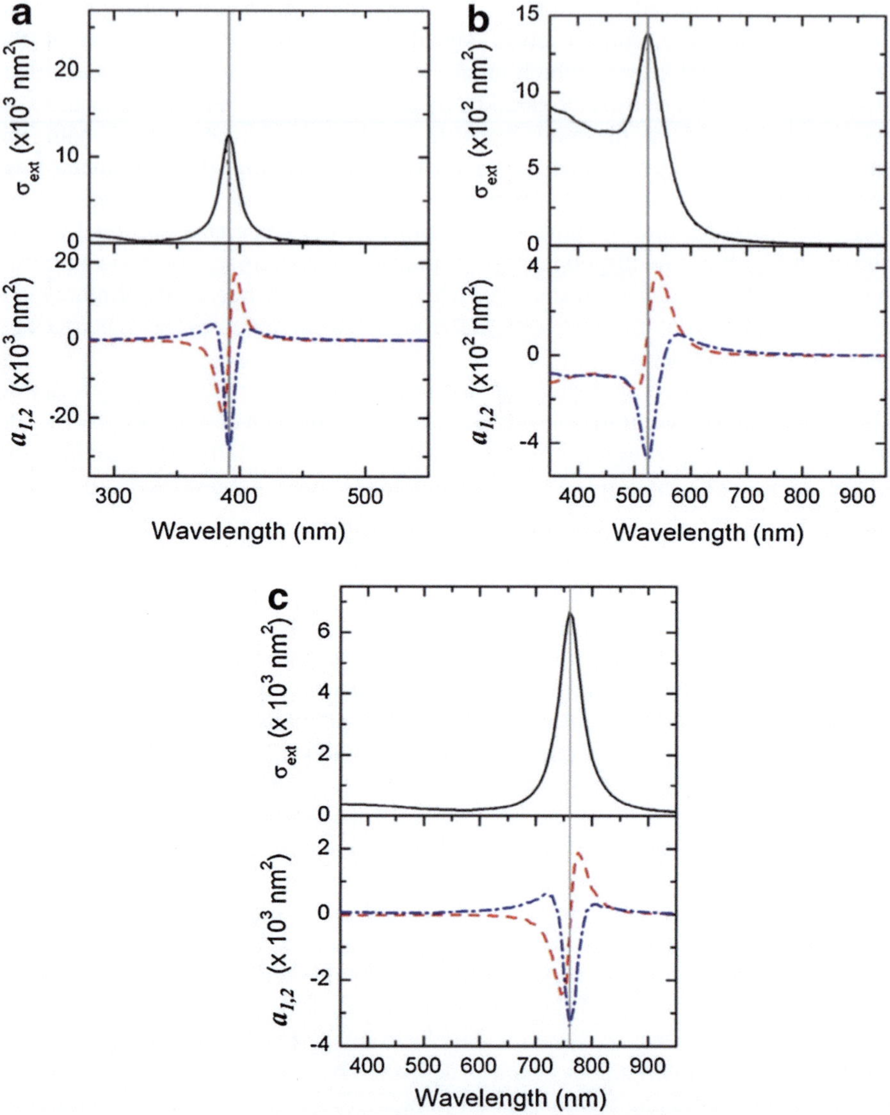

Fig. 6.5 (*Upper panel*) Extinction cross-sections, and (*lower panel*) derivatives a_1 (*red dashed*) and a_2 (*blue dash-dotted*) for a 30 nm diameter silver nanosphere (**a**), 30 nm gold nanosphere (**b**) and 43 × 12 nm gold nanorod (**c**), calculated using Mie theory. The *vertical lines* indicate the surface plasmon resonance in each case (Adapted from Ref. [24] with permission)

Following the internal thermalization by electron–electron scattering, the subsequent dynamics of the electrons and phonons can be described by the two temperature model (TTM) which assumes that both the conduction electrons and the phonons are internally thermalized to have attained a constant coupling term

G [34]. Assuming T_e and T_L as the temperatures of the internally thermalized electrons and lattice, and C_e and C_L as their specific heats per unit volume, the system dynamics follows a coupled rate equation given by

$$C_e \frac{dT_e}{dt} = -G(T_e - T_L); \quad C_L \frac{dT_L}{dt} = G(T_e - T_L) \tag{6.10}$$

The final equilibrium temperature T_{eq} of the nanoparticles common to the electron gas and the lattice, generally much smaller than T_{exc} is determined from the analytical solution as $T_{eq} = T_0 + (T_{exc}{}^2 - T_0{}^2)/(2C_L/a)$ where $a = 65$ J/m^3/K^2 for silver and gold [35]. Since the electron-lattice temperature T_{eq} is higher than the initial temperature T_0, energy is subsequently transferred to the surrounding medium through the nanoparticle interface. The heat transfer out of the nanoparticles occurs typically with 10–500 ps timescales and is limited by the thermal impedance at the interface and by the thermal diffusion within the surrounding medium [36].

First time-resolved experimental results on a single 30 nm silver nanosphere obtained by Muskens et al., [29] using 140 fs pump excitation at 850 nm are reproduced in Fig. 6.6. The probe pulses were close to the surface plasmon resonance at ~425 nm. The transient differential transmission $\Delta T/T$ data for different pump powers as presented in Fig. 6.6a show a fast rise followed by a slower decay corresponding to energy injection into the electron gas by the pump pulse and subsequent electron energy loss by thermalization with the lattice, respectively. Experiments on single nanoparticles allow precise measurement of the ΔT_{exc} which is not the case for optically excited ensemble of nanoparticles. Dotted, dashed and solid lines in Fig. 6.6a are for 180, 280 and 480 μW pump power, corresponding to $\Delta T_{exc} = 190$, 275 and 415 K, respectively. Figure 6.6b shows the simulated signals computed using Eq. (6.9) for which the derivatives a_1 and a_2 were determined from fitting the measured extinction cross-section by Mie theory (shown in the inset). The ultrafast kinetics is mainly governed by the time-dependent $\Delta\varepsilon_1^{ib}$ while $\Delta\varepsilon_2$ contribution is negligible as the probe-photon energy is far from the interband transition energy in silver.

More conventionally, time-resolved experiments have been performed on nanoparticle ensembles, either embedded in a dielectric matrix or dispersed in a liquid. In this case, due to the inhomogeneous distributions of shapes and sizes, and polarization-independent pump absorption, the measurements are less quantitative; however, the intrinsic dynamical characteristics of the nanosystems are captured. Neglecting the absorption by the surrounding matrix, for an ensemble of nanoparticles within an effective optical length l_{op}, the differential transmission signal of the probe can be written as:

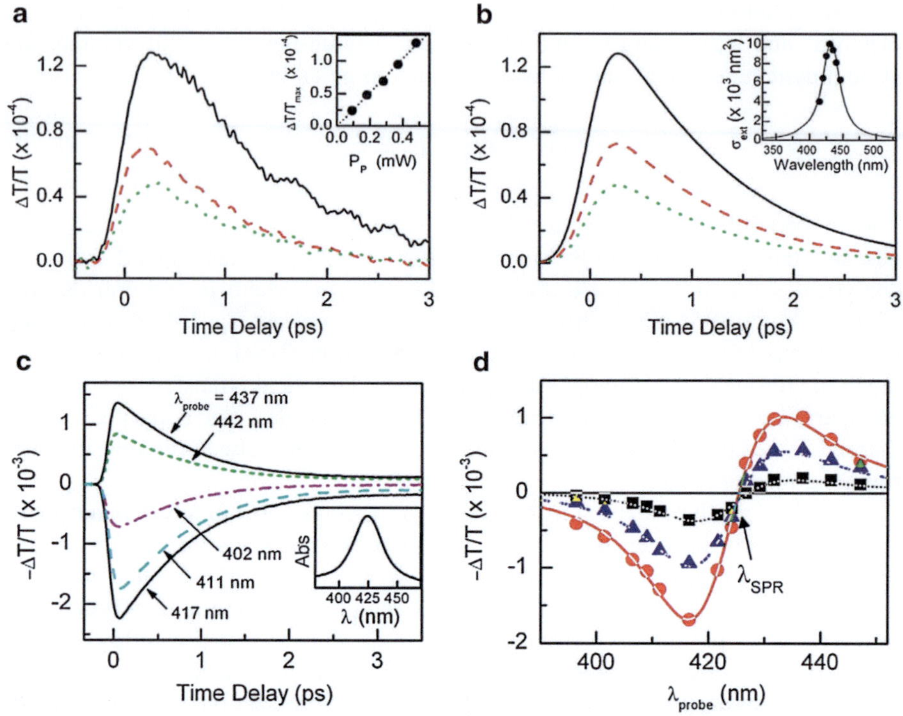

Fig. 6.6 (**a**) Transient differential transmission spectra measured for a single 30 nm Ag nanosphere supported on a glass substrate, following excitation by 140 fs pump pulses centred at 850 nm and probe pulses centred at 425 nm near the surface plasmon resonance of the nanoparticle. *Dotted*, *dashed* and *solid lines* correspond to 180, 280 and 480 μW pump power, corresponding to $\Delta T_{exc} = 190$, 275 and 415 K, respectively. The inset shows the maximum of the ΔT/T signal as a function of pump power. (**b**) Simulated signals computed with Eq. (6.9), a_1 and a_2 derivatives being determined after fitting the measured extinction cross-section by Mie theory as shown in the inset. (**c**) Similar experimental results on ensemble of silver nanospheres of average diameter 26 nm dispersed in a glass matrix using 30 fs pump pulses cantered at 850 nm and variable probe pulses between 400 and 450 nm. The inset shows the sample absorption spectrum. (**d**) Probe-wavelength dependence of the sample transmission change for probe delay of 400 fs (*circles*), 1 ps (*triangles*) and 2 ps (*squares*). Lines are fits assuming pump-induced SPR frequency shift and the broadening (Adapted from Ref. [24] with permission)

$$\frac{\Delta T}{T}(\lambda, t) \sim -\Delta\alpha(\lambda, t).l_{op} \sim -n_{np}\Delta\overline{\sigma}_{abs}(\lambda, t).l_{op} \tag{6.11}$$

where $\Delta\overline{\sigma}_{abs}$ is the mean value of the absorption cross-section of the ensemble with the volume density n_{np} of the nanoparticles. Results from one of the earliest investigations on silver nanospheres [37] are presented in Fig. 6.6c where 30 fs pump pulses at 850 nm and probe pulses varying between 400 and 450 nm were used to study silver nanospheres of mean diameter 26 nm dispersed in a glass matrix. The dispersion-like dependence on the probe-wavelength around the surface plasmon resonance at 425 nm is shown in Fig. 6.6d for probe delay of 400 fs (circles), 1 ps (triangles) and 2 ps (squares). Clearly, the nature of the transient response of the nanoparticles near zero probe-delay depends on the relative position of the probe wavelength with respect to the surface plasmon resonance as a consequence of the pump-induced transient frequency-shift and -broadening of the SPR. Following the pump-excitation, the $\Delta\varepsilon_1^{ib}$ increases very fast leading to a red-shift of the SPR and then decays to its initial value within a few ps by electron-lattice coupling. The SPR broadening at ultrafast time-scales arises due to change in $\Delta\varepsilon_2^{ib}$ modifying the state occupation number during and immediately after the excitation, and at slower time-scales due to changes in the strength of the scattering processes.

For Au nanospheres, the SPR lies in the range of 500–550 nm depending on the size and the dielectric environment which overlaps with the onset of the interband transition wavelength of ~650 nm. Selective investigation of the SPR dynamics requires spectral separation of the SPR and the interband transitions as satisfied for Ag nanospheres and LSP resonance of Au nanorods. Hence the nature of the ultrafast response probed around the SPR of the Ag nanospheres and the LSP resonance of Au nanorods is expected to be similar. Indeed this is the case as seen from Fig. 6.7 where experimental results from [31] have been presented for a single Au nanorod excited by femtosecond pump pulses at 850 nm and probe pulses with varying wavelength λ_{probe} on either side of the LSP resonance at 810 nm far from the interband transition wavelength of gold. The transient response can be interpreted well on the whole timescale as the combination of the dynamical response of the bulk metal amplified by plasmonic effects as shown by simulations of the $\Delta\sigma_{ext}^{ib}$ or the $\Delta\sigma_{abs}^{ib}$ dynamics [24], i.e., variations of $\Delta\sigma_{ext}$ induced by the interband term $\Delta\varepsilon_{1,2}^{ib}$ which reproduce the experiments, both for signal amplitude and time dependence on the first picoseconds after excitation very well. Also transient signals for short delays reflect the a_1 shape profile computed from the experimental linear σ_{ext} (Fig. 6.7c, d). This is because $\Delta\varepsilon_1^{ib}$ is approximately undispersed far from $\lambda = \lambda_{\text{ib}}$ and $\Delta\varepsilon_2^{ib}$ is small. For $\lambda \approx \lambda_{\text{LSP}}$ (black line in Fig. 6.7a, b), a_1 vanishes and $\Delta\sigma_{\text{ext}}$ reflects $\Delta\varepsilon_2^{ib}$ dynamics. Conversely, for longer delays ($t \approx 4$ ps), $\Delta\sigma_{\text{ext}}$ is not correctly reproduced any more by the $\Delta\varepsilon_{1,2}^{ib}$ terms alone, and inclusion of the $\Delta\varepsilon_2^{D}$ variation due to lattice heating becomes necessary (experiments and model prediction are represented by squares and dashed line in Fig. 6.7d).

6.3.1 *Tuning Between Ultrafast PB and PA in Gold Nanorods by Selective Probing Near LSP Resonance*

As discussed above, the nature or polarity of the initial transient response immediately after photo-excitation of metal nanoparticles depends on the probe wavelength relative to the surface plasmon resonance and the condition that the later is spectrally separated from the interband transition wavelength. We present below our experimental observations on gold nanorods where the pump excitation at

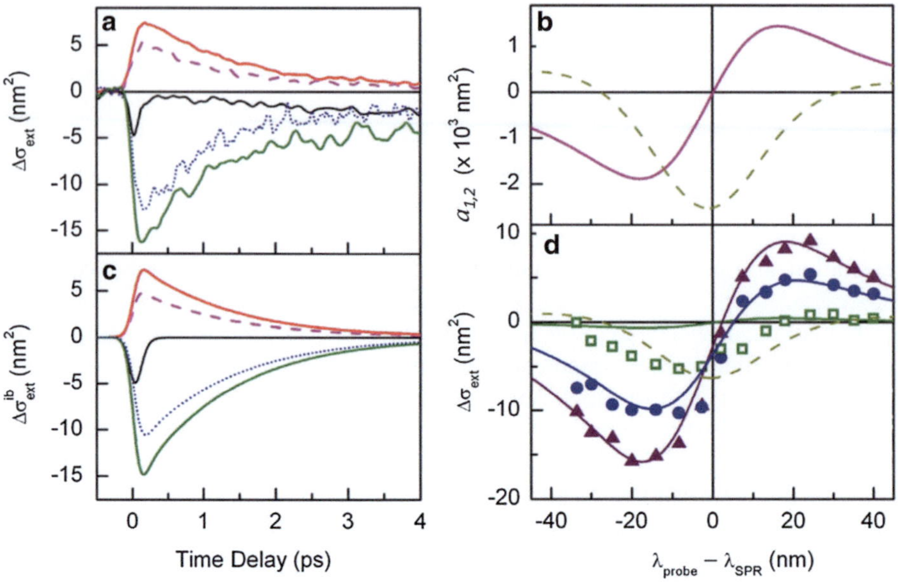

Fig. 6.7 (**a**) Ultrafast extinction or absorption cross-section changes $\Delta\sigma_{ext}$ measured for a single 43×12 nm Au nanorod following pump-excitation at 400 nm and probed using pulses with varying wavelengths around nanorod LSP resonance ($\lambda_{LSP} = 810$ nm): from *top* to *bottom*, $\lambda_{probe} - \lambda_{LSP} = 30$, 40, 0, −30 and −20 nm. (**b**) Computed interband contributions (Eq. (6.9)). (**c**) a_1 and a_2 coefficients (*solid* and *dashed lines*, respectively) computed from the linear extinction cross-section. *Vertical line* corresponds to λ_{LSP}. (**d**) Measured transient extinction spectra for probe-delay of 0 fs (*circles*), 200 fs (*triangles*) and 4 ps (*squares*), and the corresponding lines are computed from interband contributions. *Dashed line* represents Drude contribution for the longest delay Adapted from Ref. [24] with permission)

either wavelength (energy) 790 nm (1.57 eV) or its second harmonic at 395 nm (3.15 eV) creates athermal population of electrons whose relaxation is monitored by probe pulses of fixed wavelength 790 nm. To prove the point, one can either vary the probe wavelength continuously across the LSP resonance of the nanorods or equivalently, use different gold nanorod samples for a fixed probe wavelength. The later has been followed here. In these experiments, transient differential transmission spectra $\Delta T(t)/T$ have been measured where positive change in the probe transmission at the zero probe-delay ($t = 0$) is due to photo-bleaching (PB) of the LSP band and negative change in the probe transmission is due to photo-induced absorption (PA) or excited state absorption induced by the pump of either same wavelength 790 nm (degenerate pump-probe configuration) or different wavelength 395 nm (non-degenerate pump-probe configuration) for both of which the interband transition wavelength of gold ($\lambda_{ib} \sim 650$ nm) is far off.

The linear absorption spectra of four representative gold nanorod colloidal suspensions in water [38, 39], are presented in Fig. 6.8a. The gold nanorods have aspect ratio ζ (length/diameter) ranging from 3 to 5 such that the LSP resonance wavelength (energy) varies from 660 nm to 860 nm ($E_{LSP} = 1.88$–1.42 eV). According to the LSP wavelength, the samples are named as AuNR660 to

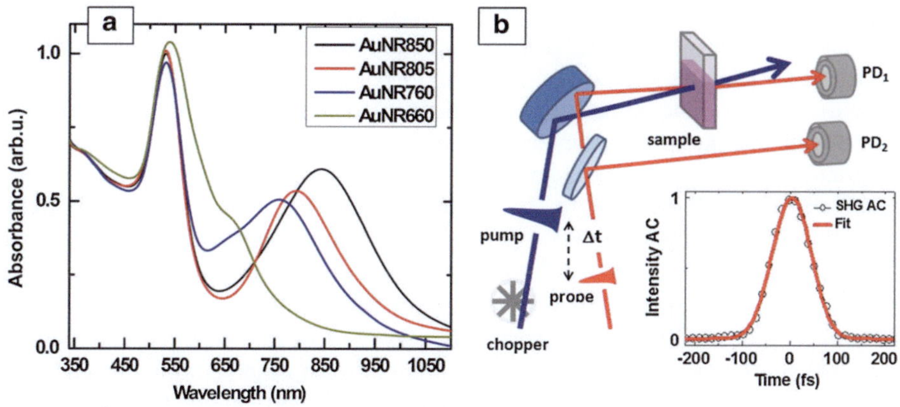

Fig. 6.8 (**a**) Linear optical absorption spectra of colloidal suspensions of the gold nanorods in water shown for four representative samples named according to the longitudinal surface plasmon wavelength. (**b**) Typical experimental arrangement for time-resolved pump-probe spectroscopy. The inset in (**b**) shows the second harmonic generation intensity autocorrelation (SHG AC) trace of the femtosecond laser pulses centred at 790 nm

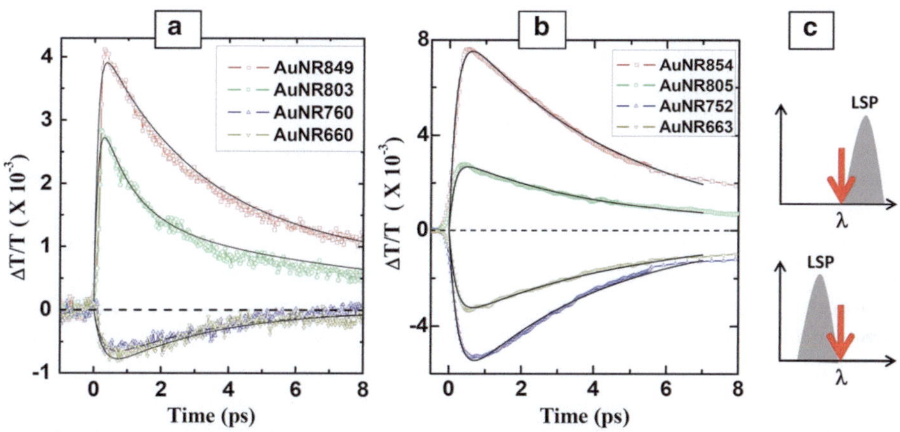

Fig. 6.9 Transient differential transmission data for gold nanorods taken using (**a**) degenerate pump-probe spectroscopy at 790 nm, and (**b**) non-degenerate pump-probe spectroscopy with 395 nm pump and 790 nm probe pulses (Adapted from Ref. [38, 39] with permission). (**c**) A depiction to emphasize that the ultrafast response near zero probe-delay is sensitive only to the relative position of the probe wavelength (marked by *red arrow*) with respect to the LSP wavelength

AuNR860 and the same nomenclature will be used in the discussion here. The typical experimental setup used for either degenerated or non-degenerated pump-probe experiments is shown in Fig. 6.8b. The differential transmission data ($\Delta T/T$) from four nanorod samples using degenerate pump-probe spectroscopy at 790 nm taken at pump-fluence of ~130 μJ/cm^2 and probe-fluence of ~8 μJ/cm^2 are shown in Fig. 6.9a. On similar nanorods, results from non-degenerate pump-probe spectroscopy at 395 nm pump and 790 nm probe taken at pump-fluence of ~600 μJ/cm^2 and probe-fluence of ~1.3 μJ/cm^2 on are presented in Fig. 6.9b. Clearly, PB is observed for samples with $\lambda_{LSP} > \lambda_{probe}$ and PA for samples with $\lambda_{LSP} < \lambda_{probe}$. A depiction in Fig. 6.9c emphasizes the observation that irrespective of the pump-wavelength and pump-flux, the response of the gold nanorods, either the PB (upper diagram) or the PA (lower diagram) is sensitive only to the probe-wavelength relative to the LSP wavelength. The solid curves in Fig. 6.9a, b are results from numerical simulations assuming a pump-induced time-dependent excess athermal electron density $n(t) = \left(1 - e^{-t/\tau_R}\right) n_1 e^{-t/\tau}$, τ_R and τ being the initial rise time-constant due to thermalization and fast relaxation time-constant, respectively. Essentially, $n(t)$ modifies the instantaneous plasma frequency $\omega_p\{n(t)\}$ and thereby the dielectric constant of the particle, resulting into the observed time-dependence of the differential transmission signal at the probe wavelength for the sample optical path length of l_{op} given by [39],

$$\Delta T(t)/T = -\Delta\alpha\{n(t)\}.l_{op} \tag{6.12}$$

6.3.2 Light Controlled Reversible Switching Between Ultrafast PB and PA in Gold Nanorods

Interestingly, reversible switching between ultrafast PB and PA effects in gold nanorods can also be achieved by using the probe photon-flux as the control parameter [39]. This occurs only for those gold nanorods which have the surface plasmon wavelength above the probe wavelength and the pump wavelength such that it excites the interband transitions. Experimental results obtained on gold nanorods using nondegenerate pump-probe spectroscopy at 395 nm pump-wavelength and 790 nm probe-wavelength, are presented in Fig. 6.10. For a fixed pump-fluence of ~580 μJ/cm^2, the results for AuNR854 are shown in Fig. 6.10a where it can be seen that the amount of the PB signal at low probe-fluences continuously decreases with the increasing probe-fluence and changes to PA at higher probe-fluences. The threshold value of the probe-fluence for which switching from PB to PA occurs is ~80 μJ/cm^2, independent of the sample and the value of the pump-fluence as can be noted from the results summarized in Fig. 6.10b, c. For the reverse, i.e., decreasing the probe-fluence from high to low, the signal continuously changes from PA to PB without any measurable hysteresis loss. This unique observation is independent of the pump-fluence and occurs even at high pump-fluences but much below the nanorod melting threshold. On the other hand, for the AuNR663 and AuNR752 samples for which the LSP wavelength is below the probe wavelength, we always observe PA as seen in Fig. 6.10b irrespective of the pump or the probe-fluences.

In Fig. 6.10c we have plotted the maximum of the ΔT/T signals near zero-delay as a function of the probe-fluence for the AuNR854 sample. Each curve is for a particular value of the pump-fluence as mentioned. It can be noted that all the curves

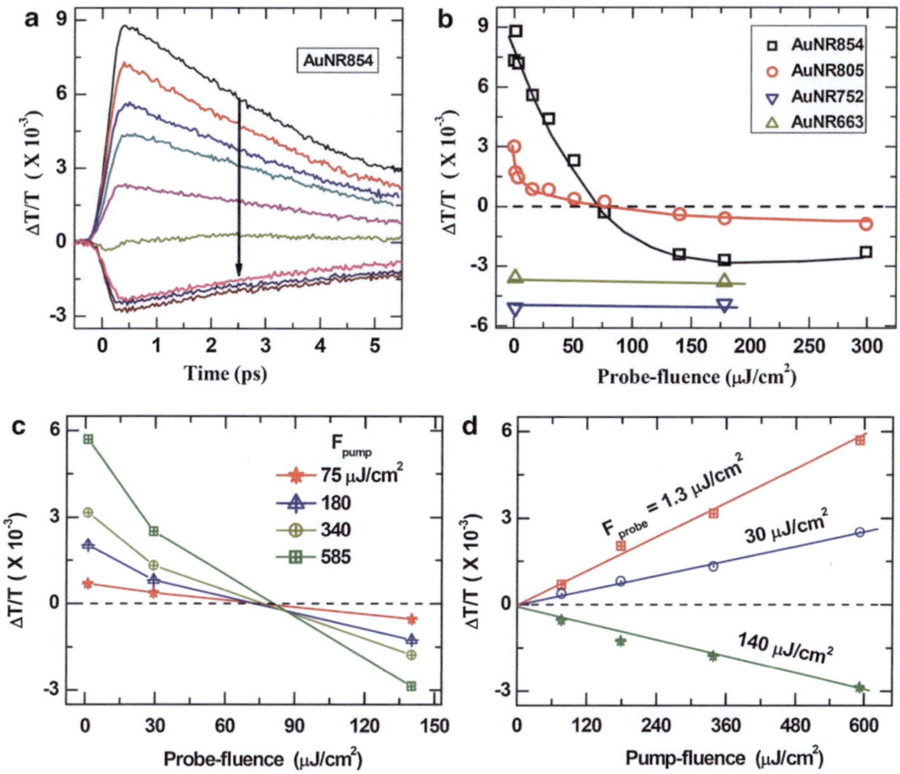

Fig. 6.10 (**a**) Transient differential transmission spectra for AuNR854 taken at a fixed pump-fluence of ~580 μJ/cm^2 and increasing the probe-fluence in the direction of the *arrow*. (**b**) Maximum amplitude of the signals near zero time-delay ($t = 0$) for all the four nanorod samples as a function of the probe-fluence. (**c**) Probe-fluence dependence of the $\Delta T(t = 0)/T$ measured at various fixed values of the pump-fluence as mentioned, and (**d**) probe-fluence dependence of the $\Delta T(t = 0)/T$ measured at various values of the probe-fluence as mentioned, for the AuNR854 sample. *Solid lines* in (**b**), (**c**) and (**d**) are guide to the eyes (Adapted from Ref. [39] with permission)

pass through a common point on the horizontal axis (probe-fluence) irrespective of the value of the pump-fluence and the change in polarity from PB to PA occurs at the probe-fluence of ~80 μJ/cm^2. If laser heating induced melting of the gold nanorods has to be invoked then it is unusual to have all the curves pass through a single point. Rather the pump and the probe-fluences should have acted in a constructive way to heat up the system additively (function of the total laser-fluence incident on the sample at time delay $t = 0$). This is, of course, not the case here.

Furthermore, the linear-dependence of the $\Delta T(t = 0)/T$ on the pump-fluence for a fixed probe-fluence as seen in Fig. 6.10d strengthens the point further that the reversible switching between PB and PA as a function of the probe-fluence is intrinsic to the nanorods and is not due to nanorod shape change due to melting. Our results clearly show that the pump-fluence dependent results at any given fixed value of the probe-fluence are usual, i.e., the magnitude of the transient absorption (either PB at low probe-fluences or PA at high probe-fluences) as shown in Fig. 6.10d, increases linearly with the pump-fluence without change in the polarity.

To explain the observed effect due to the probe-fluence in our experiments, we have to invoke a cascaded two-photon absorption process. The pump at 315 nm (photon energy larger than the TSP and LSP energies) excites electrons from the d-band to tail states of the continuum band above the interband transition energy in gold (~2.4 eV) or the TSP band (~2.7 eV) in the nanorods. The electrons from these excited states quickly thermalize and populate the LSP band within a time $\tau_R \sim 300$ fs, as reflected in the rise-time of our differential transmission signals at probe wavelength 790 nm, irrespective of the gold nanorod aspect ratio, and the pump and probe fluences. At low values of the probe fluence, the usual photo-bleaching is observed when the pump excited carriers relax to the LSP band and block the normal probe absorption to that band. However, at high probe-fluences, beyond the threshold value of ~80 μJ/cm^2, the excited state absorption from the LSP band to higher energy states dominates the signal leading to the observed photoinduced absorption.

We consider a special two-photon absorption coefficient $\beta_{(2)}$ which represents the cascaded two-photon absorption, one photon from the pump and one from the probe pushing the electrons from the d-band to the excited states. The action of the pump-fluence is to create an initial excess density n_1 of photoexcited carriers which thermalize by electron–electron scattering within time τ_R and subsequently relax with a time constant τ by electron–phonon scattering. Assuming that the time-dependence of the contribution arising from the $\beta_{(2)}$ process is the same as before in Eq. (6.12), i.e., $\left(1 - e^{-t/\tau_R}\right) n_1 e^{-t/\tau}$, the total transmission change is suggested to be of the form:

$$\left(\frac{\Delta T(t)}{T}\right)_{total} = \frac{\Delta T(t)}{T} - \beta_{(2)}\left(I_{pump} \times I_{probe}\right)^{1/2}\left(1 - e^{-t/\tau_R}\right) n_1 e^{-t/\tau} \tag{6.13}$$

where the first term on the right side is same as discussed before in Eq. (6.12) for the usual case. Using Eq. (6.13), simulations were carried out to fit the experimental data. We start with a negligibly small $\beta_{(2)}$ at the lowest probe-fluence of 1.3 μJ/cm^2 (intensity ~ 16 MW/cm^2). Then by adjusting only the value of $\beta_{(2)}$, we simulate the results at various probe-fluences to obtain fits as shown by dark continuous lines in Fig. 6.11a, b. The resultant $\beta_{(2)}$ values as a function of the probe-intensity are shown in Fig. 6.11c, d for AuNR854 and AuNR805 samples, respectively. We can clearly see from Fig. 6.11c, d that the resultant $\beta_{(2)}$ is largely independent of the probe-intensity and a simple cascaded two-photon absorption process is critical to explain the observed reversible transition between PB and PA for the AuNR854 and AuNR805 nanorods.

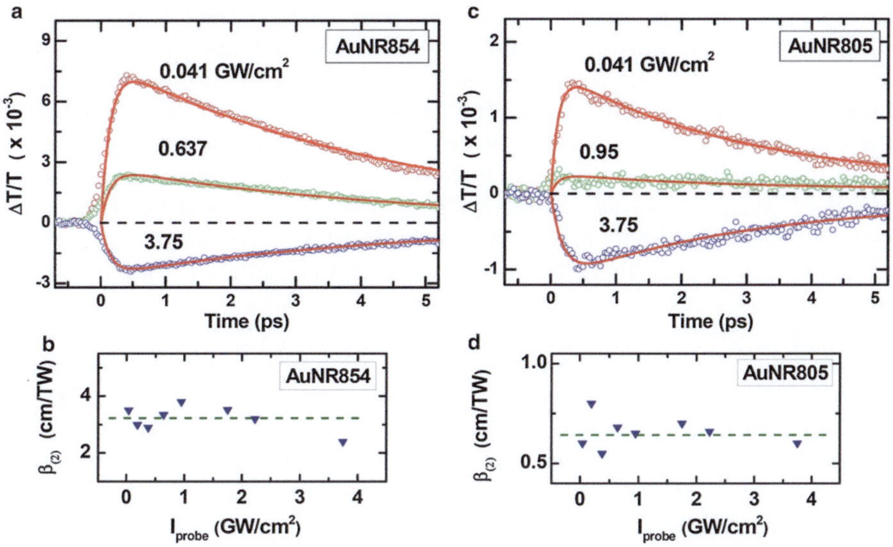

Fig. 6.11 Transient differential transmission spectra along with the results from simulations (dark curves) shown for (**a**) AuNR854, and (**c**) AuNR805 nanorods at a few probe-intensities (GW/cm^2) as mentioned after incorporating the cascaded two-photon absorption coefficient $\beta_{(2)}$ at the corresponding probe-intensity from (**b**) and (**d**), respectively. Dashed lines in (**b**) and (**d**) are guide to the eyes (Adapted from Ref. [39] with permission)

6.4 Ultrafast Optical Nonlinearities of Metal Nanoparticles

6.4.1 Surface Plasmon Resonance Tuned Optical Nonlinearities

We have already seen that the linear response, i.e., one-photon absorption and scattering cross-sections of metal nanoparticles with sizes typically in the range of ~2–100 nm, are enhanced near the surface plasmon resonance. It is obvious that the surface plasmon resonance in metal nanoparticles would also play a vital role in determining their nonlinear optical properties. It has been seen in many studies that near the surface plasmon resonance, metal nanoparticles exhibit immensely enhanced optical nonlinearities as compared to their bulk counterparts [40–43]. Generally, the optical nonlinearities can be due to electronic transitions or thermal effects. Electronic nonlinearities have ultrafast response, i.e., in the femtosecond to picosecond range while the thermal processes induced nonlinearities are slow, i.e., in the nanosecond or slower time scales. In either case, the nonlinear optical properties of materials can be measured using a single laser beam technique popularly known as the z-scan technique introduced by Sheikh Bahae et al. [44]. In the open aperture (OA) configuration, all the transmitted light from a sample is collected onto a photodiode which is a function of the complex optical susceptibility of the material. On the other hand, using close aperture (CA) configuration where only part of the laser beam transmitted through the sample is collected using an aperture before the photodiode, the nature and magnitude of the detected signal depends on the type and magnitude of the real part of the complex optical susceptibility of the sample. Therefore, z-scan technique is useful for measuring the effective cross-sections of various intensity-dependent processes such as second harmonic generation, third harmonic generation, two-photon and multi-photon absorption, and Kerr-effect, to name a few.

The optical nonlinearities in metal nanoparticles are ultrafast in nature, occurring in the range of femtosecond to picoseconds [45]. Very often, the two-photon absorption coefficient β and an intensity-dependent coefficient of refraction γ arising from third-order optical susceptibility $\chi^{(3)}$ are reported in the literature. Nonlinear optical properties of metal nanostructures are closely related to the surface plasmon resonance, i.e., the size and shape of the nanostructures. For silver nanostructures of various shapes, Luo et al., [46] observed that with the increasing incident intensity of 87 fs pulses centred at 800 nm, the reverse saturable absorption or optical limiting behaviour related to $\chi^{(3)}$ increases for nano-size flower-shaped structures but decreases for microrods. However, for silver nanowires, saturable absorption response is observed initially at low incident intensities which changes to reverse saturable absorption with higher nonlinear absorption coefficient at high incident intensities. These results have been reproduced in Fig. 6.12 where the scanning electron microscopy (SEM) images and the corresponding nonlinear transmission properties of the three samples have been presented.

The tunability of the SPR in metal nanostructures makes them unique for various potential applications in biology and medicine. It has been shown in many studies

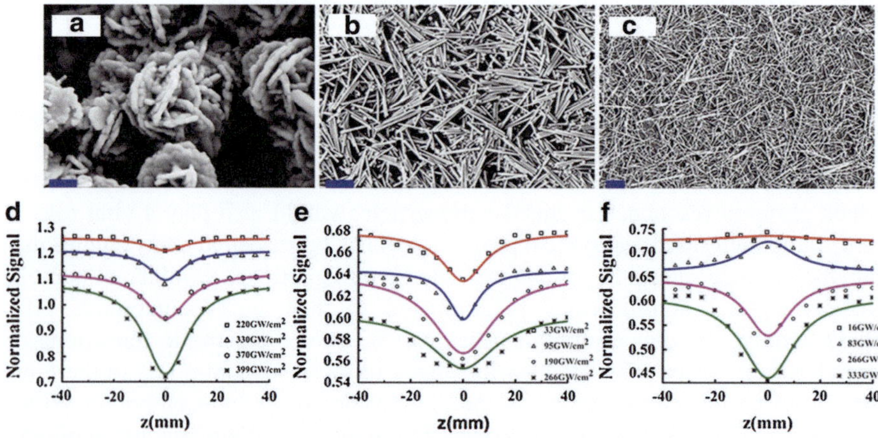

Fig. 6.12 Scanning electron microscopy images of silver nanoparticles with different shapes, (**a**) nanoflowers, (**b**) microrods, and (**c**) nanowires. The scale bars are 500 nm, 1 micron and 2 microns, respectively. The corresponding nonlinear optical transmission data obtained using open aperture z-scan experiments with 87 fs laser pulses centred at 800 nm, are presented in (**d**), (**e**) and (**f**), respectively. Adapted from reference [46] with permission

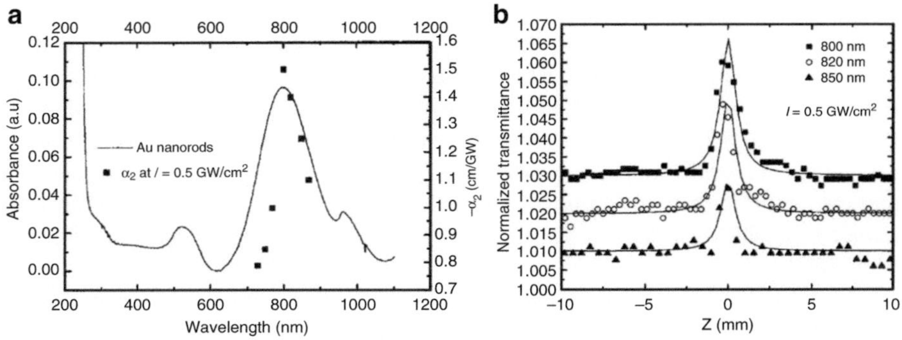

Fig. 6.13 (**a**) Spectral-dependence of the linear and nonlinear absorption coefficients of the gold nanorods, (**b**) open aperture z-scan results on the gold nanorods measured at different excitation wavelengths using 220 fs laser pulses for a fixed incident irradiance of 0.5 GW/cm^2 (Adapted from Ref. [47] with permission)

that the optical nonlinearities or the coefficients β and γ measured at the surface plasmon resonance are much enhanced, sometimes by many folds of magnitude as compared to those measured at off-resonance. For example, in gold nanorods, femtosecond z-scan experiments revealed that the saturable absorption behaviour can be continuously changed to optical limiting by increasing the amount of the incident irradiance [47]. These results have been reproduced in Fig. 6.13. In this study gold nanorods with their TSP resonance at ~520 nm and LSP resonance at ~800 nm were used. The laser excitation wavelength was in resonance with the LSP wavelength. It can be seen from Fig. 6.13 that the two-photon absorption coefficient ($-\alpha_2$ cm/GW) related to saturable absorption for smaller values of the incident intensity, is maximum for excitation wavelength in resonance with the LSP wavelength.

6.4.2 Metal Nanoclusters with Improved Ultrafast Nonlinearity

Smaller metal nanoparticles, namely, the nanoclusters containing a few atoms are a new class of materials with characteristics in between those of atoms and nanoparticles [16–18, 48–50]. The nanoclusters with size of the order of the de Broglie wavelength of conduction electrons (~0.5 nm) exhibit discrete energy levels while larger nanoparticles or nanocrystals (>2 nm) exhibit quasi-continuous electronic bands with the additional surface plasmon resonance as discussed before. The optical absorption spectra of such subnanometric clusters lack the surface plasmon resonance but rather show a distinct absorption onset at the electronic gap between the highest occupied molecular orbital (HOMO) and the lowest unoccupied molecular orbital (LUMO). Noble metal nanoclusters have been shown to exhibit several novel properties [18, 50–52]. Ultrafast photoexcited electron relaxation dynamics in 28-atom gold (Au_{28}) nanoclusters were found to show bi-exponential decay with a fast sub-picosecond and another slow nanosecond time-constant, independent of both the laser photon-energy and the pump-fluence [53]. Gold nanoclusters do not show saturable absorption at the surface plasmon wavelength of larger gold nanocrystals [54]. On the other hand, these ultra-small particles exhibit very good optical power limiting performance with significantly reduced threshold for the optical limiting and higher nonlinear optical absorption coefficient.

Open-aperture z-scan experiments by Philip et al., [54] as reproduced in Fig. 6.14 showed a transition from pure optical limiting behaviour for gold nanoclusters with 25, 38, and 144 atoms to saturable absorption behaviour for nanocrystals of size ~4 nm, all excited with same 5 ns laser pulses centred at 532 nm near the surface plasmon resonance of the nanocrystals. The valley-shaped z-scan curves for the 25- and 38-atom clusters indicate pure optical limiting behaviour due to the nonlinear absorption throughout the z-scan range whereas that of the nanocrystals show a central valley flanked by two symmetric peaks on

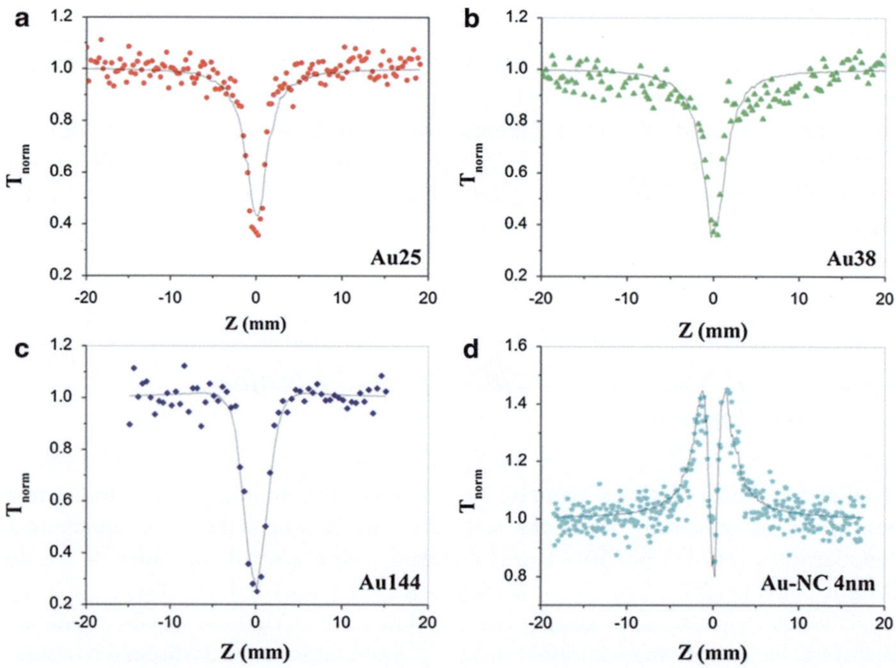

Fig. 6.14 Open-aperture z-scans measured for gold clusters, (**a**) Au25, (**b**) Au38, (**c**) Au144, and (**d**) gold nanocrystals of size ~4 nm, Au-NC 4 nm. Samples are excited using 5 ns laser pulses at 532 nm. T_{norm} is the measured transmission normalized by the linear transmission of the samples. *Solid curves* correspond to numerical fits to the data (Taken from Ref. [54] with permission)

either side. The onset of the side peaks signifying the onset of saturable absorption is visible in the form of two humps flanking the valley for the 144-atom clusters.

In the following we discuss a case study of 15-atom gold clusters (Au_{15}) whose ultrafast nonlinear optical response was found to be immensely enhanced for the clusters deposited on indium-tin-oxide (ITO) metal film as compared with that on SiO_2 glass plate [55]. Systematic experiments were carried out using femtosecond z-scan at 395 nm and time-resolved non-degenerate pump-probe spectroscopy (395 nm pump and 790 nm probe) on the gold-clusters deposited on the glass plate (Au_{15}-glass) and on the ITO (Au_{15}-ITO) as well as separately on bare ITO and glass plates. The pulse-width of the laser pulses was ~80 fs. The optical limiting performance of the Au_{15}-ITO system was found to be higher by almost an order than that of the Au_{15}-glass system. Concurrently, excited state absorption from transient transmission measurements was also observed to be enhanced by approximately the same order.

Stable Au_{15} nanoclusters were synthesized involving processes like core-etching of larger clusters and simultaneous trapping of the clusters inside cyclodextrin (CD) cavities [18] shown schematically in the inset of Fig. 6.15a. The decorating glutathione (GSH) and CD molecules are optically transparent in the 200–1000 nm window. The absorption spectrum of the gold clusters is shown in Fig. 6.15a where the experimentally obtained wavelength-dependent intensity I(λ) has been converted into energy-dependent intensity I(E) by dividing by the popular Jacobian factor $\partial E/\partial \lambda$. We note that the absorption profile of the clusters does not

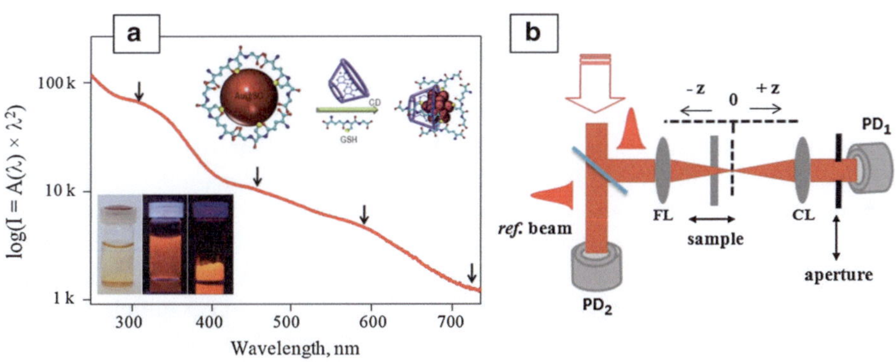

Fig. 6.15 (**a**) UV–vis absorption spectrum of Au_{15} clusters inside cyclodextrin (CD) cavities plotted as the natural logarithm of the Jacobian factor. Well defined absorption features related to molecular character of the clusters are marked with *arrows*. Lower inset shows fluorescence from the sample in suspension and solid forms. Upper inset shows the schematic illustration of CD-assisted one-pot synthesis of the Au_{15} clusters via the core etching reaction [18]. GSH indicates glutathione used as the core-etching agent and as the ligand to protect the Au_{15} core (Adapted from Ref. [55]). (**b**) Experimental setup for open and close aperture z-scan measurements. Convex lenses for focusing the light onto the sample (FL) and collecting the transmitted light from the sample (CL) are used while the sample mounted on a linear stage is moved along the z-axis on both the sides of the focal point $z = 0$. The transmitted light, either all of it or part of it allowed by an aperture, is detected at a photodiode PD_1. Another photodiode PD_2 is used as a reference to nullify the noises in the detected signal which are present in the laser beam

show surface plasmon resonance (SPR) of bigger nanoparticles. Instead, molecule-like characteristic features marked by arrows in Fig. 6.15a at about 318, 458 and 580 nm are clearly visible. The distinct absorption onset occurs near 710 nm which is indicative of an energy gap of 1.75 eV between the highest occupied molecular orbital (HOMO) and the lowest unoccupied molecular orbital (LUMO) gap.

A layer of the Au_{15} nanoclusters embedded in dense hydrogel matrix of thickness ~100 micron was sandwiched between a glass plate (~500 micron thick) and a coverslip (~100 micron thick) making the Au_{15}-SiO_2 sample and another between an ITO thin film-coated glass plate and a cover slip to make the Au_{15}-ITO sample. The commercial ITO plates having film thickness of ~100 nm and a sheet resistance of ~10 ohm per square area were used. When excited with 395 nm laser pulses, a yellowish glow is seen at the back of the samples, which was rejected by using a colour glass filter from reaching the photo-detector in all the experiments described here.

The typical z-scan experimental setup is shown in Fig. 6.15b. A femtosecond laser beam is divided into two parts, one to excite the sample and the other for a reference to be used later for correcting the laser noise in the experimental data. A lens focuses the light onto the sample which is movable across the focal point along the optical axis so that for each z-position, the sample experiences a different incident laser intensity thereby varying the laser intensity from ~2 MW/cm^2 to 70 GW/cm^2. In Fig. 6.16a, b, normalized transmittance for the Au_{15}-ITO and the Au_{15}-glass samples is presented. The signal from the OA z-scan shows optical limiting, i.e., reduction in transmission as the input beam intensity is increased, whereas the signal from CA z-scan shows a positive refractive nonlinearity (self-focusing effects), i.e., decrease in the transmitted intensity due to refraction as the sample approaches the focal point ($z = 0$) followed by an increase in intensity as it moves away from the focal point and towards the detector. We note from these results that the magnitude of change in transmission near the focal point is about 10 times larger for the Au_{15}-ITO sample as compared to the Au_{15}-glass sample. The continuous lines in Fig. 6.16a, b are theoretical fits to the data [44, 56] providing the two-photon absorption coefficient β of ~0.3 (3.0) cm/GW and the nonlinear refraction coefficient γ of $\sim 2 \times 10^{-5}$ (6×10^{-5}) cm^2/GW for the Au_{15}-glass (Au_{15}-ITO) system. Clearly, the value of β (γ) for the Au_{15}-ITO sample is about 10 (3) times higher than that for the Au_{15}-glass sample. The z-scan experiments performed on bare ITO and glass plates under the same experimental conditions resulted in comparatively negligibly small values of the nonlinear coefficients.

The results for the transient differential transmission ($\Delta T/T$) from pump-probe spectroscopy using pump-fluence of ~16 μJ/cm^2 and probe-fluence of ~1 μJ/cm^2 are presented in Fig. 6.16c, d, on a linear-linear and log-linear scale to distinguish the regions of various dynamical time-constants. The following observations can be made from these results. Firstly, the polarity of the $\Delta T/T$ signals is negative which is in contrast to the positive signal (photo-bleaching) usually seen for larger colloidal gold nanoparticles at the surface plasmon resonance wavelength. In the present case, the negative signal corresponds to excited-state absorption and not the ground state bleaching [53]. The magnitude of the signal at zero probe-delay ($t = 0$)

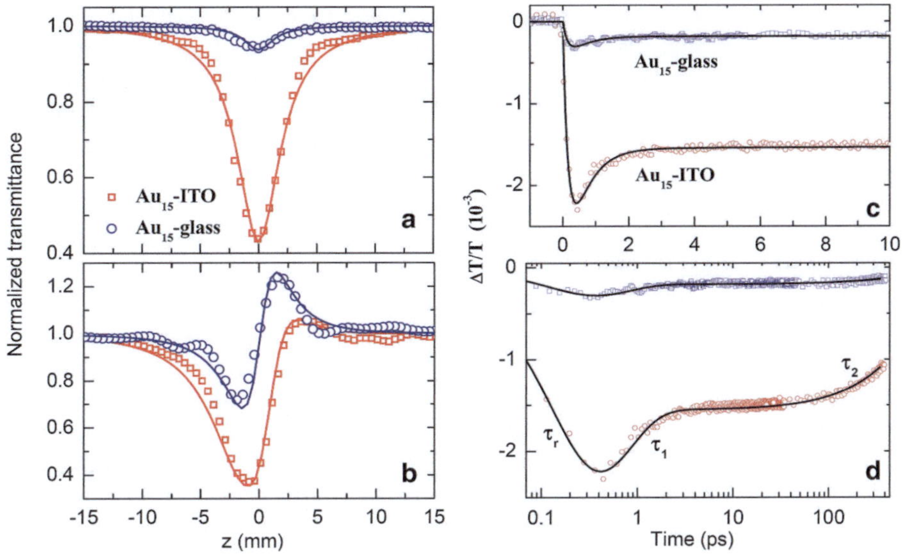

Fig. 6.16 Normalized transmittance of the Au_{15} clusters in (**a**) OA z-scan and (**b**) CA z-scan, shown for clusters deposited on ITO plate (sample Au_{15}-ITO) and on glass plate (sample Au_{15}-glass) having same molar concentration and layer thickness taken at excitation wavelength of 395 nm and maximum incident power of ~70 GW/cm^2 (at the focus). The continuous lines are numerical fits to determine nonlinear absorption and refraction coefficients. Time-resolved differential transmission data for the two samples shown on linear-linear scale (**c**) and log-linear scale (**d**), taken using 395 nm pump and 790 nm probe pulses to show the difference in the signal magnitude for the two samples and the related carrier relaxation dynamics involving initial thermalization time-constant τ_r, fast and slow decay time-constants τ_1 and τ_2, respectively, estimated from fitting the data with bi-exponentially decaying function (Adapted from Ref. [55])

is a direct consequence of the magnitude of the excited state absorption at the probe wavelength of 790 nm. We can clearly see from Fig. 6.16c that the excited state absorption in the gold clusters is enhanced by a factor of ~8 for the clusters deposited on ITO as compared to those on the glass plate. This enhancement is similar to that observed from open aperture z-scan for the optical limiting performance of the Au_{15}-ITO system. The magnitude of $\Delta T(t=0)/T$ can be used to quantify the cascaded two-photon absorption process (one photon of 395 nm and the other of 790 nm) by estimating the corresponding nonlinear absorption coefficient $\beta_{(2)}$ via the relation $|\Delta T(t=0)/T| = \beta_{(2)}\sqrt{I_{pump} \times I_{probe}}$. The estimated value is $\beta_{(2)} \sim 0.05$ cm/GW for Au_{15}-ITO and about 8 times smaller value for Au_{15}-glass.

The life-time of the photogenerated carriers in the excited state has been determined by fitting a bi-exponentially decaying function to the photo-induced absorption as shown in linear-linear and log-linear plots in Fig. 6.16c, d, respectively. It was found that the data for both the samples can be fitted nicely with an initial build-up time $\tau_r \sim 200$ fs, fast relaxation time-constant $\tau_1 \sim 700$ fs and a slow relaxation time-constant $\tau_2 \sim 1$ ns. The fast decay is attributed to the initial carrier relaxation from optically coupled states to the LUMO level followed by the slow decay via radiative and/or nonradiative processes from the LUMO level to the ground state. We find (data not shown) that the magnitude of the pump-induced absorption linearly increases with the excitation power (pump-fluence) indicating one-photon absorption processes at 3.15 eV. The carrier life-times remain pump-fluence independent. These results are similar to those reported for Au28 clusters [53].

The important outcome from this study was the observation of enhanced optical nonlinearity for the Au_{15}-ITO coupled system. A possible reason due to photoinduced energy transfer between the overlapped excited states of Au_{15} and ITO has been outlined here. In the schematic energy level construction shown in Fig. 6.17, the molecular energy level diagram of the gold clusters [57] has been shown in the left side of the figure considering the absorption spectrum given in Fig. 6.15. The HOMO-LUMO gap of ~1.75 eV is very close to our probe photon energy of 1.57 eV. Three molecular like levels have been drawn at energies (wavelengths) of 2.1 eV (580 nm), 2.7 eV (458 nm) and 3.9 eV (318 nm). Similarly, on the right side in Fig. 6.17, we have drawn schematically, the energy bands in ITO as inferred from the indirect and direct band gap energies of ~2.4 eV (515 nm) and 3.6 eV (345 nm), respectively [58]. Since the enhancement in the excited state absorption (Fig. 6.16c) is similar to that in the optical limiting (Fig. 6.16a), hence it is natural to

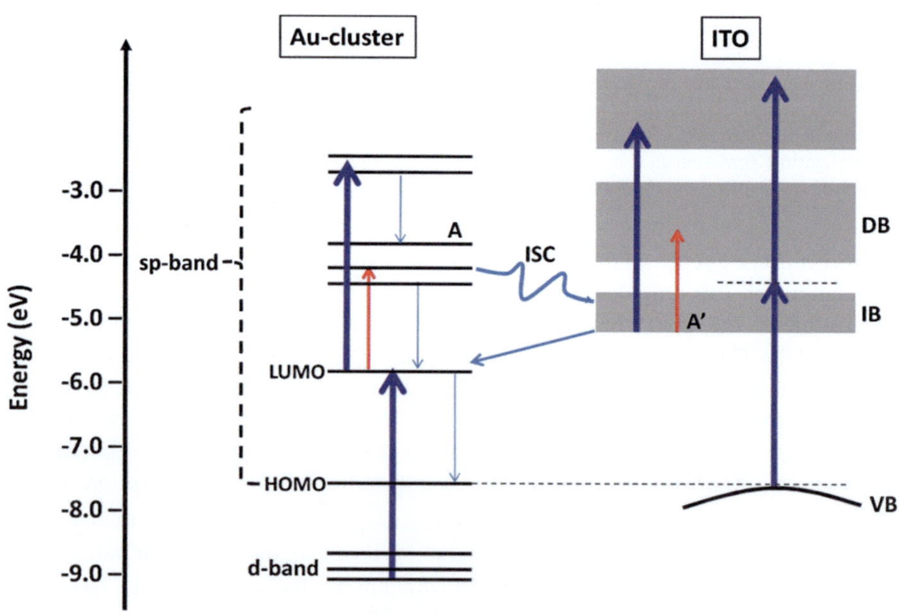

Fig. 6.17 Schematic of the electronic energy-diagram of Au_{15} clusters coupled with the electronic energy-band diagram of ITO. The *blue* and *red arrows* represent the optical transitions by the pump and probe photon energies of ~3.15 and 1.57 eV, respectively. Possible excited-state intersystem coupling between Au_{15} clusters and ITO is represented by the curved arrow and out of many possible interstate relaxation processes few are marked by thin *downward arrows* (Taken from Ref. [55])

assume that the optical limiting in Au_{15} clusters at 395 nm is also a two-step cascaded single photon absorption process. Whereas, a direct two-photon absorption in ITO at 3.15 eV (shown by two thick blue arrows in Fig. 6.17) from the top of the valance band (VB) to the continuum states bypassing the indirect band (IB) but via an intermediate virtual state is more probable. In the Au_{15}-ITO coupled system, due to close proximity of the energy levels of the molecular Au_{15} clusters and energy bands of ITO, excited state electronic-coupling can occur. In that case the electrons created in the excited state of Au_{15} (marked by A) can cross-over to the IB state of ITO which can later readily absorb single photons of either 3.15 or 1.57 eV photons as shown by vertical blue and red arrows in Fig. 6.17 leading to an enhanced combined two-photon absorption or photo-induced excited state absorption in the coupled system.

6.5 Direct Observation of Surface Vibrations in Nanoparticles

Plasmonic nanostructures' large absorption cross-sections, high sensitivity to geometry and refractive index changes, and the ability to localize electromagnetic fields into a subwavelength volume make them an ideal candidate for controlling nanoscale mechanical motion and sensing applications. By manipulating the geometry of these nanostructures, multiple vibrational states can be tailored and dynamically selected by acousto-plasmonic coherent control [59]. The shape-dependent confined vibrations of metal nanoparticles can be studied using various optical techniques, such as Raman spectroscopy, transient absorption spectroscopy and terahertz time-domain spectroscopy. In ultrafast time-resolved pump-probe spectroscopy, an incident femtosecond laser pulse initially excites the plasmons which quickly dephase in time and the hot electrons diffuse throughout the nanostructure at the Fermi velocity transferring their energy to the lattice. This results into rapid thermal expansion of the nanostructure generating coherent acoustic phonons by impulsive thermal excitation, [60, 61] which are imprinted as damped oscillations in the experimentally measured transient optical response signal.

The vibrational modes of a freely vibrating elastic sphere [62] are labelled by two integers, n, the harmonic order, i.e., the number of radial nodes, and l, the angular momentum number, which represents the angular dependence of the mode. As shown in Fig. 6.18, interferometric pump-probe measurements on single gold nanospheres using probe wavelength close to the surface plasmon resonance [28] revealed coherent excitation of a nonspherically symmetric mode of vibration $(n, l) = (0, 2)$ along with the usual radial breathing mode of vibration $(n, l) = (0, 0)$ that is observed very often for ensemble of particles where particle's environment is isotropic. The former can be generated by an isotropic heat pulse only if the spherical symmetry of the particle's expansion is broken either by the substrate or by the particle's shape.

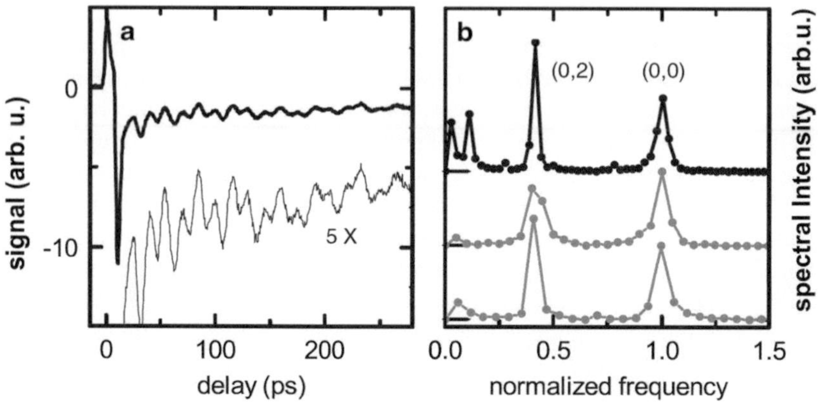

Fig. 6.18 (**a**) Delay scan of a single gold nanosphere with 45 nm diameter using interferometric pump-probe spectroscopy. (**b**) Power spectra of this particle's oscillations (*top spectrum*) and of two other particles of slightly different size. Frequencies and amplitudes are normalized to those of the (n, l) = (0, 0) mode (the absolute frequencies are 67, 59, and 63 GHz, from *top* to *bottom*). The low-frequency peak at ~28 GHz corresponds to the (n, l) = (0, 2) mode of the vibrations (Taken from Ref. [28] with permission)

For a rod-shaped nanoparticle, both the fundamental breathing mode of vibration along the short axis and the extensional mode of vibration along the long axis of the particle can be impulsively excited by laser-induced heating [63]. For gold nanorods in ensembles, these vibrational modes have been observed in many ultrafast pump-probe studies [63, 64], however, for a single nanorod, polarization of the pump-pulse can be used for selective excitation. Coherent acoustic vibrations of an optically trapped single gold nanorod as shown in Fig. 6.19 were reported by Ruijgrok et al. [30], which are very useful for determining the accurate elastic constants from the observed frequencies of the oscillations. The breathing mode involves a pure expansion and contraction (with change in volume) along the transverse radial directions of the rod, and depends on both the bulk and shear elastic moduli. Whereas the extensional mode is along the length of the rod involving dimensional changes in both transverse and longitudinal directions (without change in volume), and depends on the Young's modulus along the long axis of the rod.

Both the extensional and breathing modes of gold nanorods having large aspect-ratio can be excited by femtosecond laser impulse heating in ultrafast time-resolved experiments. Experimentally, it is observed that for thinner rods (diameter < 20 nm), the breathing mode cannot be excited because the period of oscillations is comparable to the time-scale of laser heating [63]. Moreover, the detection of the extensional mode can be greatly improved by tuning the probe wavelength near the longitudinal surface plasmon resonance.

Our time-resolved pump-probe experiments performed upto probe-delays of $\sim$400 ps on the colloidal suspensions of gold nanorods as discussed before, namely

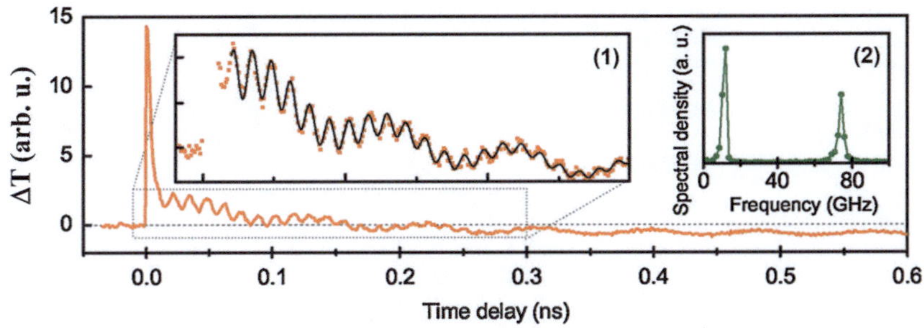

Fig. 6.19 Acoustic vibrations of a single gold nanorod. (**a**) Coherent oscillations in the transient transmission signal of a single 25× 60 nm gold nanorod, optically trapped in water. The pump and the probe wavelengths were 785 nm and 610 nm, respectively (probe wavelength in the *blue wing* of the plasmon resonance). Inset (*1*): Zoomed-in of a part of the trace, fitted with a sum of two damped oscillations and slow cooling to extract the frequencies and damping times of the oscillations, inset (*2*): power spectral density of the oscillatory part of the vibrational trace (Taken from Ref. [30] with permission)

AuNR663, AuNR752, AuNR805 and AuNR854, revealed long time-period oscillations as shown in Fig. 6.20a. These nanorods have diameter D of ~10 nm but growing length L so that the aspect ratios (ζ = L/D) are 2.8, 4.2, 4.7, and 5.2, respectively [39]. As can be seen from Fig. 6.20b, the time-period of oscillations for the four gold nanorod samples is linearly dependent on the LSP wavelength. Since the LSP wavelength for these samples also varies linearly with the aspect ratio (Fig. 6.20c), it was confirmed that the observed coherent oscillations were due to extensional mode of vibrations (time-period τ_{ext}) along the length of the nanorods. For the fundamental extensional mode, the relation between τ_{ext} and the Young's modulus is given [63] as $\tau_{ext} = 2L\sqrt{\rho/Y}$ where ρ is the material density. Taking Y = 65 GPa and ρ = 19.3 g/cm^3 for gold, [63] the average value of the nanorod length for the longest nanorods in our experiments, i.e., AuNR854 sample is estimated to be ~55 nm which is very close to that expected from D ~ 10 nm and ζ ~ 5.2.

The nanoparticle vibrational modes can also be characterized using time-domain terahertz (THz) spectroscopy as shown by Kumar et al., for silver nanoparticles [65]. Nearly spherical silver nanoparticles of mean diameter ~3.4 nm were prepared in a poly(vinyl alcohol) matrix; Fig. 6.21a, b show the transmission electron microscopy image and the absorption spectrum, respectively. Experimentally measured real and imaginary parts of the complex dielectric function ε are shown in Fig. 6.21c where the resonances due to the nanoparticle vibrations have been marked by arrows. A broad feature in the THz spectrum at frequency of ~1.1 THz is related to the crystalline lamellae in the thin polymer film.

The observed resonances at frequencies of ~0.6 and 2.12 THz are due to the spheroidal and toroidal vibrational modes of the nanospheres. These results clearly demonstrated for the first time that THz time-domain spectroscopy technique can be a complementary tool to the well established Raman scattering spectroscopy for

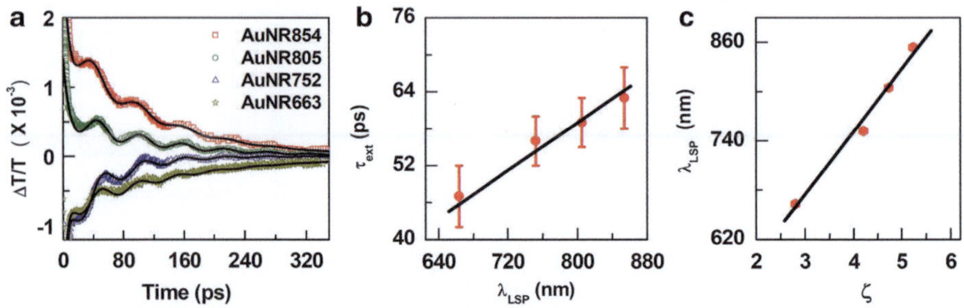

Fig. 6.20 Extensional mode of vibrations of gold nanorods observed using time-resolved pump-probe spectroscopy. (**a**) Coherent oscillations in the transient differential transmission spectra of colloidal suspensions of gold nanorods due to the extensional mode of vibrations, (**b**) correspondence between the time-period of the oscillations with LSP wavelength for the four samples, and (**c**) the linear dependence of the LSP wavelength on the nanorod aspect ratio (Adapted from Ref. [39] with permission)

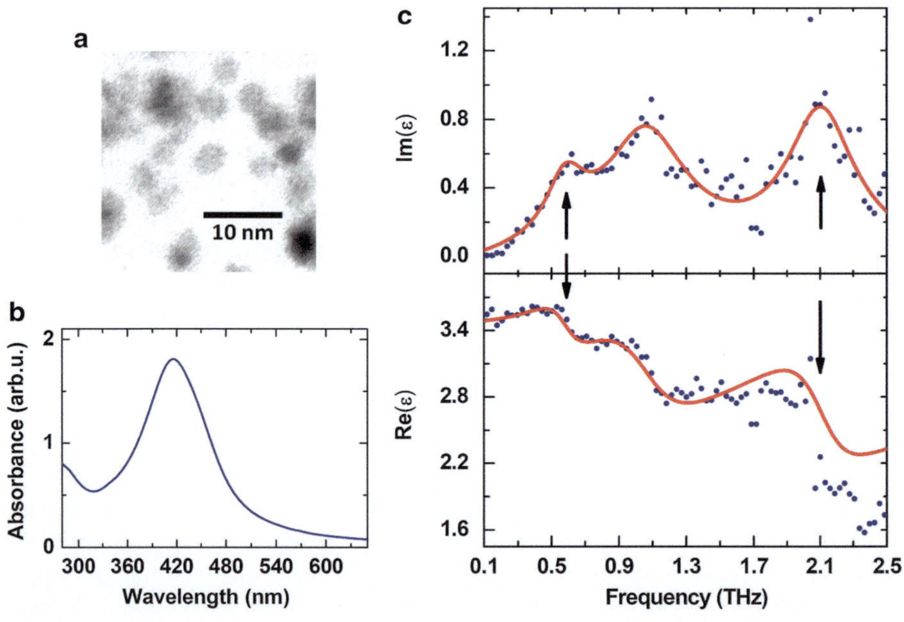

Fig. 6.21 Elastic vibrations of a silver nanospheres observed using terahertz time-domain spectroscopy. (**a**) Transmission electron micrograph of the silver nanoparticles dispersed inside thin film of poly(vinyl) alcohol, and (**b**) UV-Visible absorption spectrum of the nanoparticles. (**c**) Real and imaginary parts of the experimentally measured complex dielectric function ε in the THz frequency range showing vibrational resonances of the nanoparticles at frequencies marked by *arrows*. The resonance at ~1.1 THz is due to the crystalline lamellae in the polymer matrix (Adapted from Ref. [65] with permission)

characterizing nanoparticles, particularly in the low-frequency regime where Raman spectroscopy is difficult to achieve the desired results due to strong Rayleigh scattering. We note that the phonons observed in the THz time-domain spectroscopy are the equilibrium phonons of the system unlike those created coherently by impulsive stimulated Raman scattering in the time-resolved pump-probe spectroscopy.

6.6 Summary and Outlook

To summarize, ultrafast photoresponse of plasmonic nanostrucutres was described in this chapter by taking some of the interesting examples in the literature. The surface plasmon effect in metal nanostrucutres was discussed starting from the fundamental principles. We then showed that the ultrafast time-resolved spectroscopy of plasmonic nanostructures reveals the fundamental processes of electron scattering which help in releasing the excess energy of photoexcited electrons and the system returns to its equilibrium. For gold nanorods, we showed that the ultrafast photo-bleaching and photoinduced absorption can be tuned and reversibly switched between the two by varying the probe-wavelength across the LSP resonance and using the probe-fluence as the control parameter, respectively. A possible application of this novel tunability between the PB and PA could be an ultrafast optical switch where switching is controlled by the fluence of the second laser pulse. In other words, it shows potential in applications where ultrafast light controls the light itself. We also discussed the effect of surface plasmon resonance in enhancing the nonlinear optical properties of metal nanostructures. For a comparison, we presented a study on nonplasmonic gold nanoclusters where the enhanced optical limiting or photoinduced absorption property was discussed. It was found that about an order enhancement in the third order optical susceptibility is observed for the Au_{15} clusters in contact with the ITO as compared to the clusters deposited on a SiO_2 glass plate. These results indicate a role of the excited state inter-system crossing between the electronic states of the clusters and the ITO to enhance the nonlinear response in the coupled system. Such coupled systems which are, of course, different from the extensively studied polymer-nanoparticle composites, need to be exploited much more in the future to have better understanding of the underlying physics of very small metal clusters in proximity of the tunable planar surface plasmon mode of a metal film or having an excited state overlap. Finally, we discussed time-reolved pump-probe spectroscopy and terahertz time-domain spectroscopy as efficient experimental tools for characterizing the confined acoustic phonons in the metal nanoparticles.

In recent years, there has been a growing interest to understand coherent vis-a-vis incoherent energy transfer between the surface plasmon polaritons in a metal and excitons in molecule assemblies [66, 67]. The interplay between coherent and incoherent pathways leads to a significant modification of the radiative damping, opening up unexplored avenues in coherent active plasmonics to be studied by time-resolved spectroscopies. We expect that the studies related to the ultrafast energy transfer between molecular assemblies and surface plasmons in hybrid nanomaterials like two dimensional graphene or transition metal dichalcogenides (e.g., MoS_2 layers) with metal nanostructures will attract increasing interest in coming years.

Acknowledgements We acknowledge financial assistance from Nanomission Project of Department of Science and Technology, Government of India.

References

1. Krasavin AV, Zheludev NI (2004) Active plasmonics: controlling signals in Au/Ga waveguide using nanoscale structural transformations. Appl Phys Lett 84(8):1416–1418
2. Renger J, Quidant R, van Hulst N, Palomba S, Novotny L (2009) Free-space excitation of propagating surface plasmon polaritons by nonlinear four-wave mixing. Phys Rev Lett 103:266802
3. Rotenberg N, Betz M, van Driel HM (2010) Ultrafast all-optical coupling of light to surface plasmon polaritons on plain metal surfaces. Phys Rev Lett 105:017402
4. Zia R, Schuller JA, Chandran A, Brongersma ML (2006) Plasmonics: the next chip-scale technology. Mater Today 9:20–27
5. MacDonald KF, Zheludev NI (2010) Active plasmonics: current status. Laser Photonics Rev 4(4):562–567
6. Ru ECL, Blackie E, Meyer M, Etchegoin PG (2007) Surface enhanced Raman scattering enhancement factors: a comprehensive study. J Phys Chem C 111:13794–13803
7. Lal S, Link S, Halas NJ (2007) Nano-optics from sensing to waveguiding. Nat Photonics 1:641–648
8. Khlebtsov NG, Dykman LA (2010) Optical properties and biomedical applications of plasmonic nanoparticles. J Quant Spectrosc Radiat Transf 111:1–35
9. Clavero C (2014) Plasmon-induced hot-electron generation at nanoparticle/metal-oxide interfaces for photovoltaic and photocatalytic devices. Nat Photonics 8:95–103
10. Maier SA (2007) Plasmonics: fundamentals and applications. Springer, New York
11. Garcia MA (2011) Surface plasmons in metallic nanoparticles: fundamentals and applications. J Phys D Appl Phys 44:283001–283020
12. Bohren CF, Huffman DR (1998) Absorption and scattering of light by small particles. Wiley, New York
13. Mishchenko MI, Travis LD, Lacis AA (2004) Scattering, absorption, and emission of light by small particles. Cambridge University Press, New York
14. Burda C, Chen X, Narayanan R, El-Sayed MA (2005) Chemistry and properties of nanocrystals of different shapes. Chem Rev 105:1025–1102
15. Noguez C (2007) Surface plasmons on metal nanoparticles: the influence of shape and physical environment. J Phys Chem C 111:3806–3819
16. Schmid G (2008) The relevance of shape and size of Au_{55} clusters. Chem Soc Rev 37:1909–1930
17. Qian H, Zhu M, Wu Z, Jin R (2012) Quantum sized gold nanoclusters with atomic precision. Acc Chem Res 45(9):1470–1479
18. Shibu ES, Pradeep T (2011) Quantum clusters in cavities: trapped Au_{15} in cyclodextrins. Chem Mater 23(4):989–999
19. Mie G (1908) Articles on the optical characteristics of turbid tubes, especially colloidal metal solutions. Ann D Phys 25:377

20. Gans R (1912) Uber Die Form Ultra Mikroskopischer Goldteilchen. Ann Phys 342:881
21. Johnson PB, Christy RW (1972) Optical constants of the noble metals. Phys Rev B 6:4370–4379
22. Link S, El-Sayed MA (1999) Spectral properties and relaxation dynamics of surface plasmon electronic oscillations in gold and silver nanodots and nanorods. J Phys Chem B 103:8410–8426
23. Novo C, Gomez D, Perez JJ, Zhang ZY, Petrova H, Reismann M, Mulvaney P, Hartland GV (2006) Contributions from radiation damping and surface scattering to the linewidth of the longitudinal plasmon band of gold nanorods: a single particle study. Phys Chem Chem Phys 8:3540–3546
24. Stoll T, Maioli P, Crut A, Fatti ND, Vallee F (2014) Advances in femto-nano-optics: ultrafast nonlinearity of metal nanoparticles. Eur Phy J B 87:260–278
25. Voisin C, Fatti ND, Christofilos D, Vallee F (2001) Ultrafast electron dynamics and optical nonlinearities in metal nanoparticles. J Phys Chem B 105:2264–2280
26. Hartland GV (2011) Optical studies of dynamics in noble metal nanostructures. Chem Rev 111(6):3858–3887
27. Kiel M, Mohwald H, Bargheer M (2011) Broadband measurements of the transient optical complex dielectric function of a nanoparticle/polymer composite upon ultrafast excitation. Phys Rev B 84:165121–165126
28. van Dijk MA, Lippitz M, Orrit M (2005) Detection of acoustic oscillations of single gold nanospheres by time-resolved interferometry. Phys Rev Lett 95:267406–267409
29. Muskens OL, Fatti ND, Vallee F (2006) Femtosecond response of a single metal nanoparticle. Nano Lett 6(3):552–556
30. Ruijgrok PV, Zijlstra P, Tchebotareva AL, Orrit M (2012) Damping of acoustic vibrations of single gold nanoparticles optically trapped in water. Nano Lett 12(2):1063–1069
31. Baida H, Mongin D, Christofilos D, Bachelier G, Crut A, Maioli P, Fatti ND, Vallee F (2011) Ultrafast nonlinear optical response of a single gold nanorod near its surface plasmon resonance. Phys Rev Lett 107:057402
32. Masia F, Langbein W, Borri P (2013) Polarization-resolved ultrafast dynamics of the complex polarizability in single gold nanoparticles. Phys Chem Chem Phys 15(12):4226–4232
33. Voisin C, Christofilos D, Loukakos P, Fatti ND, Vallee F, Lerme J, Gaudry M, Cottancin E, Pellarin M, Broyer M (2004) Ultrafast electron–electron scattering and energy exchanges in noble-metal nanoparticles. Phys Rev B 69(19):195416–195428
34. Allen PB (1987) Theory of thermal relaxation of electrons in metals. Phys Rev Lett 59:1460–1463
35. Ashcroft NW, Mermin ND (1976) Solid state physics. Holt, Rinehart and Winston, New York
36. Cahill DG, Ford WK, Goodson KE, Mahan GD, Majumdar A, Maris HJ, Merlin R, Phillpot SR (2003) Nanoscale thermal transport. J Appl Phys 93(2):793–818
37. Fatti ND, Vallee F, Flytzanis C, Hamanaka Y, Nakamura A (2000) Electron dynamics and surface plasmon resonance nonlinearities in metal nanoparticles. Chem Phys 251:215–226
38. Anija M, Kumar S, Kamaraju N, Tiwari N, Kulkarni SK, Sood AK (2011) Ultrafast dynamics of gold nanorods: tuning between photo-bleaching and photo-induced absorption. Int J Nanosci 10:687–691
39. Kumar S, Anija M, Sood AK (2013) Tuning ultrafast photoresponse of gold nanorods. Plasmonics 8:1477–1483
40. Chemla DS, Herritage JP, Liao PF, Isaacs ED (1983) Enhanced four-wave mixing from silver particles. Phys Rev B 27(8):4553–4558
41. Ricard D, Roussignol P, Flytzanis C (1985) Surface-mediated enhancement of optical phase conjugation in metal colloids. Opt Lett 10(10):511–513
42. Bloemer MJ, Haus JW, Ashley PR (1990) Degenerate four-wave mixing in colloidal gold as a function of particle size. J Opt Soc Am B 7(5):790–795

43. Philip R, Kumar GR, Sandhyarani N, Pradeep T (2000) Picosecond optical nonlinearity in monolayer-protected gold, silver and gold-silver alloy nanoclusters. Phys Rev B 62(19):13160–13166
44. Bahae MS, Said AA, Wei T-H, Hagan DJ, Stryland EWV (1990) Sensitive measurement of optical nonlinearities using a single beam. IEEE J Quant Electron 26(4):760–769
45. Tokizaki T, Nakamura A, Kaneko S, Uchida K, Omi S, Tanji H, Asahara Y (1994) Subpicosecond time response of third-order optical nonlinearity of small copper particles in glass. Appl Phys Lett 65(8):941–943
46. Luo S, Chen Y, Fan G, Sun F, Qu S (2014) Saturable absorption and reverse saturable absorption on silver particles with different shapes. Appl Phys A 117:891–894
47. Elim HK, Yang J, Lee J-Y, Mi J, Ji W (2006) Observation of saturable and reverse saturable absorption at longitudinal surface plasmon resonance in gold nanorods. Appl Phys Lett 88:083107
48. Schaaff TG, Knight G, Shafigullin MN, Borkman RF, Whetten RL (1998) Isolation and selected properties of a 10.4 kDa gold: glutathione cluster compound. J Phys Chem B 102(52):10643–10646
49. Schaaff TG, Whetten RL (2000) Giant gold–glutathione cluster compounds: intense optical activity in metal-based transitions. J Phys Chem B 104(12):2630–2641
50. Link S, Beeby A, FitzGerald S, El-Sayed MA, Schaaff TG, Whetten RL (2002) Visible to infrared luminescence from a 28-atom gold cluster. J Phys Chem B 106(13):3410–3415
51. Jin R (2010) Quantum sized, thiolate-protected gold nanoclusters. Nanoscale 2(3):343–362
52. Yadav BD, Kumar V (2010) Gd@Au15: a magic magnetic gold cluster for cancer therapy and bioimaging. Appl Phys Lett 97(13):133701–133703
53. Link S, El-Sayed MA, Schaaff TG, Wetten RL (2002) Transition from nanoparticle to molecular behavior: a femtosecond transient absorption study of a size-selected 28 atom gold cluster. Chem Phys Lett 356(3–4):240–246
54. Philip R, Chantharasupawong P, Qian H, Jin R, Thomas J (2012) Evolution of nonlinear optical properties: from gold atomic clusters to plasmonic nanocrystals. Nano Lett 12:4661–4667
55. Kumar S, Shibu ES, Pradeep T, Sood AK (2013) Ultrafast photoinduced enhancement of nonlinear optical response in 15-atom gold clusters on indium tin oxide conducting film. Opt Express 21:8483–8492
56. Kamaraju N, Kumar S, Sood AK, Guha S, Krishnamurthy S, Rao CNR (2007) Large nonlinear absorption and refraction coefficients of carbon nanotubes estimated from femtosecond z-scan measurements. Appl Phys Lett 91(25):251103–251105
57. Zhu M, Aikens CM, Hollander FJ, Schatz GC, Jin R (2008) Correlating the crystal structure of a thiol-protected Au_{25} cluster and optical properties. J Am Chem Soc 130(18):5883–5885
58. Matino F, Persano L, Arima V, Pisignano D, Blyth RIR, Cingolani R, Rinaldi R (2005) Electronic structure of indium-tin-oxide films fabricated by reactive electron-beam deposition. Phys Rev B 72(8):085437–085445
59. O'Brien K, Kimura NDL, Rho J, Suchowski H, Yin X, Zhang X (2014) Ultrafast acousto-plasmonic control and sensing in complex nanostructures. Nat Commun 5:4042–4047
60. Perner M, Gresillon S, März J, von Plessen G, Feldmann J, Porstendorfer J, Berg K-J, Berg G (2000) Observation of hot-electron pressure in the vibration dynamics of metal nanoparticles. Phys Rev Lett 85:792–795
61. Zijlstra P, Tchebotareva AL, Chon JWM, Gu M, Orrit M (2008) Acoustic oscillations and elastic moduli of single gold nanorods. Nano Lett 8:3493–3497
62. Nishiguchi N, Sakuma T (1981) Vibrational spectrum and specific heat of fine particles. Solid State Commun 38:1073–1077
63. Hu M, Wang X, Hartland GV, Mulvaney P, Juste JP, Sader JE (2003) Vibrational response of nanorods to ultrafast laser induced heating: theoretical and experimental analysis. J Am Chem Soc 125:14925–14933

64. Hartland GV, Hu M, Wilson O, Mulvaney P, Sader JE (2002) Coherent excitation of vibrational modes in gold nanorods. J Phys Chem B 106(4):743–747
65. Kumar S, Kamaraju N, Karthikeyan B, Tondusson M, Freysz E, Sood AK (2010) Direct observation of low frequency confined acoustic phonons in silver nanoparticles: terahertz time domain spectroscopy. J Chem Phys 133:014502–014505
66. Vasa P, Wang W, Pomraenke R, Lammers M, Maiuri M, Manzoni C, Cerullo G, Lienau C (2013) Real-time observation of ultrafast Rabi oscillations between excitons and plasmons in metal nanostructures with J-aggregates. Nat Photonics 7:128–132
67. Wang W, Vasa P, Pomraenke R, Vogelgesang R, Sio AD, Sommer E, Maiuri M, Manzoni C et al (2014) Interplay between strong coupling and radiative damping of excitons and surface plasmon polaritons in hybrid nanostructures. ACS Nano 8:1056–1064

Chapter 7
Graphene-Based Ultra-Broadband Slow-Light System and Plamonic Whispering-Gallery-Mode Nanoresonators

Weibin Qiu

Abstract In this chapter, the graphene monolayer is modeled as surface current density without thickness in order to explore the properties of the plasmons propagating along it. The proposed model is verified analytically in terms of the propagation constant of the plasmons. By using this zero-thickness model, the nanofocusing properties of the mid infrared electromagnetic wave and the slow-light system of infrared wave are investigated. A focus width of 1.6 nm for a wavelength of 56 THz is achieved on a gradient chemical potential distribution graphene. Variant frequency components of the plasmons are stopped at the different locations of the gradient chemical potential distributed graphene monolayer forming "rainbow" capture effect. Additionally, the tunable whispering-gallery mode properties of the graphene monolayer coated semiconductor nano wire are also studied. An effective mode area smaller than $3.75 \times 10^{-5}(\lambda_0)^2$, and a quality factor of 195 are obtained in a nanowire with a diameter of 5 nm. The proposed model and the plasmonic structures may find broad implications in the fields of high density plasmonic integrated circuits technique and transformation plasmonics.

Keywords Graphene • Plasmon • Zero-thickness • Nanofocusing • Slow light • Whispering-Gallery Mode

7.1 Introduction

Graphene, a 2-dimentional (2D) material composed of a single carbon atom layer, has caught intensive research since it was discovered. This awesome material offers promising fantastic applications in electronic and photonic devices [1–5], metametarials [6], optical cloaking [7], femtosecond lasers [8], nonlinear optics [9], and solar cells [10], due to its unique zero-gap band structure with linear dispersion relation, In a suitable range of chemical potential and photon frequency,

W. Qiu (✉)
College of Information Science and Engineering, Huaqiao University, Xiamen, China
e-mail: wbqiu@hqu.edu.cn

C.D. Geddes (ed.), *Reviews in Plasmonics 2015*,
DOI 10.1007/978-3-319-24606-2_7

graphene behaves like a semi-metallic material and consequently is able to support surface plasmonic polariton (SPP) wave [11]. Furthermore, comparing with noble materials, graphene supported SPP waves possess outstanding advantages, such as relatively low loss, high confinement, and most significantly, and tunability [12]. This application of graphene in the field of plasmonics includes sensors [13], nanoribbon waveguides [14, 15], resonators [16], splitters [17], filter [18, 19], polarization convertor [20], modulators [21], and antenna [22, 23], Also, peculiar optical behaviors such as slow light system [24], and anormalous reflection [25] of the graphene supported SPP have been reported. The propagation of SPP wave along the graphene sheet can be tailored by modifying the chemical potential. Thus, graphene offers another freedom to engineer the SPP wave without any geometric engineering. In this chapter, by using the finite element method (FEM) technique, the nanofocusing of MIR on a graphene monolayer which has spatially gradient chemical potential without complicated device geometry is numerically analyzed. In addition, the dynamic group velocity and tunable rainbow capture and relieving of plamsons along graphene sheet is discussed. Also, the optical properties of the graphene monolayer-coated dielectric nanowire resonators are discussed.

This chapter is organized as the following, besides the introduction section, the validity of the zero-thickness graphene monolayer model is reviewed analytically in the second section. In Sect. 7.3, the surface conductivity of graphene is discussed in terms of the chemical potential. The propagation constant of the plasmon along graphene is discussed in the Sect. 7.4. Nanofocusing properties of the MIR EM field are reviewed in the Sect. 7.5. Later on, the dynamic group velocity control and rainbow capture and relieving mechanism of plasmon along graphene sheet is discussed in the Sect. 7.6. In the Sect. 7.7, the tunable WGM properties of the graphene monolayer coated dielectric nanowire resonator are discussed (Fig. 7.1).

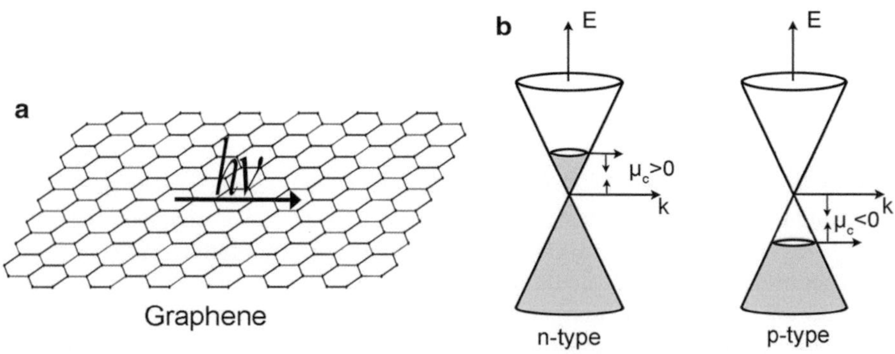

Fig. 7.1 The schematic molecule structure and energy band structure of graphene monolayer

7.2 Validity of the Zero-Thickness Graphene Monolayer Model

Among most of the literatures, the graphene monolayer sheet was modeled as a metallic layer with a finite "effective thickness" d_{eff}, where the corresponding "effective dielectric constant" $\varepsilon_{eff}(d_{eff})$. Then the thin layer was treated similarly to the noble metal materials. The dispersion relation, i.e. the propagation constant as a function of the frequency, is obtained from the common three-layer planar waveguide system. They also demonstrated that when the effective thickness below certain value, the dispersion relation kept unchanged. However, it should be pointed out that the concept of dielectric constant is based on bulk materials. Nevertheless, graphene is composed of a single layer of carbon atom. Substantially, it is not a bulk material, and of course, it is unsuitable to use the concept of dielectric constant. The similar case is the concept of temperature, one cannot define a temperature of a molecule, or a finite number of molecules.

In order to avoid the embarrassment described above, the concepts of "effective thickness" and "effective dielectric constant" should be abandoned. Alternatively a concise model is adapted, where the graphene monolayer is treated as a conductive layer without thickness. So it is characterized by surface conductivity σ. It could be shown that dispersion relation got from the proposed model is identical that got from the "effective thickness" model when the "effective thickness" approaches to zero.

The graphene ultra-thin film is supposed with a thickness $\Delta = a$. And it is characterized by an equivalent permittivity of $\varepsilon_{g,eq}(\Delta)$, which is dependent upon the thickness Δ.

$$\varepsilon_{g,eq} = 1 + \frac{i\sigma_g\eta_0}{k_0\Delta} \tag{7.1}$$

In Fig. 7.2a, the graphene sheet is located on ZOY plane and the plasmon propagates along z direction.

Only the TM polarized SPPs supported by graphene is considered. Assume the EM field propagates along the y-axis. Since the graphene sheet is infinite in y direction, $\frac{\partial}{\partial y} \to 0$, the non-zero components of the TM polarization are (E_y, E_z, H_x),

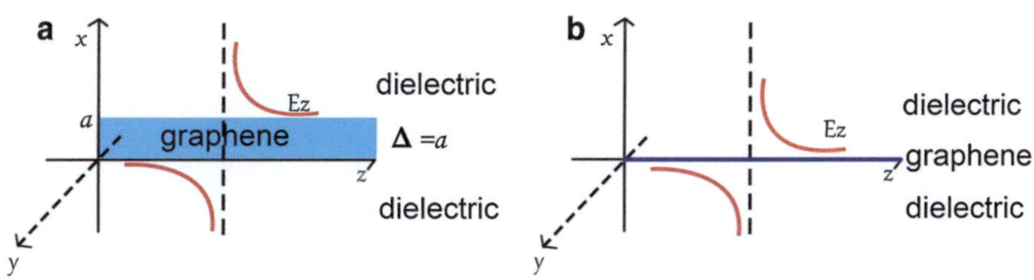

Fig. 7.2 The schematic diagram of (**a**) three layers non-zero thickness model (**b**) zero-thickness model of graphene monolayer

According to the traditional waveguide theory, the electric field in the different regions can be written as

$$E_x = \begin{cases} Ae^{-k_h(x-a)}\exp(i(\omega t - \beta z)) & (x > a) \\ (Be^{k_g(x-a)} + Ce^{-k_g x})\exp(i(\omega t - \beta z)) & (0 < x < a) \\ De^{k_h(x+b)}\exp(i(\omega t - \beta z)) & (x < 0) \end{cases} \quad (7.2)$$

Where A-D are unsolved mode amplitudes, $k_h = \left(\beta^2 - \varepsilon_h k_0{}^2\right)^{1/2}$, $k_g = \left(\beta^2 - \varepsilon_{g,eq} k_0{}^2\right)^{1/2}$, β is the propagation constant. ε_h is the relative permittivity of the free space.

$$\begin{cases} H_y = \dfrac{\omega\varepsilon}{\beta} E_x \\ E_z = -\dfrac{j}{\omega\varepsilon}\dfrac{\partial H_y}{\partial x} \end{cases} \rightarrow E_z = -\frac{j}{\beta}\frac{\partial E_x}{\partial x}$$

and

$$E_y = H_x = H_z = 0$$

For the different regions,

$$z > a \qquad \begin{cases} E_x = Ae^{-k_h(x-a)} \\ H_y = \dfrac{\omega\varepsilon_r\varepsilon_0}{\beta} Ae^{-k_h(x-a)} \\ E_z = \dfrac{i}{\beta} k_h A e^{-k_h(x-a)} \end{cases}$$

$$0 < z < a \qquad \begin{cases} E_x = Be^{k_g(x-a)} + Ce^{-k_g x} \\ H_y = \dfrac{\omega\varepsilon_g\varepsilon_0}{\beta}\left(Be^{k_g(x-a)} + Ce^{-k_g x}\right) \\ E_z = -\dfrac{i}{\beta} k_g \left(Be^{k_g(x-a)} - Ce^{-k_g x}\right) \end{cases}$$

$$z < 0 \qquad \begin{cases} E_x = De^{-k_h x} \\ H_y = \dfrac{\omega\varepsilon_r\varepsilon_0}{\beta} De^{-k_h x} \\ E_z = -\dfrac{i}{\beta} k_h D e^{-k_h x} \end{cases} \quad (7.3)$$

Take the boundary conditions into account,

$$\text{z = a} \quad \begin{cases} A\varepsilon_r = \varepsilon_g\left(B + Ce^{-k_g a}\right) \\ Ak_h = -k_g\left(B - Ce^{-k_g a}\right) \end{cases} \qquad \text{z = 0} \quad \begin{cases} \varepsilon_g\left(Be^{-k_g a} + C\right) = \varepsilon_r D \\ k_g\left(Be^{-k_g a} - C\right) = k_h D \end{cases} \quad (7.4)$$

$$\begin{cases} \varepsilon_{g,eq} = 1 + \dfrac{i\sigma_g\eta_0}{k_0\Delta} \\ k_h = \left(\beta^2 - \varepsilon_h k_0^2\right)^{1/2} \\ k_g = \left(\beta^2 - \varepsilon_{g,eq} k_0^2\right)^{1/2} \end{cases} \quad (7.5)$$

then $k_h = -\dfrac{\varepsilon_h}{1 + \frac{i\sigma_g\eta_0}{k_0 a}}\dfrac{2}{a}$

When $a \rightarrow 0$,

$$k_h = -\frac{2\varepsilon_h k_0}{i\sigma_g \eta_0} \tag{7.6}$$

substitute Eqs. (7.6, 7.5), the propagation constant (wave vector) of the SPPs is given by

$$k_{SPP} = k_g = k_0 \sqrt{\varepsilon_h - \left(\frac{2\varepsilon_h}{\sigma_g \eta_0}\right)^2} \tag{7.7}$$

When it comes to the zero-thickness model, the grahene sheet also locates on the surface of YOZ, and the plasmon propagates along z direction. Also, only TM mode is considered. The non-zero components of the EM field are identical with the non-zero-model. The whole space is divided into two parts, z > 0, and z < 0. The non-zero components of the EM fields in the different regions are written as

$$E_x = \begin{cases} A\exp(-k_h x)\exp(i(\omega t - \beta z)) & x > 0 \\ B\exp(k_h x)\exp(i(\omega t - \beta z)) & x < 0 \end{cases}$$

According to Maxwell's equations $\nabla \times H = \frac{\partial D}{\partial t} = \varepsilon_0 \varepsilon_r \frac{\partial E}{\partial t}$ and $\begin{cases} \frac{\partial H_y}{\partial x} = \varepsilon_0 \varepsilon_r \frac{\partial E_z}{\partial t} \\ E_x = \frac{\beta}{\omega \varepsilon_0 \varepsilon_r} H_y \end{cases}$ we can get H_y E_z in different regions. Here, one can ignore $\exp(i(\omega t - \beta z))$.

$$\overset{x>0}{\begin{cases} H_y = \frac{\beta}{\omega \varepsilon_0 \varepsilon_r} A\exp(-k_h x) \\ E_z = \frac{i\beta}{(\omega \varepsilon_0 \varepsilon_r)^2} k_h A\exp(-k_h x) \end{cases}} \qquad \overset{x<0}{\begin{cases} H_y = \frac{\beta}{\omega \varepsilon_0 \varepsilon_r} B\exp(-k_h x) \\ E_z = -\frac{i\beta}{(\omega \varepsilon_0 \varepsilon_r)^2} k_h B\exp(-k_h x) \end{cases}} \tag{7.8}$$

Where A-B are unsolved mode amplitudes which can be eliminated. k_1 and k_2 is wave vector of the dielectric in the different regions, β is the propagation constant. ε_r is the relative permittivity. ω is the angular frequency.

$$k_h = \left(\beta^2 - \omega^2 \varepsilon_0 \varepsilon_r \mu_0\right)^{1/2} = \left(\beta^2 - \varepsilon_r {k_0}^2\right)^{1/2} \tag{7.9}$$

According to the boundary conditions

$$\begin{cases} \lim\limits_{x \to 0^+} H_y - \lim\limits_{x \to 0-} H_y = J_s \\ J_s = \sigma_g E_z(x = 0) \\ E_z(x = 0) = \lim\limits_{x \to 0^-} E_z = \lim\limits_{x \to 0^+} E_z \end{cases}$$

Where σ_g is the surface conductivity.

A, B, k_1 and k_2 can be solved by solving the boundary equation,

$$\begin{cases} A = -B \\ k_1 = k_2 = \dfrac{2\omega\varepsilon_0\varepsilon_r}{i\sigma_g}. \end{cases}$$

Substitute them to Eq. (7.11), the propagation constant can be got

$$\beta = \frac{\omega}{c}\sqrt{\varepsilon_r - \frac{4{\varepsilon_r}^2\varepsilon_0}{\mu_0{\sigma_g}^2}} \tag{7.10}$$

Where vacuum wave impedance $\eta_0 = \sqrt{\frac{\mu_0}{\varepsilon_0}}$ and $k_0 = \frac{\omega}{c}$, the propagation constant β can be written as

$$\beta = k_0\sqrt{\varepsilon_r - \frac{4{\varepsilon_r}^2}{\eta_0{\sigma_g}^2}} \tag{7.11}$$

So, the validity of the zero-thickness is conformed. It is the limit condition when the effective thickness approaches to 0.

7.3 Surface Conductivity of Graphene

Graphene's surface conductivity σ_g is composed of the contribution from interband electron transition $\sigma_{\mathrm{int}\,er}$ and intraband electron-photon scattering $\sigma_{\mathrm{int}\,ra}$, $\sigma_g = \sigma_{\mathrm{inter}} + \sigma_{\mathrm{intra}}$, According to Kubo formula,

$$\sigma_{\mathrm{inter}} = i\frac{e^2}{4\pi\hbar}\ln\left[\frac{2|\mu_c| - \hbar\left(\varpi + i\tau^{-1}\right)}{2|\mu_c| + \hbar\left(\varpi + i\tau^{-1}\right)}\right], \tag{7.12}$$

$$\sigma_{\mathrm{intra}} = i\frac{e^2 k_B T}{\pi\hbar^2\left(\varpi + i\tau^{-1}\right)}\left[\frac{\mu_c}{k_B T} + 2\ln\left(\exp\left(-\frac{\mu_c}{k_B T} + 1\right)\right)\right] \tag{7.13}$$

Where ω is the angular frequency of the plasmon, τ is the electron momentum relaxation time. Experimental works demonstrated that more than 2 eV of chemical potential [26] and more 3 ps [27] of relaxation time have been achieved. Here the adapted chemical potential is less than 1.3 eV and the relaxation time is less than 1.4 ps. So it is conservative enough to ignore the additional damping channels due to the plamson-phonon interaction [28]. Figure 7.3 reveals the schematic of the interband transitions and intraband scattering.

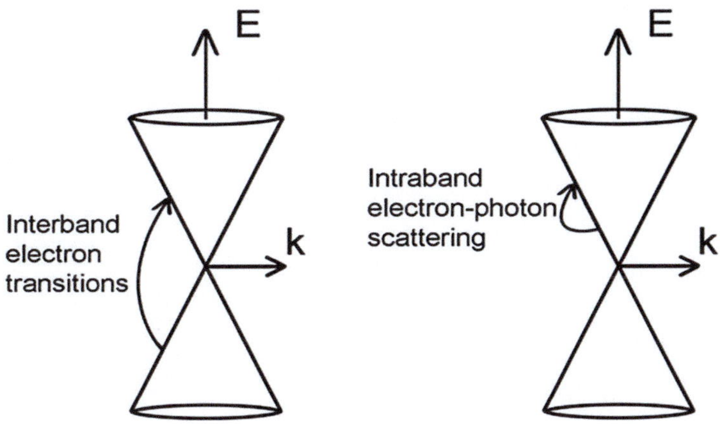

Fig. 7.3 The schematic diagram of interband electron transitions and intraband electron-photon scattering

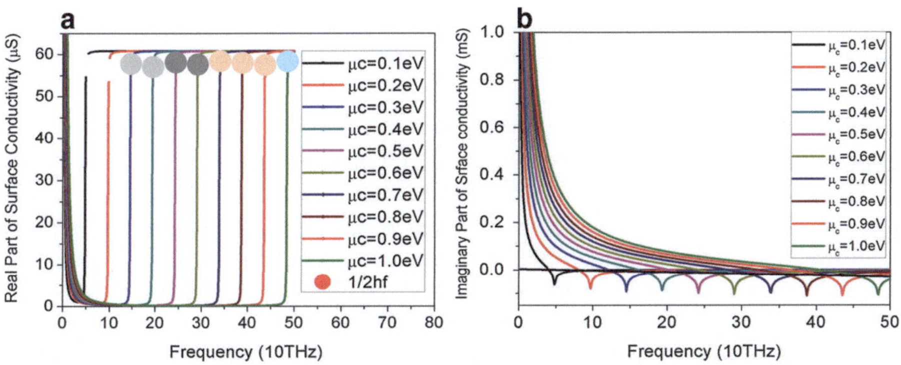

Fig. 7.4 The real and imaginary parts of the surface conductivity of graphene as a function of photon frequency with respect to various chemical potentials. (**a**) real parts, (**b**) imaginary parts

Furthermore, σ_g can be rewritten as $\sigma_g = \sigma_g^{'} + i\sigma_g^{''}$, Fig. 7.4a demonstrate the real and imaginary parts of surface conductivity as function of frequency respective to various chemical potentials. SPP wave can be supported when $\sigma_g^{''} > 0$, while the real part of the conductivity $\sigma_g^{'}$ denotes the absorption loss of the SPP wave on graphene sheet. In Fig. 7.4b, some special points are highlighted, which correspond to the relation $\mu_c = 1/2\hbar\varpi$. when $\mu_c < 1/2\hbar\omega$, $\sigma_g^{'}$ is stable at 60.5 μS, which is substantially high and constrains the propagation distance of the SPP wave. When $\mu_c > 1/2\hbar\omega$, $\sigma_g^{'}$ is low and ensures the long propagation distance of the SPP. Interestingly, when $1/2\hbar\omega$ approaches to μ_c, σ_g^{\prime\prime} reduces and even becomes negative when μ_c is very close to $1/2\hbar\omega$, where TM SPP is no longer supported. Albeit TE surface mode might be supported when $\sigma_g^{''} < 0$, it suffers high

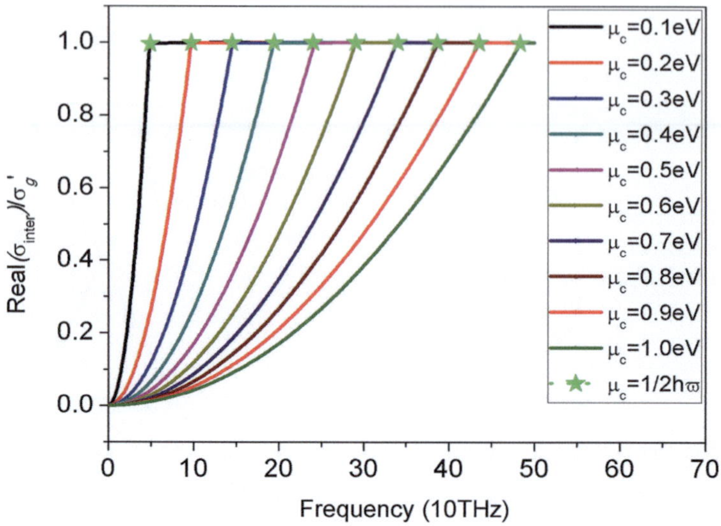

Fig. 7.5 Percentage of the contribution from interband electron transitions on the total surface conductivity as a function of the photon frequency with respect to various chemical potentials

absorption loss due to the high $\sigma_g^{'}$ value at this region. So $\mu_c = 1/2\hbar\varpi$ is very close to the critical point of the graphene supported TM SPP wave.

Figure 7.5 reveals the ratio between the real part of $\sigma_{\text{int}\,er}$ and the real part of the total surface conductivity σ_g as a function frequency with respect to various chemical potentials. In addition, the points corresponding to $\mu_c = 1/2\hbar\varpi$ are highlighted. For a given chemical potential, in the low frequency region, the contribution from electron interband transition is low, which means the propagation loss is high. Moreover, when $1/2\hbar\omega$ approaches to μ_c the contribution grows as the increasing of the frequency, corresponding to a weaker contribution from intraband electron-photon scattering. Particularly, More than 99 % percent of the total surface conductivity comes the contribution from interband electron transition when $\mu_c < 1/2\hbar\omega$. So, $\mu_c = 1/2\hbar\varpi$ is the critical point not only for the absolute value of the surface conductivity, and also for the contribution radio for different mechanisms.

7.4 Analysis of the Propagation Constant of the Plasmons Along the Graphene Monolayer

When the plasmons propagate along an infinite graphene monolayer sheet with a uniform chemical potential μ_c, the dispersion relation is given by Eq. (7.13). In more detail, the propagation constant β is rewritten by $\beta = \beta^{'} + \beta^{''}$, The wavenumber β' and the group velocity of SPP waves on graphene sheet with spatially homogeneous μ_c are shown in Fig. 7.6a as a function of frequency f.

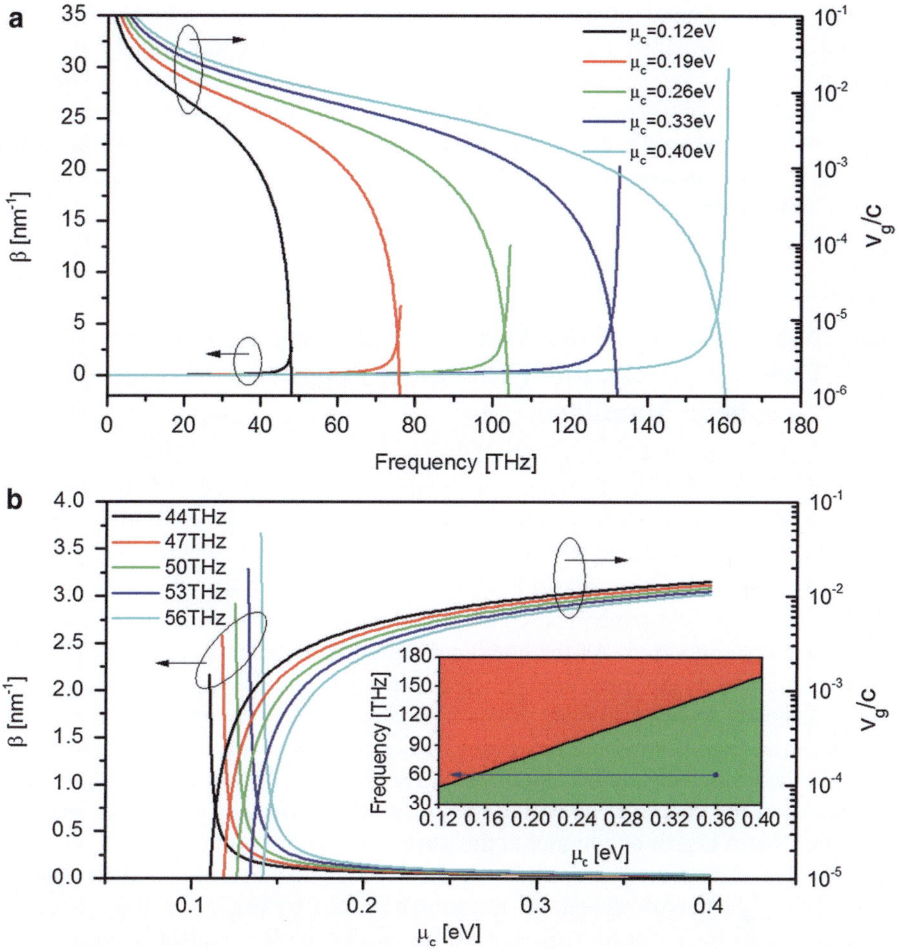

Fig. 7.6 (**a**) The dispersion relation and group velocity of SPP waves propagating along free-standing graphene with homogeneous chemical potential. (**b**) The wavenumber and group velocity of SPP waves as a function of chemical potential. The inset in (**b**) shows the critical frequency as a function of chemical potential. The *green regime* indicates that SPP waves are allowed and the *red regime* indicates the SPP waves are fully damped or not supported. For each chemical potential, calculation is not performed for frequencies higher than its critical frequency

Here we define that for a certain chemical potential, the critical frequency f_c is the frequency that gives equal real part and imaginary part of the complex wavenumber, i.e. $\beta' = \beta''$. In other words, the SPP wave is fully damped. The curve of the inset of Fig. 7.6b shows critical frequency corresponding to different chemical potentials. As the frequency approaches the critical frequency f_c for a certain chemical potential, β' increases drastically. Since $2\pi/\beta'$ is the wavelength of the SPP wave and implies the spatial extension of EM waves, the confinement of the EM waves along graphene is very strong and nanofocusing effect is expected around this frequency f_c. For example, the critical frequencies for chemical potentials of 0.12 and 0.40 eV are 48.1 and 160.9 THz, respectively. Furthermore, around critical frequency, the group velocity v_g becomes very low. Ideally if β' goes to infinite as the frequency increases, zero group velocity of the SPP wave would be achieved and the wavelength of the SPPwave would shrink to zero as well, which means that the EM field is concentrated on a singularity point on the graphene sheet. Of course, there is loss accompanying wave propagation and such ideal case is hardly achievable.

7.5 Nanofocusing of the Mid Infrared Electromagnetic Field on the Gradient Chemical Potential Distributed Graphene Monolayer

Similar behavior will occur when the frequency is kept fixed and the chemical potential is gradually reduced, which means that the SPP wave of a certain frequency propagate along a graphene sheet with spatially decreasing gradient chemical potential. This scenario is indicated by the blue trajectory in the inset of Fig. 7.2b. The green regime means $\beta' > \beta''$ and the red regime means $\beta' < \beta''$ or SPP waves are not supported. As the chemical potential decreases, β' becomes larger, the confinement of the light field becomes stronger, and the group velocity slows down. Together with the slowing down of group velocity, the intensity accumulates and becomes increasingly higher, which is essentially a result of energy conservation. As the point follows the trajectory and approaches the red regime in the inset of Fig. 7.6b, the chemical potential goes sufficiently low and the loss of the SPP waves becomes higher and higher at the same time. Therefore, nanofocusing effect actually depends on the competition between energy accumulation and loss during propagation. As long as the energy accumulation rate far exceeds energy loss rate, nanofocusing effect can be expected. Additionally, if the incident wave is composed of several frequency components, different frequencies would be focused at different positions on a graphene sheet with appropriately designed chemical potential distribution.

To numerically verify the discussion above, Finite element method simulation is used COMSOL Multiphysics to compute the light field of SPP waves propagating along a graphene monolayer with gradient chemical potential μ_c. The graphene sheet is in the XOZ plane at y = 250 nm and SPP waves are launched at x = 0 nm. The distribution of chemical potential μ_c over the graphene sheet is a function of position. In the simulation, it is defined to be

$$\mu_c = \mu_{c0}[1 - \exp(0.015(x - 1000))] \tag{7.14}$$

where $\mu_{c0} = 0.4eV$ is the chemical potential at x = 0 nm. The purpose of choosing this distribution is nothing more than having a spatially rapidly varied chemical potential. The frequencies of simulated SPP waves are 44, 47, 50, 53, and 56 THz.

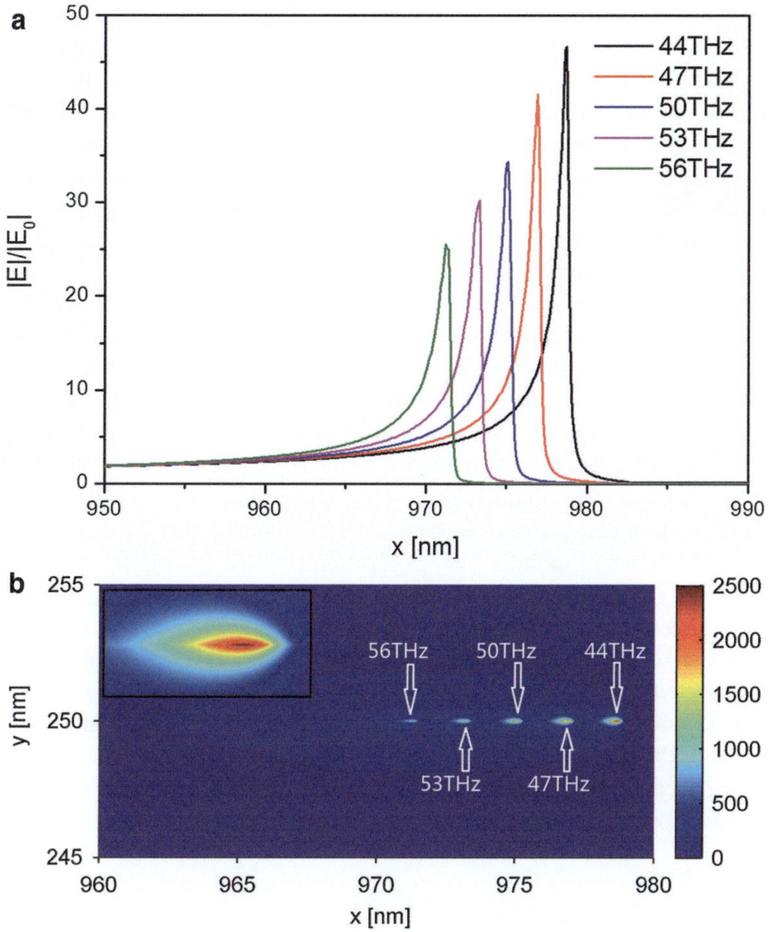

Fig. 7.7 (**a**) The evolution of electric field norm around the focusing spots along the graphene sheet at y = 250 nm. The values are normalized by the electric field norm at the launching point. (**b**) The time averaged energy density distribution near the focusing spots. The values are normalized by the time averaged energy density at the launching point. If five frequencies are launched together, they will be focused at different positions. The inset in (**b**) is a magnification of the focusing spot of 44 THz

The corresponding wavelength in free space is 6.81 6.38, 6.00, 5.66, and 5.35 μm. The evolution of $|\mathbf{E}|$ during propagation at y = 250 nm, i.e. along the graphene sheet, is shown in Fig. 7.7a and the values have been normalized with the norm of electric field at the launching point $|\mathbf{E}_0|$. The intensity enhancement factor $R = |\mathbf{E}_{\text{peak}}|^2/|\mathbf{E}_0|^2$ reaches 654 for 56 THz and even 2178 for 44 THz. The full width at half maximum (FWHM) of the focusing spot of 56 THz is 1.6 nm. It is obvious that other frequencies have spot size on the same order, which indicates that nanofocusing effect is achieved for different frequencies on the same structure.

The five distinct focusing spots of the frequencies are clearly shown in Fig. 7.7b, where the time averaged energy density of each frequency is normalized by the time averaged energy density at the launching point. Since the position of the focusing spot depends on the chemical potential distribution, it can be tuned by modifying the distribution of chemical potential. Meanwhile the intensity enhancement factor R might be changed consequently as well. Passing the focusing spot, the SPP waves vanish due to EM loss [29]. Before SPP waves reach the focusing spot, the local chemical potential is far above $\hbar\omega/2$, σ_{inter} has a small real part and the loss is dominated by intraband electron-photon scattering, thereby bringing very limited loss to the propagation of SPP waves. However, when the chemical potential continues decreasing and reaches around $\hbar\omega/2$, the contribution of loss from interband electron transition indicated by Eqs. (7.14) and (7.15) becomes more significant and eventually dwarfs the counterpart from intraband electron-photon scattering, resulting the SPP waves to disappear. In a way similar to defining critical frequency, the critical chemical potential μ_c for a certain frequency can be defined as the chemical potential that gives $\beta' = \beta''$. It is worthwhile to point out that the position of the focusing point is determined by the location of the critical μ_c, while the size of the focusing pointis fully controlled by how rapidly μ_c approaches the critical value. Practically, one should keep the chemical potential far from the threshold to avoid propagation loss and then reduce it to the critical value as rapidly as possible at desired position to get nanofocusing. In our chemical potential distribution governed by Eq. (7.16), the group velocity remains almost unchanged before the chemical potential falls down rapidly. Take 56 THz as an example. At the FWHM of the focusing spot, the group velocity finally drastically slows down to 5×10^{-5} times the light speed in vacuum, effectively enabling energy accumulation in the vicinity of the focusing spot.

7.6 Ultra-Broadband Rainbow Capture and Releasing Along Gradient Chemical Potential Distributed Graphene Monolayer

In order to dynamically control the group velocity of the plasmon, one can consider a graphene sheet where the chemical potential gradually decreases along the SPP wave propagation direction. From the analysis of Sect. 7.4, it is already known that the group velocity of a wave with radian frequency ω will be trapped at the position where the critical condition discussed above is satisfied. Furthermore, different frequencies would be trapped at different positions on the gradient chemical potential graphene. This actually means broadband light trapping. In contrast to the nanofocusing effect, it does require the chemical potential decreasing abruptly. The above arguments can be verified by simulation based on COMSOL multiphysics Finite Element Method (FEM). In the simulation, the length of the

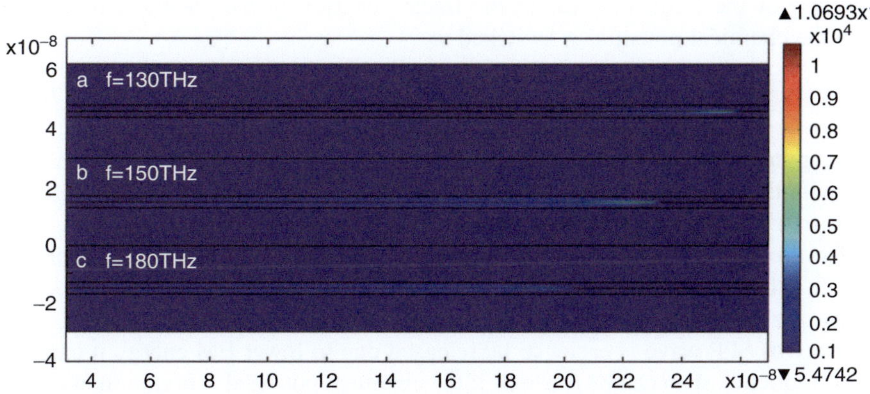

Fig. 7.8 The |E| field distribution of 130, 150, and 180 THz plasmons along the graphene sheet whose chemical potential is governed by $\mu_{c0}(x)$. The 130, 150, and 180 THz plasmons are trapped at 279.45, 231 and 200.5 nm, respectively

graphene sheet is 300 nm. The chemical potential decreases linearly from 1.0 to 0.2 eV, so the chemical potential is expressed by

$$\mu_{c0}(x) = (1.0 - 0.00267x)eV \tag{7.15}$$

Propagation of 3 frequency components, 130, 150 and 180 THz, corresponding to the wavelength of 2.32, 1.99 and 1.66 μm, is simulated. The trapping position is 279.45, 231 and 200.5 nm, respectively. If we incident a broadband SPP wave of which all frequencies have the same amplitude, we will observe broadband light trapping, which is, in other words, rainbow trapping. Figure 7.8 shows the 2 dimensional |**E**| field contours of the three frequencies of SPP wave trapped at different locations. Besides broadband frequency trapped, gradient chemical potential distribution structure leads the small momentum mismatching between the incident wave and the eignmode of the local chemical potential and reduces the loss due to the mode evolution.

The other issue arises here is how to release the trapped SPP wave. A. Vakil and N. Engheta proposed a feasible way to tune the chemical potential of the graphene by applied the external gate voltage [30]. In this work, we follow their scheme and make a little modification of the electrode to taper shape to obtain the linear gate voltage distribution along the graphene sheet, and finally dedictate the chemical potential distribution according to the design. Here we still use the chemical potential distribution function of Eq. (7.17), then one can calculate the trapped frequency distribution along the graphene sheet by COMSOL multiphysics. If the incident SPP is composed of a broadband 80 to 350 THz with the same amplitude. Each frequency component is trapped on the graphene sheet according to the chemical potential. The coordinate of the trapped position shifts linearly with the increasing of the frequency due to the linear function of the chemical potential. When the chemical potential is switched to

$$\mu_{c1}(x) = (1.10 - 0.00267x)eV \quad (7.16)$$

by tuning the external gate voltage. If the impinging frequency span is kept as the same with the above case, The trapped frequency span is from 120 to 350 THz, as shown in Fig. 7.9, which means that the frequency span from 80 to 120 THz transfers from the trapped mode to the propagation mode and leaves the graphene sheet. In other words, this frequency span is released from the graphene sheet as long as the gate voltage increased uniformly along the graphene sheet. Lower frequency of SPP wave is easier to be released due to lower EM energy it carries. Wider frequency span can be released if the chemical potential is higher than $\mu_{c1}(x)$ uniformly. Furthermore, the releasing speed is relied on how fast the external electric field is established and modified and the speed of the reconstruction of the chemical potential distribution. It should be much faster than that due to the thermal-optical effect [31], even it is possible to get real time control of the group velocity of the SPP wave along the graphene sheet.

It is impossible to get the group velocity equal to zero exactly if the SPP wave suffers loss during propagation [32]. Even it is possible to get the SPP mode vanished before the critical condition is reached. So, estimating the lifetime of the SPP wave at the minimum group velocity becomes significant. The lifetime of the SPP τ_{spp} is expressed as $\tau_{spp} = 1/v_g\alpha$, where v_g is the group velocity and α is the attenuation coefficient of the SPP mode including inherent materials absorption, reflectance, scattering loss and the coupling loss due to the mode momentum mismatching during propagation. However, the graphene monolayer sheet is flatten and smooth is our case, the loss from reflectance and scattering is negligible. Consequently, the total propagation loss which is given by the imaginary part of the propagation constant β. v_g and α are calculated from the dispersion relation Eq. (7.9) combining Kubo formula and plotted the lifetime as a function of frequency at various conditions in Fig. 7.10. In the first step, the relaxation time was kept at 0.5 ps, the lifetime as a function of wavelength was calculated at temperature of 270, 300 and 330 K, respectively and the results are plotted in Fig. 7.10a. Interestingly, the lifetime is stable at 3.14 ps in the calculated temperature and frequency spectrum, which is one order higher than other publications [33]. It is rooted from the fact that the electron momentum relaxation time of graphene is orders higher than the counterpart of the noble metals [30]. Furthermore, when we tuned the relaxation time, the lifetime was modified significantly. Figure 7.10b plots the lifetime of 150 THz as a function of relaxation time τ, where the lifetime increases linearly from 2.51 ps with a relaxation time of 0.4 ps to 6.29 ps with a relaxation time of 1 ps. So, the lifetime of the trapped SPP wave is mainly determined by the relaxation time, which also determines the metal absorption loss.

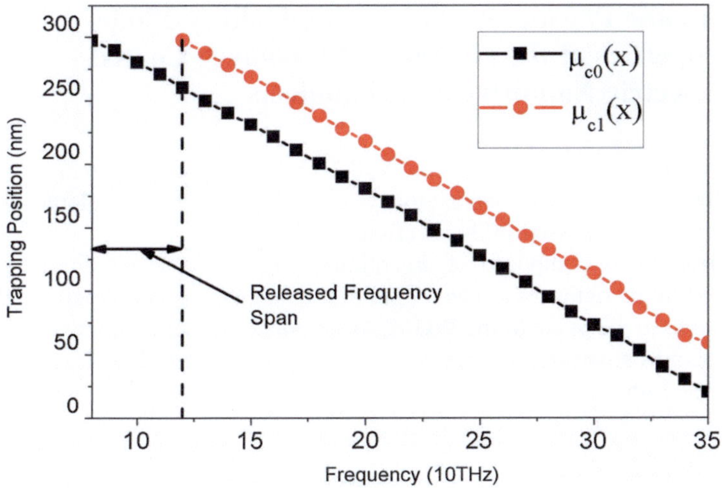

Fig. 7.9 The trapping position as a function of frequency with various chemical potential distribution of $\mu_{c0}(x)$ and$\mu_{c21}(x)$. The frequency span from 80 to 120 THz transfers from the trapped mode to the propagation mode and leaves the graphene sheet

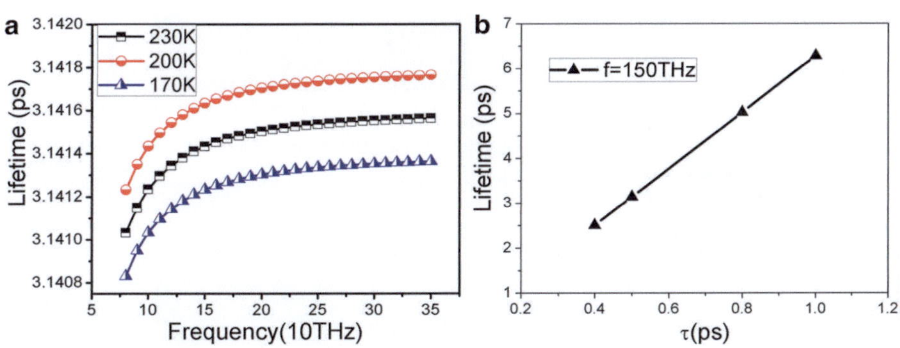

Fig. 7.10 The lifetime of the trapped plasmons as a function of frequency at a temperature ranged from 170 to 230 K with a relaxation time of 0.5 ps. The lifetime of the trapped plasmon is almost independent from both temperature and frequency. (**b**). The lifetime of the trapped 150 THz plasmon as a function of relaxation time. The lifetime increases linearly with the increasing of the relaxation time

7.7 Tunable Plasmonic Whispering-Gallery-Mode Properties of the Graphene Monolayer Coated Dielectric Nanowire and Nanodisks

In order to take full advantages of the WGMs and the high confinements of the SPP waves, dielectric nanowires resonators coated by graphene monolayer instead of noble metals are discussed in this section.

The structure is composed of an infinite long dielectric nanowire coated by graphene with the radius of r. The length of the nanowire is infinite in the direction perpendicular to the plane of the WGM, so we only consider 2 dimensional pictures of the proposed structure, and the schematic is as shown in Fig. 7.11a. Here, $n_{air} = 1$ and $n = 3.45$.

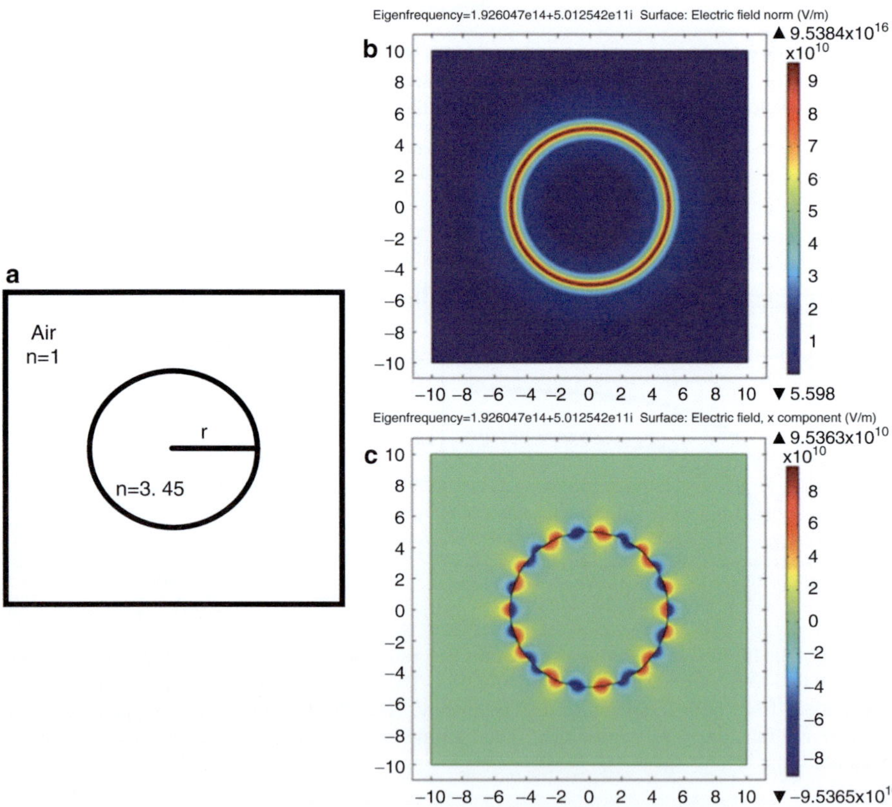

Fig. 7.11 (**a**) Schematic diagram of InGaAs nanowire cavity coated with graphene and corresponding COMSOL finite-element computational window. r indicates the radius of the cavity. The cross-sectional views of the Electrical field (**b**) (Ex) and (**c**) (|E|) distribution of the plasmonic whispering gallery mode with the azimuthal number of 9 at the wavelength of 1.55 um. The cavity radius is 5 nm and the chemical potential is 0.9 eV

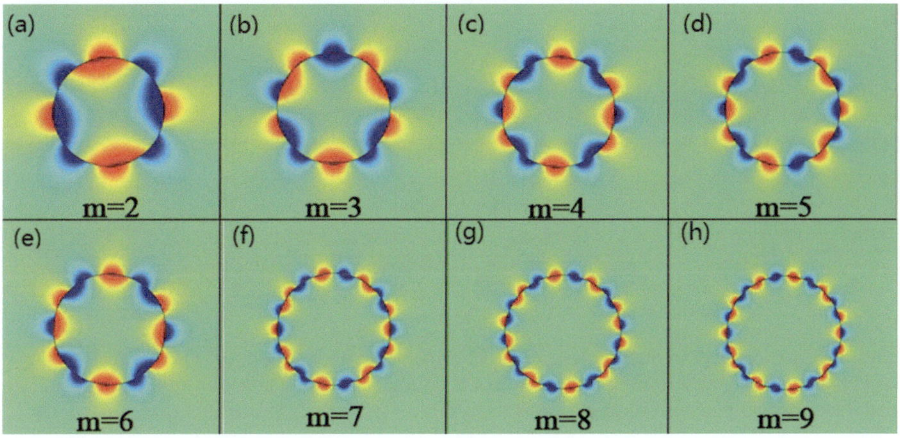

Fig. 7.12 **(a–h)** The cross-sectional views of the Electrical field (E_x) distribution of the plasmonic whispering-gallery mode with the different azimuthal number (m = 2,3,4,5,6,7,8,9) for the cavity with radius r = 5 nm and chemical potential $\mu_c = 0.9$ eV

In a dielectric nanowire cavity with r = 5 nm and chemical potential of graphene $\mu_c = 0.9$ eV, plasmonic WGM with an azimuthal number (m) of 9 can be excited at a resonant wavelength of 1550 nm in free space. The electric field (E_x) profiles are shown in Fig. 7.11b, c. It is obvious in Fig. 7.11b that the electric field is tightly localized on the surface of the graphene-coated cavity. Also, the E_x field profiles of various azimuthal numbers from 2 to 9 are revealed in Fig. 7.12a–h.

The electric field becomes tighter and tighter along the circumference of the nanocavity as the azimuthal mode number increases. The E_x component is anti-symmetric inside and outside the dielectric resonator due to the TM field nature, while the azimuthal number is different in dielectric resonator from that in air region owing to the different refractive index and the non-zero current density in the graphene layer.

According to the eigenvalue analysis section in COMSOL, the eigenvalue is given by $\beta = \delta - i\gamma$,which has an imaginary part γ representing the eigen frequency of the resonator, and a real part δ representing the damping. The Q factor is defined as

$$Q = \gamma / (2|\delta|), \tag{7.17}$$

For the SPP WGM, besides the quality factor and the azimuthal mode number, another key parameter is the effective mode area A_{eff}, which is defined by the ratio of a mode's total energy density per unit length to its peak energy density [34],

$$A_{eff} = \int_{all} W(r) ds / \max[W(r)], \tag{7.18}$$

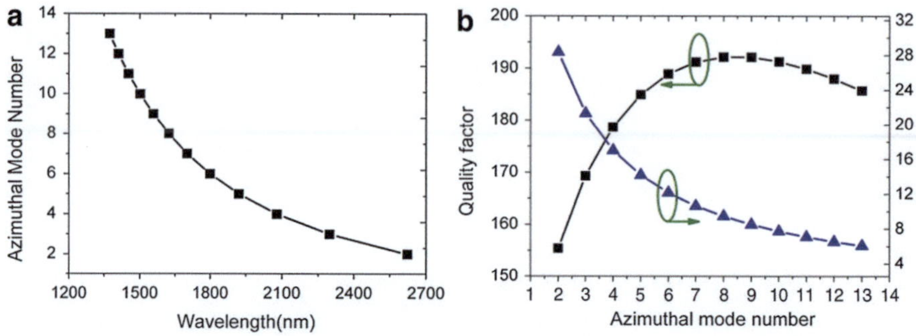

Fig. 7.13 (**a**) Azimuthal number as a function of resonant wavelength, (**b**) Quality factor (*black*) and effective mode area (*blue*) as a function of azimuthal number. The radius of the cavity is 5 nm and the chemical potential of graphene is 0.9 eV

where $W(r) = \frac{1}{2}\mathrm{Re}\left\{\frac{d[\omega\varepsilon(r)]}{d\omega}\right\}|E(r)|^2 + \frac{1}{2}\mu_0|H(r)|^2$, representing the mode energy density. $|E(r)|^2$ and $|H(r)|^2$ are the intensity of electric and magnetic fields, respectively, $\varepsilon(r)$ and μ_0 are the electric permittivity and magnetic permittivity in the free space, respectively.

Q factors and the effective mode area of the nanowire resonators are calculated as a function of the azimuthal mode number. The radius of the cavity is fixed to r = 5 nm and chemical potential of graphene is kept constant $\mu_c = 0.9$ eV for comparison. The azimuthal mode number falls down from 13 to 2 with the increasing of the resonance wavelength from 1320 to 2680 nm in Fig. 7.13a. Figure 7.13b plots the Q factor and the effective area as a function of the azimuthal number m. The Q-factor increases with the increasing the azimuthal number before it reaches the maximum 192.1 at m = 8, then it falls down slightly when m keeps going up. However, the effective area reduces monotonously with the increasing of the azimuthal number. This trend means that with the increasing of the azimuthal number, the confinement of the EM field becomes tighter, and it would enhance the EM field density at the interface. In general, the loss of plasmonic resonator is composed of radiation loss and metallic absorption loss. With the increasing of the resonance wavelength, the metallic absorption loss decreases, leading to the Q factor going up. While the resonance wavelength reaches to a certain degree, radiation loss begins rising gradually and even exceeds the reduction of absorption loss, resulting in the falling of Q factor [35]. In the graphene-coated dielectric nanowire structure, the effective mode area is typically smaller than 3.75×10^{-5} $(\lambda_0)^2$ with the resonance wavelength λ_0 from 1370 to 2620 nm, much smaller than the conventional optical whispering-gallery mode with the same wavelength.

As a comparison, one can simulate the dielectric nanowire cavity coated by Au and Ag thin films with a thickness of 2 nm. If the resonance wavelength is kept at 1.55 μm, it is found that the quality factor of the Au-coated nanocavity is around 0.33 with an azimuthal mode number of 2, and a radius of 54.5 nm. The quality

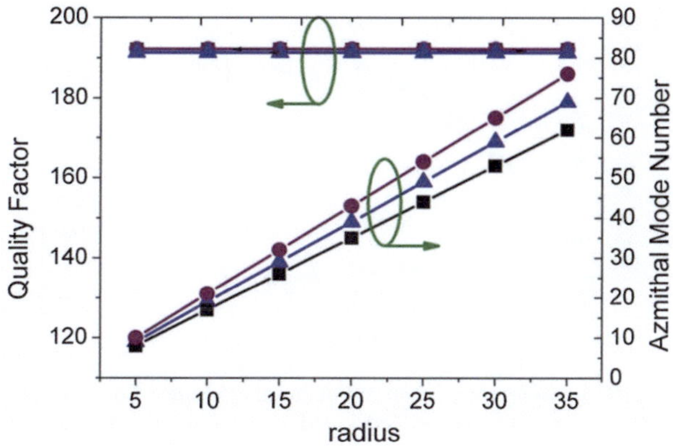

Fig. 7.14 Quality factor and azimuthal number as a function of the radius. The chemical potential is 0.9 eV, and the resonant wavelength is 1.50 um (*black*), 1.56 um (*blue*), 1.62 um (*purple*), respectively

factor of the Ag-coated dielectric nanocavity is 0.14 with the same azimuthal number. For the graphene coated dielectric nanowire cavity, the azimuthal number m is 109 with the same radius in which the chemical potential is 0.9 eV. This is in stark contrast to the cavity coated by the conventional metal films. Therefore, graphene is much more effective than Au and Ag in achieving a high quality factor for the nanowire resonators.

One can also study the Q factor and the azimuthal number with the change of the radius, when the chemical potential is 0.9 eV. The resonance wavelengths are fixed to 1500 nm (black), 1550 nm (blue) and 1620 nm (purple), respectively. It is shown in Fig. 7.14 that the quality factor is stable at around 192 with all the three resonance wavelengths as the radius increases from 5 to 35 nm, while the azimuthal mode number increases linearly. The absorption loss of the SPP mode decreases with the decreasing of the radius due to the reduction of the travelling distance along the circumference. This effect would result in increasing of the quality factor. On the other hand, the confinement capability of the dielectric cavities without coating degrades as the radius decreases, which would induce a decreasing quality factor. When these two effects balance each other, the total loss during one period along the circumference becomes stable, which finally results in stable quality factors of the various wavelengths and radius. In a SPP WGM cavity, the azimuthal number is roughly satisfied by $m = 2n_{eff}\pi r/\lambda_0$, where λ_0 is the wavelength of the EM wave in free space. So it is obvious that the azimuthal mode number is proportional to the radius of the nanowire cavity when the resonant wavelength is kept constant. Nevertheless, if the cavity radius is small, the azimuthal mode number is roughly independent upon the radius and is reverse proportional to the wavelength in high m region, as shown in Fig. 7.14.

Finally, rhe quality factors and azimuthal mode numbers of the nanowire cavities are studied with the radius r = 5 nm versus the chemical potential at the resonant wavelength λ = 1.56 μm (blue), 1.61 μm (black), which are shown in Fig. 7.15a. It is obvious that the Q factor increases with the increasing of the chemical potential while the azimuthal mode number decreases. When the chemical potential reduces 0.5 eV from higher level, the azimuthal mode number rises

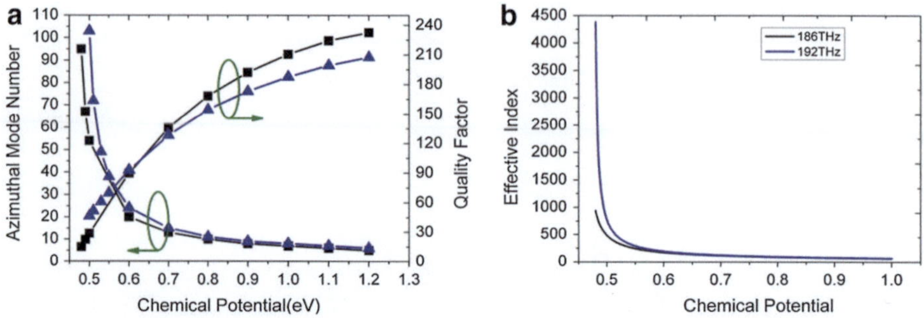

Fig. 7.15 (**a**) Quality factor and azimuthal mode number as a function of chemical potential. The radius is 5 nm, (**b**) The effective index as a function of the chemical potential in flatted graphene. The resonant wavelength is 1.56 um (*blue*), 1.61 um (*black*), respectively

sharply and the quality factor is less than 20. As it is well known, the propagation loss of graphene supported SPPs is determined by the real parts of the surface conductivity of graphene, and finally determined by the value comparison between the chemical potential μ_c and half of the energy of the SPP $\hbar\varpi/2$ [36]. When the chemical potential is much higher than $\hbar\varpi/2$, the surface conductivity is intraband electron-photon scattering dominated, which results in low propagation loss of SPP mode on graphene. Otherwise, the surface conductivity is interband electron-electron transition dominated and the propagation loss is high which would lead to low quality factor or even no SPP WGM supported in this category of nanocavity [11]. One should make the chemical potential much higher than half energy of the SPP modes to avoid the propagation loss, which would increase the quality factor of the nanocavity, as shown in Fig. 7.15a. A quality factor as high as 235 is achieved when the chemical potential is 1.2 eV. The effective index of SPP wave as a function of the chemical potential of flattened graphene is plotted in Fig. 7.15b. For a certain frequency, as the chemical potential approaches to $\hbar\varpi/2$, which is around 0.4 eV in our case, the effective index increases drastically, which means the wavelength of the SPP decreases accordingly. The relationship between the effective refractive index and the chemical potential is not exactly the same with that in the WGM case. The trend is adoptable to understand Fig. 7.15a.

References

1. Grigorenko AN, Polini M, Novoselov KS (2012) Graphene plasmonics. Nat Photonics 6:749
2. Wu Y, Lin Y, Bol AA, Jenkins KA, Xia F, Farmer DB, Zhu Y, Avouris P (2011) High-frequency, scaled graphene transistors on diamond-like carbon. Nature 472:74
3. Xia F, Farmer DB, Lin Y, Avouris P (2010) Graphene field-effect transistors with high on/off current ratio and large transport band gap at room temperature. Nano Lett 10:715
4. Bao Q, Loh KP (2012) Graphene photonics, plasmonics, and broadband optoelectronic devices. ACS Nano 6:3677

5. Low T, Avouris P (2014) Graphene plasmonics for terahertz to mid-infrared applications. ACS Nano 8:1086
6. Wang J, Lu W, Liu J, Cui T (2015) Digital metamaterials using graphene. Plasmonics 10:1141
7. Chen PY, Alù A (2011) Atomically thin surface cloak using graphene monolayers. ACS Nano 5:5855
8. Bao Q, Zhang H, Wang Y, Ni Z, Yan Y, Shen ZX, Loh KP, Tang DY (2009) Atomic-layer graphene as a saturable absorber for ultrafast pulsed lasers. Adv Funct Mater 19:3077
9. Bao Q, Zhang H, Wang B, Ni Z, Haley C, Lim YX, Wang Y, Yuan DT, Loh KP (2011) Broadband graphene polarizer. Nat Photonics 5:411
10. Kanade P, Yadav P, Kumar M, Tripathi B (2015) Plasmon-induced photon manipulation by Ag nanoparticle-coupled graphene thin-film: light trappng for photovoltaics. Plamsonics 10:157–164
11. Jablan M, Buljan H, Soljačić M (2009) Plasmonics in graphene at infrared frequencies. Phys Rev B 80:245435
12. Farhat M, Rockstuhl C, Bağcı H (2013) A 3D tunable and multi-frequency graphene plasmonic cloak. Opt Express 21:12592
13. Dash JN, Jha R (2015) On the performance of graphene-based D-shaped photonic crystal fibre biosensor using surface plasmon resonance. Plasmonics 10:1123
14. Christensen J, Manjavacas A, Thongrattanasiri S, Koppens FHL, de García Abajo F (2012) Graphene plasmonic waveguiding, and hybridization in individual paired nano-ribbons. ACS Nano 6:431
15. He S, Zhang X, He Y (2013) Graphene nano-ribbon waveguides of recordsmall mode area and ultra-high effective refractive indices for future VLS. Opt Express 21:30664
16. Brar VW, Jang MS, Sherrott MJ, Atwater H (2013) Highly confined tunable mid-infrared plasmonics in graphene nanoresonators. Nano Lett 13:2541
17. Zhu X, Yan W, Mortensen N, Xiao S (2013) Bends and splitters in graphene nanoribbon waveguides. Opt Express 21:3486
18. Li H, Wang L, Liu J, Huang Z, Sun B (2014) Tunable, mid-infrared ultra-narrowband filtering effect induced by two coplanar graphene stripes. Plamsonics 9: 1239–1243
19. Danaeifar M, Granpayeh N, Mohammadi A, Setayesh A (2013) Graphene-based tunable terahertz and infrared band-pass filter. Appl Opt 52:E68
20. Cheng H, Chen S, Yu P, Li J, Deng L, Tian J (2013) Mid-infrared tunable optical polarization converter composed of asymmetric graphene nanocrosses. Opt Lett 38:1567
21. Sensale-Rodriguez B, Yan R, Kelly M, Fang T, Tahy K, Hwang W, Jena D, Liu L, Xing H (2011) Broadband graphene terahertz modulators enabled by intraband transitions. Nat Commun 3:1787
22. Yao Y, Kats MA, Genevet P, Yu N, Song Y, Kong J, Capasso F (2013) Broad electrical tuning of graphene-loaded plasmonic antennas. Nano Lett 13:1257
23. Liu H, Sun S, Wu L, Bai P (2015) Optical near-field enhancement with graphene bowtie antennas. Plamonics 9:845
24. Chen L, Zhang T, Li X, Wang G (2013) Plasmonic rainbow trapping by a graphene monolayer on a dielectric layer with a silicon grating substrate. Opt Express 21:28628
25. Nikitin AY, Low T, Martin-Moreno L (2014) Anomalous reflection phase of graphene plasmons and its influence on resonators. Phys Rev B 90:041407
26. Efetov DK, Kim P (2010) Controlling electron-phonon interactions in graphene at ultrahigh carrier densities. Phys Rev Lett 105:256805
27. Dean CR, Young AF, Meric I, Lee C, Wang L, Sorgenfrei S, Watanabe K, Taniguchi T, Kim P, Shepard KL, Hone J (2010) Boron nitride substrates for high-quality graphene electronics. Nat Nanotechnol 5:722
28. Yan H, Low T, Zhu W, Wu Y, Freitag M, Li X, Guinea F, Avouris P, Xia F (2013) Damping pathways of mid-infrared plasmons in graphene nanostructures. Nat Photonics 7:394
29. Lindquist NC, Nagpal P, Lesuffleur A, Norris DJ, Oh SH (2010) Three-dimensional plasmonic nanofocusing. Nano Lett 10:1369

30. Vakil A, Engheta N (2010) Transformation optics using graphene. Science 332:1291–1294, 2011
31. Gan Q, Ding YJ, Bartoli FJ (2009) "Rainbow" trapping and releasing at telecommunication wavelengths. Phys Rev Lett 102:56801
32. Reza A, Dignam MM, Hughes S (2009) Can light be stopped in realistic metamaterials? Nature 455(7216):E10–E11
33. Chen L, Wang GP, Gan Q, Bartoli FJ (2009) Trapping of surface-plasmon polaritons in a graded bragg structure: frequency-dependent spatially separated localization of the visible spectrum modes. Phys Rev B 80:161106
34. Falk AL, Koppens FH, Chun LY, Kang K, de Snapp Leon N, Akimov AV, Jo M, Lukin MD, Park H (2009) Near-field electrical detection of optical plasmons and single-plasmon sources. Nat Phys 5:475
35. Chen Y, Zou C, Hu Y, Gong Q (2013) High-Q plasmonic and dielectric modes in a metal-coated whispering-gallery microcavity. Phys Rev A 87:23824
36. Hanson GW (2008) Dyadic Green's functions and guided surface waves for a surface conductivity model of graphene. J Appl Phys 103:64302

Chapter 8
Fano Resonance in Plasmonic Optical Antennas

Siamak Dawazdah Emami, Richard Penny, Hairul Azhar Abdul Rashid, Waleed S. Mohammed, and B.M. Azizur Rahman

Abstract This chapter describes an analytical model developed to study the Fano resonance effect in clusters of spherical plasmonic nanoparticles under local excitation. The model depicted the case of a parallel single dipole emitter that was near-field coupled to a pentamer or heptamer cluster of nanospheres. Spatial polarization and field distributions of the optical states and resonance spectra for these cluster configurations were calculated. The directivity calculation was analyzed in order to qualify the redirection of emission. Performances of various nanoantennae were investigated and fully characterized in terms of spatial geometric differences and the Fano resonance effect on plasmonic nanoparticles in the optical domain. Light radiation patterns were found to be significantly affected by nanosphere sizes and positioning of nanospheres with respect to the dipole. A coupling capacitor is calculated as an equivalent component in the proposed circuit model in order to describe the coupling effect between subradiant and superradiant mode in the Fano resonance. The circuit impedances of tetramer, pentamer, and broken symmetry pentamer are simulated, with resultant circuit models in agreement with the calculated results based on S-parameters. The analytical treatment of these modeled nanoantennae yielded results that are applicable to physical design and utilization considerations for clusters in nanoantennae mechanisms.

Keywords Plasmonic nanoparticles • Optical nanoantennae • Fano resonance • Circuit model

S.D. Emami (✉) • R. Penny
Photonic Research Center, University of Malaya, 50603 Kuala Lumpur, Malaysia
e-mail: s.d.emami@gmail.com

H.A. Abdul Rashid
Faculty of Engineering, Multimedia University, 63100 Cyberjaya, Selangor, Malaysia

W.S. Mohammed
School of Engineering, Bangkok University, Klong Luang, Patumthani 12120, Thailand

B.M.A. Rahman
Electrical, Electronic and Information Engineering Department, City University, London, UK

C.D. Geddes (ed.), *Reviews in Plasmonics 2015*,
DOI 10.1007/978-3-319-24606-2_8

8.1 Introduction

The unique optical properties of surface Plasmon have attracted tremendous interest due to important applications in biological sensors [1], subwavelength optical imaging [2], metamaterial research [3], fluorescence generation [4], nonlinear optical effects enhancement [5] and optical antennas [6]. In the vicinity of metal nanoparticles, a resonant coupling occurs between incident light and the collective oscillation mode of the conduction electrons within noble metals. This effect is responsible for converging incident light into sub-wavelength scale regions with very intense local fields at the near surfaces of nanoparticles [7]. The interaction between these intense fields and surrounding molecules will increase the Raman cross-section of the molecules, and consequently trigger surface-enhanced Raman scattering (SERS) [8]. In SERS, the maximum obtainable field enhancement is generally larger for narrower resonances. The linewidth of plasmon resonance is a key element in explaining several phenomena and applications, and a narrow plasmon linewidth implies a long plasmon lifetime, large field enhancement, and high detection sensitivity [9]. Propagation lengths in plasmonic waveguides are typically proportional to the plasmon lifetime. Localized surface plasmon resonance (LSPR) sensing applications require use of substrates with narrow plasmon resonance [10].

The Fano resonance effect on plasmonic nanoparticle materials results in such materials possessing a number of unique optical properties, and the potential applicability for sensing [11], lasing [12], switching [13], nonlinear devices [14] and slow-light devices [15–17] has generated significant attention. A Fano resonance is a consequence of coherent interference between superradiant and subradiant hybridized plasmon modes [18]. Incident light on subradiant modes will initiate excitation that results in superradiant modes, and these superradient modes possess zero or finite dipole moments alongside a negligible coupling with light. The superradiant mode is strongly lifetime-broadened with a decay rate $\Gamma_{\text{Superradiant}}$ dependent on radiative and nonradiative losses, whereas the subradiant mode is weakly lifetime-broadened and has a decay rate of $\Gamma_{\text{Subradiant}} << \Gamma_{\text{Superradiant}}$, caused by nonradiative losses [19, 20]. This difference in decay rates allows for a significantly broadened spectral feature in superradiant modes in comparison to subradiant modes, mirroring the electromagnetic induced transparency (EIT) phenomenon that is omnipresent in atomic systems [21, 22].

Fano resonance has been recently observed in a variety of complex plasmonic nanostructures such as nanoshells [23], nanodisks [24], nanocubes [25], nanoparticle chains [26], carbon nanotubes [14] and nanoclusters [19, 20]. Research on nanoclusters has shown that the Fano resonance wavelength and amplitude spectrum is strongly dependent on the dimensions, geometry, and relative size of the individual nanocluster and the dielectric environment [27]. Such findings allow for the Fano resonance effect to be optimally enhanced, and a maximum local density of optical states (LDOS) can be achieved by sufficiently increasing the overlap of the radiation pattern between subradiant and superradiant modes. LDOS affects practically all features of optical antennas, including excitation and emission rates [28]. Experimental observations of nanoclusters [28] allow for the Fano resonance superradiant mode to be considered as a promising candidate to enhance LDOS effects in nanoantennas.

This chapter details the derivation of an electrodynamics coupling model for the interaction of dipolar transitions and radiation via plasmonic nanoclusters such as quadrimers [29] pentamers [30, 31] and heptamers [28, 32]. The quantitative evolution of antenna modes, particularly the gradual emergence of subradiant and superradiant as antennas become increasingly bounded, is discussed. Geometrical properties of nanostructure antennas, such as sphere radius, inter-sphere distance, and distance of the model from the dipole, were evaluated over an appropriate range for the purpose of optimizing the directivity and bandwidth of the model [33]. Despite the analogy of a coupled oscillator model for Fano resonance, there is a deficit of models for accurately describing the impedance and induced currents that chiefly determine the impedance matching of resonators with propagated light. A representation circuit diagram for Fano resonance is presented in this chapter to address this shortfall. The starting point is an intuitive explanation of the physical nature of Fano resonance using the three-level quantum system. The basic idea of Fano resonance arising in a circuit model for pentamer nanoparticles is demonstrated. It is found that symmetry breaking in shells [34, 35], disks [8, 36] and rings [37–39] gives rise to the appearance of an additional subradiant that potentially can be used in LSPR sensing. Effects of symmetry breaking of pentamers and consequent second Fano resonances are investigated by circuit modeling. Also shown is that the offset of the central nanosphere from the nearest neighbor nanosphere leads to the appearance of the second Fano resonance in the same spectrum. Considering the similarity of Fano resonance phenomena in all plasmonic nanostructures, the proposed circuit model can be employed in studies for other coupled plasmonic nanostructures such as ring-disk cavities [40], dolmen structures [40], multilayer nanoshells [41] and stacked optical metamaterials [42]. This analytical model was simulated using Matlab commercial software and was validated using the finite integration technique (FIT) of the CST studio suite commercial product [43]. The numerical simulations and analytical results described within this work were sufficient to allow determination of the optimum transmission directivity, intensity and polarization using the basis of optimized nanosphere radius, inter-sphere distance, and distance between emitter and cluster.

8.2 Fano Resonance in Optical Nanoantennas

A Fano resonance plasmonic system can be described by the three-level quantum system [44] depicted in Fig. 8.1. The $|1\rangle$, $|2\rangle$ and $|3\rangle$ levels represent the incident photon at the ground, subradiant, and superradiant mode levels respectively [21, 45]. A system pumped at superradiant wavelengths causes the superradiant mode to be engaged via two pathways, these pathways being $|1\rangle \rightarrow |2\rangle$, as

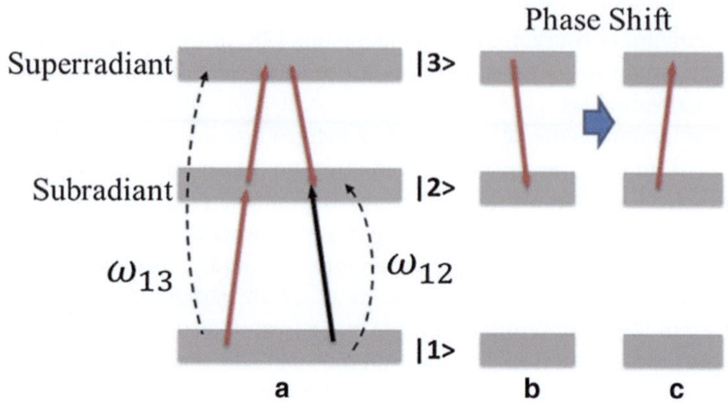

Fig. 8.1 Fano resonance in a three-level quantum system

indicated by the black arrow in Fig. 8.1a, and |1> → |2> → |3> → |2>, as depicted by the red arrow in Fig. 8.1b. Figure 8.1c illustrates the coupling amid subradiant and superradiant modes that occurs between levels |2> and |3>, and this coupling contributes to phase shifting across a 0 to π range. The coupling mechanism invokes a delay between the incidence of the external electric field and the response of the plasmonic nanostructures. As ω increases, the phase increases from 0 at values below resonance wavelength, to π/2 at resonance wavelength, and to π as resonance wavelength increases to infinity. Consequently, plasmonic nanostructures and the external electric field that were in phase before the resonance become out of phase after the resonance [46]. The Fano resonance effect appears when the combined phase shift from |2> → |3 > and |3> → |2> is equal to π, since in this situation the two paths cancel the polarization of the superradiant mode. This phenomenon results in a narrow window of transparency in the transmission spectrum [45].

8.3 Analytical Model for Optical Nanoantenna Clusters

8.3.1 Electrodynamics Coupling Model for Nanoclusters

An electrodynamics coupling model was utilized for calculation of the scattered dipole field matrix (reflection and transmission) in the vicinity of nanoclusters. The analyzed configuration consisted of arrays of metallic nanospheres that were irradiated by an E-field emerging from a single dipole emitter. This emitter possessed a dipole moment with magnitude $\mathbf{P_0}$ located at the origin and along the X-axis. The induced dipole moment was described by Mei theory [47] as

$$\mathbf{P}_0 = \alpha_{ee}.\mathbf{E}_{exc}(\mathbf{r}_0.\mathbf{k}_B) \tag{8.1}$$

where α_{ee} is the electric polarizability of the nanoparticle, $\mathbf{r}_0$ and $\mathbf{k}_\mathrm{B}$ are the distance from the emitter and the background wavevectors respectively, and $\mathbf{E}_{exc}$ is the exciting E-field produced cumulatively by the nanoparticles and any externally induced fields. The electric polarizability (α_{ee}) of a generic isolated particle may be expressed in terms of its Mie scattering coefficient c_1^{TM} (corresponding to a_n in the notation used in [48]). Such an expression arises by comparing the Mie TM^r spherical harmonic for $n=1$ with the dipolar field generated by

$$\alpha_{ee} = -\mathbf{i}\frac{6\pi\varepsilon_0}{\mathbf{k}_B^3}\mathbf{a}_1 \tag{8.2}$$

in which wave vector $\mathbf{K}_B = k_x\hat{\mathbf{X}} + k_y\hat{\mathbf{Y}} + k_z\hat{\mathbf{Z}}$ has amplitude $|\mathbf{K}_B| = (\omega/c)\sqrt{\varepsilon_B}$ for the exterior medium, where ε_B is the relative dielectric constant and c is the speed of light. Excitation by a periodic field results in each nanosphere forming an electric dipole moment of $\mathbf{P}_n = \mathbf{P}_0 e^{ik_B.d_n}$, where $d_n = d_x\hat{\mathbf{X}} + d_y\hat{\mathbf{Y}} + d_z\hat{\mathbf{Z}}$ and represents the index of nanoparticle positions [47].

Consideration of the model geometry allowed for the dipole moments (represented in plane wave form) to be expanded into spherical harmonics, which facilitated a solution for the scattered light. The expansion permitted the electric field produced by an electric dipole moment $\boldsymbol{p}_j$, to be expressed as [49]

$$\mathbf{E}_{scat}(\mathbf{r}) = \frac{e^{i\mathbf{k}_B r}}{4\pi\varepsilon_b\varepsilon_0 r^3}\left\{\mathbf{k}_B^2 r^2(\hat{\mathbf{r}}\times\hat{\mathbf{p}}_0)\times\hat{\mathbf{r}} + (1 - i\mathbf{k}_B r)[3(\hat{\mathbf{r}}\cdot\hat{\mathbf{p}}_0)\hat{\mathbf{r}} - \hat{\mathbf{p}}_0]\right\} \tag{8.3}$$

where $\mathbf{k}_\mathrm{B}$ is wave vector and $\hat{\mathbf{p}}_0$ is induced dipole moment. For an incident field polarized in the $\hat{\mathbf{Z}}$ direction, the excitation field $E_{exc,j}$ at the center of a sphere j also lies along the $\hat{\mathbf{Z}}$ direction. $E_{exc,j}$ is the sum of the incident field and the field scattered by the other spheres, and is represented in Eq. (4) wherein the unit vector $\hat{\mathbf{r}}$ points from sphere j to sphere l [50]: in which

$$\gamma_{j,l} \equiv -\mathrm{e}^{\mathrm{i}k_b R_{j,l}}\left(\frac{\mathrm{a}_l}{R_{\mathrm{j,1}}}\right)^3\left(1 - \mathbf{i}\mathbf{k}_B R_{\mathrm{j,1}} - \mathbf{k}_B^2 R_{\mathrm{j,1}}^2\right) \tag{8.4}$$

and where the dimensionless field coupling factors $R_{n,l} = |\mathbf{R}_{n,l}| = |\mathbf{r}_n - \mathbf{r}_l|$ are distance between different particle centers, and $R_{n,e} = |\mathbf{R}_{n,\mathbf{e}}| = |\mathbf{r}_n - \mathbf{r}_\mathbf{e}|$ represents distance of particles from the emitter in the Cartesian plane. The factor $\widetilde{\alpha}_n(\omega)$ is the dimensionless polarizability function that can be simplified to:

$$\widetilde{\alpha}_n(\omega) = \frac{\alpha_n(\omega)}{4\pi R_n^3} \tag{8.5}$$

where R_n is the radius of the nth sphere and $\alpha_n(\omega)$ is particle polarizability as defined in [51]. Figure 8.2 illustrates the pentamer model structure as single-layered and located perpendicular to the YZ plane at a distance of $x = R_{0,e} = |\mathbf{R}_{0,\mathbf{e}}| = |\mathbf{r}_0 - \mathbf{r}_\mathbf{e}|$

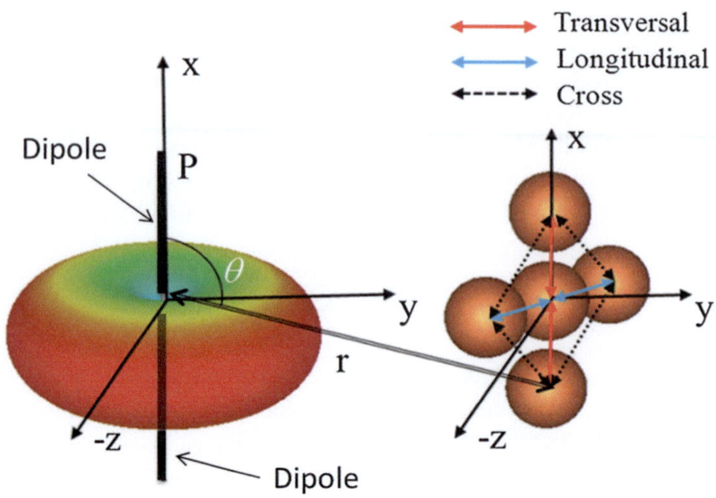

Fig. 8.2 Spatial schematic for a pentamer influenced by a dipole along the X-axis

from the dipole. The response of each particle to the local incident electric field invoked a dipole moment $\mathbf{p}(\omega) = \varepsilon_0\varepsilon_b(\omega)\alpha(\omega)\mathbf{E}_{exc}(\omega)$, where $\varepsilon_b(\omega)$ is dielectric response of metal. The dielectric response $\varepsilon_b(\omega)$ was calculated using the Drude fit D + 2CP model for gold metal as shown in Eq. (8.6) [52]:

$$\varepsilon_b(\omega) = \varepsilon_\infty - \frac{\omega_B^2}{\omega(\omega + i\delta)} + \sum_{p=1}^{2} A_p\Omega_p\left(\frac{e^{i\phi p}}{\Omega_p - \omega - i\Gamma_p} + \frac{e^{-i\phi p}}{\Omega_p + \omega + i\Gamma_p}\right) \tag{8.6}$$

where $\varepsilon_\infty = 1.1431$ is background material, $\omega_B = 1.32 \times 10^{16}(rad.s^{-1})$ is Bulk Plasmon energy, $\delta = 1.805 \times 10^{14}(rad.s^{-1})$ represents the damping ratio, $A_1 = 0.26698$, $A_2 = 3.0834$, $\phi_1 = -1.2371$, $\phi_2 = -1.0968$, and $\Omega_2 = 4.16 \times 10^{15}$. $\Omega_2 = 4.16 \times 10^{15}$. The real and imaginary parts of gold dielectric permittivity as functions of wavelength are depicted in Fig. 8.3.

In this model, the field strength of an arbitrary $\hat{z}$ -polarized incident field at each particle position (as shown in Fig. 8.2) and the excitation fields was described as $\mathbf{E}_{exc,n} = \mathrm{E}_{exc,n}.\hat{\mathbf{r}}$, which represented the local excitation field created by external sources and all other particles.

Derivation of the dimensionless coupling coefficient between the particles and dipole and particles using Mei theory in Eq. (8.7) was necessary to calculate the coupling between particles and the coupling between the dipole and particles. This derivation process involved a dimensionless coefficient a_n expressed by the Riccati-Bessel function as follows [53–55]:

$$\mathrm{a_n} = \frac{\psi_\mathrm{n}(\mathrm{x})\psi_\mathrm{n}'(\mathrm{mx}) - \mathrm{m}\psi_n'(x)\psi_\mathrm{n}(\mathrm{mx})}{\xi_n(\mathrm{x})\psi_\mathrm{n}'(\mathrm{mx}) - \mathrm{m}\xi_n'\psi_\mathrm{n}(\mathrm{mx})} \tag{8.7}$$

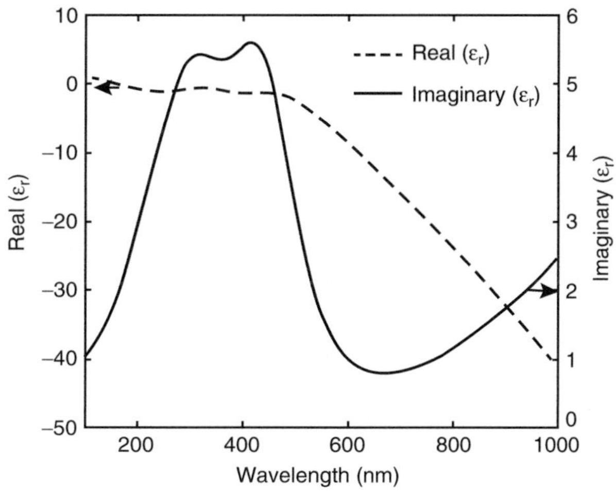

Fig. 8.3 Gold dielectric permittivity as a function of wavelength

where $\psi_n(x)=xj_n(x), \chi_n(x)=xy_n(x)$, and $\xi_n(x) = xh_n^{(1)}(x)$ are Riccati–Bessel, Riccati–Neumann and Riccati–Hankel functions respectively, $x=\mathbf{k}_B R_{n,e}$ and the prime denotes derivation with respect to the argument.. Couplings between a nanoparticle and dipole, and between nanoparticles, were expressed after consideration of particle polarizability Eq. (8.2) in conjunction with introduced dimensionless field coupling factors $\gamma_{j,l}$ and $\gamma_{j,e}$. These latter factors represented the coupling between particles *l* and *j*, and between the emitter and a particle *j* respectively. Based on the positions of the nanospheres and consideration of Eq. (8.4), the transverse, longitudinal and cross-section coupling factors were described as follows [50].

For the transverse case where $\hat{\mathbf{r}}_l \cdot \mathbf{p} = 0$ and $(\hat{\mathbf{r}}_l \times \mathbf{p}) \times \hat{\mathbf{r}}_l = \mathbf{p}$ the coupling factor was determined as:

$$\gamma_{j,l,T} \equiv -\mathrm{e}^{\mathrm{i}k_b R_{j,l}} \left(\frac{\mathrm{a_l}}{R_{j,l}}\right)^3 \left(1 - \mathrm{i}\mathbf{k}_B R_{\mathrm{j,1}} - \mathbf{k}^2{}_B R_{\mathrm{j,1}}^2\right) \tag{8.8}$$

$$\gamma_{j,e,T} \equiv -\mathrm{e}^{\mathrm{i}k_b R_{j,e}} \left(\frac{\mathrm{a_j}}{R_{j,e}}\right)^3 \left(1 - \mathrm{i}\,\mathbf{k}_B R_{\mathrm{j,e}} - \mathbf{k}^2{}_B R_{\mathrm{j,e}}^2\right) \tag{8.9}$$

For the longitudinal case where $\hat{\mathbf{r}}_l \cdot \mathbf{p} = p$ and $(\hat{\mathbf{r}}_l \times \mathbf{p}) \times \hat{\mathbf{r}}_l = \mathbf{0}$, the coupling factors were:

$$\gamma_{j,l,L} \equiv -2\mathrm{e}^{i\mathbf{k}_B R_{j,l}} \left(\frac{\mathrm{a_l}}{R_{\mathrm{j,1}}}\right)^3 \left(\mathrm{i}\mathbf{k}_B R_{j,l} - 1\right) \tag{8.10}$$

$$\gamma_{j,e,L} \equiv -2\mathrm{e}^{\mathrm{i}k_B R_{\mathrm{j,e}}} \left(\frac{\mathrm{a_j}}{R_{\mathrm{j,e}}}\right)^3 \left(\mathrm{i}\mathbf{k}_B R_{\mathrm{j,e}} - 1\right) \tag{8.11}$$

For the cross-section case where $\hat{\mathbf{r}}_l \cdot \mathbf{p} = \cos\theta$ and $(\hat{\mathbf{r}}_l \times \mathbf{p}) \times \hat{\mathbf{r}}_l = -\cos\theta$, the coupling factors were derived as:

$$\gamma_{j,l,C} \equiv -e^{ik_B R_{j,l}} \left(\frac{a_l}{R_{j,l}}\right)^3 \left(1 - i\mathbf{k}_B R_{j,l} - \mathbf{k}_B{}^2 R_{j,l}^2\right) \cos\theta \qquad (8.12)$$

$$\gamma_{j,e,C} \equiv -e^{i\mathbf{k}_B R_{j,e}} \left(\frac{a_j}{R_{j,e}}\right)^3 \left(1 - i\mathbf{k}_B R_{j,e} - \mathbf{k}_B{}^2 R_{j,e}^2\right) \cos\theta \qquad (8.13)$$

Examination of Eqs. (8.8, 8.9, 8.10, 8.11, 8.12 and 8.13) revealed that the coupling coefficient factors were strongly proportional to the inverse cubic distance between two particles. Furthermore, the form of these coupled equations indicated that the dipole moment formula $\mathbf{p}(\omega) = \varepsilon_0 \varepsilon_b \alpha(\omega) \mathbf{E}_{exc}(\omega)$ could be solved using matrix inversion. The developed model described in this paper was able to provide moment derivations for quadrumer, pentamer, hexamer and heptamer clusters with fully electrodynamically coupled dipoles [28].

8.3.2 *Superradiant and Subradiant Coupling Matrix Elements*

Calculations related to the coupling matrix elements were simplified by using values obtained from observed results of the E-field distribution [30, 31]. Figure 8.4a and e illustrate the E-field distribution in superradiant mode for a pentamer and a heptamer respectively. Figure 8.4b and c provide a comparison of E-field distribution results for a pentamer with the Fano-resonance peak at vertical and horizontal polarized angles respectively. Figure 8.4f and g depict a vertical and a 45° induced Fano resonance peak E-field from a dipole source in a heptamer cluster respectively. Consideration of external E-field distributions for pentamer and heptamer clusters led to a hypothesis that the superradiant and subradiant modes, amongst all antennae radiant modes, were sufficient to explain all observations within the wavelength range of 400–1000 nm [56]. The E-field distribution between nanospheres at superradiant and subradiant modes wavelength in pentamers and heptamers resulted in an absence of interactions between nanospheres in the outer ring of the cluster structure. Such behavior was due to the coupling factor (γ) between particles j and l (Eqs. (8.8, 8.9, 8.10, 8.11, 8.12 and 8.13) being inversely proportional with the cube of the distance between two particles. In the case of E-field distribution at the Fano resonance wavelength, the couplings among nanospheres in all E-field polarizations caused excitation in those nano-spheres along the E-field, having similar interaction with the central nano-sphere so that the effects of the other nanospheres on the generation of sub-radiant mode can be ignored. The symmetry broken pentamer and heptamer are depicted in Fig. 8.4d and h respectively. In this latter figure, at 685 nm wavelength, force is more localized in the 10 nm gap between central nanosphere 5 and topmost nanosphere 1 while it is reduced between central nanosphere 5 and rightmost nanosphere 3 due to the larger 30 nm inter-nanosphere gap. E-field distribution between the nanospheres at Fano resonance wavelength results in no interaction between the outer nanospheres 1, 2, 3 and 4, and so the springs between these spheres in the oscillator system can be neglected.

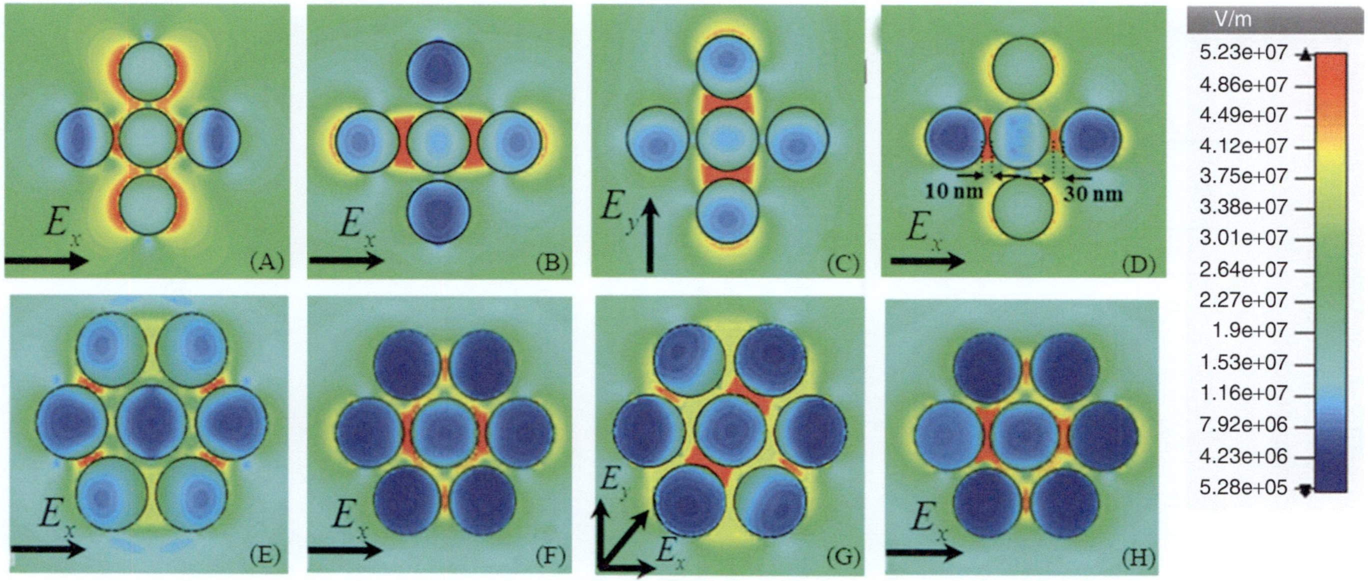

Fig. 8.4 External E-field distribution at Fano resonance wavelength. (**a**) E-field located along X-axis and the cluster plane located in XZ plane at 560 nm (superradiant mode) for pentamer. (**b**) E-field located along X-axis and the cluster plane located in XZ plane at 600 nm (Fano dip) for pentamer. (**c**) E-field located along Y-axis and the cluster plane located in XY plane at 600 nm (Fano dip). (**d**) E-field located along Y-axis and the cluster plane located in XY plane at 600 nm for symmetry break pentamer (Fano dip). (**e**) E-field located along X-axis and the cluster plane located in XZ plane at 560 nm (superradiant mode) for heptamer. (**f**) E-field located along X-axis and the cluster plane located in XZ plane at 600 nm (Fano dip) for heptamer. (**g**) E-field located along 45° and the cluster plane located in XZ plane at 600 nm (Fano dip) for heptamer. (**h**) E-field located along Y-axis and the cluster plane located in XY plane at 600 nm for symmetry-broken heptamer (Fano dip)

Consideration of the E-field distribution observed results and the stated hypotheses facilitated a simplification of the dipole analogy of the coupling matrix elements for the superradiant and subradiant modes, and this simplified calculation resulted in Eqs. (8.14) and (8.15) respectively:

$$\begin{bmatrix} \tilde{p}_1 \\ \\ \vdots \\ \tilde{p}_N \end{bmatrix} = \begin{bmatrix} \tilde{\alpha}_1^{-1} & 0 & \cdots & 0 \\ 0 & \tilde{\alpha}_2^{-1} & \cdots & 0 \\ \vdots & \vdots & \ddots & \vdots \\ 0 & 0 & \cdots & \tilde{\alpha}_N^{-1} \end{bmatrix}^{-1} \begin{bmatrix} E_{inc,1} \\ \\ \vdots \\ E_{inc,N} \end{bmatrix} \tag{8.14}$$

$$\begin{bmatrix} \tilde{p}_1 \\ \\ \vdots \\ \tilde{p}_N \end{bmatrix} = \begin{bmatrix} 0 & -\gamma_{1,2,X} & \cdots & -\gamma_{1,N,X} \\ -\gamma_{2,1,X} & 0 & \cdots & -\gamma_{2,N,X} \\ \vdots & \vdots & \ddots & \vdots \\ -\gamma_{N,1,X} & -\gamma_{N,2,X} & \cdots & 0 \end{bmatrix}^{-1} \begin{bmatrix} E_{inc,1} \\ \\ \vdots \\ E_{inc,N} \end{bmatrix} \tag{8.15}$$

where the notation X refers to the case of T (transverse), L (longitudinal) or C (cross-sectional). The matrix inversion in Eq. (8.14) is the dipole analog of the multi-pole T-matrix and incorporates both subradiant and superradiant modes. $\mathbf{E}_{inc}$ is the incident field that interacts with each particle, while N represents the number of particles in both the pentamer and the heptamer model. The interaction matrix has an inverse of α on the diagonal, while the off-diagonal elements describe the interaction between dipoles as determined by the electrodynamics coupling factor γ. For any dipole assembly, α can be inverted to find the polarization state $\tilde{p}$ induced by any interacting $\mathbf{E}_{inc}$.

Superposition of the superradiant and subradiant modes (Eqs. (8.14) and (8.15)) resulted in coupled linear equations that can be written as:

$$\begin{bmatrix} \tilde{p}_1 \\ \\ \vdots \\ \tilde{p}_N \end{bmatrix} = \begin{bmatrix} \tilde{\alpha}_1^{-1} & -\gamma_{1,2,X} & \cdots & -\gamma_{1,N,X} \\ -\gamma_{2,1} & \tilde{\alpha}_2^{-1} & \cdots & -\gamma_{2,N,X} \\ \vdots & \vdots & \ddots & \vdots \\ -\gamma_{N,1,X} & -\gamma_{N,2,X} & \cdots & \tilde{\alpha}_N^{-1} \end{bmatrix}^{-1} \begin{bmatrix} E_{inc,1} \\ \\ \vdots \\ E_{inc,N} \end{bmatrix} \tag{8.16}$$

8.3.3 Analysis of Optical Directivity Properties

Antennae are characterized by specific properties such as directivity (D_{MAX}), spectral efficiency bandwidth and radiation efficiency (ε_{rad}). The near-field coupling of quantum emitters to particle plasmon resonances provides a mechanism to redirect the light emission, while far-field radiation is completely determined by an antenna's mode [57, 58]. This redirection of emission is quantified by a directivity expression [57], while directivity properties of developed nanoantennae in the vicinity of subradiant and superradiant mode spectra were investigated in the current work. The assumption that the accumulation of all nanoparticle dipole moments within given spherical angles θ and φ can be referenced as $P(\theta, \phi)$ permitted the simplified Eq. (8.17). The P_{rad} and P_{loss} factors are indicative of radiated and absorbed powers respectively [59]. The maximum directivity of the antenna (D_{MAX}) and radiation efficiency (η_{rad}) are given by:

$$D_{MAX} = \left(\frac{4\pi}{\mathrm{P_{rad}}}\right) MAX[p(\theta, \phi)] \quad \text{(a)}, \qquad \varepsilon_{rad} = \frac{P_{rad}}{P_{rad} + P_{loss}} \quad \text{(b)} \tag{8.17}$$

Figure 8.5a and b show in 3D the directivity $D(\theta, \phi)$ of a dipolar emitter in free space when coupled with a pentamer nanoantenna oriented along the y-axis. The dipole antenna acted solely as the feed element and was included for purposes of a reference situation. The 2D results in the two major planes for the dipolar emitter in free space are depicted in Fig. 8.5c and d. Dipole-coupled results in the vicinity of a pentamer nanoantenna are demonstrated in Fig. 8.5e and f. In this scenario the emitter was placed 200 nm from the antenna element, while in contrast the angular emission of the emitter coupled to the pentamer was strongly directed along the +y-axis. The maximum directivity of the emitter was 5.2, which about twice that of the dipole antenna maximum directivity.

8.4 Transmission and Reflection Analysis

8.4.1 Superradiant Mode Analysis

To demonstrate the control of the superradiant mode wavelength, single gold nanospheres of various sizes were initially investigated in the vicinity of a dipole. Figure 8.6 illustrates the reflected spectra and radiation pattern of single nanospheres with radius size varying from 20 to 100 nm in increments of 20 nm. It is clear that the superradiant mode peak wavelength was shifted significantly to a longer wavelength with increasing particle size. The effect of particle size on the peak resonant wavelength results from two different mechanisms depending on the particle size range. For small particles with diameters of less than 10 nm, known as the quasi-static regime, the effects of phase retardation and multiple modes can be neglected. While Mie theory gives a constant resonant frequency independent of particle size if the bulk dielectric constant is used, the size effect in the quasi-static regime comes from the dependence of the permittivity on particle size because of quantum confinements. For larger particles with diameters greater than 10 nm, such as in the investigated examples, the quantum confinement effect or size dependence of dielectric coefficients becomes negligible, and the aspect of phase retardation effects comes into play. The peak red shift observed in our experiments was a fully electrodynamic effect due to phase retardation. In order to have better insight regarding red shift due to phase retardation, electric fields elongated along the

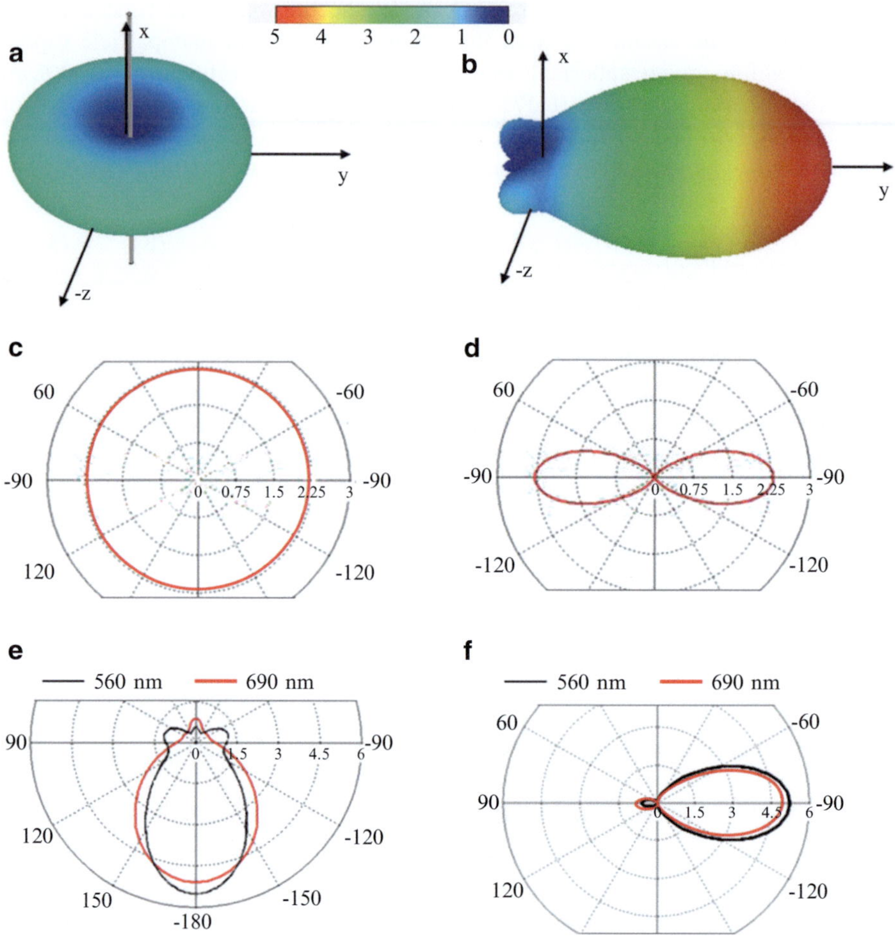

Fig. 8.5 Directivity for a y-oriented dipolar emitter in (**a**) free space and (**b**) coupled to the pentamer antenna, (**c**) and (**d**) the directivity in the two major planes for dipole emitter. (**e**) and (**f**) is the directivity in the two major planes for dipole emitter coupled to the pentamer antenna. The schematics show the location of the emitter and the orientation of the emitter and pentamer antenna relative to the image plane. The distance between the emitter and the antenna cluster is 200 nm. The emission of the emitter coupled to the pentamer antenna is highly directed

direction of the dipole moment for different nanospheres size were considered, and are shown in the inset of Fig. 8.6. Phase retardation in plasmonic nanospheres lead changes to electric polarizability. The 500, 550, and 600 nm wavelengths elevation ($\Phi = 0$) and azimuth ($\Phi = 90$) radiation patterns for different nanosphere sizes are demonstrated in the right-hand side of Fig. 8.6. As can be seen from Fig. 8.6(C.3) and (D.3) for the elevation plane and 6(C.4) and 6(D.4) for the azimuth plane, high radiation is observed at wavelength ranges from 500 to 600 nm for sphere radius 60 and 80 nm. Thus, a sphere with radius from 60 to 80 nm could be considered as an optimized candidate for an optical antenna in these ranges.

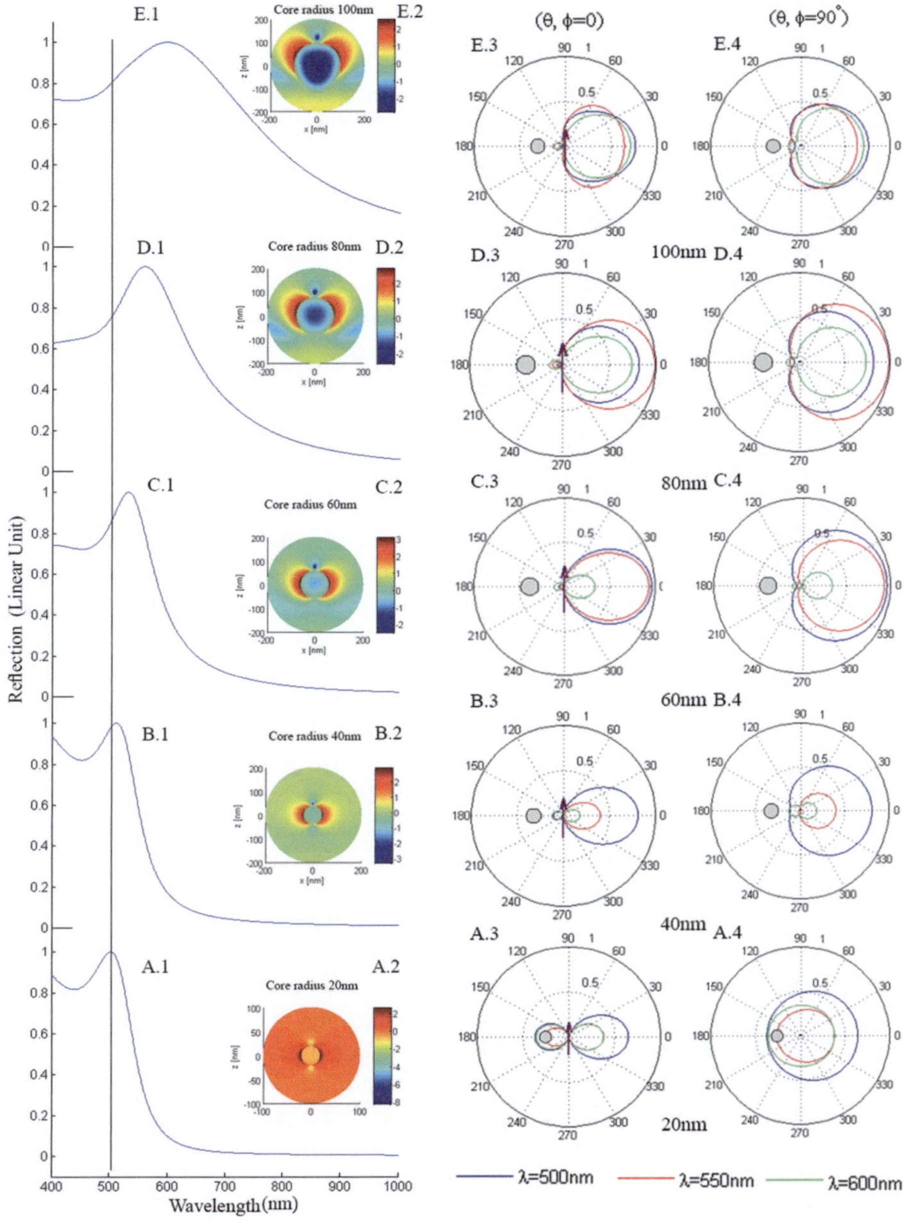

Fig. 8.6 (*Left side*) Monomer reflection of (A.1) 20 nm sphere radius (B.1) 40 nm sphere radius (C.1) 60 nm sphere radius (D.1) 80 nm sphere radius (E.1) 100 nm sphere radius (*left-side* figure inset) Monomer E-field distribution of (A.2) 20 nm sphere radius (B.2) 40 nm sphere radius (C.2) 60 nm sphere radius (D.2) 80 nm sphere radius (E.2) 100 nm sphere radius (*right side*) elevation ($\Phi = 0$) radiation pattern of (A.3) 20 nm sphere radius (B.3) 40 nm sphere radius (C.3) 60 nm sphere radius (D.3) 80 nm sphere radius (E.3) 100 nm sphere radius (*right side*) azimuth ($\Phi = 90$) radiation pattern of (A.4) 20 nm sphere radius (B.4) 40 nm sphere radius (C.4) 60 nm sphere radius (D.4) 80 nm sphere radius (E.4) 100 nm sphere radius

8.4.2 Fano Resonance Analysis

The origin and polarization-independent nature of Fano resonance has been previously reported [60], wherein the authors provided formal and rigorous definitions of the distinct electric and magnetic eigenmodes present in their investigated systems. Verification of the Fano resonance model introduced here was aided by the transmission and reflection profiles of a dipole in the vicinity of pentamer and heptamer with spectra as plotted in Figs. 8.7(A.1) and (B.1) respectively. Analogously to nanoshell symmetry breaking [31], the signature of the subradiant mode became more apparent when the distance between two nanospheres in a pentamer or heptamer was reduced to 20 ~ 35 nm. The calculated subradiant mode reflection (Eq. (8.15)) and superradiant mode reflection (Eq. (8.14)) are demonstrated by the green dash-dotted and blue dotted line respectively. The superposition of the calculated radiated field at the subradiant and superradiant modes (Eq. (8.16)) was responsible for the resulting transmission and reflection spectra. Reflection and transmission profiles due to dipole radiation are shown by the red dashed line and solid black line respectively, for which sphere radii was set at 70 nm. The existence of Fano resonance in a pentamer or heptamer also led to the formation of the subradiant mode, and this mode could be observed from reflection spectra in an asymmetric line shape at around 600 nm in Figs. 8.7(A.1) and (B.1) for pentamer and heptamer respectively. As detailed in the previous section, coupling between subradiant and superradiant modes contributed to phase shifting across a 0 to π range, and Fano resonance appeared when the combined phase shift was equal to π. Characteristics of the Fano resonances in the pentamer are explained by examination of the electric field distributions at the respective spectral positions as shown in Figs. 8.7(A.2) and (A.3). All of the nanoparticles oscillated in phase as a result of the charge distributions in the nanoparticles at the superradiant mode, as observed by reference to Fig. 8.7(A.2). The E-field distribution pattern at the Fano resonance wavelength of 600 nm is plotted in Fig. 8.7(A.3). Phase shifts of spheres 1, 3, and 5 with respect to spheres 2 and 4 led to the formation of the superradiant and subradiant modes [20]. As in the case with pentamers, heptamers having a superradiant mode experienced the oscillating plasmon in the seven nanoparticles as in-phase, and exhibited significant mode broadening due to radiative damping. The trend of the transmission and reflection profiles in Fig. 8.7(B.1) for the developed model agreed well with those of an earlier report [60].

In order to analyze the behavior of Fano resonance, the curves of phase change versus the E-field wavelength were established, and are shown in Fig. 8.7(A.4) for quadrumers and pentamers, and in Fig. 8.7(B.4) for hexamers and heptamers. For a quadrumer or hexamer, the phase shifting across a 0 to π range that contributed to the external electric field was a $-\pi$ phase before the resonance, then a 0 phase after the resonance [46]. As shown by the red dashed line in Fig. 8.7(A.4) and (B.4) when the subradiant and superradiant modes were being invoked near the Fano resonance dip, a phase shift of π appeared at the Fano resonance wavelength. This phenomenon was a consequence of the subradiant mode canceling the polarization of the

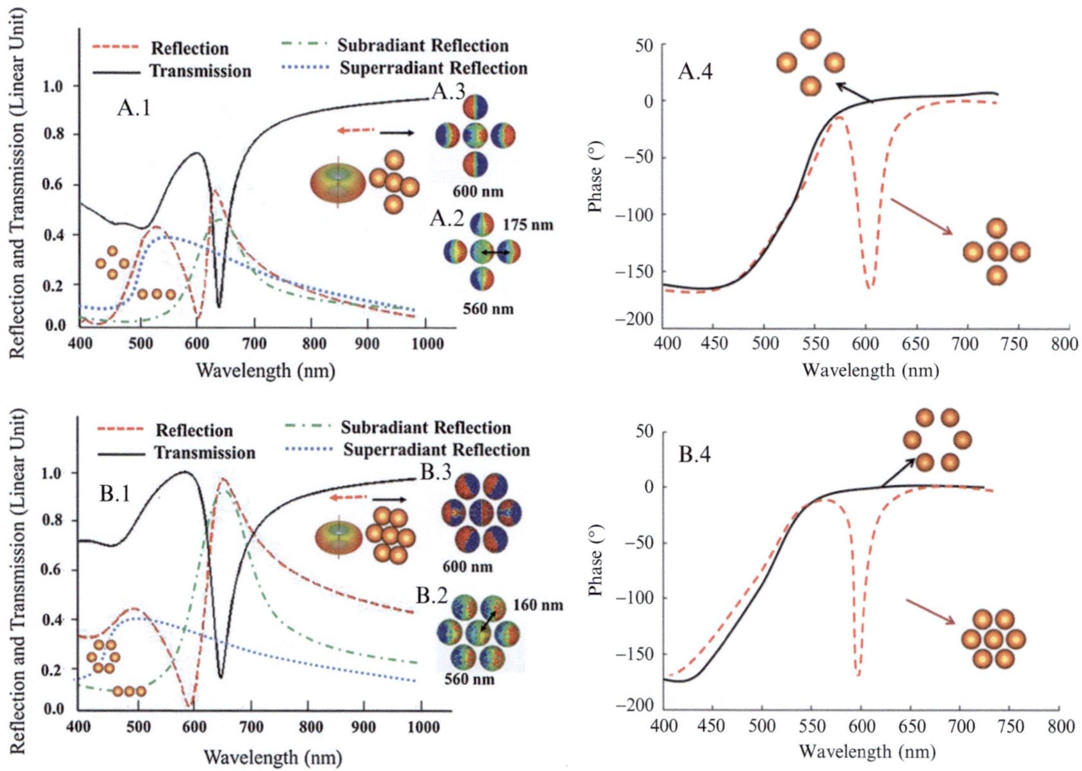

Fig. 8.7 (A.1) Reflection and transmission spectrum for a pentamer (A.2) E-field distribution pattern at 600 nm for a pentamer (Fano dip) (A.3) E-field distribution pattern at 560 nm (superradiant mode) for a pentamer (A.4) Phase spectrum for a quadrumer and a pentamer (B.1) Reflection and transmission spectrum for a heptamer (B.2) E-field distribution pattern at 600 nm for a heptamer (Fano dip) (B.3) E-field distribution pattern at 560 nm (superradiant mode) for a heptamer (B.4) Phase spectrum for a heptamer and an hexamer

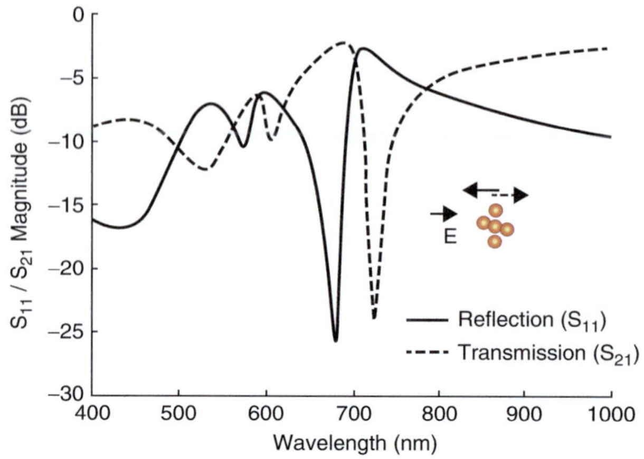

Fig. 8.8 Magnitude transmission and reflection spectra of a symmetry-broken pentamer

superradiant mode, and resulted in a narrow window of transparency in the transmission spectrum [45].

In order to see the effect of geometrical symmetry breaking on the optical properties of a pentamer, the distance between nanospheres 1 and 5 was decreased to 10 nm with a simultaneous increase of 30 nm in distance between nanospheres 3 and 5. This alteration caused the appearance of the second Fano resonance in the reflection spectrum. This phenomenon resulted from an increased local E-field strength between nanospheres 1 and 5 (Fig. 8.8).

8.5 Element Optimization

8.5.1 Directivity Analysis

Determining the optimum directivity of the subradiant mode in a pentamer nanoantenna required attention on three variables, these being nanoparticle radius, distance between the dipole resonator and nano-particles, and gap between each nanoparticle in the model structure as shown in Fig. 8.9. The resonance wavelength was set at maximum directivity of the subradiant mode. Figure 8.9a illustrates the variations of directivity at the subradiant mode when the nanoparticle radii was varied from 50 to 100 nm and gap between two adjacent nanoparticles changed from 120 to 220 nm, while the distance from the dipole remained constant at 200 nm. The highest directivity could be observed at 5.8 dBi, which corresponded to a distance of 160 nm between nanoparticles of 70 nm radii. As the pentamers moved apart from each other (i.e. the outer ring became larger), the coupling factor between the nanospheres decreased drastically. Figure 8.9b shows the effects of varying nanoparticle radius when the pentamer was repositioned away from the dipole and gap distance between the nanoparticles remained at 20 nm. Observation

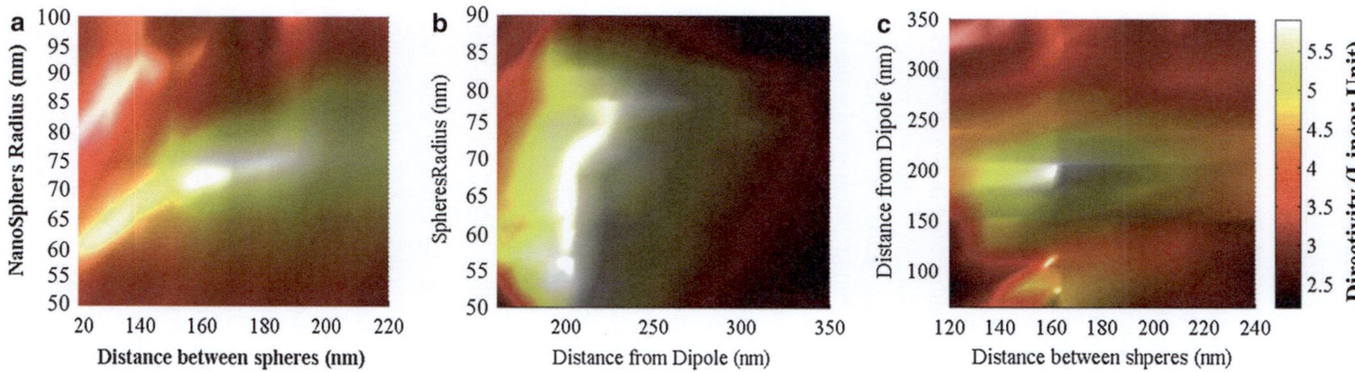

Fig. 8.9 Directivity of the pentamer nanoantenna (**a**) nanosphere radius (nm) versus space between nanospheres (nm) (**b**) nanosphere radius (nm) versus distance form dipole (nm) (**c**) distance from dipole (nm) versus space between nanospheres (nm)

of Fig. 8.9b confirms that directivity at the changes in subradiant mode was rather subtle when nanoparticle size was varied within a 50–75 nm range. This phenomenon could be explained by consideration of the coupling factor (γ), in which the subradiant mode was seen as less affected by a variation in sphere radius than variation of the gap distance.

Figure 8.9c shows the change of directivity at subradiant mode for distance between the nanospheres when varied from 120 to 240 nm and for the pentamer when moved away from the dipole by 100–350 nm. As expected from Eqs. (8.8) to (8.13), the effect of the subradiant mode became less effective as the nanoparticles became increasingly separated from each other. The effect of subradiant mode vanished when the nanosphere gap distance was fixed while the dipole distance from nanoparticles increased; the situation was a consequence of the lower local intensity of the optical state. This phenomenon resulted in the directivity decreasing when the pentamer was moved away from the dipole. Consideration of Eqs. (8.14) and (8.15) along with the results illustrated in Fig. 8.9 led to a conclusion: changes of nanoparticles radii within a reasonable range was not as effective in comparison to variation of gap between the nanoparticles. The most effective dipole distance from the nanoparticles was measured at 200 nm while the optimum size was 70 nm and distance of nanoparticles in a state of subradiant directivity was 160 nm. Based on Fig. 8.9b and c when the dipole distance is lower than 130 nm (hot spot), the effect of subradiant mode on directivity vanished and the cluster functioned as a nanosphere. This outcome at a short dipole distance from the cluster is due to the effect of the dipole E-field at the center sphere polarization being stronger than at the outer spheres, and led to a low coupling effect between nanospheres and lower subradiant mode at those regions.

8.5.2 Bandwidth Analysis

The S11 parameter, i.e. the ratio of the reflected wave to the incident wave as a function of the operation frequency, was investigated for dipole optical nanoantennas in order to find the frequency bandwidth of each model. The frequency bandwidth for pentamer nanoantennas was calculated for conditions of $S_{11} < -10$ dB while varying three parameters within reasonable ranges. These parameters consisted of the gap between the nanoparticles, the distance of the cluster from the dipole, and the radii of the involved nanoparticles. Figure 8.10a–c illustrate the variations of bandwidth (THz) when these aforementioned parameters were varied with respect to each other. Figure 8.10a shows the bandwidth variation of the antenna when the pentamer was moved away from the dipole and nanoparticle radius was increased from 50 to 90 nm. Nanosphere radius was observed to have a minor effect on the bandwidth of the antenna, which is in line with the directivity results. The highest peak of bandwidth was achieved at a 200 nm dipole distance for different nanosphere sizes. Figure 8.10b shows the bandwidth of pentamer structures when distance from the dipole and the gap between nanospheres were varied from 100 to 350 nm and 10 to 100 nm respectively. The

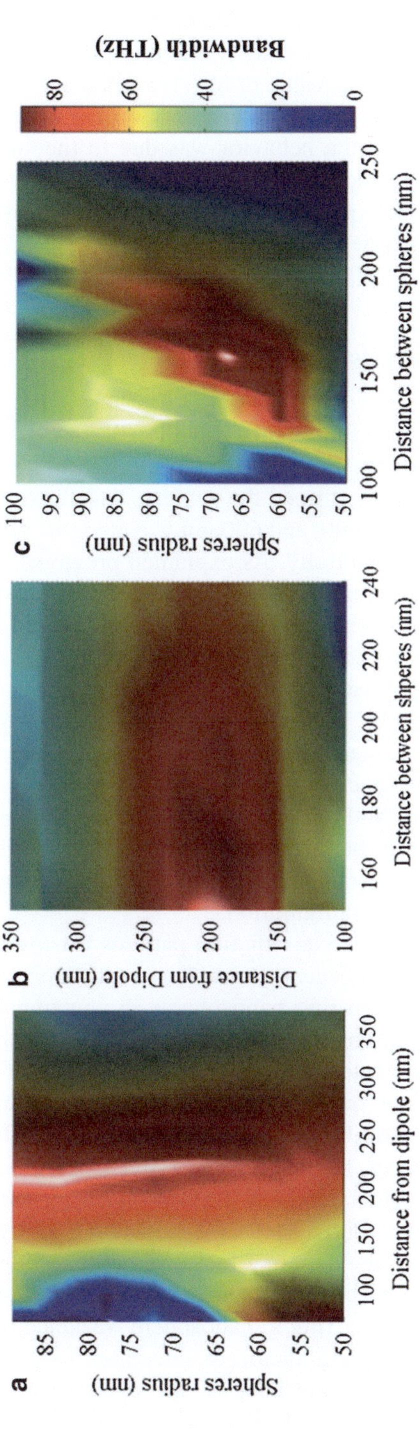

Fig. 8.10 Bandwidth of the pentamer nanoantenna (**a**) nanosphere radius (nm) versus space between nanospheres (nm) (**b**) nanosphere radius (nm) versus distance form dipole (nm) (**c**) distance from dipole (nm) versus space between nanospheres (nm)

radii of these nanoparticles were fixed at 70 nm. In agreement with Fig. 8.10a, the highest peak of bandwidth was achieved at a 200 nm dipole distance. Bandwidth could be observed at a maximum value of 80 THz when nanoparticle gap distance was 165 nm. Bandwidth dramatically decreased as the gap distance of nanoparticles increased above 175 nm, and the behavior was due to the smaller effects of the subradiant mode at higher nanosphere gap distances. The highest bandwidth was obtained when the nanosphere gap distances and sphere radii were 170 and 70 nm respectively, and this case is depicted in Fig. 8.10c.

8.6 Directivity Analysis

The maximum directivity as a function of wavelength for a quadrumer is depicted in Fig. 8.11a and for a pentamer structure with four different nanosphere gap values in Fig. 8.11b–e. Such results confirmed the analysis described prior to this section. Emission patterns at 490, 560, 600, and 690 nm wavelengths were established with the results of this adaptive directivity. As explained previously, the geometry of a configured array of nanostructures strongly influenced the transmission and reflection properties, which subsequently resulted in the directivity of each antenna being related to the nanosphere size and gap distances between the nanospheres in each model's structure. A well-separated configuration of nanospheres resulted in the structure behaving similarly to monomers, with spectra peaks of a broad superradiant mode being centered within the vicinity of 560 nm wavelength as seen in Fig. 8.11(A.3). The spectral range of calculation did not indicate directivity resonance at the vicinity of 560 nm (Fig. 8.11(A.2)) and 690 nm (Fig. 8.11(A.4)), and this situation was due to the absence of a subradiant mode in the quadrumer dipole. The optimized sphere radius of 70 and 200 nm dipole distance from the nano-particles was set in this configuration.

Pentamer directivity and relative radiation patterns for four different distances between the centers of two adjacent nanospheres are depicted in Fig. 8.11b–e. Reducing the distance between ring nanospheres in pentamers and heptamers to 20 ~ 60 nm caused the coupling factor and charge distributions to increase and invoke a subradiant mode centered within the vicinity of 690 nm. The pentamer models, in comparison to the quadrumer model, have the directivity at subradiant spectra increased significantly by optimizing the distances between two adjacent nanospheres. As can be seen from Fig. 8.11b and c, the subradiant mode was dominated by the superradiant mode at 160 and 175 nm distance between the outer ring and centered core (Fig. 8.11(B.2)–(B.5)) and Fig. 8.11(C.2)–(C.5)), and was due to higher values of the coupling factor rather than polarizability function $\alpha_n(\omega)$. This conclusion was reached by examination of Eqs. (8.4) and (8.5) respectively. Increasing the outside ring's nanosphere distance from the centered nanospheres caused the coupling factor to decrease, with the coupling factor being inversely proportional to the cube of the distance between nanospheres. This phenomenon led to a dramatic decrease of the superradiant mode at a

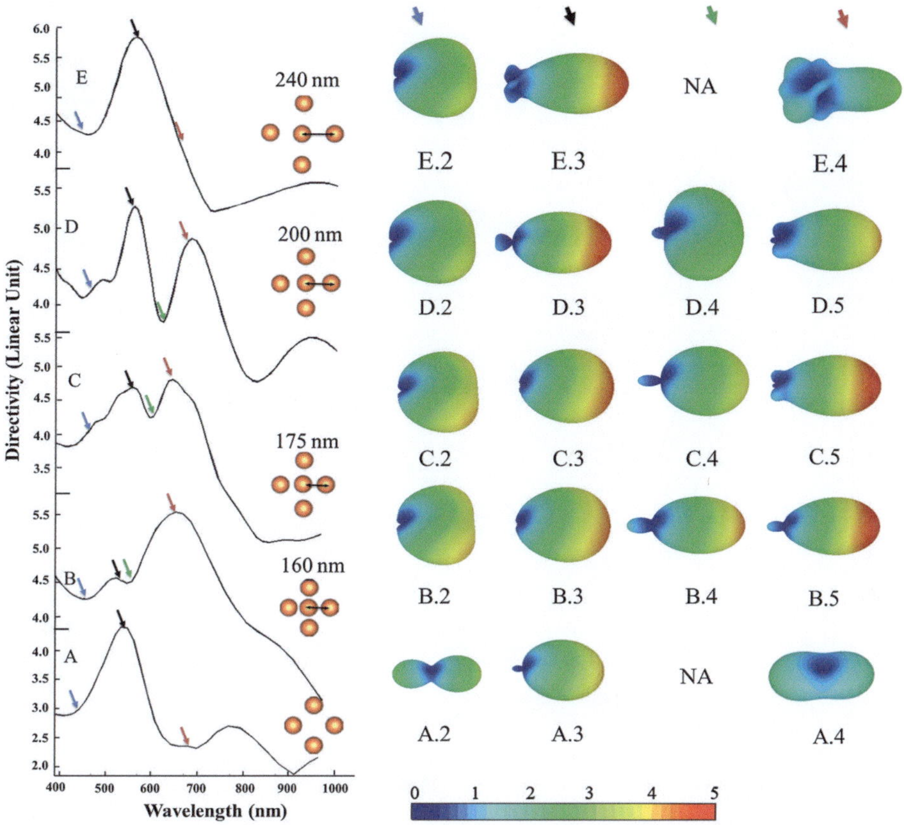

Fig. 8.11 (*Left side*) directivity of (**a**) quadrumer and (**b**) pentamer with 160 nm gap space from center of each nanosphere (**c**) pentamer with 175 nm gap space from center of each nanosphere (**d**) pentamer with 160 nm gap space from center of each nanosphere (**e**) pentamer with 175 nm gap space from center of each nanosphere (*right side*) radiation pattern of (A.1) monomer at 490 nm (A.2) monomer at 560 nm (A.3) monomer at 690 nm (B.1) pentamer (160 nm) at 490 nm (B.2) pentamer (160 nm) at 560 nm (B.3) pentamer (160 nm) at 690 nm (C.1) pentamer (175 nm) at 490 nm (C.2) pentamer (175 nm) at 560 nm (C.3) pentamer (175 nm) at 690 nm (D.1) pentamer (160 nm) at 490 nm (D.2) pentamer (160 nm) at 560 nm (D.3) pentamer (160 nm) at 690 nm (E.1) heptamer (175 nm) at 490 nm (E.2) pentamer (175 nm) at 560 nm (E.3) pentamer (175 nm) at 690 nm

200 nm distance between the central nanosphere and outer rings, as depicted in Fig. 8.11d and Fig. 8.11(D.2)–(D.5). Effects from a superradiant mode vanished beyond a 240 nm distance between central and outer ring nanospheres, and in such a case the cluster could be regarded as an array of monomers (Fig. 8.11e and (E.2)–(E.5)). The destructive effect of Fano resonance is demonstrated in Fig. 8.11(B.4), (C.4) and (D.4). A low coupling factor at a 200 nm distance between central nanospheres led to a dramatic increase of the Fano dip and week emission patterns, as depicted in Fig. 8.11(D.4).

The same explanation could be extended to hexamer and heptamer structures as shown in Fig. 8.12. The maximum directivity of hexamer (sub-figure A) and

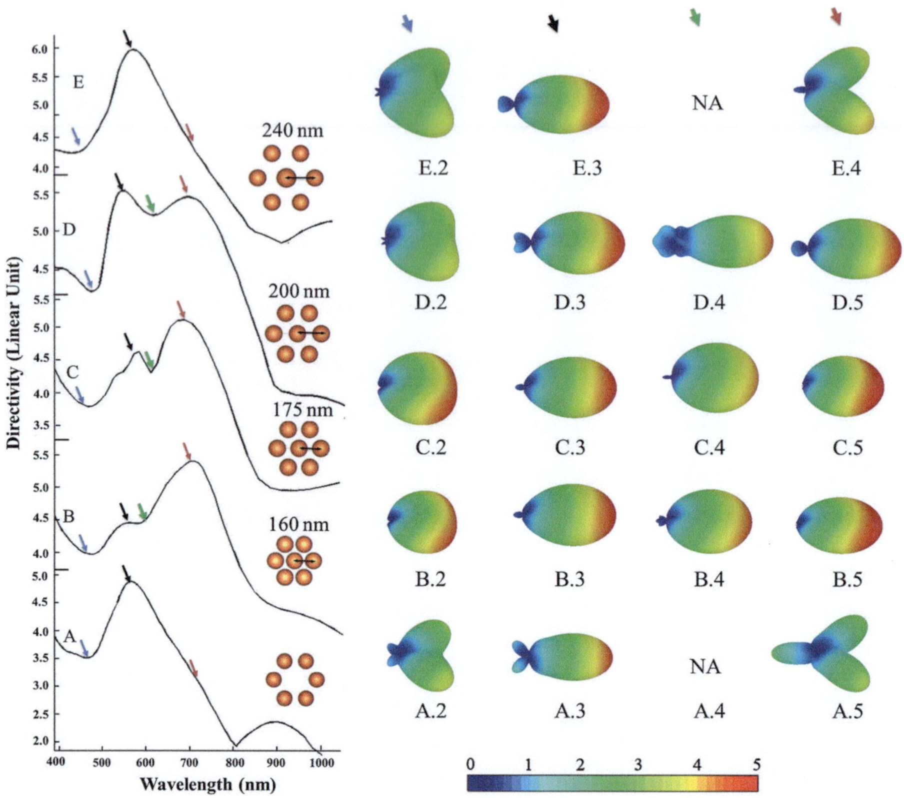

Fig. 8.12 (*Left side*) directivity of (**a**) hexamer (**b**) heptamer with 160 nm gap space from center of each nanospher (**c**) heptamer with 175 nm gap space from center of each nanosphere (**d**) heptamer with 160 nm gap space from center of each nanosphere (**e**) heptamer with 175 nm gap space from center of each nanosphere (*right-side*) radiation pattern of (A.1) hexamer at 490 nm (A.2) hexamer at 560 nm (A.3) hexamer at 690 nm (B.1) heptamer (160 nm) at 490 nm (B.2) heptamer (160 nm) at 560 nm (B.3) heptamer (160 nm) at 690 nm (C.1) heptamer (175 nm) at 490 nm (C.2) heptamer (175 nm) at 560 nm (C.3) heptamer (175 nm) at 690 nm (D.1) heptamer (160 nm) at 490 nm (D.2) heptamer (160 nm) at 560 nm (D.3) heptamer (160 nm) at 690 nm (E.1) heptamer (175 nm) at 490 nm (E.2) heptamer (175 nm) at 560 nm (E.3) heptamer (175 nm) at 690 nm

heptamer structures with four different geometry structure values as a function of wavelength is depicted in Fig. 8.12 sub-figures B, C, D and E. Similarly to the quadrumer case, a well-separated configuration of nanospheres in an hexamer resulted in the structure behaving similarly to monomers with spectra peaks of a broad superradiant mode (Fig. 8.12(A.2)–(A.4)). Reducing the distance between ring nanospheres to 20 ~ 60 nm increased both the coupling factor and the charge distributions, and furthermore invoked and increased a sub-radiant mode centered within the vicinity of 690 nm (Fig. 8.12b–d). As can be seen from Fig. 8.12b and c, the subradiant mode was dominated by the superradiant mode at 160 and 175 nm distance between the outer ring and centered core. An increase in the outside ring's nanosphere distance from the centered nanospheres caused a dramatic decrease of the superradiant mode at a 200 nm distance between the central nanosphere and outer rings, as depicted in Fig. 8.12d. The effect of a superradiant mode vanished beyond a 240 nm distance between central and outer ring nanospheres, and in such a case the cluster could be regarded as an array of monomers (Fig. 8.12e and (E.2)–(E.4)). The destructive effect of Fano resonance at a 600 nm wavelength (a Fano resonance dip) is demonstrated in Fig. 8.12(B.4), (C.4) and (D.4).

By comparison to published numerical results regarding the plane wave effect on pentamer and heptamer clusters, there is a minor change on Fano dip wavelength and superradiant mode peak wavelength observed in the dipole-coupled model. Based on the authors' founding, increasing the outer nanospheres distance from center nanosphere, causes the dipole polarization effect on the outer nanospheres to decrease accordingly. This results in a different Fano resonance wavelength whereby the π phase shift related to subradiant and superradiant modes is achieved.

8.7 Mass Spring Model

A mass spring model provides an explanation of the optical responses of Fano resonance in plasmonic nanoparticles. Masses represent the superradiant and subradiant modes while a spring represents the coupling between them. Figure 8.13 illustrates the basic model of Fano resonance in a pentamer plasmonic material by using the analogy of masses connected to bases by springs with a fractional constant of γ. The spring connecting the masses together has an oscillation coupling of Ω [46]. This model is an extension of the classical two-oscillator system used to model electromagnetic induced transparency. Oscillator $|1\rangle$ is driven by a periodic harmonic force $(F(t) = Fe^{i\omega t})$ and is analogous to optical excitation of the superradiant mode.

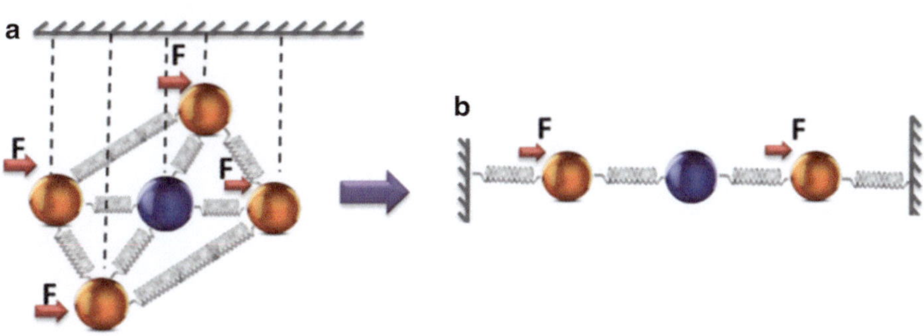

Fig. 8.13 Mass spring analogy of Fano resonance (**a**) Five coupled interacting oscillators representing the pentamer optical responses (**b**) simplified three coupled oscillators

The mass spring model of a pentamer can be simplified to a three-coupled interacting oscillators system as shown in the Fig. 8.13 sequence. A number of pentamer simplifications are applied on the basis of the following observed results [30, 31], and these considerations allow the motion equations of the pentamer to be written as:

$$\ddot{x}_1(t) + \gamma_1\dot{x}_2(t) + \omega_1^2 x_2(t) - \Omega_1^2 x_1(t) = 0 \tag{8.18}$$

$$\ddot{x}_2(t) + \gamma_2\dot{x}_2(t) + \omega_2^2 x_2(t) - \Omega_2^2 x_1(t) = ae^{i\omega t} \tag{8.19}$$

$$\ddot{x}_3(t)^3 + \gamma_3\dot{x}_3(t) + \omega_3^2 x_3(t) - \Omega_3^2 x_1(t) = ae^{i\omega t} \tag{8.20}$$

where ω_1, ω_2 and ω_3 correspond to 1, 2, and 3 oscillator frequencies respectively that are defined as subradiant and superradiant mode wavelengths. Ω represents coherent coupling frequencies between interconnected oscillators. γ_1, γ_2 and γ_3 are friction coefficients relating to the energy dissipation between nanospheres for |1>, |2> and |3> respectively. The motion equations (Eqs. (8.18), (8.19) and (8.20)) can be solved by a steady-state function $x_i(t)$ whereby the term $ae^{-i\omega t}$ is incorporated into these equations. In order to show the accuracy of the mass spring model, a calculation of the normalized extinction power versus wavelength of the oscillator system was performed and subsequently compared with CST Microwave Studio results [30, 31, 44]. A Plasmon superradiant mode was modeled with $\omega_{12} = 2.2$ eV as representative of the resonance frequencies of the 1, 2, 3 and 4 nanospheres, and the resonance frequency $\omega_{13} = 1.87$ eV represented the central nanosphere frequency. Substitute coupling between modes ($\Omega^2 = 0.85$) and other parameters ($\gamma_1 = 2.1$, $\gamma_2 = 0.43$, $\gamma_3 = 0.52$ and $a = 1.4$) allowed the normalized absorption power versus energy of the oscillator system to be calculated and compared with CST Microwave Studio result, as shown in Fig. 8.14.

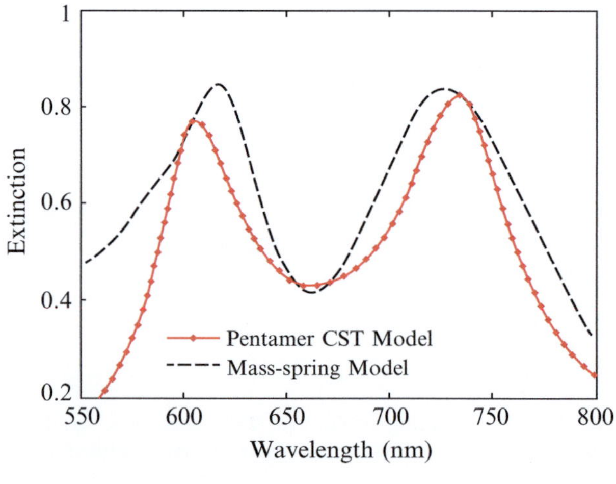

Fig. 8.14 Extinction power versus wavelength

8.8 Circuit Model

8.8.1 *Plasmonic Nanosphere Circuit Model*

Several circuit models of plasmonic nanoparticles as circuit elements have been recently proposed [61–69]. A complete paradigm and theoretical frame work for the quantities description regarding the complex electromagnetic response of nanospheres is described by Nader Engheta's group [61, 62, 67, 70]. This theory is based on the observation that the small size of the particle in comparison to the wavelength leads to lumped-impedance representations under the quasi-static approximation. Such a method stipulates the wavelength of light must be larger than the size of the particle; thus the method only works for very small particles lower than 15 nm radii whereupon phase changes are considered negligible. A quasi-static limitation on the nanoparticle size has led to a proposal of a time-varying approach to circuit modeling of plasmonic nanospheres based on radial considerations [69]. This time-varying approach uses vector wave functions to derive accurate resonance frequency and impedance expressions for the metallic nanospheres at and around the plasmon resonance. In comparison to the quasi-static model, this proposed approach can more accurately predict the dependence of the resonance frequency on nanoparticle size, and yield more precise expressions for the equivalent capacitance and inductance lumped elements. The derived circuit model, with the constituent equivalent internal and external impedances and resonance frequency for a nanosphere, is demonstrated in Fig. 8.15:

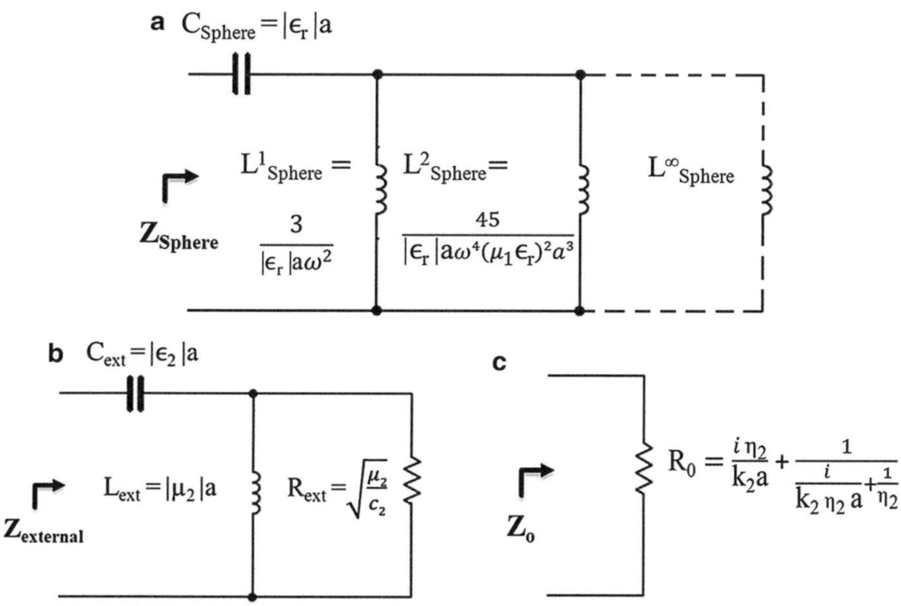

Fig. 8.15 Plasmonic nanosphere circuit model. (**a**) Internal circuit diagram (**b**) External circuit diagram impedance (**c**) Source impedance

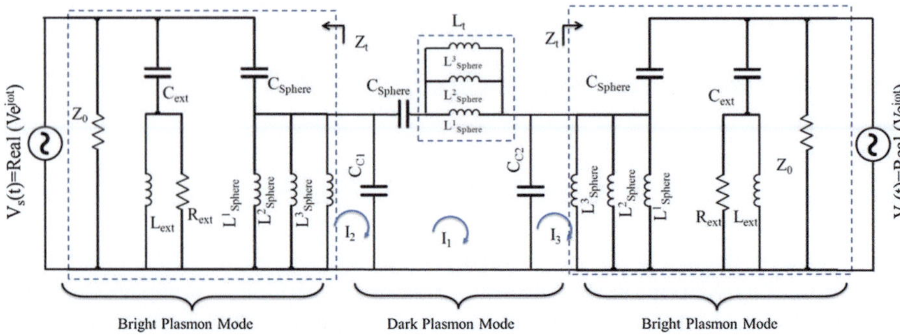

Fig. 8.16 Circuit diagram of pentamer

In these circuit diagrams, the value of *(a)* is spherical radius, which is characterized by a material of complex permittivity $\varepsilon_r(\omega)$ and embedded in a dielectric medium. Parameters $k_1 = \sqrt{\varepsilon_r(\omega)\mu_1}$, $k_2 = \sqrt{\varepsilon_2\mu_2}$ and $\eta_2 = \sqrt{\mu_2/\varepsilon_2}$. ε_2 and μ_2 are the material constants of the external medium and possess positive real values.

8.8.2 Fano Resonance Circuit Model

Different circuit diagrams for Fano resonance effects are generated by analogy to quantum EIT in [45, 63]; the phenomenon is based on the Stark effect in a DC-field, and constitutes a design in the form of 2 electric circuits coupled by constant mutual inductance k as developed via [71]. An analogy of single and double EIT three-level quantum atoms circuit using three coupled RLC circuits is discussed in [72]. Designing the accurate circuit model requires calculation of the equivalent superradiant RLC circuits coupled by a so-called coupling capacitor to the subradiant mode circuit. An equivalent circuit diagram for a pentamer is demonstrated in Fig. 8.16, with response of the left and right loops being analogous to the superradiant plasmon mode. The basis for this analogy is broadness due to the resistive loss, and a direct connection to the excitation source. The middle circuit diagram is analogous to the dark plasmonic mode, since no radiative loss occurs and there is no direct connection to the excitation source.

The two equal impedance Z_S refer to nanospheres with 73 nm diameter, in parallel with external impedance (Z_{ext}) and internal impedance (Z_{sphere}), are shown in the bright plasmon mode circuit diagram. The external impedance consisted of an external capacitor, $C_{ext} = 70$ aF in series with parallel external inductor $L_{ext} = 10$ pH representing a large cross sectional area of a large sphere, and free space impedance $R_{ext} = 376.8\ \Omega$. The sphere internal components consisted of internal sphere capacitor series with three parallel inductors: L^1 sphere, L^2 sphere and L^3 sphere. The electromagnetic field incident on the designs was modeled as a time harmonic voltage source, $V_s(t) = \mathrm{Re}\{Ve^{j\omega t}\}$, which represented the dipolar mode that was excited by the incident E-field. The dark plasmon mode circuit model consists of coupled capacitances C_{C1} and C_{C2} that corresponded to a coupling between dark and bright plasmon modes. The coupled capacitance coefficients two or more conductors are hypothesized by the John Lekner model [73], whereupon capacitance coefficients between spheres of radii a and b are defined as:

$$-C_c = \frac{ab}{c}\sinh u \sum_{n=1}^{\infty} [\sinh nu]^{-1} \tag{8.21}$$

where c = a + b + s, in which s is distance between surface of spheres and u is an auxiliary dimensionless positive parameter defined by:

$$\cosh u = \frac{c^2 - a^2 - b^2}{2ab} \tag{8.22}$$

When $a = b$ this parameter u is:

$$u \approx 2\left(\frac{s}{a}\right)^{1/2} \tag{8.23}$$

The equivalent capacitance (assuming other loops are open) is labeled as C_{eq} and given by:

$$\frac{1}{C_{eq_1}} = \frac{1}{C_S} + \frac{1}{C_{C_1}} \tag{8.24}$$

$$\frac{1}{C_{eq_2}} = \frac{1}{C_S} + \frac{1}{C_{C_2}} \tag{8.25}$$

Figure 8.17 shows the 3D graph of the equal capacitance between two-closed conduct spheres (C_{eq1} and C_{eq2}), where sphere radii *(a)* and *(b)* varies from 60 to 100 nm and spheres surface distance s varies from 10 to 50 nm. As expected from published works [73], distance *s* has a greater effect on capacitance than spherical radius. Increasing distance results in higher capacitance and hence a lower subradiant mode rather than superradiant mode.

Applying the Kirchhoff law for voltage to each of the loops in the direction of the arrows as shown in Fig. 8.16 yielded a coupled system of equations:

$$\frac{1}{j\omega C_S}I_1 + j\omega L_1 I_1 + \frac{1}{j\omega C_{C_1}}I_1 - \frac{1}{j\omega C_{C_1}}I_2 + \frac{1}{j\omega C_{C_2}}I_1 - \frac{1}{j\omega C_{C_2}}I_3 = 0 \tag{8.26}$$

$$Z_T I_2 + \frac{1}{j\omega C_{C_1}}I_2 - \frac{1}{j\omega C_{C_1}}I_1 = V \tag{8.27}$$

$$Z_T I_3 + \frac{1}{j\omega C_{C_2}}I_3 - \frac{1}{j\omega C_{C_2}}I_1 = V \tag{8.28}$$

where I_i represents the current on loop i, with i = 1, 2 or 3.

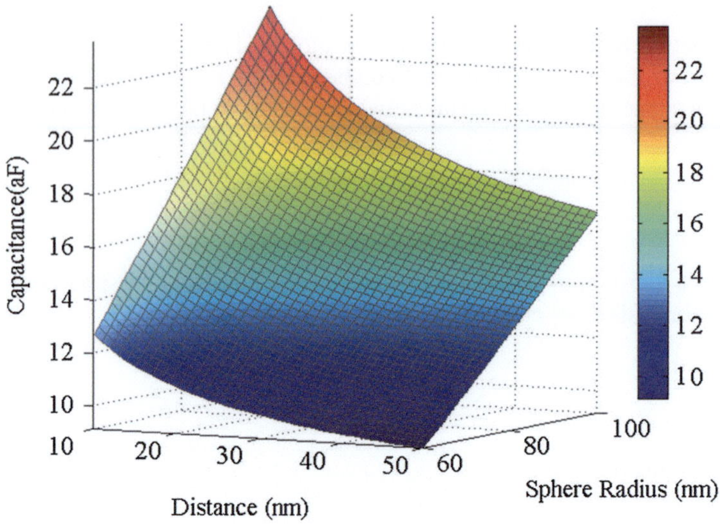

Fig. 8.17 Variations of coupling capacitance for variations of both distance and spherical radius

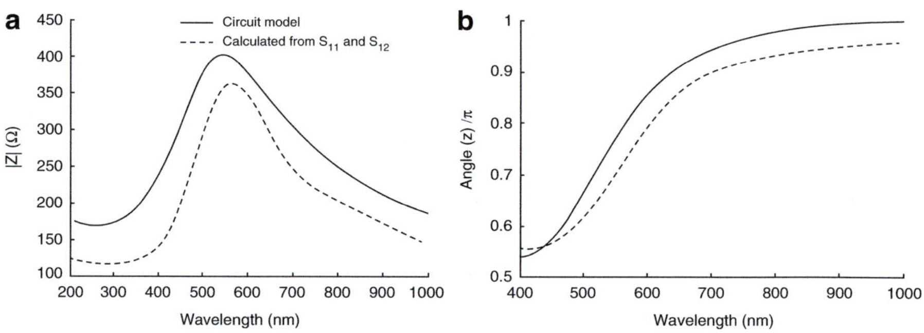

Fig. 8.18 Tetramer impedance spectra (**a**) Real value of Z (**b**) Imaginary value of Z

Accuracy of the proposed model was evaluated by comparing these results to the calculated impedance from S_{11} and S_{21} parameters. The circuit model impedance is presented alongside calculated impedance from reflection and transmission of tetramer for real and imaginary magnitudes in Fig. 8.18a and b respectively. Significant separation of nanospheres caused the value of C_{c1} to increase to 17 aF and behave as a short circuit. The effect of the coupling on the Fano resonance was represented solely as a broad superradiant mode resonance peak, as presented in Fig. 8.18a.

Figure 8.17 indicates that reducing the distance between two nanospheres in pentamer down to 20 nm causes the symmetry condition ($Ceq_1 = Ceq_2$), coupling capacitors to decrease to 10 aF. The superradiant mode circuits were coupled to the subradiant mode loop, and a pronounced Fano dip appeared in the impedance spectra due to destructive interference between the two resonant loops. In order

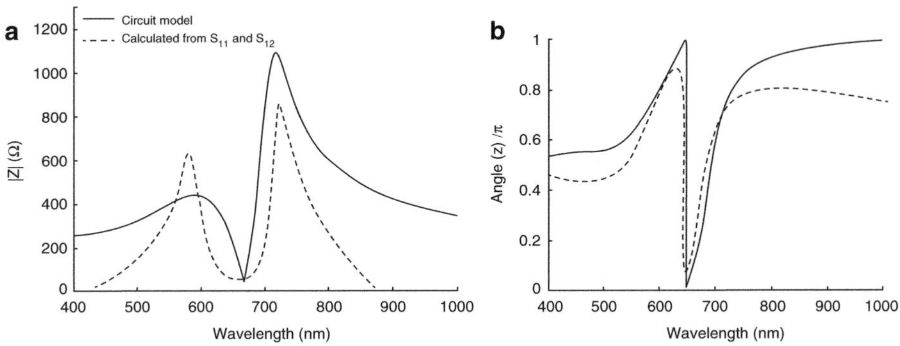

Fig. 8.19 Pentamer impedance spectra Z. (**a**) Magnitude value of Z (**b**) Phase value of Z

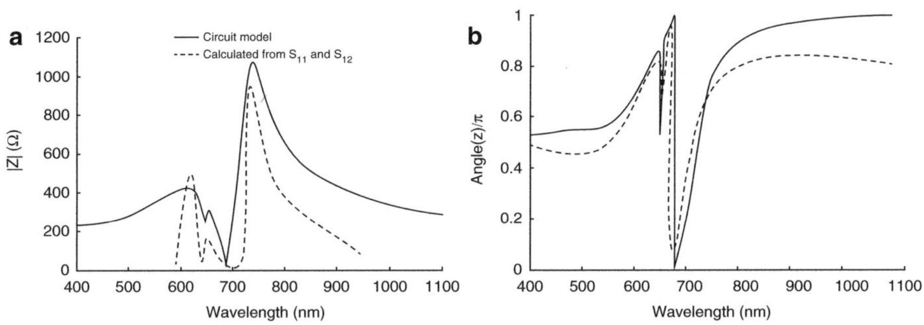

Fig. 8.20 Symmetry-broken pentamer impedance spectra Z. (**a**) Magnitude value of Z (**b**) Phase value of Z

to find the optimum distance and radius, the genetic algorithm Matlab toolkit was used. Figure 8.19a and b show the pentamer impedance magnitude and phase value respectively. The optimum radius of spheres was determined as 73 nm, with a 20 nm distance between the spheres.

In order to see the effect of geometrical symmetry breaking on the optical properties of a pentamer, the distance between nanospheres 1 and 5 was decreased to 10 nm along with a simultaneous increase of 30 nm in distance between nanospheres 3 and 5. This offset led to different equivalent capacitor of 10 and 12 aF for Ceq_1 and Ceq_2 respectively. The magnitude and phase functions of impedance for a symmetry-broken pentamer are depicted in Fig. 8.20a and b respectively. As expected from the S_{11} and S_{12} magnitude spectrum, the second Fano resonance in the same reflection spectrum appeared when an offset was applied to the center nanosphere. Approximations on S parameters measurements and impedance conversion were mainly responsible for the discrepancies between circuit model and S parameter calculated values.

The calculated capacitances from S_{11} and S_{21} parameters are depicted in Fig. 8.21, with the equivalent capacitance for tetramers, pentamers and

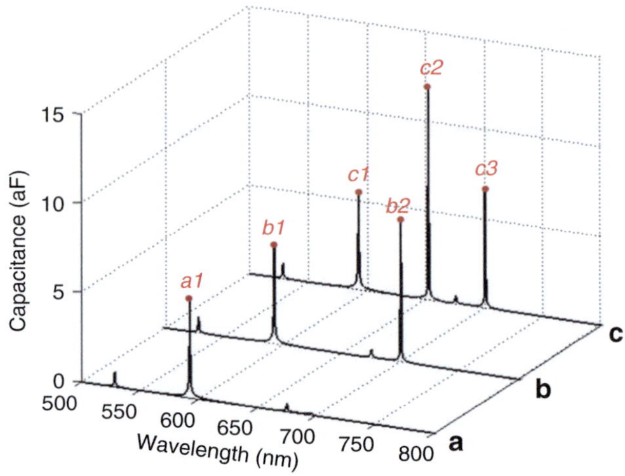

Fig. 8.21 Calculated capacitances of (**a**) tetramer (**b**) pentamer (**c**) symmetry-broken pentamer

symmetry-broken pentamers demonstrated in curves **a, b** and **c** respectively. Curve **a** concerned a single $\mathbf{a_1}$ capacitor which was related to sphere impedance. Curve **b** was in complete agreement with the calculated *Ceq* from Eq. (8.4), wherein a sphere capacitance at 600 nm and a 10 aF capacitance at Fano dip of 680 nm emerged in the pentamer. The symmetry breaking produced a third capacitor as shown by $\mathbf{c_2}$ in curve **c**. This offset led to different equivalent capacitors of 12 and 10 aF at 650 and 700 nm respectively. As can be observed, these results were in total agreement with the circuit model presented in Fig. 8.8.

8.9 Conclusion

This paper describes a detailed analysis of the various aspects of the directivity properties of arrays of metamaterial or plasmonic particles, with a specific focus on pentamer and heptamer nanoparticle models. In-depth analysis was performed on the transmission, reflection, and directive properties of pentamer and heptamer structures, these being key properties for nanoantenna functionality. The performance properties of analytically-designed nanoantennas were numerically simulated using Matlab. Results for the pentamer modeling described in this paper showed directivity was at a peak when the gap between the nanospheres was 30 nm and the space between the pentamer and dipole was 200 nm, whilst bandwidth was observed to be at a peak when the gap between nanospheres was 35 nm and the space between the pentamer and dipole was 200 nm. The reflection and transmission properties of the pentamer were improved with respect to other models by placing the nanospheres 30 nm apart and the cluster 200 nm from the emitter dipole. These findings are potentially valuable in considerations of functionality for far-field plasmonic gold nanoparticle array configurations, especially in the context of light radiation patterns, emission, reflection and control of arbitrary ranges for directivity and bandwidth. Space variation between nanoparticles in pentamers and heptamers allowed for the radiation pattern governed by overlap coupling between two involved modes to be optimized, so as to ensure an enhanced LDOS phenomenon by constructive interference. The parameter of distance from dipole emitter was utilized to optimize the Fano resonance and interaction intensity for far-field plasmonic gold molecules. Simultaneous adjustment and optimization of a model's geometrical parameters was required for the model to achieve best performance. The mass spring model was used as a basis for the presentation of the circuit diagram of a pentamer with coupled capacitance in order to couple the subradiant and superradiant modes. The effect of symmetry breaking, invoked by offsetting a central nanosphere from a nearest neighbor nanosphere, for a pentamer was investigated by means of accurate circuit modeling. Results showed such symmetry breaking led to the spectral appearance of a second Fano resonance. Circuit model impedance results for tetramers were in good agreement with the calculated impedance from S parameters and mass spring model. The approach described in this article for modeling the numerical and analytical representation of pentamer and heptamer nanoantennae array configurations resulted in a strong agreement between the numerical and analytical aspects, along with good alignment with existing literature.

References

1. Anker JN, Hall WP, Lyandres O, Shah NC, Zhao J, Van Duyne RP (2008) Biosensing with plasmonic nanosensors. Nat Mater 7:442–453
2. Ono A, Kato J, Kawata S (2005) Subwavelength optical imaging through a metallic nanorod array. Phys Rev Lett 95:267407
3. Garcia-Vidal FJ, Martin-Moreno L, Pendry JB (2005) Surfaces with holes in them: new plasmonic metamaterials. J Opt A Pure Appl Opt 7:S97–S101
4. Ha T, Enderle T, Ogletree DF, Chemla DS, Selvin PR, Weiss S (1996) Probing the interaction between two single molecules: fluorescence resonance energy transfer between a single donor and a single acceptor. Proc Natl Acad Sci U S A 93:6264–6268
5. Cesca T, Calvelli P, Battaglin G, Mazzoldi P, Mattei G (2012) Local-field enhancement effect on the nonlinear optical response of gold-silver nanoplanets. Opt Express 20:4537–4547
6. Dregely D, Taubert R, Dorfmuller J, Vogelgesang R, Kern K, Giessen H (2011) 3D optical Yagi-Uda nanoantenna array. Nat Commun 2:267
7. Hutter E, Fendler JH (2004) Exploitation of localized surface plasmon resonance. Adv Mater 16:1685–1706
8. Sonnefraud Y, Verellen N, Sobhani H, Vandenbosch GAE, Moshchalkov VV, Van Dorpe P et al (2010) Experimental realization of subradiant, superradiant, and Fano resonances in ring/disk plasmonic nanocavities. ACS Nano 4:1664–1670
9. Christ A, Martin OJF, Ekinci Y, Gippius NA, Tikhodeev SG (2008) Symmetry breaking in a plasmonic metamaterial at optical wavelength. Nano Lett 8:2171–2175
10. Hao F, Sonnefraud Y, Van Dorpe P, Maier SA, Halas NJ, Nordlander P (2008) Symmetry breaking in plasmonic nanocavities: subradiant LSPR sensing and a tunable Fano resonance. Nano Lett 8:3983–3988

11. Tetz KA, Pang L, Fainman Y (2006) High-resolution surface plasmon resonance sensor based on linewidth-optimized nanohole array transmittance. Opt Lett 31:1528–1530
12. Chua SL, Chong Y, Stone AD, Soljacic M, Bravo-Abad J (2011) Low-threshold lasing action in photonic crystal slabs enabled by Fano resonances. Opt Express 19:1539–1562
13. Samson ZL, MacDonald KF, De Angelis F, Gholipour B, Knight K, Huang CC et al (2010) Metamaterial electro-optic switch of nanoscale thickness. Appl Phys Lett 96:143105
14. Kim J, Kim J-R, Lee J-O, Park JW, So HM, Kim N et al (2003) Fano resonance in crossed carbon nanotubes. Phys Rev Lett 90:166403
15. Papasimakis N, Fedotov VA, Zheludev NI, Prosvirnin SL (2008) Metamaterial analog of electromagnetically induced transparency. Phys Rev Lett 101:253903
16. Zhang S, Genov DA, Wang Y, Liu M, Zhang X (2008) Plasmon-induced transparency in metamaterials. Phys Rev Lett 101:047401
17. Chiam SY, Singh R, Rockstuhl C, Lederer F, Zhang WL, Bettiol AA (2009) Analogue of electromagnetically induced transparency in a terahertz metamaterial. Phys Rev B 80:153103
18. Luk'yanchuk B, Zheludev NI, Maier SA, Halas NJ, Nordlander P, Giessen H et al (2010) The Fano resonance in plasmonic nanostructures and metamaterials. Nat Mater 9:707–715
19. Fan JA, Bao K, Wu C, Bao J, Bardhan R, Halas NJ et al (2010) Fano-like interference in self-assembled plasmonic quadrumer clusters. Nano Lett 10:4680–4685
20. Lassiter JB, Sobhani H, Knight MW, Mielczarek WS, Nordlander P, Halas NJ (2012) Designing and deconstructing the Fano lineshape in plasmonic nanoclusters. Nano Lett 12:1058–1062
21. Abdumalikov AA Jr, Astafiev O, Zagoskin AM, Pashkin YA, Nakamura Y, Tsai JS (2010) Electromagnetically induced transparency on a single artificial atom. Phys Rev Lett 104:193601
22. Boller KJ, Imamolu A, Harris SE (1991) Observation of electromagnetically induced transparency. Phys Rev Lett 66:2593–2596
23. Prodan E, Radloff C, Halas NJ, Nordlander P (2003) A hybridization model for the plasmon response of complex nanostructures. Science 302:419–422
24. Fang Z, Cai J, Yan Z, Nordlander P, Halas NJ, Zhu X (2011) Removing a wedge from a metallic nanodisk reveals a Fano resonance. Nano Lett 11:4475–4479
25. Zhang S, Bao K, Halas NJ, Xu H, Nordlander P (2011) Substrate-induced Fano resonances of a plasmonic nanocube: a route to increased-sensitivity localized surface plasmon resonance sensors revealed. Nano Lett 11:1657–1663
26. Solis D Jr, Willingham B, Nauert SL, Slaughter LS, Olson J, Swanglap P et al (2012) Electromagnetic energy transport in nanoparticle chains via dark plasmon modes. Nano Lett 12:1349–1353
27. Lassiter JB, Sobhani H, Fan JA, Kundu J, Capasso F, Nordlander P et al (2010) Fano resonances in plasmonic nanoclusters: geometrical and chemical tunability. Nano Lett 10:3184–3189
28. Frimmer M, Coenen T, Koenderink AF (2012) Signature of a Fano resonance in a plasmonic metamolecule's local density of optical states. Phys Rev Lett 108:077404
29. Rahmani M, Tahmasebi T, Lin Y, Lukiyanchuk B, Liew TY, Hong MH (2011) Influence of plasmon destructive interferences on optical properties of gold planar quadrumers. Nanotechnology 22:245204
30. Rahmani M, Lukiyanchuk B, Ng B, Tavakkoli KGA, Liew YF, Hong MH (2011) Generation of pronounced Fano resonances and tuning of subwavelength spatial light distribution in plasmonic pentamers. Opt Express 19:4949–4956
31. Rahmani M, Lukiyanchuk B, Nguyen TTV, Tahmasebi T, Lin Y, Liew TYF et al (2011) Influence of symmetry breaking in pentamers on Fano resonance and near-field energy localization. Opt Mater Express 1:1409–1415
32. Bao K, Mirin NA, Nordlander P (2010) Fano resonances in planar silver nanosphere clusters. Appl Phys A Mater Sci Process 100:333–339
33. Emami SD, Soltanian MRK, Attaran A, Abdul-Rashid HA, Penny R, Moghavvemi M et al (2015) Application of Fano resonance effects in optical antennas formed by regular clusters of nanospheres. Appl Phys A Mater Sci Process 118:139–150

34. Bao K, Mirin N, Nordlander P (2010) Fano resonances in planar silver nanosphere clusters. Appl Phys A 100:333–339
35. Mukherjee S, Sobhani H, Lassiter JB, Bardhan R, Nordlander P, Halas NJ (2010) Fanoshells: nanoparticles with built-in Fano resonances. Nano Lett 10:2694–2701
36. Hao F, Nordlander P, Sonnefraud Y, Dorpe PV, Maier SA (2009) Tunability of subradiant dipolar and Fano-type plasmon resonances in metallic ring/disk cavities: implications for nanoscale optical sensing. ACS Nano 3:643–652
37. Liu H, Wang N, Liu Y, Zhao Y, Wu X (2011) Light transmission properties of double-overlapped annular apertures. Opt Lett 36:385–387
38. Singh R, Al-Naib IAI, Koch M, Zhang W (2011) Sharp Fano resonances in THz metamaterials. Opt Express 19:6312–6319
39. Dong Z-G, Liu H, Xu M-X, Li T, Wang S-M, Cao J-X et al (2010) Role of asymmetric environment on the dark mode excitation in metamaterial analogue of electromagnetically-induced transparency. Opt Express 18:22412–22417
40. Verellen N, Sonnefraud Y, Sobhani H, Hao F, Moshchalkov VV, Van Dorpe P et al (2009) Fano resonances in individual coherent plasmonic nanocavities. Nano Lett 9:1663–1667
41. Hu Y, Noelck SJ, Drezek RA (2010) Symmetry breaking in gold-silica-gold multilayer nanoshells. ACS Nano 4:1521–1528
42. Tuniz A, Lwin R, Argyros A, Fleming SC, Pogson EM, Constable E et al (2011) Stacked-and-drawn metamaterials with magnetic resonances in the terahertz range. Opt Express 19:16480–16490
43. Attaran A, Emami SD, Soltanian MRK, Penny R, Behbahani F, Harun SW et al (2014) Circuit model of Fano resonance on tetramers, pentamers, and broken symmetry pentamers. Plasmonics 9:1303–1313
44. Emami SD, Ahmad H, Harun SW, Mirnia SE, Soltanian MRK, Rashid HAA (2012) Fano resonance on plasmonic nanostructures. In: 2012 I.E. 3rd international conference on photonics (ICP), Kuala lumpur. pp 144–148
45. Litvak AG, Tokman MD (2002) Electromagnetically induced transparency in ensembles of classical oscillators. Phys Rev Lett 88:095003
46. Joe YS, Satanin AM, Kim CS (2006) Classical analogy of Fano resonances. Phys Scr 74:259–266
47. Campione S, Steshenko S, Albani M, Capolino F (2011) Characterization of the optical modes in 3D-periodic arrays of metallic nanospheres. In: General assembly and scientific symposium, 2011 XXXth URSI, Istanbul. pp 1–4
48. Aden AL, Kerker M (1951) Scattering of electromagnetic waves from two concentric spheres. J Appl Phys 22:1242–1246
49. Jackson J (1998) Classical electrodynamics, vol 3. Wiley, New York
50. Stout ADB, Rolly B, Bonod N (2011) Multipole methods for nano-antennas design: applications to Yagi-Uda configurations. J Opt Soc Am B 28:1213–1223
51. Evlyukhin A, Reinhardt C, Seidel A, Luk'yanchuk B, Chichkov B (2010) Optical response features of Si-nanoparticle arrays. Phys Rev B 82:045404
52. Vial A, Laroche T (2007) Description of dispersion properties of metals by means of the critical points model and application to the study of resonant structures using the FDTD method. J Phys D Appl Phys 40:7152–7158
53. Bohren CF, Huffman DR (2007) Absorption and scattering of light by small particles. Wiley-VCH Verlag GmbH, United States. pp 1–11
54. Quinten M (2011) Mie's theory for single spherical particles. In: Optical properties of nanoparticle systems. Wiley-VCH Verlag GmbH & Co. KGaA, Weinheim. pp 75–122
55. Quinten M (2011) Beyond Mie's theory II – the generalized Mie theory. In: Optical properties of nanoparticle systems. Wiley-VCH Verlag GmbH & Co. KGaA, Weinheim. pp 317–339
56. Taminiau TH, Stefani FD, van Hulst NF (2011) Optical nanorod antennas modeled as cavities for dipolar emitters: evolution of sub- and super-radiant modes. Nano Lett 11:1020–1024

57. Taminiau T, Stefani F, Segerink F, Van Hulst N (2008) Optical antennas direct single-molecule emission. Nat Photonics 2:234–237
58. Pakizeh T, Käll M (2009) Unidirectional ultracompact optical nanoantennas. Nano Lett 9:2343–2349
59. Krasnok AE, Miroshnichenko AE, Belov PA, Kivshar YS (2012) All-dielectric optical nanoantennas. Opt Express 20:20599–20604
60. Hopkins B, Poddubny AN, Miroshnichenko AE, Kivshar YS (2013) Revisiting the physics of Fano resonances for nanoparticle oligomers. Phys Rev A 88:053819
61. Engheta N, Salandrino A, Alù A (2005) Circuit elements at optical frequencies: nanoinductors, nanocapacitors, and nanoresistors. Phys Rev Lett 95:095504
62. Engheta N (2007) Circuits with light at nanoscales: optical nanocircuits inspired by metamaterials. Science 317:1698–1702
63. Harden J, Joshi A, Serna JD (2011) Demonstration of double EIT using coupled harmonic oscillators and RLC circuits. Eur J Phys 32:541–558
64. Song K, Mazumder P (2009) An equivalent circuit modeling of an equispaced metallic nanoparticles (MNPs) plasmon wire. IEEE Trans Nanotechnol 8:412–418
65. Engheta N, Salandrino A, Alù A (2011) Erratum: circuit elements at optical frequencies: nanoinductors, nanocapacitors, and nanoresistors [Phys Rev Lett 95:095504 (2005)]. Phys Rev Lett 106:089901
66. Zareie HM, Morgan SW, Moghaddam M, Maaroof AI, Cortie MB, Phillips MR (2008) Nanocapacitive circuit elements. ACS Nano 2:1615–1619
67. Alu A, Engheta N (2011) Optical metamaterials based on optical nanocircuits. Proc IEEE 99:1669–1681
68. Alam M, Massoud Y (2006) RLC ladder model for scattering in single metallic nanoparticles. IEEE Trans Nanotechnol 5:491–498
69. Alam M, Massoud Y, Eleftheriades GV (2011) A time-varying approach to circuit modeling of plasmonic nanospheres using radial vector wave functions. IEEE Trans Microwave Theory Tech 59:2595–2611
70. Alu A, Engheta N (2008) Tuning the scattering response of optical nanoantennas with nanocircuit loads. Nat Photonics 2:307–310
71. Lamb WE Jr, Retherford RC (1951) Fine structure of the hydrogen atom. Part II. Phys Rev 81:222–232
72. Amin M, Bağci H (2012) Investigation of Fano resonances induced by higher order plasmon modes on a circular nano-disk with an elongated cavity. Prog Electromagn Res 130:187–206
73. Lekner J (2011) Capacitance coefficients of two spheres. J Electrost 69:11–14

Chapter 9
Elongated Nanostructured Solar Cells with a Plasmonic Core

Marcel Di Vece

Abstract In this chapter the effects of the plasmonic response in an elongated nano-scale solar cell with a silver nanoneedle core are explored by measuring photocurrents. The silver nanoneedles formed the support of a conformally grown hydrogenated amorphous silicon (a-Si:H) n-i-p junction around it. A spherical morphology of the solar cell functions as a nano-lens, focusing incoming light directly on the plasmonic silver nanoneedle. We found that plasmonics, geometric optics, and Fresnel reflections affect the nanostructured solar cell performance, depending strongly on light incidence angle and polarization. Besides the plasmonic effects, nano-focusing, and orthogonalization of carrier and photon pathways are simultaneously present at illumination of this structure. In this chapter the photovoltaics characterization techniques and simulations are explained and discussed as well. This work provides valuable insight in solar cell processes in which novel concepts such as plasmonics, elongated nanostructures, and nano-lenses are used.

Keywords Nanostructures • Photovoltaics • Nanoneedles • Thin films • a-Si:H • Finite difference time domain simulations • Nanolensing • Light orthogonalisation

9.1 Introduction

Light management forms a crucial component in high efficiency solar cells. Textured surfaces at the front or back contact of a solar cell that cause increased light trapping yield important improvement of the over-all efficiency. Plasmonics, the electronic response of light interacting with metal nanostructures, is a promising novel concept in the quest for photovoltaic devices with high efficiencies with low cost materials [1, 2]. Interest in plasmonics, which merges optics with electronics [3, 4] at very small scales, has increased considerably since about a decade. The potential of plasmonics in photovoltaics is recognized and there has been a

M. Di Vece (✉)
Nanophotonics–Physics of Devices, Debye Institute for Nanomaterials Science, Utrecht University, P.O. Box 80000, 3508 TA Utrecht, The Netherlands
e-mail: m.divece@uu.nl

C.D. Geddes (ed.), *Reviews in Plasmonics 2015*,
DOI 10.1007/978-3-319-24606-2_9

considerable research effort in this field [1–4]. Five main effects are distinguished with the metal plasmon and optically active neighbouring materials [5–14]: **(1) Plasmonic field enhancement** [6, 7, 11, 12], (a near-field effect) increases the electromagnetic field close to the metal surface by orders of magnitude. This is the result of dipole radiation of the metal nanoparticle in response to illumination. This is particularly interesting for thin film solar cells as light concentration has a significant effect on the overall absorption. **(2) Scattering** (a far field-effect) at/by the metal nanoparticle [15] increases the light path length in a solar cell and is therefore an important mechanism to increase efficiency. Incident light is scattered off the object into a distribution of optical modes within the semiconductor. **(3) Coupling** of the plasmon resonance to waveguide modes in thin semiconductor slabs. **(4) The coupling of an emitter to plasmon modes** also affects both the radiative and non-radiative decay rates. This phenomenon is based on the principle that the strength with which an emitter couples with an electromagnetic field depends on its environment [16–18]. **(5) Energy transfer** from the absorber/emitter to plasmon modes of a nearby nanoparticle can occur by for example Forster Energy Transfer (FRET) [19] or vice versa. This energy will be partially dissipated or coupled to radiation [20, 21]. The balance between these effects depends strongly on the geometry, shape, size and distance [22]. One of the first experimental investigations of plasmonic enhancement for photovoltaics involved silver nanoparticles in an organic solar cell [23, 24]. An absorption increase of 10–15 % was found to be feasible by using such nanoparticles [25]. In the system (ITO/metal-clusters/CuPc/In) it has been shown that the incorporation of copper or gold clusters increases the photocurrent by a factor of more than two. The first papers to explore the potential of localized surface plasmon resonance (LSPR) in solid state photovoltaics were published in the late 1990s by Stuart and Hall [14]. Their device consisted of a thin silicon-on-insulator wafer, where the Si was 165 nm thick, and since the absorbing Si layer was separated from the Si substrate by a layer of SiO_2, it acts as a thin waveguide. Metal island formation was achieved by depositing a thin metallic film and annealing under N_2 so that the film coalesced into discrete islands. A recent work by Barnard et al. [26] shows the enhancing effect of silver nanostructures on photocurrent in silicon. Underneath a metal nanoantenna the field intensity is enhanced by a factor of 2.8 while locally the field enhancement can be significantly higher (up to 20) in "hotspots" under the corners of the antenna. A theoretical study of nanoparticles for solar cells shows that an absorption increase of 10–15 % is easily feasible [25].

In nano-engineered plasmonic back reflectors that scatter strongly, light is coupled into guided modes of an ultra-thin a-Si:H solar cell [27] which results in an increase of the external quantum efficiency (EQE) between 600 and 800 nm of about 8 %. The spatial order of such patterned nanostructures has a strong effect on its plasmonic performance [28].

With systematically designed pseudorandom arrays of nanostructures based on their power spectral density, the spatial frequencies are correlated with measured and simulated photocurrent. Integrated cell designs consisting of patterned plasmonic back reflectors and a nanostructured semiconductor top interface give broadband and isotropic photocurrent enhancement. An example of a viable nanostructured three-dimensional solar cells employing an ultrathin hydrogenated amorphous silicon a-Si:H n-i-p junction uses a zinc oxide nanorod arrays as substrate [29]. The ZnO nanorods were prepared by aqueous chemical growth at 80 °C. The photovoltaic performance of the nanorod/a-Si:H solar cell with an ultrathin absorber layer of only 25 nm yields an efficiency of 3.6 % and a short-circuit current density of 8.3 mA/cm^2, significantly higher than values achieved for planar or even textured counterparts with three times thicker absorber layers.

Because metals have electric resistance, the electron movement of a plasmon experiences such resistance by which optical energy can be converted into heat. A study of these Ohmic losses in various materials shows that silver has relatively low loss compared to metals such as aluminum and copper [30].

The promising approach in photovoltaics by using elongated nano-structured solar cells has the advantage of enhanced charge collection by short charge carrier paths and increased light absorption due to orthogonalization of light and carrier paths [31]. This is the category of nanocoax [32], nanocone [33, 34], nanodome [35], nanopillar [36, 37], nanorod [29, 38–41], and nanowire [42–47] solar cells. Employing hydrogenated amorphous silicon (a-Si:H) as light absorbing material, an efficiency of up to ~9 % [18] has been reported for such nanostructured solar cells.

To obtain the best of both worlds, the elongated solar cell can be combined with a plasmonic feature. A plasmonic core inside the elongated solar cell could provide effects such as described above to enhance the absorption efficiency. Here we discuss an optical nano-antenna in the form of a silver nanoneedle which is embedded in an a-Si:H solar cell [48]. This is achieved by remarkably conformal growth using hot wire chemical vapor deposition (HWCVD) [49]. A semi-sphere on top of the nanoneedle is formed, which acts like a plano-convex lens, focusing light directly on the silver nanoneedle antenna. The pointy shape of the nano-antenna provides a wide range of plasmonic resonance conditions which satisfy excitation by light of different wavelengths, i.e. increasing the optical mode density. The elongated metal nanostructure potentially enables transport of surface plasmon polaritons (SPP) [17] towards the bottom of the solar cell. In this device the photocurrents are used to measure the effect of the nano-lens combined with plasmonic nano-antenna inside the protruded solar cell. Illumination under an angle, which provides an asymmetric geometry for s- and p-polarized light, provides detailed information. S-polarized light affects the free electrons in the radial direction while p-polarized light has a main resonance along the nanoneedle length. The experimental results are explained by using finite-difference time-domain (FDTD) calculations, which provide insight into the plasmonic and photonic properties. In this work the unique combination of plasmonics, nano-focusing, and orthogonalisation in an elongated solar cell are explored.

9.2 Experimental Procedures and Setups

9.2.1 Sample Fabrication

Silver nanoneedles are formed by thermal evaporation on polycarbonate nucleopore track-etch membranes (Whatman) with a pore size of 200 nm. The polycarbonate is dissolved in chloroform after which the nanostructured silver foil was transferred on a glass substrate [50]. During the silver deposition the pores are partially filled, resulting in the nanoneedle structure as shown in Fig. 9.1. The silver nanoneedles have a wide range of values for the length and width but are on average 650 nm high and have a base width of 90 nm. The silver film with needles is covered with a 20 nm ZnO:Al (2 %) transparent conductive layer by magnetron sputtering to prevent inter diffusion with the semiconductor and optimize the reflective properties. Then, the deposition of an a-Si:H n-i-p layer stack was carried out in a multichamber deposition system which is described elsewhere [51]. To provide conformal coverage, the intrinsic (i-) layer was deposited by HWCVD [52] using SiH_4:H_2 (30:60) as source gasses, whereas plasma-enhanced CVD was employed for the deposition of the p- and n-layers. $B(CH_3)_3$ and PH_3 were utilized for p- and n-doping, respectively. The n-layer, i-layer, and p-layer had a nominal thickness of 60, 400, 50 nm respectively. The arrays were sputter-coated with 2.1 mm diameter circles of transparent conducting indium tin oxide (ITO) layer with a thickness of 80 nm after the deposition of the a-Si:H n-i-p layers to provide window contacts to the cells. In Fig. 9.2 the nanoneedle solar cell is schematically depicted (left) as well as the final real solar cells (right).

High resolution scanning electron microscopy (HRSEM) imaging was performed with a Philips XL30SFEG microscope.

To obtain a working solar cell by deposition on the pointy silver nanoneedles (Fig. 9.3a), the active film thickness (i-layer thickness) has to be carefully chosen. When the active layer is too thin, shunting prevents normal operation; when the active film is too thick, light will not reach the silver nanoneedle. By using HWCVD, a conformal growth of the n-i-p solar cell around the silver nanoneedle was achieved. A minimal thickness of the i-layer of 400 nm was required to prevent shunting and still have sufficient transparency. Conformal growth of all layers is demonstrated in the HRSEM micrographs in Fig. 9.3b. The silver nanoneedles are uniformly covered with the ZnO, n-i-p and ITO layers. From the cross sections of purposely broken nanoneedle solar cells the diameter after deposition was found to be about 400 nm while the height is about 1 micrometer. As the position of nanoneedles is randomly distributed the distances to each other vary. Independent of this distance, the nanoneedles were always fully covered. Patches of flat solar cell cover the area in between the nanoneedles that are further apart.

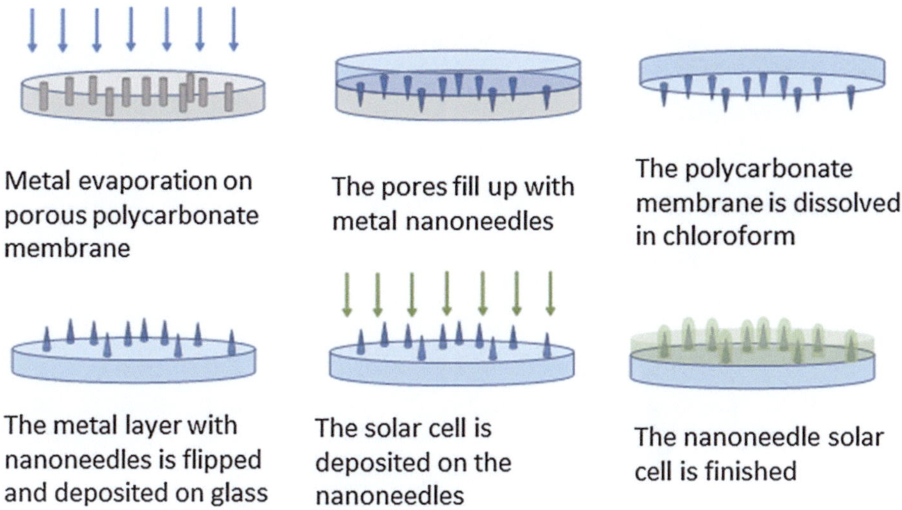

Fig. 9.1 The different stages to fabricate silver nanoneedles and the conformal deposition of a solar cell

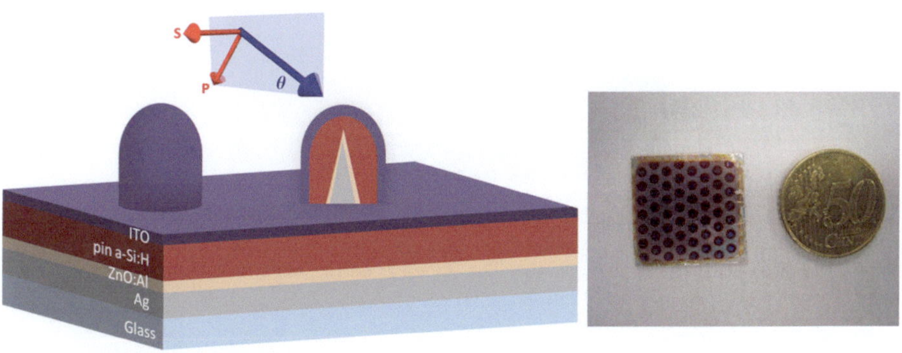

Fig. 9.2 (*left*) Schematically depicting the nanoneedle solar cell with angle of incidence and polarization directions. (*right*) Photograph of the nanoneedle sample with solar cells (*pink dots*)

9.2.2 Solar Simulator

Characterization of the solar cells is initially performed by using a solar simulator (Wacom WXS-140-S). It is equipped with a halogen lamp and a xenon arc lamp to be able to simulate the power of one sun corresponding to the AM1.5 spectrum (1000 W/m^2). Dark and light current measurements are measured through a computer controlled Keithley 238 ammeter which is used as voltage source and current meter. Losses occur in different forms such as shunts, bad contacts and pinholes. These losses are represented as the parallel and series resistance, R_p. and R_s

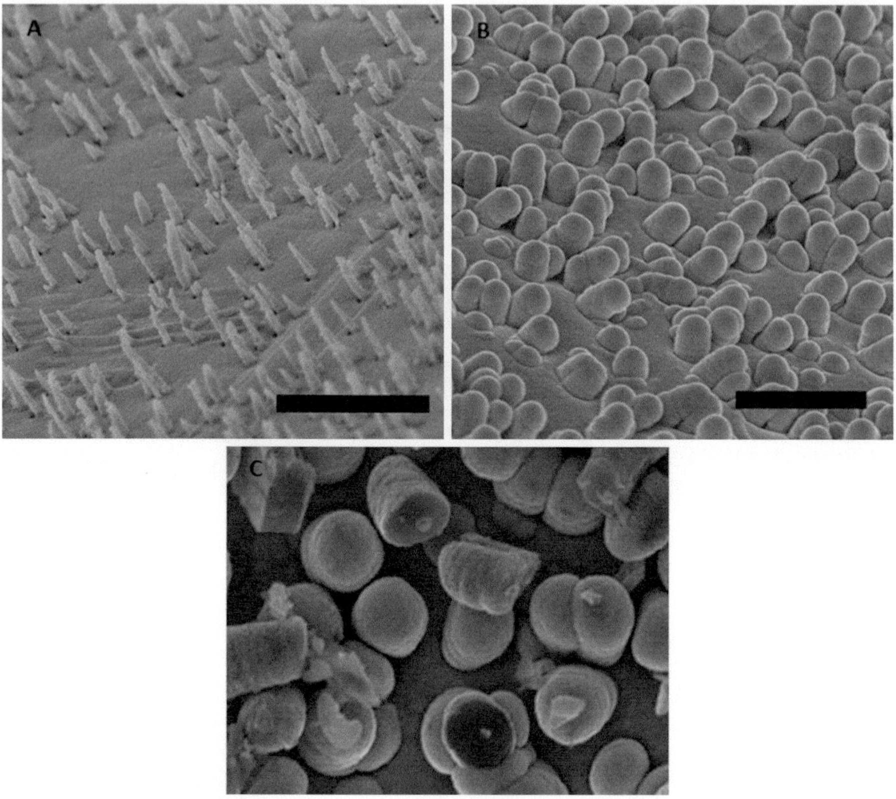

Fig. 9.3 High resolution SEM micrographs of (**a**) silver nanoneedles on a silver film and (**b**) silver needles and film plus ZnO, n-i-p and ITO. (**c**) after purposely damaging the sample the inside of the nanoneedle solar cell becomes visible. The scale bar corresponds to 1 micrometer

respectively, as indicated in Fig. 9.4. This results in a current density-voltage (J-V) dependency:

$$J(V) = -J_{ph} + J_0\Big(exp\Big(\frac{e\big(V - JR_s}{nk_bT} - 1\Big) + \frac{V - JR_s}{R_p}$$

With:

J_0 = dark saturation current
J_{ph} = photo current density
n = diode quality factor
k_b = Boltzmann constant
T = absolute temperature
e = elementary charge
V = applied voltage

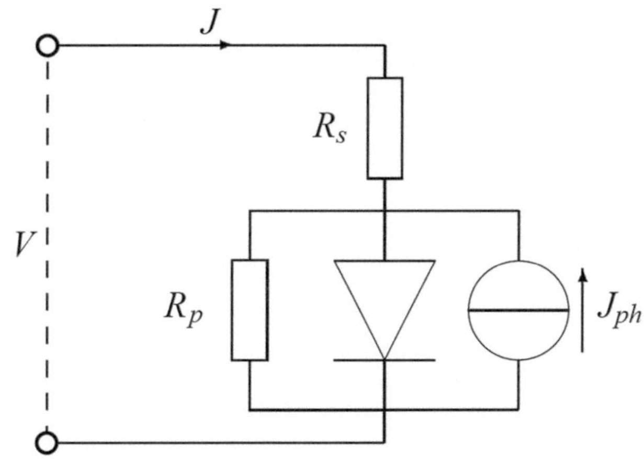

Fig. 9.4 Schematic figure of a solar cell with indicated resistances, current, and voltage

J_0 and n are obtained from dark J-V measurements ($J_{ph} = 0$). Figure 9.5 shows the IV-characteristics of a typical amorphous silicon solar cell. The short-circuit current density (J_{sc}) is measured at AM1.5. The slope at this point is $1/R_{sc}$, associated with the parallel resistance R_p and the slope at the V_{oc} is $1/R_{oc}$, associated with the series resistance R_s. The value of the diode quality factor is 1 if the cell is assumed to be an ideal diode with only band to band recombination. When recombination also occurs in the depletion region (junction) a value of 2 is applicable.

The dark characteristic (Fig. 9.5, right) illustrates the contribution of the parallel and series resistances by three red lines. The resistances influence the current and cause a further deviation of the J-V curve from the ideal shape. The contribution can be explained as follows:

1. Voltages below 0.3 V, the current density are dominated by the parallel resistance.
2. Voltages above 0.4 V, the current density behavior corresponds with exponential, diode-like behavior.
3. Voltages above 0.9 V, the curve starts to deviate from the diode line, due to the increasing influence of the series resistance and space charges.

9.2.3 Spectral Response

In a spectral response measurement the external quantum efficiency (EQE) of the solar cell is determined at each wavelength. The EQE is defined as the number of collected electrons per incident photon with reflection and transmission losses included. A monochromatic light beam from a Xenon arc lamp is used to illuminate the solar cell. Backlight is filtered using a lock-in amplifier. By measuring the photocurrent of a solar cell and comparing it with the reference photodiode, the EQE is calculated with the following equation (at 0 V):

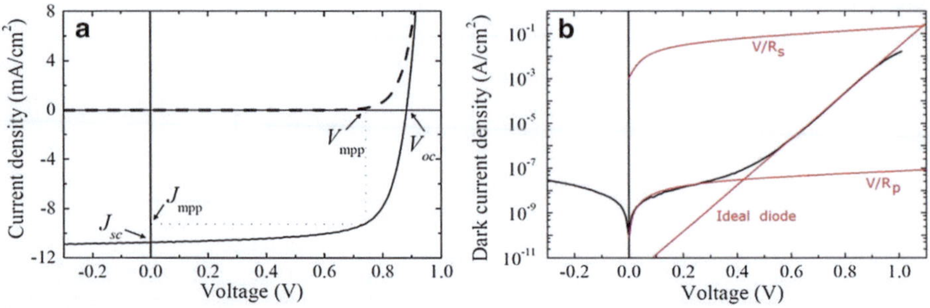

Fig. 9.5 Characterization plots of a typical amorphous silicon solar cell in the dark (*right*) and under AM1.5 conditions (*left*). The dark characteristics include the different contributions of the parallel and series resistances (Adapted from [53])

$$\eta_{col}(\lambda) = \frac{J_{ph}(\lambda)}{e\phi_{ph}(\lambda)} \tag{9.1}$$

With η the EQE, ϕ_{ph} the photon flux, e electron charge and J_{ph} the photo current. Absorption and reflection losses can be analyzed from spectral response measurements. Short wavelengths are absorbed in the top layers. A low response indicates losses in the front contact, p-layer and/or p-i-interface. The band gap, thickness of the absorber layer and light scattering properties at the metal back contact determine the spectral response at longer wavelengths, red light in particular.

9.2.4 Angle and Polarization Resolved Measurements

In order to obtain isolated effects of plasmonic resonances, for example along or perpendicular to the nanoneedle axis, a distinction between different angles of incidence and polarization of the light has to be made. Dependencies on the polarization angle and light intensity are well known in plasmonics, therefore angle and polarization dependent measurements are appropriate. A schematic illustration of the experiment is depicted in Fig. 9.6. On the sample, which is attached to the holder, each purple dot is an individual solar cell. A monochromatic polarized light beam illuminates the sample. The current is measured by placing one probe at the back contact, top left of the sample, while the front contact is positioned at the edge of a solar cell.

9.2.5 Specular Reflection and Scattering

The optical reflection and scattering of a surface provides information about how light is absorbed by the solar cell. Reflection and scattering experiments are

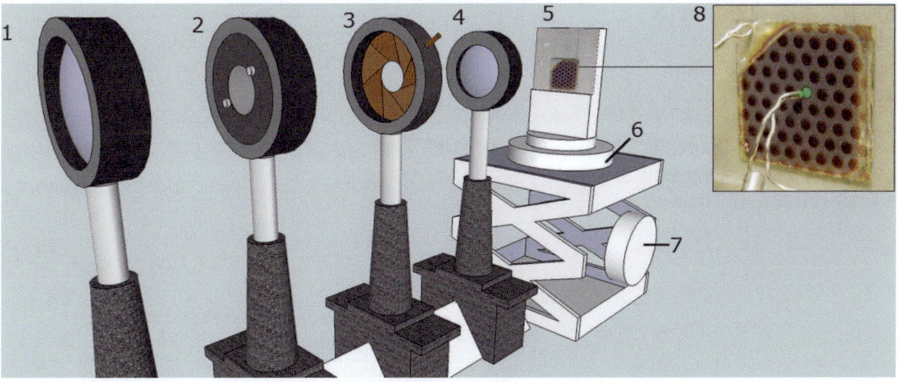

Fig. 9.6 Experimental setup for angle and polarization dependent measurements of the EQE. With two focus lenses (*1*, *4*), polarizer (*2*), diaphragm (*3*), sample holder (*5*), sample angle adjuster (*6*) and sample height adjuster (*7*). A sample inset (*8*) is also shown with attached probes. The projection of a 550 nm beam is visible on the surface of the cell. Probes are attached to measure the induced current

performed with an integrating sphere, which enables the distinction between specular and diffusive reflection. The total reflection is measured by placing a sample at a small angle. With a perpendicular mounted sample diffusive reflected light is detected. In this case scattered reflection (R_s) is measured. A mirror, evaporated aluminium on glass, is used to calibrate for total reflection and a rough white reflector for scattering. It is important to measure with the spectrometer (Perkin-Elmer Lambda 2S UV–VIS) in a wide range (from 200 to 1000 nm).

9.3 Experimental Study of a Silver Nanoneedle Plasmonic Core Inside an Elongated a-Si:H Solar Cell

Optical scattering measurements of the bare and n-i-p covered nanoneedles show a strong effect due to the presence of nanoneedles as shown in Fig. 9.7. The silver nanoneedles scatter much stronger than the flat silver film after the onset of the plasmon resonance wavelength (340 nm). The scattering of the silver nanoneedles has a maximum around 600 nm, which is attributed to the plasmon resonance corresponding to the specific silver nanoneedle size and shape. The scattering spectrum of the nanoneedle n-i-p a-Si:H solar cell exhibits a significant reduction in scattering intensity around 600 nm, which is caused by absorption at the a-Si:H band gap. At longer wavelengths the scattering intensity is much stronger because most of the light in the red and infra-red is not absorbed and reaches the nanoneedles. An important condition for observing a significant plasmonic effect in the solar cell is that the plasmon resonance energy is at or larger than the a-Si:H band gap. Only then the radiation absorbed and re-emitted by the plasmonic silver nanoneedle can be detected by measuring photocurrents.

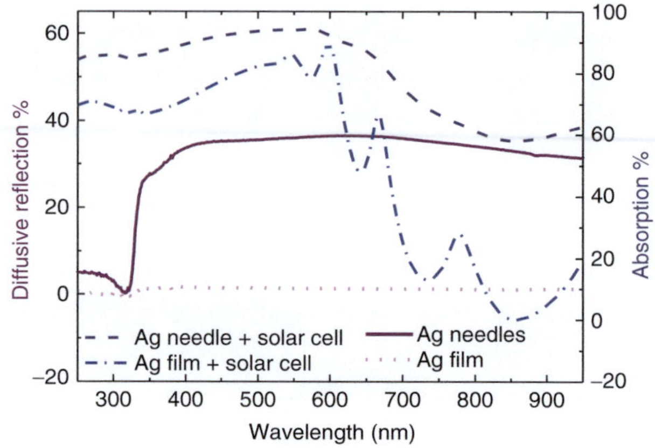

Fig. 9.7 *Left axis*: diffusive scattering of the silver nanoneedles (*solid*), flat silver film (*dotted*). *Right axis*: optical absorption of nanoneedles and solar cell (*dashed*) and flat silver film + solar cell (*dashed dotted*)

The current density-voltage (*J-V*) characteristics of the nanoneedle solar cells were measured using a solar simulator under one sun illumination (AM1.5G, 100 mW/cm^2) as shown in Fig. 9.8. The nanoneedle solar cells have a higher J_{SC} as compared to the flat reference. This can be the result of plasmonic interaction or simply by increased surface area. Simulations can provide more insight into the origin of this increased current. The V_{OC} values of the nanoneedle solar cells are significantly lower as a result of an increase in the defect density. These defects can be cured by anneal treatment which will increase the V_{OC}. However, annealing may also create shunting in the solar cell, making experiments impossible.

The overall electrical quality of the nanoneedle solar cells as indicated by the fill factor (FF) is lower for the nanoneedle solar cells, probably caused by more shunting pathways. The growth around the nanostructure may not be completely uniform due to the presence of sharp points and deep corners. The energy conversion efficiencies are somewhat lower for the nanoneedle solar cells than for the flat reference solar cells taking into account the inaccuracy of 10 % due to unavoidable error margins in the calibration of the simulator and the solar cell active areas. The lower V_{OC} neutralizes the gains in J_{SC}, which leaves room for improvement.

The effectiveness of the silver nanoneedles under illumination depends strongly on light absorption within a-Si:H. At low wavelengths most light will be absorbed near the upper silicon surface of the nanoneedle solar cell, not reaching the silver nanoneedle, which renders it inactive. At the large wavelengths most light passes undisturbed by a-Si:H, which makes the detection of the nanoneedle interaction with light immeasurable. Therefore, only between those spectral ranges in the window between about 550 and 670 nm a measurable effect of the nanoneedle will be possible.

To investigate the effect of the plasmonic properties of the silver nanoneedles on the surrounding solar cell, s- and p-polarized light are used to excite different plasmon resonance modes in the nanoneedle (as shown in Fig. 9.2), which can be considered as a Fabry-Perot resonator. Particularly under glancing incidence, the plasmon resonance modes for s- polarized light is radial while for p-polarized light

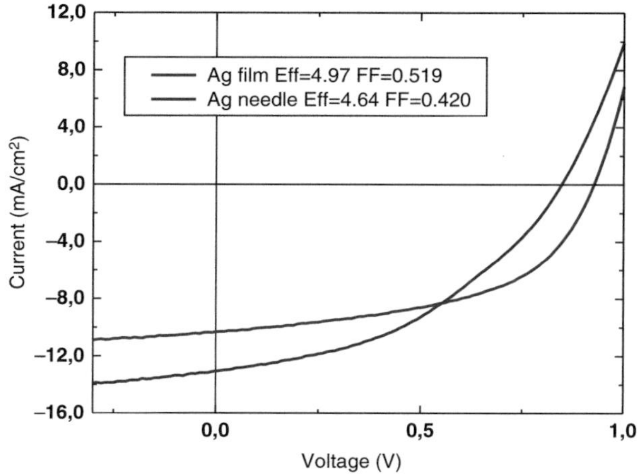

Fig. 9.8 IV curves of nanoneedle and flat film solar cell under AM1.5 illumination

they have a strong component along the nanoneedle length. In Fig. 9.9. the experimental spectral response of the silver nanoneedle and flat solar cell is shown for s- and p-polarized light at normal incidence and under an incidence angle of 30°. For normal incidence the spectral response profile on the Ag nanoneedle and the flat solar cells does not depend on the polarization of light (blue curve in Fig. 9.9a, b) due to symmetry. The flat solar cell has two shoulders at 560 and 610 nm, which can be explained by constructive interference inside the layer. This is not present in the nanoneedle solar cell because the optical path length within the layers varies considerably because of the wide range of incidence angles with respect to the solar cell surface. Because the nanoneedle solar cell surface angle is varying so strongly, the light passing through experiences different thicknesses, which obscures such spectral features.

The spectral response curves of the nanoneedle and flat reference solar cells taken at an incidence angle of 30° depend strongly on the polarization of light. The dependence of the EQE on light polarization of the flat reference solar cells is explained by the Fresnel reflection and Brewster angle (62° for air to ITO), which results in a reflection coefficient of s-polarized light of about 0.3 while it is close to 0 for p-polarized light. This results in different light transmission for different polarizations at the solar cell surface leading to the observed difference in EQE. The flat solar cell has a higher EQE value for s-polarized light as compared to p-polarized light at 600 and 670 nm. Because of the stronger transmission for s-polarized light due to Fresnel reflection this light interferes constructively. The absorption spectrum of the flat solar cell as calculated by FDTD (Fig. 9.11a) confirms this polarization dependence at 30° incidence. At the higher wavelengths s- and p-polarized absorption spectra have optima at alternating wavelength positions due to constructive interference.

At 30° incidence angle on the nanoneedle solar cell the blue and red part of the EQE spectrum are not significantly different between the two polarizations. The

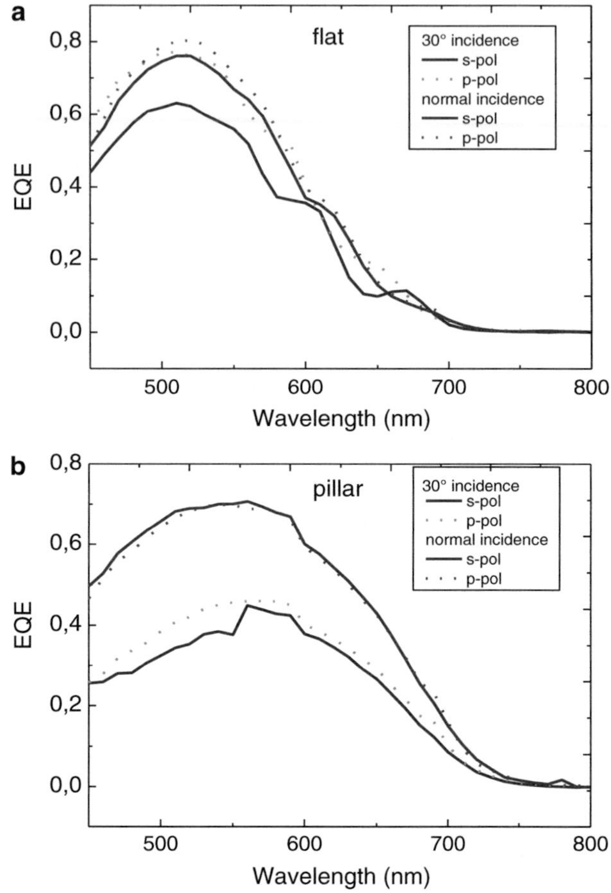

Fig. 9.9 EQE of (**a**) flat film and (**b**) nanoneedle (pillar) solar cell at normal incidence and 30° incidence angle for s- and p-polarized light

smaller EQE values of the nanoneedle solar cell for s-polarized light between 460 and 570 nm must be related to the presence of the silver nanoneedle.

In Fig. 9.9b the EQE maximum at normal incidence is positioned approximately 26 nm blue-shifted as compared to that at 30° incidence. This indicates that relatively less light is absorbed in the red at normal incidence due to a loss mechanism. To clarify this loss, the optical absorption in the a-Si:H layer of a nanoneedle solar cell was calculated with FDTD as a function of wavelength for s- and p-polarization at normal and 30° incidence for a nanoneedle solar cell (Fig. 9.11c).

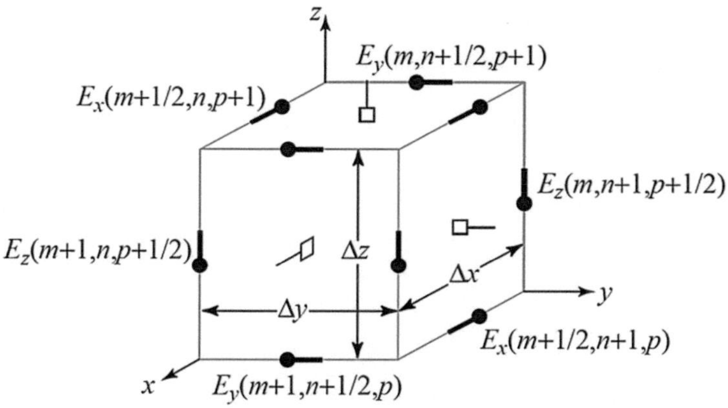

Fig. 9.10 Yee cell. In FDTD simulations the Maxwell's Equations at the Yee cell positions are solved. For each **cell**, the Electric field is calculated at four positions per axis, namely one for each cell edge. The magnetic fields are calculated at two positions for each axis, namely the cell faces

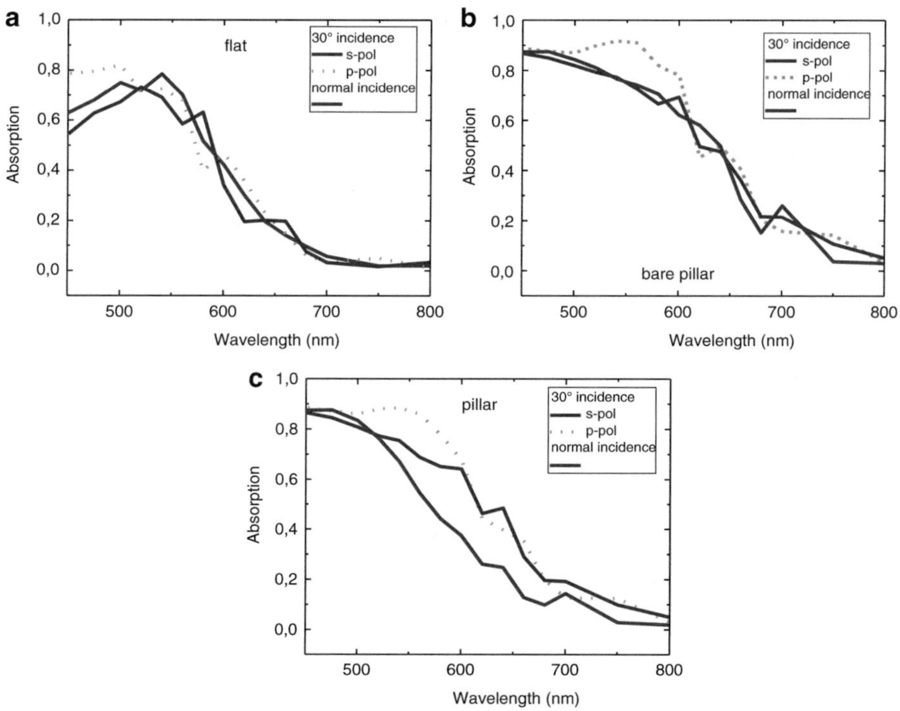

Fig. 9.11 Optical absorption spectra in a-Si:H (**a**) flat film (**b**) without nanoneedle and (**c**) nanoneedle solar cell at normal and 30° incidence for s- and p-polarized light

9.4 FDTD Simulations on the Ag Nanoneedle Inside a a-Si: H Core-Shell Structure

9.4.1 The Finite-Difference Time Domain Method

The FDTD method utilizes the Maxwell's equations to calculate the properties of electromagnetic waves as they propagate through space and interact with the object of interest. In principle the FDTD method can provide accurate results to problems where complex structures, small structures, and complex materials like metals are involved. Classical ray-based method cannot achieve such versatility and accuracy [54]. Yee proposed an algorithm to solve the propagation of electromagnetic waves in a certain configuration over time numerically [55] The FDTD method is developed further since then, but it still works based on above principle.

A FDTD simulation setup contains energy (light) sources, structures of interest and monitors. Once these are defined, the simulation space is divided in to small cells (Yee Cell) [56]. The values and flows of the electric and magnetic fields are calculated by including the refractive index of the material in each cell. When the values for all cells are calculated, a step forward in time is taken and the process of calculating the electric and magnetic fields of all the cells is restarted using the new starting values. This cycle repeats itself until certain conditions are met. For example, these conditions can be a certain amount of simulation or a minimum or maximum amount of energy present in the simulation box.

In the Yee algorithms the Maxwell's Equations are used as a basis. These equations are modified in such a way that they can be used for the simulation:

Maxwell's Equations are:

$\nabla \cdot \vec{D} = \rho_{free}$	Gauss' Law for electric fields
$\nabla \cdot \vec{B} = 0$	Gauss' Law for magnetic fields
$\nabla \times \vec{E} + \frac{\partial \vec{B}}{\partial t} = \vec{M}$	Faraday's Law
$\nabla \times \vec{H} - \frac{\partial \vec{D}}{\partial t} = \vec{J}$	Ampere's Law

Where

$\vec{M}$	Equivalent magnetic current density $[V\ m^{-2}]$

These Maxwell equations are modified as follows [57]:

1. "*Replace all the derivatives in Ampere's Law and Faraday's Laws with finite differences. Discretize space and time so that the electric and magnetic fields are staggered in both space and time.*"
2. "*Solve the resulting difference equations to obtain "update equations" that express the (unknown) future fields in terms of (known) past fields.*"
3. "*Evaluate the magnetic fields one time-step into the future so they are now known (effectively they become past fields).*"

4. "*Evaluate the electric fields one time-step into the future so they are now known (effectively they become past fields).*"
5. "*Repeat the previous two steps until the fields have been obtained over the desired duration.*"

The algorithm makes certain assumptions concerning constructive relations, sources and losses. The materials are considered linear, isotropic and non-dispersive so their electric and magnetic properties are independent of field, direction and frequency. In formulas, this results in the following:

$\vec{D} = \varepsilon\vec{E} = \varepsilon_0\varepsilon_r\vec{E}$
$\vec{B} = \mu\vec{H} = \mu_0\mu_r\vec{H}$
$\vec{J} = \vec{J}_{source} + \sigma\vec{E}$
$\vec{M} = \vec{M}_{source} + \sigma^m\vec{H}$

Where

ε	Electric permittivity	$[F/m]$
ε_0	Electric permittivity of free space	$8.854E^{-12}\,[F/m]$
ε_r	Relative electric permittivity	
μ	Magnetic permittivity	$[H/m]$
μ_0	Magnetic permittivity of free space	$4.0\pi E^{-7}\,[H/m]$
μ_r	Relative electric permittivity	
σ	Electric conductivity	$[S/m]$
σ^m	Equivalent magnetic loss	$[\Omega/m]$

Including the assumptions to Maxwell's equations and rewriting them in such a way that they are direction dependent, result in the following equations for the electric and magnetic flux. The y and z directions are obtained by replacing the indexes.

$$\frac{\partial H_x}{\partial t} = \frac{1}{\mu}\left(\frac{\partial E_y}{\partial z} - \frac{\partial E_z}{\partial y} - \sigma^m H_x - M_{source\_x}\right) \tag{9.2}$$

$$\frac{\partial E_x}{\partial t} = \frac{1}{\varepsilon}\left(\frac{\partial H_z}{\partial y} - \frac{\partial H_y}{\partial z} - \sigma E_x - J_{source\_x}\right) \tag{9.3}$$

These formulas are applied to all cells as given in Fig. 9.11. All cells together form a 3D grid, which is the simulation area. It is important to keep the cells relatively small since no effects can be calculated within the cell. This is an important limitation, for instance, when the cell is located at the edge of the skewed surface of a nanostructure.

9.4.2 Simulation Details

The structure of interest can be exposed to energy sources such as the Dipole, Gaussian, Plane wave, Total-field, scattered-field or a custom source. The response monitors include the refractive index, field time, movie, frequency-domain field profile, frequency-domain field and power, mode expansion, and/or several analysis groups build from these components. A selection of these sources and monitors should be inserted in the defined simulation region. The boundaries of this region can be set to absorbing all of the radiation, using the Perfect Absorbing Layer (PML) boundary, or use periodicity. There are several options for periodic boundary conditions: Periodic, Symmetric, Anti-Symmetric and Bloch. Periodic boundaries simulate an infinite periodicity of the structure within the boundaries. Thus fields that pass through a periodic boundary at the one side come back into the simulation at the other side. Symmetric and Anti-Symmetric boundaries are boundary conditions that make use of symmetry in the simulation. It simulates only one half and due to symmetry it calculates the other half. This is useful for reducing the required simulation time and resources. Which one should be used depends on the direction of the electric and magnetic field. Bloch boundary conditions are the most advanced boundary conditions. They do not feature symmetry, but are able to handle periodic EM fields. The most common type of simulation that requires Bloch BC is periodic structures with a plane wave source at non-normal angles of incidence.

An optimum has to be found between simulation accuracy and calculation time which results in a few important options:

- Size of the simulation region: the size should be as small as possible, but cannot always be so small that all the components precisely fit in the region, e.g. a monitor should not be located against a PML layer. Near PML layers artefacts of practical implementations of the FDTD method occur, resulting in inaccurate results when a monitor is placed in that region. Therefore, there should be enough distance between structures and PML layers. 'Enough' is generally half the maximum wavelength involved. Using a larger simulation region does however result in more Yee cells and thus longer calculation time.
- Size of the Yee cells: the smaller the Yee cells, the more spacially precise the simulation can be performed.
- Simulation time and auto-shutoff levels: The auto shutoff level is an indicator of the field strength with respect to the maximum field strength of the source. When this level becomes too high, the simulation should be stopped due to an error. When this level becomes very low, typically 10^{-5}, the simulation is considered finished, at the point where more calculations do not contribute significantly any

more. In addition, a maximum simulation time can be set. The shut-off condition which is first reached will terminate the simulation.

The optimum parameters are determined by using convergence tests which consist of a series of simulations where the value of one parameter is changed until an optimum is reached.

9.4.3 FDTD Simulation Results and Comparison with Experiment

Finite-difference time-domain (FDTD) calculations were performed with commercial software (Lumerical Solutions Inc.) on a cluster super computer consisting of several hundreds of multi-core nodes. In line with the averaged nanoneedle dimensions as obtained from SEM, a silver nanoneedle of 600 nm high and 80 nm wide at the base, covered with 20 nm ZnO, 400 nm a-Si:H and 80 nm ITO was modeled. The optical absorption is calculated by integrating the optical absorption over the a-Si:H structure and over the time a monochromatic light pulse resides in the structure.

In agreement with the experimental results at normal incidence the nanoneedle solar cell has considerably lower absorption in the red as compared to 30° incidence. A likely mechanism is Ohmic loss by which the moving free electrons in the metal result in the dissipation of electromagnetic energy into heat [58]. This is confirmed by comparing the FDTD calculated optical absorption of the nanoneedle solar cell with and without silver nanoneedle inside. The optical absorption slope in Fig. 9.11c at normal incidence is positioned 75 nm to the blue for the nanoneedle solar cell as compared to the bare solar cell in Fig. 9.11b. This can only be explained by the nanoneedle presence leading to losses in the red part of the spectrum. The EQE difference at the short wavelengths between experiment (Fig. 9.9) with low EQE and FDTD calculations (Fig. 9.11) with strong absorption is explained by optical losses in the p-layer, which was not included in the calculations.

In Fig. 9.11b and c the FDTD calculations confirm that s-polarized light at 30° incidence is much less absorbed than p-polarized light between 500 and 600 nm while at other wavelengths the absorption is comparable. A slight shift of the position of this gap of about 30 nm is explained by a slight mismatch between the average nanoneedle size in the experiment and calculation. Because this lower absorption for s-polarized light is also present in the bare solar cell (without nanoneedle, Fig. 9.11b), the morphology in combination with Fresnel reflection is the most likely cause.

The cross sections of the light absorption in the a-Si:H with and without the nanoneedle are shown in Fig. 9.12. The wavelength of 580 nm is chosen as it allows a full pass through the entire active layer. Pockets of high absorption are clearly distinguishable within the solar cell. Side and top view cross sections at normal and 30° incidence show that most of the light is absorbed in the central region of the nanostructured solar cell. Comparison of the side cross sections with (Fig. 9.12a–d) and without (Fig. 9.12e–h) nanoneedle shows clearly that the light is focused onto the nanoneedle.

The distance between high absorption spots in the cross sections is about 100 nm. Since the wavelength of 580 nm light inside the a-Si:H medium is 164 nm a constructive interference pattern has a periodicity of half that wavelength: 82 nm. Therefore these strong absorption spots are the result of constructive interference caused by the incident light interacting with its own reflections and scattering. Top view cross sections at 340 nm (Fig. 9.12c, d) show that the nanoneedle forms a dipole scattering profile. At 540 nm (Fig. 9.12d) this scattering strongly influences the absorption intensity profile. This demonstrates the activity of the nanoneedles plasmonic response in the far field (scattering), which must affect overall light absorption.

In Fig. 9.13 the absorption cross section of the nanoneedle solar cell structure are shown for light with an incidence angle of 30°. At this angle of incidence the focal

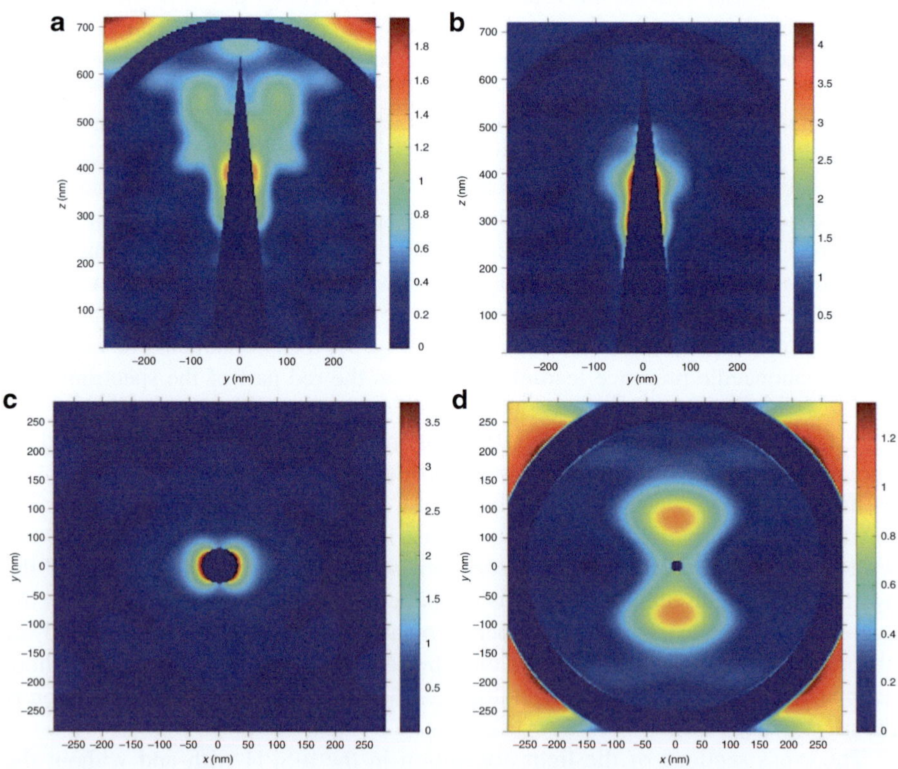

Fig. 9.12 Cross sections of the time-integrated photo-absorption in the a-Si:H part of the nanoneedle solar cell at 580 nm with nanoneedle (**a**) view perpendicular to plane of incidence (**b**) along plane of incidence (**c**) top view at z = 340 nm and (**d**) 540 nm. Without silver nanoneedle inside (**e**) view perpendicular to plane of incidence (**f**) along plane of incidence (**g**) top view at z = 340 nm and (**h**) 540 nm. The intensity scale bar units are absorption 10^{19} m^{-3}

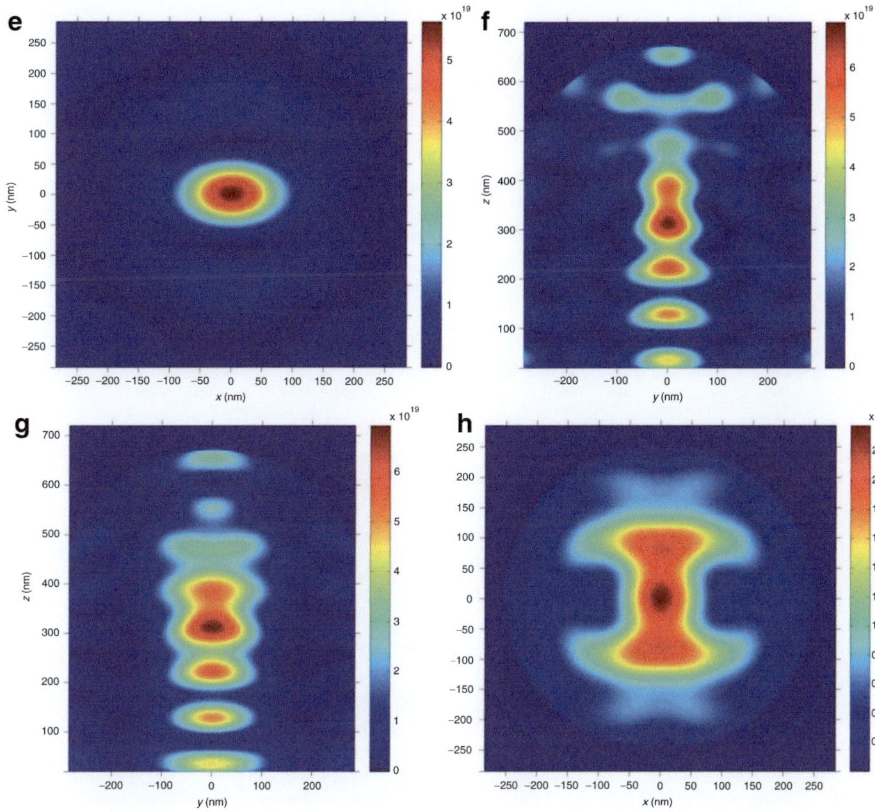

Fig. 9.12 (continued)

point is not positioned at the nanoneedle, therefore the light-plasmon interaction is much weaker. This is particularly well illustrated by Fig. 9.13b where the constructive interference pattern also has an angle of 30° with the nanoneedle and therefore is positioned mainly inside the a-Si.

To illustrate the effect of the incident light polarization, in Fig. 9.13c the difference of the optical absorption for s and p polarized light is shown along the plane of incidence. Bright red indicates a stronger p-polarized absorption while dark blue represents a very small difference in absorption between the two polarizations. In agreement with Fig. 9.13b, the main absorption occurs at an angle of 30° with the nanoneedle. However, on this line spots of a high difference (stronger p-polarisation) are present. This is in agreement with Fig. 9.9a, b where a difference in overall absorption was measured, likely caused by the effect of Fresnel reflections at the surface of the structure. In Fig. 9.13 f the loss of symmetry in the optical absorption profile is well illustrated. The complex interactions which occur under wide angle incidence, results in an absorption profile which does not strongly resemble a dipole response.

The nanoneedle has a small radius at the top and a large radius at the base resulting in different responses to incoming electromagnetic fields. At the silver

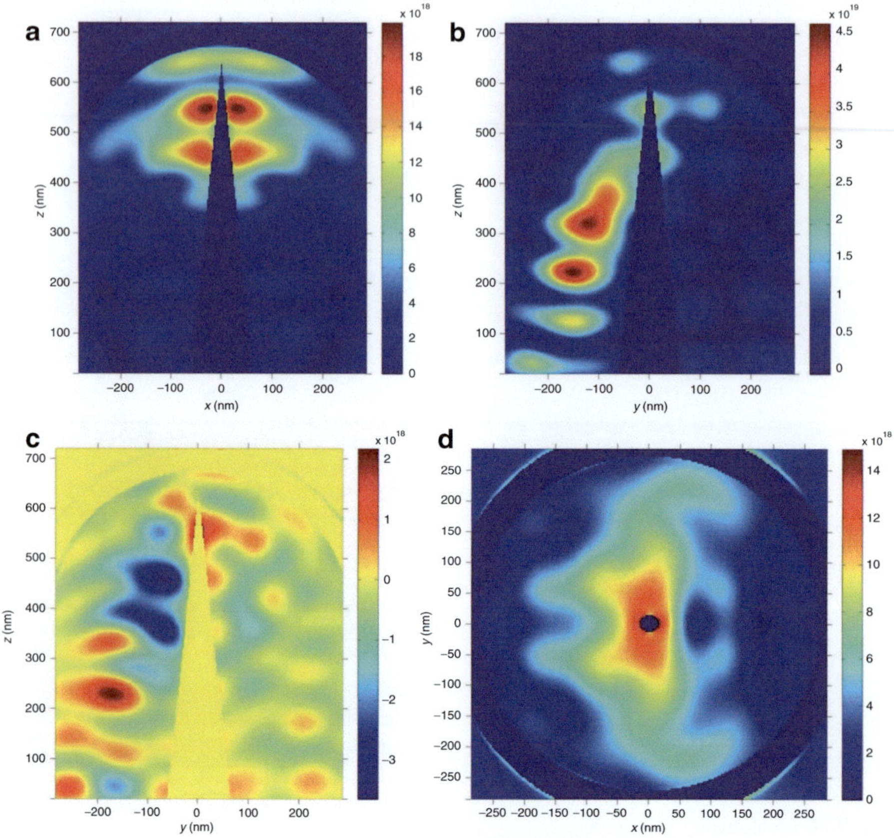

Fig. 9.13 Cross sections of the time integrated photo absorption in the a-Si:H part of the nanoneedle solar cell at 580 nm under an incidence angle of 30°. (**a**) view perpendicular to plane of incidence (**b**) along plane of incidence (**c**) difference between s-and p-polarized light along plane of incidence (**d**) top view. The intensity scale bar units are absorption m^{-3}

nanoneedle tip the size is so small that visible light fully penetrates until about 27 nm [57] (the quasi-static regime). For silver particles of comparable size, about 30 % of the incident energy is absorbed and lost as heat [59]. At the lower part of the nanoneedle the radius is too large for light to cross, instead retardation effects become important. A wide range of plasma resonance wavelengths are possible as shown by calculations on nanorods [60]. Although no significant effect is observed here, energy transport via SPPs to the bottom of the solar cell is in principle possible [17].

Concluding, an a-Si:H solar cell was conformally grown around a silver nanoneedle providing a unique device to measure plasmonic activity. The spherical top of the solar cell provided a unique nano-lens, focusing polarized light on the silver nano-antenna. The EQE for s- and p-polarized light at normal incidence and under an angle of 30° shows that the nanoneedle solar cell has a lower absorption for red light at normal incidence. This is confirmed by FDTD calculations. The electromagnetic field is captured by the silver nano-antenna and via the motion of free electrons (plasmons) converted into heat. Differences in EQE as a function of polarization are attributed to the nanostructured morphology in combination with Fresnel reflections. This work demonstrates that even in an optimized geometry, where light is directly focused on a nano-antenna, the combination of plasmonics and geometrical optics needs to be finely tuned to obtain solar cell efficiency gain.

9.5 Future Possibilities

The here described results of combining plasmonics in elongated nanostructured solar cells are only a first step. The choice of metal is important for future improvements. Silver has the advantage that it has a sharp and strong plasmon resonance. Although the plasmon resonance is positioned in the visible, by embedding silver in a solar cell medium, the plasmon resonance may shift to wavelengths which are less interesting. Instead Aluminium could be an interesting alternative [61]. Although Aluminum has its bulk plasmon resonance in the far UV around 150 nm, it is a cheap alternative since the plasmon resonance frequency can be shifted. The plasmon resonance wavelength of aluminum nanoparticles can be shifted towards the visible by selection of size, shape and configuration [62–67]. This plasmon resonance shift makes the aluminum particles suitable for solar cells. A theoretical study by Akimov at al. [68] has demonstrated that aluminum particles increases the optical absorption in a-Si:H over a wide wavelength range, better than silver particles.

The plasmonic effect of local field enhancement occurs only a few tens of nm away from the metal surface. Therefore, ultra-thin solar cell absorbers are required to optimize the benefits from that effect. Recent work on Au-Cu2O core-shell nanowires [69] with a very thin absorber shows clear photovoltaics properties.

Although the detrimental effects of plasmonics, such as Ohmic losses and stimulated radiative losses are not desired, they need to be studied in order to optimize the benefits. It is known from photoluminescence experiments that the distance between an emitter and the plasmonic structure makes the difference between enhancement or quenching [18, 70]. Therefore, thin transparent layers are likely necessary between the plasmonic structure and optical absorber in a solar cell. Besides the plasmonic benefit, this also reduces the effect of defects, introduced at the interface between the metal and optical absorber. Such defects can form strong recombination centers, reducing the photocurrent and voltage.

Extending the random silver nanoneedle pattern to regular or periodic nanoneedle patterns would change the optical response considerably. If the nanoneedles are in close proximity, an interaction will occur between the electromagnetic fields of the plasmon resonances. This will enhance or inhibit the strength of the plasmon resonance, depending on the distance and periodicity. Complicated interactions are also likely giving rise to meta properties.

Controlling the density of nanoneedles on the surface and the dimensions of the nanoneedles even further, would provide the possibility to optimize the structure for light absorption in the solar cell. In the here described work, only one such configuration has been explored, leaving ample space for more detailed studies.

References

1. Green MA (2003) Third generation photovoltaics: ultra high efficiency at low cost. Springer, Berlin
2. Conibeer G (2007) Third-generation photovoltaics. Mater Today 10:42
3. Ozbay E (2009) Merging photonics and electronics at nanoscale dimensions. Science 311:189
4. Maier SA, Brongersma ML, Kik PG, Meltzer S, Requicha AAG, Atwater H (2001) Plasmonics–a route to nanoscale optical devices. Adv Mater 13:1501
5. Maier SA (2007) Plasmonics fundamentals and applications. Springer, New York
6. Atwater HA, Polman A (2010) Plasmonics for improved photovoltaic devices. Nat Mater 9:205
7. Pala RA, White J, Barnard E, Liu J, Brongersma ML (2009) Design of plasmonic thin-film solar cells with broadband absorption enhancements. Adv Mater 21:3504
8. Pillai S, Green MA (2010) Plasmonics for photovoltaic applications. Sol Energy Mat Sol C 94:1481
9. Schaadt DM, Feng B, Yu ET (2005) Enhanced semiconductor optical absorption via surface plasmon excitation in metal nanoparticles. Appl Phys Lett 86:063106
10. Rand BP, Peumans P, Forrest SR (2004) Long-range absorption enhancement in organic tandem thin-film solar cells containing silver nanoclusters. J Appl Phys 96:7519
11. Ferry VE, Munday JN, Atwater HA (2010) Design considerations for plasmonic photovoltaics. Adv Mater 22:4794
12. Pillai S, Catchpole KR, Trupke T, Green MA (2007) Surface plasmon enhanced silicon solar cells. J Appl Phys 101:093195
13. Stuart HR, Hal DG (1998) Island size effects in nanoparticle-enhanced photodetectors. Appl Phys Lett 73:3815
14. Stuart HR, Hal DG (1996) Absorption enhancement in silicon-on-insulator waveguides using metal island films. Appl Phys Lett 69:2327–2329
15. Bohren CF, Huffman DR (2004) Absorption and scattering of light by small particles. Wiley-VCH, Weinheim
16. Purcell E (1946) Spontaneous emission probabilities at radio frequencies. Phys Rev 69:681
17. Novotny L, Hecht B (2008) Principles of nano-optics. Cambridge University Press, Cambridge
18. Anger P, Bharadwaj P, Novotny L (2006) Enhancement and quenching of single-molecule fluorescence. Phys Rev Lett 96:113002
19. Govorov AO, Bryant GW, Zhang W, Skeini T, Lee J, Kotov NA, Slocik JM, Naik RA (2006) Exciton-plasmon interaction and hybrid excitons in semiconductor–metal nanoparticle assemblies. Nano Lett 6:984–994
20. Clapp AR, Medintz IL, Mattoussi H (2005) Förster resonance energy transfer investigations using quantum-dot fluorophores. ChemPhysChem 7:47
21. Michalet X, Pinaud FF, Bentolila LA, Tsay JM, Doose S, Li JJ, Sundaresan G, Wu AM, Gambhir SS, Weiss S (2005) Quantum dots for live cells, in vivo imaging, and diagnostics. Science 307:538–544
22. Haes AJ, Zou S, Schatz GC, Van Duyne RP (2004) Nanoscale optical biosensor: short range distance dependence of the localized surface plasmon resonance of noble metal nanoparticles. J Phys Chem B 108:109

23. Stenzel O, Stendal A, Voigtsberger K, von Borczyskowski C (1995) Enhancement of the photovoltaic conversion efficiency of copper phthalocyanine thin film devices by incorporation of metal clusters. Sol Energy Mat Sol C 37:337
24. Westphalen M, Kreibig U, Rostalski J, Lüth H, Meissner D (2000) Metal cluster enhanced organic solar cells. Sol Energy Mat Sol C 61:97–105
25. Akimov YA, Koh WS, Ostrikov K (2009) Enhancement of optical absorption in ... nanoparticle plasmon modes. Opt Express 17:10195–10205
26. Barnard ES, Pala RA, Brongersma ML (2011) Photocurrent mapping of near-field optical antenna resonances. Nat Nanotechnol 6:588–593
27. Ferry VE, Verschuuren MA, Li HBT, Verhagen E, Walters RJ, Schropp REI, Atwater HA, Polman A (2010) Light trapping in ultrathin plasmonic solar cells. Opt Express 18:A237–A245
28. Ferry VE, Verschuuren MA, van Lare MC, Schropp REI, Atwater HA, Polman A (2011) Optimized spatial correlations for broadband light trapping nanopatterns in high efficiency ultrathin film a-Si:H solar cells. Nano Lett 11:4239–4245
29. Kuang Y, van der Werf KHM, Houweling ZS, Schropp REI (2011) Nanorod solar cell with an ultrathin a-Si:H absorber layer. Appl Phys Lett 98:113111
30. Tassin P, Koschny T, Kafesaki M, Soukoulis CMA (2012) Comparison of graphene, superconductors and metals as conductors for metamaterials and plasmonics. Nat Photonics 6:259–264
31. Kuang Y, Di Vece M, Rath JK, van Dijk L, Schropp REI (2013) Elongated nanostructures for radial junction solar cells. Rep Prog Phys 76:106502
32. Naughton MJ, Kempa K, Ren ZF, Gao Y, Rybczynski J, Argenti N, Gao W, Wang Y, Peng Y, Naughton JR, McMahon G, Paudel T, Lan YC, Burns MJ, Shepard A, Clary M, Ballif C, Haug FJ, Söderström T, Cubero O, Eminian C (2010) Efficient nanocoax-based solar cells. Phys Status Solidi R 4:181–183
33. Zhu J, Yu Z, Burkhard GF, Hsu CM, Connor ST, Xu Y, Wang Q, McGhee M, Fan S, Cui Y (2009) Optical absorption enhancement in amorphous silicon nanowire and nanocone arrays. Nano Lett 9:279–282
34. Lu Y, Lal A (2010) High-efficiency ordered silicon nano-conical-frustum array solar cells by self-powered parallel electron lithography. Nano Lett 10:4651–4656
35. Zhu J, Hsu CM, Yu Z, Fan S, Cui Y (2010) Nanodome solar cells with efficient light management and self-cleaning. Nano Lett 10:1979–1984
36. Fan Z, Razavi H, Do J, Moriwaki A, Ergen O, Chueh YL, Leu PW, Ho JC, Takahashi T, Reichertz LA, Neale S, Yu K, Wu M, Ager JW, Javey A (2009) Three-dimensional nanopillar-array photovoltaics on low-cost and flexible substrates. Nat Mater 8:648–653
37. Li J, Yu HY, Wong SM, Zhang G, Sun X, Lo PGQ, Kwong DL (2009) Si nanopillar array optimization on Si thin films for solar energy harvesting. Appl Phys Lett 95:033102
38. Kayes BM, Atwater HA, Lewis NS (2005) Comparison of the device physics principles of planar and radial p-n junction nanorod solar cells. Appl Phys 97:114302
39. Spurgeon JM, Atwater HA, Lewis NSJ (2008) A comparison between the behavior of nanorod array and planar Cd (Se, Te) photoelectrodes. Phys Chem C 112:6186–6193
40. Kieven D, Dittrich T, Belaidi A, Tornow J, Schwarzburg K, Allsop N, Lux-Steiner M (2008) Effect of internal surface area on the performance of ZnO/In[sub 2]S[sub 3]/CuSCN solar cells with extremely thin absorber. Appl Phys Lett 92:153107
41. Belaidi A, Dittrich T, Kieven D, Tornow J, Schwarzburg K, Lux-Steiner M (2008) Influence of the local absorber layer thickness on the performance of ZnO nanorod solar cells. Phys Status Solidi R 2:172–174
42. Sivakov V, Andrä G, Gawlik A, Berger A, Plentz J, Falk F, Christiansen SH (2009) Silicon nanowire-based solar cells on glass: synthesis, optical Properties, and cell parameters. Nano Lett 9:1549–1554
43. Tsakalakos L, Balch J, Fronheiser J, Korevaar BA, Sulima O, Rand J (2007) Silicon nanowire solar cells. Appl Phys Lett 91:233117
44. Law M, Greene LE, Johnson JC, Saykally R, Yang P (2005) Nanowire dye-sensitized solar cells. Nat Mater 4:455
45. Kelzenberg MD, Boettcher SW, Petykiewicz JA, Turner-Evans DB, Putnam MC, Warren EL, Spurgeon JM, Briggs RM, Lewis NS, Atwater HA (2010) Enhanced absorption and carrier collection in Si wire arrays for photovoltaic applications. Nat Mater 9:239–244

46. Tian B, Zheng X, Kempa TJ, Fang Y, Yu N, Yu G, Huang J, Lieber CM (2007) Coaxial silicon nanowires as solar cells and nanoelectronic power sources. Nature 449:885–889
47. Garnett E, Yang P (2010) Light trapping in silicon nanowire solar cells. Nano Lett 10:1082–1087
48. Di Vece M, Kuang Y, van Duren S, Charry JM, van Dijk L, Schropp REI (2012) Plasmonic nano-antenna a-Si:H solar cell. Opt Express 20:27327
49. Schropp REI (2001) Status of Cat-CVD (hot-wire CVD) research in Europe. Thin Solid Films 395:17–24
50. Di Vece M, Kelly JJ (2003) Synthesis and characterization of Pd nano-pillar arrays in the metal hydride switchable mirror. MRS Proc 776:Q11.7
51. Schropp REI, Feenstra KF, Molenbroek EC, Meiling H, Rath JK (1997) Device-quality polycrystalline and amorphous silicon films by hot-wire chemical vapour deposition. Philos Mag B 76:309–321
52. van Veen MK, van der Werf CHM, Rath JK, Schropp REI (2003) Incorporation of amorphous and microcrystalline silicon in n–i–p solar cells. Thin Sol Film 430:216–219
53. van Veen MK (2003) Tandem solar cells deposited using hot-wire chemical vapor deposition, PhD thesis, Utrecht
54. Shlager KL, Schneider JB (1995) A selective survey of the finite-difference time domain literature. IEEE Antenna Propag M 37:39–57
55. Yee K, IEEE T (1966) Numerical solution of initial boundary value problems involving Maxwells equations in isotropic media. Antenna Propog 14:302–307
56. FDTD (2012) Lumerical Solutions, Inc.
57. Schneider J (2012) Understanding the finite-difference time-domain method, unpublished. www.eecs.wsu.edu/~schneidj/ufdtd, 2010
58. Kreibig U, Vollmer M (1995) Optical properties of metal clusters. Springer, Berlin
59. Messinger BJK, von Raben U, Chang RK, Barber PW (1981) Local fields at the surface of noble-metal microspheres. Phys Rev B 24:649–657
60. Encina ER, Perassi EM, Coronado EA (2009) Near-field enhancement of multipole plasmon resonances in Ag and Au nanowires. J Phys Chem A113:4489–4497
61. Ramadurgam S, Lin TG, Yang C (2014) Aluminum plasmonics for enhanced visible light absorption and high efficiency water splitting in core-multishell nanowire photoelectrodes with ultrathin hematite shells. Nano Lett 14:4517–4522
62. Ekinci Y, Solak HH, Löffler JF (2008) Plasmon resonances of aluminum nanoparticles and nanorods. J Appl Phys 104:083107
63. Langhammer C, Schwind M, Kasemo B, Zorić I (2008) Localized surface plasmon resonances in aluminum nanodisks. Nano Lett 8:1461–1471
64. Tan SJ, Zhang L, Zhu D, Goh XM, Wang YM, Kumar K, Qiu CW, Yang JKW (2014) Plasmonic color palettes for photorealistic printing with aluminum nanostructures. Nano Lett 14:4023–4029
65. Olsona J, Manjavaca A, Liub L, Changa WC, Foerstera B, King NS, Knight MW, Nordlander P, Halas NJ, Link S (2014) Vivid, full-color aluminum plasmonic pixels. Proc Natl Acad Sci U S A 111:14348–14353
66. Knight MW, Liu L, Wang Y, Brown L, Mukherjee S, King NS, Everitt HO, Nordlander P, Halas NJ (2012) Aluminum plasmonic nanoantennas. Nano Lett 12:6000–6004
67. Temple TL, Bagnall DM (2011) Optical properties of gold and aluminium nanoparticles for silicon solar cell applications. J Appl Phys 109:084343
68. Akimov YA, Koh WS (2011) Design of plasmonic nanoparticles for efficient subwavelength light trapping in thin-film solar cells. Plasmonics 6:155–161
69. Oener SZ, Mann SA, Sciacca B, Sfiligoj C, Hoang J, Garnett AC (2015) Au-Cu2O core-shell nanowire photovoltaics. Appl Phys Lett 106:023501
70. Biteen JS, Pacifici D, Lewis NS, Atwater HA (2005) Enhanced radiative emission rate and quantum efficiency in coupled silicon nanocrystal-nanostructured gold emitters. Nano Lett 5:1768–1773

Chapter 10
Controlled Assembly of Plasmonic Nanostructures Templated by Porous Anodic Alumina Membranes

Xingce Fan, Qi Hao, and Teng Qiu

Abstract Non-lithographic template-based approach, especially the porous anodic alumina (PAA) template, owing to its long-range ordering self-assembly hexagonal cells, facile controllability of various configuration shapes, excellent reproducibility, easy fabrication and modest cost, have attracted much attentions these days. Recent research interests of our group are based on this promising template. Several plasmonic nanostructures based on the PAA membranes were successfully fabricated, which were applied to surface-enhanced Raman scattering sensing, the process of tailoring fluorescence as well as surface-enhanced fluorescence cellular imaging. Our recent plasmonic research results based on the PAA membranes and their applications are reviewed here.

Keywords Porous anodic alumina • Plasmonics • Surface-enhanced Raman scattering • Surface-enhanced fluorescence • Cellular imaging

10.1 Introduction

Localized surface plasmons resonance (LSPR) is collective oscillations of free electrons confined to metallic nanoparticles and nanostructures [1]. LSPR is usually excited by an electric field (light) at an incident wavelength where resonance occurs, resulting in strong light scattering, in the appearance of intense surface plasmon absorption bands, and an enhancement of the local electromagnetic fields [2]. The frequency and intensity of the SP absorption bands are characteristic of the type of material (typically, gold, silver, platinum and aluminum) and are highly sensitive to the size, size distribution, shape of the nanostructures, as well as their surrounding environments [3, 4]. The extraordinarily intense local electromagnetic field enhancements at certain interstitial sites, so-called 'hot-spots' or 'hot-junctions' in the nanostructures, consist of two or more coupled metallic nanoparticles or nanostructures with closely spaced features [5].

X. Fan • Q. Hao • T. Qiu (✉)
Department of Physics, Southeast University, Nanjing 211189, People's Republic of China
e-mail: tqiu@seu.edu.cn

C.D. Geddes (ed.), *Reviews in Plasmonics 2015*,
DOI 10.1007/978-3-319-24606-2_10

The fact that particle plasmons allow the direct coupling of light to resonant electron plasmon oscillations has spurred tremendous research activities in the design and fabrication of highly plasmonic active substrates in metallic nanostructured films and nanoparticles [6, 7]. The conventional plasmonic substrates were mostly produced by electrochemical oxidation and reduction cycles [8, 9], vacuum deposition [10] as well as colloidal metal particles [11]. The substrate surface produced by such methods are either non-uniform or irreproducible, thus the hot-spots are unevenly distributed. However, highly ordered plasmonic substrates are required for the application of Surface-enhanced Raman Scattering (SERS) sensing, Surface-enhanced Fluorescence (SEF) and others, to achieve the collection of reliable, stable and uniform signal spanning a wide dynamic range [12].

To produce highly ordered plasmonic substrates, several new strategies, including electron beam lithography [13], nanosphere lithography [14] and self-assembly template method [12, 15–22], have been proposed. However, electron beam lithography suffers from low throughput and high cost as well as nanosphere lithography inhomogeneity of close-packed nanosphere with pattern resolution less than 100 nm.

In contrast, non-lithographic template-based approach, especially the porous anodic alumina (PAA) template [12, 16–22], owing to its long-range ordering self-assembly hexagonal cells, facile controllability of various configuration shapes, excellent reproducibility, easy fabrication and modest cost, have attracted much attentions these days.

A typical structure of PAA membranes is illustrated in Fig. 10.1a (pore surface) and Fig. 10.1b (bottom surface). By using electron microscopy, Keller et al. [23] found that the pore surface of PAA membranes consisted of predominately close-packed hexagonal cells of alumina in shape and each cell contained a single cylindrical pore in the center which is perpendicular to the surface of the underlying aluminum substrate. It should be noted that there are small protrusions along the surface of pore wall, and a dent exists between every two neighboring protrusions [12]. The pore extends down to an alumina barrier layer (a layer of hemispherical alumina, composing the inner bottom of cylindrical pore, so-called barrier layer) which is between the pore bottom and the aluminum substrate, and the bottom surface of PAA consists of periodically hexagonal cells which show an array of partial hemispheres, one-by-one corresponding to the pore surface geometry.

So far, many efforts have been made to improve the arrangement of the pore patterns [24, 25], for instance by optimizing the experimental conditions such as temperature and pre-texturing to initiate pore growth on the aluminum surface and enhance the order of the patterns [26, 27]. To produce highly ordered PAA films, a two-step anodization process in conjunction with pre-texturing have been proposed by Masuda and co-workers [25, 28]. Later on, Li et al. [29] reported that self-assembly hexagonal pore arrays with a 50–420 nm interpore distance (the average distance between two neighboring pore center) in PAA membranes have been obtained by anodizing aluminum foils in oxalic, sulfuric and phosphoric acid solutions. The interpore distance increases linearly with anodic potential

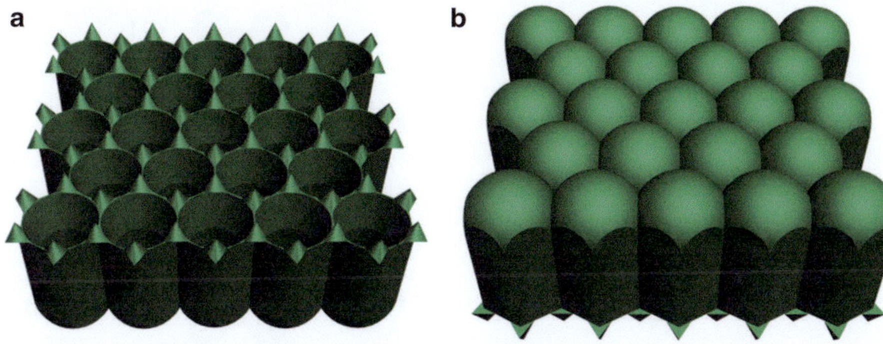

Fig. 10.1 (**a**) The pore surface of PAA membranes and (**b**) the bottom surface of PAA membranes

(or anodic voltage). The pore diameter of PAA can be enlarged via a pore enlarging process by emersion in some electrolyte solutions (typically, phosphoric acid) [28].

Recent research interests of our group are based on this promising template. Several plasmonic nanostructures based on the PAA membranes were successfully fabricated, which were applied to SERS sensing, the process of tailoring fluorescence as well as SEF cellular imaging. Our recent plasmonic research results based on the PAA membranes and their applications are reviewed here.

10.2 Highly Ordered Plasmonic Nanostructures Templated by PAA Membranes

Figure 10.2 shows the fabrication of highly ordered hemispherical silver nanocap arrays [12] templated by the PAA pore surface (as shown in Fig. 10.1a) as robust and cost-efficient plasmonic substrates, which will be detailed in Sect. 10.2.1.

The bottom surface of PAA membranes also consist of highly ordered hexagonal PAA cells, which can be utilized to fabricate plasmonic nanostructures as well. Generally, the PAA membranes obtained by previous anodic procedures do not have the underlying aluminum substrate removed. By dissolving this layer of aluminum, the bottom surface of PAA membranes (as shown in Fig. 10.1b) can be uncovered. Herein, silver nanocap superlattice arrays [30] and silver nanovoid arrays [31] templated by the bottom surface of PAA membranes were successfully fabricated by our group, which will be detailed in Sect. 10.2.2.

Later on, our group proposed a new strategy to fabricate nanostructures by using single step direct imprint process [32]. The highly ordered, close-packed, hexagonal PAA patterns can be precisely transferred to the substrate materials. This new method will be detailed in Sect. 10.2.3.

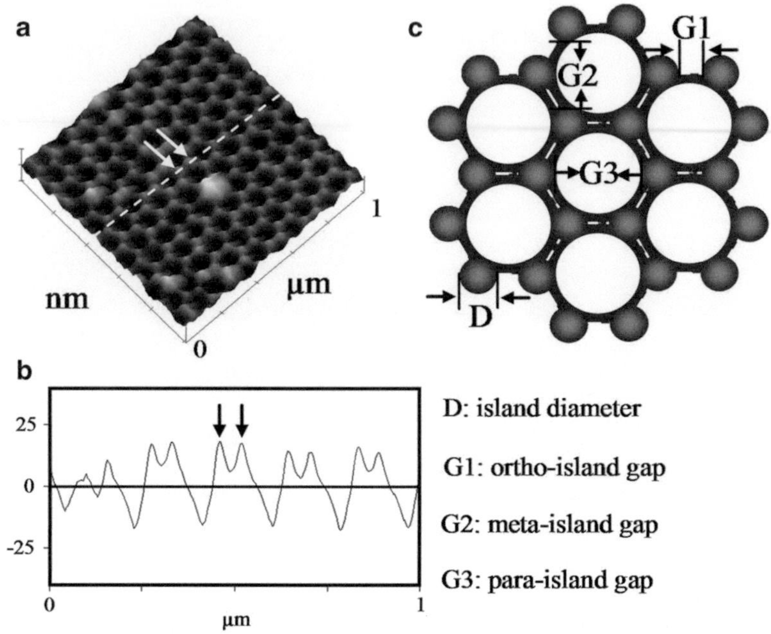

Fig. 10.2 (**a**) A 3D AFM image of the PAA surface. (**b**) Section analysis along the *white line* in (**a**). The two *arrows* correspond to those in (**a**). (**c**) Schematic representation of a planar PAA surface with dimensional parameters D, G1, G2 and G3, where D is the island diameter and G1,G2 and G3 are the ortho-, meta-, and para-island gap, respectively

10.2.1 Plasmonic Nanostructures Templated by the Pore Surface of PAA Membranes

It is well known that highly ordered PAA can be partially exposed with electrodeposition techniques to form regular nanowire and nanoelectrode [33, 34]. In a later study, Wang and colleagues adopted a similar approach to produce PAA templated silver nanoparticle arrays with a precisely controlled variation of interparticle gaps of between 5 and 25 nm [17]. Their work represented the first quantitative observation of the SERS effect on a substrate with controlled hot junctions in the sub-10-nm regime, and confirmed the theoretical prediction of interparticle coupling induced Raman enhancement. In another work, Ko and Tsukruk introduced three-dimensional (3D) SERS substrates using PAA membranes with inner walls decorated with gold nanoparticles [35]. These works have a similar feature that they all utilize uniformly aligned, vertical cylindrical pores of PAA to fabricate SERS-active nanostructures to maintain the uniformity of the nanostructures and realize reliable, stable, and uniform SERS signal.

As aforementioned, with the exception of ordered hexagonal pore arrays in PAA templates, there are small protrusions and dents one-by-one arranging along the surface of the PAA pore wall [12]. A typical structure of PAA pore surface is shown

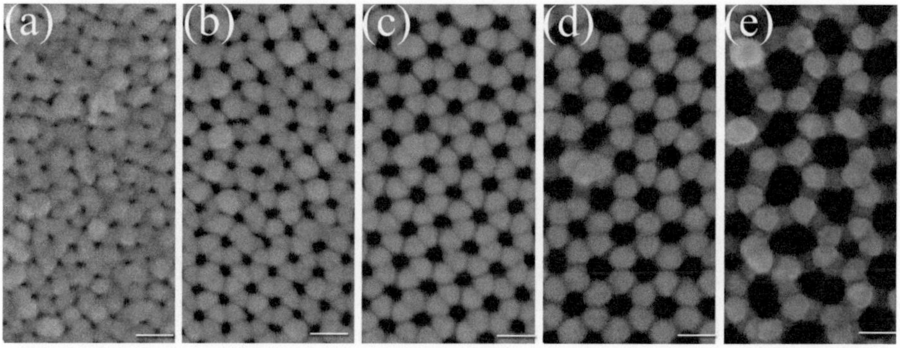

Fig. 10.3 A series of SEM images acquired from the silver-coated PAA membranes formed under different constant DC voltages: (**a**) 20, (**b**) 30, (**c**) 40, (**d**) 50, and (**e**) 60 V. The scale bar is 100 nm. The sputtering times for the silver are all set to be 10 min (Reprinted with permission from [37]. Copyright (2011) American Chemical Society)

in Fig. 10.2a (3D AFM image). A protrusion with a diameter of approximately 20 nm occurs at each corner of each hexagonal cell and a concave shape can be observed between every two adjacent protrusions. A section analysis was conducted along the white line in Fig. 10.2a and the corresponding result is presented in Fig. 10.2b. Top-to-bottom fluctuations of the cell boundary between the two protrusions are obvious. The planar geometry of the PAA substrate is shown schematically in Fig. 10.2c, in which D is the island (protrusion) diameter and G1, G2, and G3 are the ortho-, meta-, and para-island gap, respectively.

Figure 10.3 shows the procedure of fabricating arrays of hemispherical silver nanocaps separated by tunable gaps on the PAA substrates. Silver was deposited onto the pore surface of the PAA substrates using conventional direct-current magnetron sputtering system. And highly ordered silver nanocap arrays (SNAs) can grow on the protrusions along the PAA pore walls. The geometry of SNAs can be modified by tuning the anodic voltage and silver depositing parameters (as illustrated in Fig. 10.3), thus realizing the strongest coupling among adjacent silver nanocaps. The plasmonic enhancing capability of SNAs was evaluated by applying Rhodamine 6G (probe molecules: R6G) to the SNAs substrate. Enhanced R6G Raman signals can be detected by using a He-Ne laser to excite the coupled surface plasmon resonance arising from a hybrid mode at the air-silver interface. The intrinsic fluorescence of R6G molecule was quenched by coupling with SNAs, owing to the Förster resonance energy transfer mechanism [36] (this resonance energy transfer mechanism will be detailed in Sect. 10.3.2). Therefore, the plasmonic capability of SNAs can be qualitatively analyzed from SERS signals.

The strongest Raman enhancement was observed from the sample in Fig. 10.3c compared with other samples, which can be attributed to the fact that the nanocap arrays are assembled with a favorable gap configuration (G3 $\approx$ 35 nm, G2 $\approx$ 25 nm, G1 $<$ 10 nm) and a highly ordered arrangement. These, as well as a high density of both silver nanocaps ($\approx 1 \times 10^{10}$ cm^{-2}) and hot spots ($\approx 1 \times 10^{11}$ cm^{-2}), are necessary for intense SERS enhancement. Although mass particle aggregation can be observed in samples a and b (Fig. 10.3a, b), the values of G2 and G3 decrease, thus not reducing the SERS enhancement significantly.

Apart from tuning the gap between neighboring silver nanocaps, silver nanocaps size can be modified by adjusting the silver depositing time to optimize the plasmonic capabilities. The results show that there is a favorable silver depositing time to produce the most enhanced Raman signals, when the fabrication anodic potential of PAA membranes is fixed.

Therefore, this highly ordered hemispherical SNAs plasmonic substrate is very promising owing to its flexibility when spectrally tuning the SP resonance and controlling the size of interparticle gaps in sub-10-nm regime, as well as easy fabrication, excellent reproducibility, modest cost and large area production. The applications of this novel plasmonic substrate will be discussed from viewpoints of SERS trace sensing and SEF cellular imaging in Sect. 10.3.

10.2.2 Plasmonic Nanostructures Templated by the Bottom Surface of PAA Membranes

Different from the hemispherical SNAs templated by the small protrusions along the PAA pore wall, another convenient approach to fabricate plasmonic nanostructures was proposed and experimentally achieved by our group. The bottom surface of PAA membranes were acquired by the two-step anodization process, subsequently dissolving underlying aluminum substrate by immersing in saturated solution of copper sulfate with copper tape as a protective layer. Then, silver was directly deposited on the bottom surface of PAA membranes by conventional direct-current magnetron sputtering system with proper depositing time, thus silver nanocap superlattice arrays (SNSAs) were acquired (the fabrication process as shown in Fig. 10.4) [30]. If the silver depositing time was prolonged, silver nanovoid arrays (SNVAs) could be uncovered by dissolving PAA template (the fabrication process as shown in Fig. 10.5) [31].

By adjusting the anodic potential and subsequent silver deposition parameters, a series of SNSAs and SNVAs with different sizes were obtained. The shape of the silver nanocap resembles a cap based on the ratio of the height to radius of the alumina protrusions, and the silver nanovoid has the complementary shape with the nanocap. The size of each silver nanocap (or nanovoid) is consistent with the one of protuberance on the bottom surface of PAA membranes. That is, the value of D (the silver nanocap or nanovoid diameter) is almost unchanged after silver deposition. Moreover, the average size of nanocap or nanovoid can be tailored from 60 to 150 nm, and the functional relationship between D and the anodic potential is approximately linear (as shown in Fig. 10.6). Thus, combined with the adjustion of silver deposition parameters, the plasmonic properties of SNSAs and SNVAs can be tuned to achieve the strongest coupling between adjacent silver nanocaps or nanovoids.

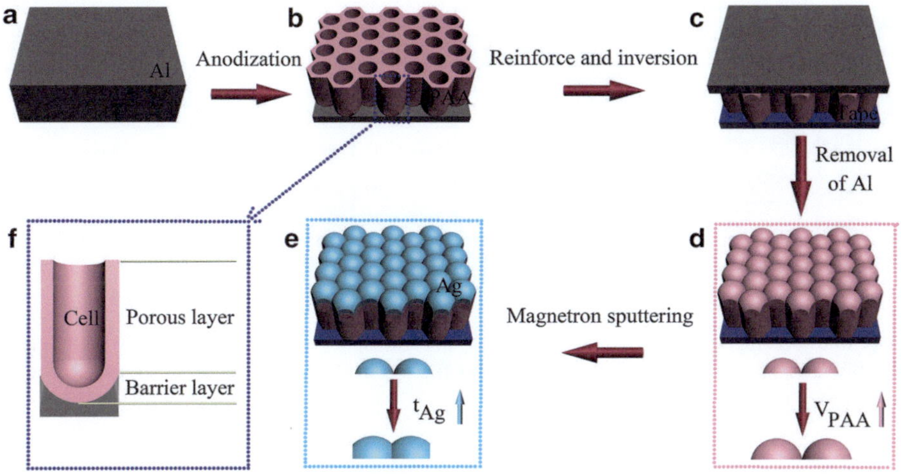

Fig. 10.4 Schematic of the fabrication method of silver nanocap superlattice arrays (SNSAs): (**a**) pretreated aluminum sheets, (**b**) PAA templates, (**c**) reinforced and inversed PAA templates, (**d**) the bottom surface of PAA, (**e**) SNSAs and (**f**) alumina cell (Reprinted with permission from [30]. Copyright (2011) American Chemical Society)

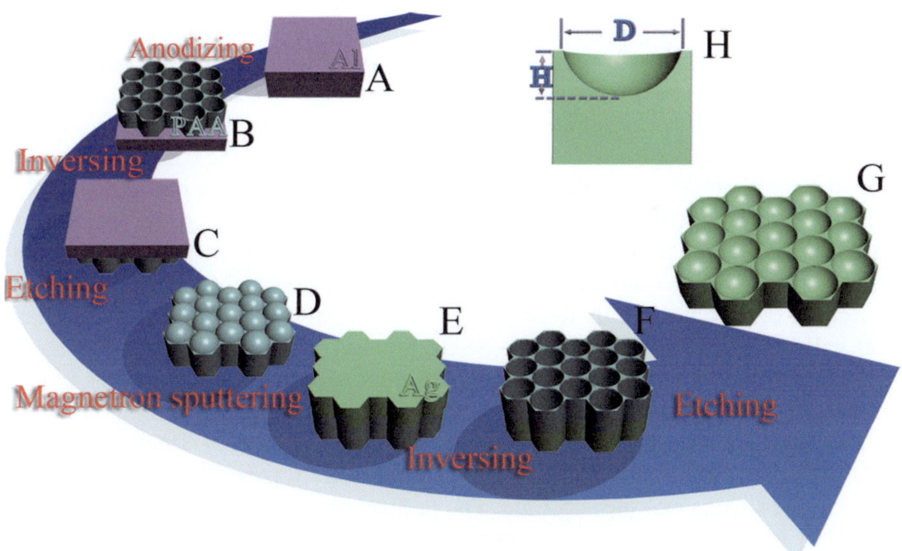

Fig. 10.5 Schematic illustrating the fabrication process for the silver nanovoid arrays (SNVAs): (**a**) pretreated aluminum sheets, (**b**) PAA templates, (**c**) reinforced and inversed PAA templates, (**d**) PAA templates with alumina protrusion arrays, (**e**) silver films which are deposited on the alumina protrusion arrays, (**f**) SNVAs protected by the barrier layer of PAA, (**g**) SNVAs and (**h**) void structure (Reprinted with permission from [31]. Copyright (2012) American Chemical Society)

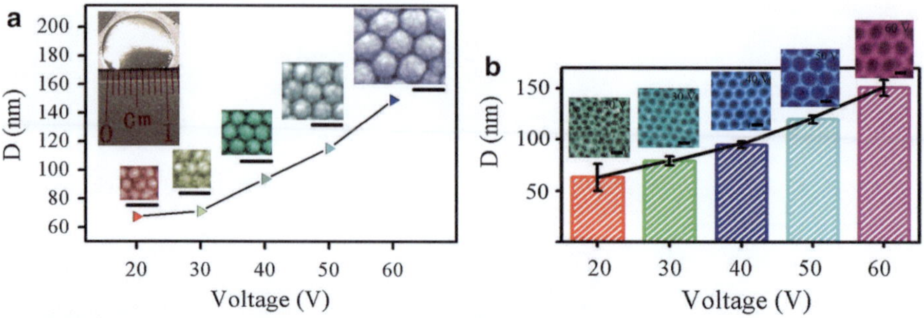

Fig. 10.6 Average size D as a function of V_{PAA} and corresponding SEM images of (**a**) SNSAs and (**b**) SNVAs. The scale bar is 200 nm in (**a**) and 100 nm in (**b**). The inset of (**a**) shows large-area SNSAs with size of about 1 cm^2

10.2.3 Plasmonic Nanostructures Patterned by Single Step Direct Imprint Process

The traditional lithography preparation methods to fabricate metallic patterns are indirect, that is, the nanoscale features are first patterned on a polymer by common lithography techniques such as photolithography, electron-beam lithography or focused ion beam, and subsequently metal deposition as well as lift-off or etching are performed [38, 39]. However these fabrication procedures tend to be complex and labor-intensive due to the multiple steps and special conditions. Consequently, alternative methods such as transfer printing [40, 41], imprinting without melting the metals [42, 43], pulsed laser melting [44], electrochemical nanoimprinting [45, 46] or direct micro-scale imprinting [47] have been proposed to produce metallic patterns directly. Although some of these techniques suffer from poor resolution, direct metal patterning has immense potential in next generation micro-scale and nanoscale manufacturing because of the low cost and high throughput.

Herein, our group recently developed a novel metal direct imprinting technique [32], called single step direct imprint (SSDI) process, which resembles stamping on a piece of paper. The "stamp" is PAA, and the "paper" is metal or other hybrid multi-layer materials.

The periodic hexagonal geometry on the pore surface of PAA membrane ("stamp") was successfully transferred to the metals ("paper") by mechanically pressing the stamps into the substrate. The protrusions along the PAA pore wall are favorable to nanoimprint process because they are too sharp to press into the substrates easily in a short time and at a small pressure. The transferred patterns show one-to-one correlation with the features of the pore surface of PAA stamp. The mean value of gaps in the transferred patterns of G1 and G2 are approximately 6 and 15 nm respectively (as shown in Fig. 10.7a), where G1 is the ortho-island gap and G2 is the size of the nanopit among adjacent three islands. This narrow size distribution (Fig. 10.7b) suggests the capability to fabricate small features (less than 10 nm), which is extremely necessary for plasmonic applications.

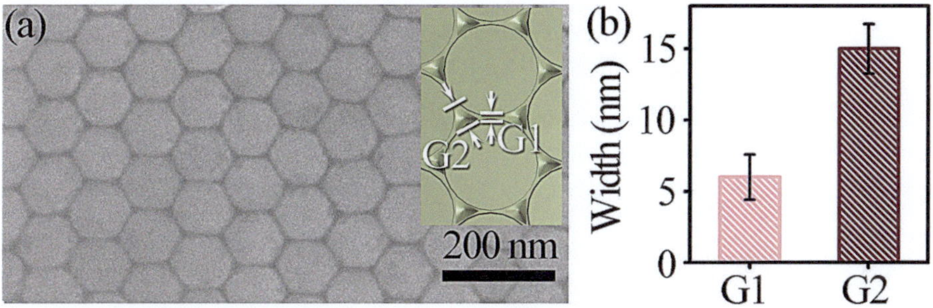

Fig. 10.7 (**a**) Representative SEM image of metallic nanostructures during pattern transfer. The inset is a schematic representation of nanopatterns with dimensional parameter G1 and G2, where G1 is the ortho-island gap and G2 is the size of the nanopit among adjacent three islands. (**b**) The width of G1 and G2

A long-range ordered hexagonal cell arrays in the PAA with dimensions of millimeters or even centimeters and pore density of 10^{10} cm^{-2} are produced. Our results illustrate that metallic nanostructures with dimensions of the order of centimeters can be constructed by this SSDI process, and it can be used to replicate large area metallic nanostructures with resolution down to a few nanometers as well as good trade-off between resolution and throughput.

With this novel SSDI process, various metal nanopatterns with different hardness and elastic moduli (Sn, Zn, Al, Ag, Mg and Cu) were precisely replicated. By adjusting the cell size and pore diameter of PAA stamps, metal nanoislands (or nanopits imprinted with relatively small pressure) with well-defined size (or spacing) between 75 and 150 nm can be readily replicated and the sizes or spacing of transferred patterns increase monotonically as the anodic voltage is increased, as shown in Fig. 10.8.

The impacts of pressure on the height of transferred patterns were investigated. We found that the height increases monotonically with pressure (as shown in Fig. 10.9, different nanopattern height with different imprint pressure), and it can be easily understood how the PAA stamp cavities are gradually filled with the substrate materials. Moreover, low error height indicates that the imprint process has good spatial uniformity.

Later on, this SSDI process was further developed by our group. Polymers and metal/polymer multi-layer structures were chosen as substrate materials ("paper"). The underlying soft polymer layer is employed to reduce the imprinting pressure in SSDI process.

Gold semi-shell arrays [48] with controllable dimensions were fabricated by using Au/PMMA bilayer structure as substrate. Figure 10.10 schematically illustrates the formation steps of close-packed gold semi-shells (both nanocaps and nanobowls); the left route corresponds to the fabrication of nanocaps by using the pore surface of PAA stamp and the right route nanobowls by using the bottom surface of PAA stamp.

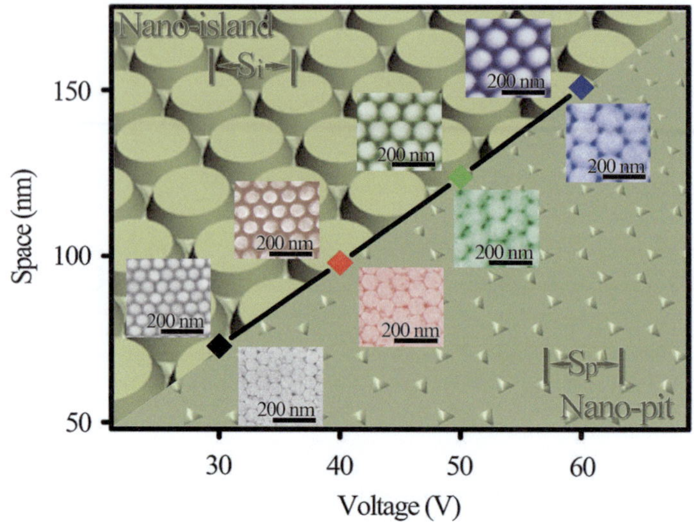

Fig. 10.8 Average spacing as a function of voltage and corresponding SEM images of aluminum nanoislands (*above*) and nanopits (*below*) templated by SSDI process

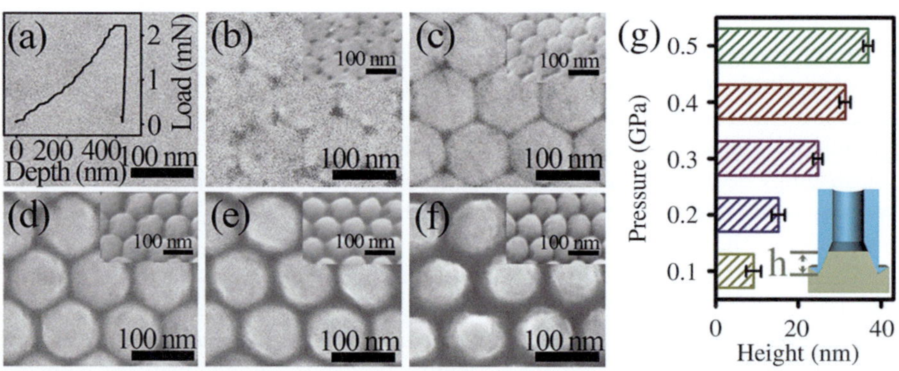

Fig. 10.9 (**a**) Typical load-displacement curve of aluminum substrate on smooth surface (inset). (**b–f**) SEM top and side views (inset) of transferred patterns under different imprinting pressures: (**b**) 0.1 GPa, (**c**) 0.2 GPa, (**d**) 0.3 GPa, (**e**) 0.4 GPa and (**f**) 0.5 GPa. (**g**) Average height as a function of working pressure. The inset is the legend to the height

The imprinting pressure of SSDI process is only about 1 % that of the previous direct imprint method [49], making it compatible with conventional imprinting processes. The low imprinting pressure results in an extended service life of PAA stamps, which can be reused several times nearly without abrasion or loss of resolution. After SSDI process, the low adhesion between noble metal and PAA stamp surface makes it easy to demold without further surface treatment.

Another approach to fabricate gold semi-shell arrays [50] was performed by only using polymer as substrate and subsequently depositing gold on the transferred

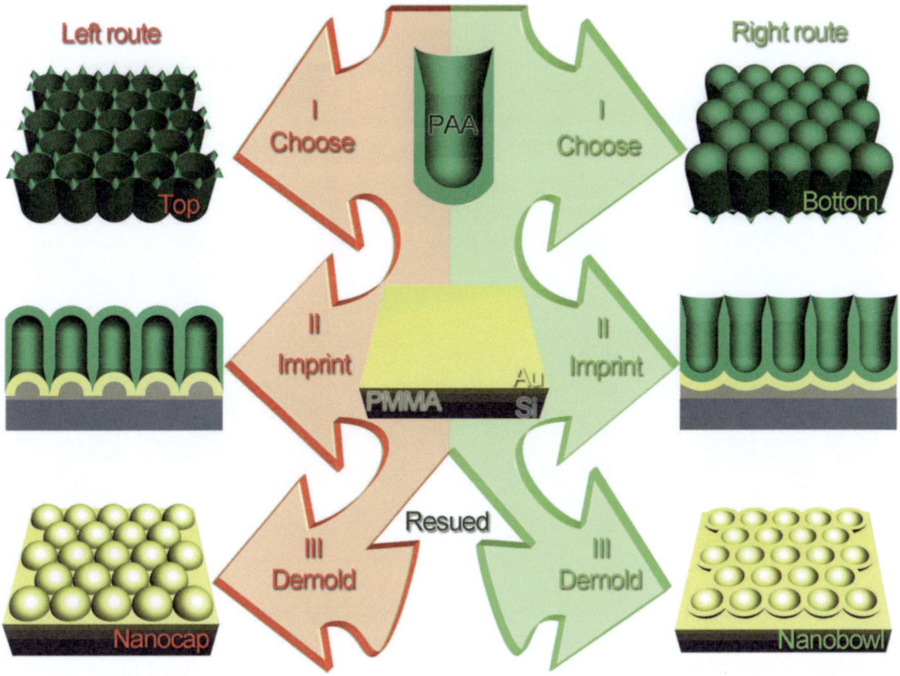

Fig. 10.10 Representation of the SSDI process for patterning gold semi-shell arrays by using PAA stamps

patterns (as shown in Fig. 10.11). Polyethylene terephthalate (PET) was chosen as substrate material ("paper"). The gap spacing between neighboring two protuberances on the patterned PET substrate can be controlled in sub-10-nm regime. The gold semi-shell arrays were templated by subsequent gold depositing procedure and the thickness of gold layer can be adjusted from 5 to 25 nm for further tailoring the geometries, thus the average gap spacing between adjacent gold semi-shells can be adjusted. Hence, the LSPR supported by these gold semi-shell arrays can be tuned to achieve the strongest coupling between adjacent gold semi-shell [30].

10.3 Applications

10.3.1 Surface-Enhanced Raman Scattering (SERS) Sensing

SERS has been demonstrated as a promising way to overcome the poor efficiency of inelastic scattering processes and weak signals inherent to normal Raman spectroscopy [51], as well as drawn widespread attention as a powerful spectroscopic technique capable of nondestructive and ultrasensitive characterization down to single molecular level [5, 52].

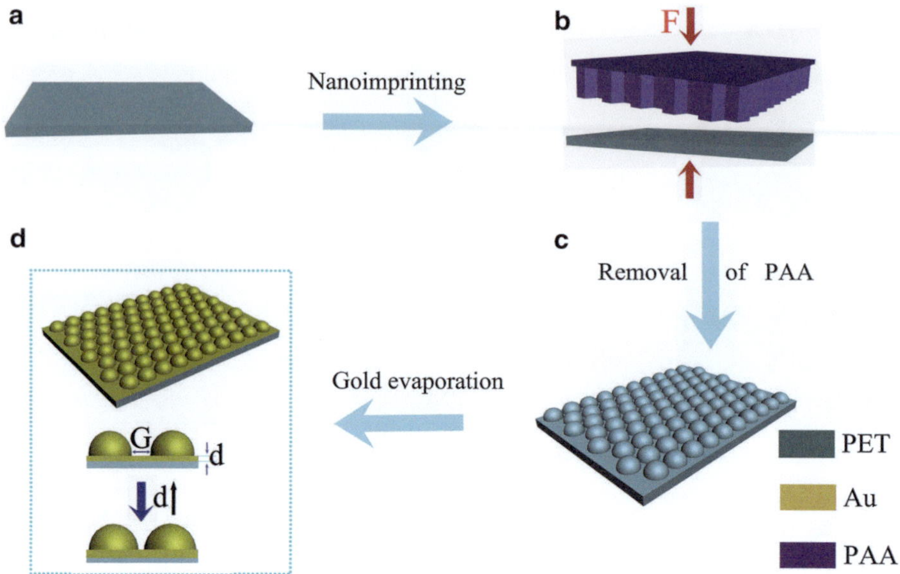

Fig. 10.11 Schematic illustrating the fabrication process for the gold semi-shell arrays: (**a**) pretreated PET substrate; (**b**) SSDI process; (**c**) patterned PET substrate; (**d**) gold semi-shell arrays (Reprinted with permission from [50]. Copyright [2013], AIP Publishing LLC)

It is important to note that, after much debate, it is currently widely accepted that the SERS can be discussed in terms of electromagnetic and chemical enhancements, in which the former term is responsible for the major portion of SERS enhancement [7]. Hence, the following discussion about SERS would be only focused on electromagnetic enhancement mechanism.

Electromagnetic enhancement is caused by the excitation of SPs, resulting in enhancement of the local field experienced by a molecule adsorbed on the surface of the nanoparticles (or nanostructures) [53]. Although the increase in the local electrical field is usually modest, the enhancement in the inelastically scattered light intensity scales to the fourth power, causing a remarkable SERS effect, that is, the enhancement factor (EF) of SERS signal $= (E_{loc}/E_0)^4$, where (E_{loc}/E_0) is the local electrical filed enhancement ratio.

For the application of routine SERS online trace sensing, SERS substrates are required to be stable with a high concentration of giant Raman cross-section hot spot and also be reproducibly prepared, inexpensive and easy to make [12]. Although conventional method, spraying Ag or Au colloids onto a substrate leads to an extremely high SERS signal at some local hot spots, due to particle aggregation, it is not easy to achieve a reliable, stable, and uniform SERS signal spanning a wide dynamic range. Popular approaches to remedy the problems of poor control over the particle aggregation states include immobilization at surfaces [54], entrapment within stable matrices [55], or fabrication of complex surface structures (e.g., with microfabrication), but these processes can be demanding in terms of labor and cost, and it is sometimes impossible to extend them to large dimensions [13, 56]. Precise control of the gaps between the nanostructures on a SERS-active substrate, with interparticle-gap dimensions in the sub-10-nm regime is extremely necessary for intense SERS enhancement.

Three approaches to fabricate plasmonic nanostructures based on PAA membranes have been described above [12, 30–32], all of which overcome the disadvantages discussed above, that is, PAA-based plasmonic nanostructure can provide a stable, reliable and reproducible platform for SERS sensing application. Besides, low costs and easy fabrication procedures make it a promising candidate for practical applications. All these three approaches can control the interparticle-gap dimensions of metallic nanostructures in sub-10-nm regime by simply adjusting the PAA fabrication conditions, metal depositing parameters or nanoimprint pressure. Additionally, the dimensions of PAA-based long-range uniform plasmonic nanostructures can be extended to the centimeter scale.

The Raman enhancing capabilities of these plasmonic nanostructures can be depicted by assessing the EF values. In our experimental researches, the probe molecules were placed near the "hot spots" by drop-coating technique, maintaining in the solution of probe molecules with subsequent rinsing, or other methods. The EF values were empirically estimated by ratios of the average SERS peak intensity (typically, one conspicuous Raman peak of the probe molecule) to the corresponding average unenhanced Raman signals (the same Raman peak with enhanced one).

For instance, the Raman enhancing capabilities of silver nanocap superlattice arrays (SNSAs) [30] were evaluated by applying an aqueous solution (10^{-5} M) of R6G to the substrates. Figure 10.12 shows the typical Raman spectrum of the R6G solution obtained on the SNSAs (V_{PAA} =40 V, where V_{PAA} is the anodic potential of PAA template) which is compared to that acquired from the flat silver sheet (unpatterned one). However, the normal Raman signature of 10^{-2} M of R6G on the flat silver sheet is barely recognizable. The empirical EF is estimated to be larger than 10^5 by comparing ratios of the average SERS peak intensity (at 1510 cm^{-1}) of R6G to the corresponding average unenhanced signals (at 1510 cm^{-1}).

By employing the finite-difference time-domain (FDTD) simulations, this intense SERS enhancement can be confirmed arising from the enhanced local electrical field. FDTD is a kind of calculation method which has recently been demonstrated to be highly useful in the study of the electromagnetic properties of metallic nanostructures for almost arbitrary complexity [57]. A typical planar and cross sectional view of the calculated radial EM field components of the SNSAs (V_{PAA} = 40 V, t_{Ag} = 15 min) [30] is displayed in the inset of Fig. 10.12. Calculation shows that these localized resonant plasmon modes are able to produce maximum SERS EF as large as 3.5×10^{10} in the Raman signal of the molecules adsorbed at the V-shaped gaps between two adjacent nanocaps. The SERS performances of other SNSAs formed under different V_{PAA} (t_{Ag} is fixed at 15 min) were also experimentally and theoretically conducted, demonstrating good correlation between experiments and theories. The results show that there exists a favorable V_{PAA} and t_{Ag} parameters for the most intense SERS enhancement.

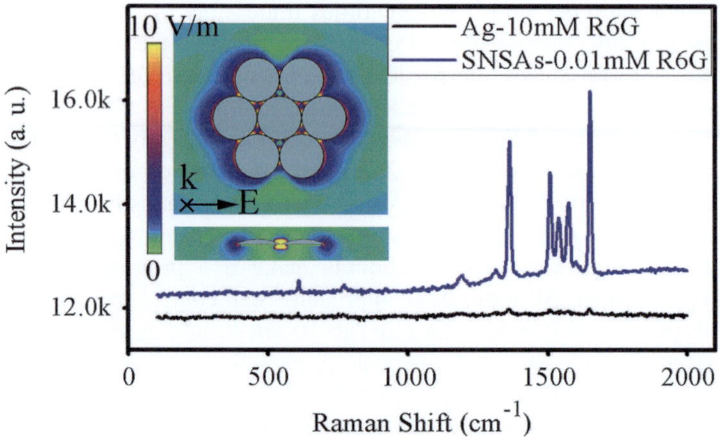

Fig. 10.12 SERS spectral comparison of R6G adsorbed on the SNSAs and the flat silver sheet. The inset shows a typical simulated EM-field distribution map of the SNSAs ($V_{PAA} = 40$ V, $t_{Ag} = 15$ min) (Reprinted with permission from [30]. Copyright (2011) American Chemical Society)

Moreover, the homogeneity of plasmonic substrates is of great significance for realizing stable, reliable and uniform signals spanning a wide dynamic range, which can be evaluated by means of 2D point-by-point Raman mapping. It can visualize the spatial distribution of Raman intensity by monitoring some finger print Raman peak of probe molecules. For instance, excellent uniformity of hemispherical SNAs [37] over a large area of 20.0×20.0 μm is confirmed by measuring the SERS signals of 400 points with a regular scanning step of 1 μm (as shown in Fig. 10.13). The map is obtained from the integrated intensity of the 612 ± 10 cm^{-1} band of 10^{-5} M R6G molecules adsorbed on the substrate. The relative SERS peak intensity of the collection spots is centered in a narrow range, and the spot-to-spot relative standard deviation is calculated to be 5 %. This relative standard deviation value can be used to evaluate the homogeneity of the plasmonic substrates, that is, relatively smaller value stands for a better homogeneity, and vice-versa.

Hence, our group proposed a brief analytical methodology to evaluate the dispersity of adsorbed multi-walled carbon nanotubes (MWNTs) with different ultra-low concentrations by 2D point-by-point Raman mapping strategy, based on a highly SERS-active substrate with excellent homogeneity [58]. Figure 10.14 shows a series of 2D Raman mappings of MWNTs on hemispherical SNAs compared with those on Si chips, with different concentrations of R6G molecules. The variation in the intensity of Raman signals around 1575 cm^{-1} can be observed according to brightness, with brighter spots representing higher Raman intensity in each Raman mapping images.

In Fig. 10.14a–c, an obvious weakening trend of intensity in the normal Raman mappings on decreasing the concentration of MWNTs can be observed, and MWNTs with concentration below 3 ppm on Si chip exhibit no visible Raman bands. However, intense SERS enhancement can be observed even at 3 ppm in

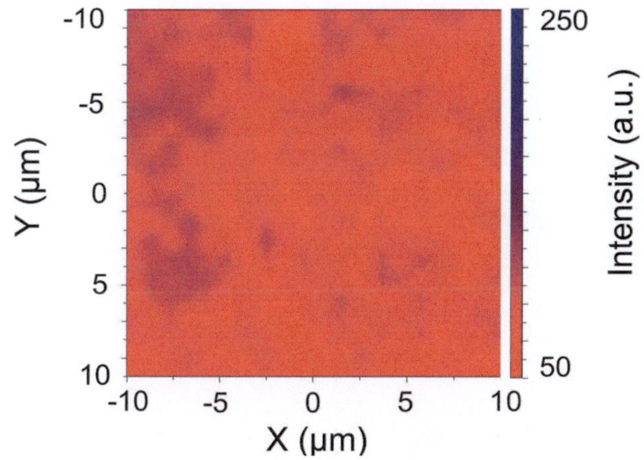

Fig. 10.13 SERS mapping (20.0 × 20.0 μm) obtained from the hemispherical SNAs with $V_{PAA} = 40$ V and $t_{Ag} = 15$ min. The map is obtained from the integrated intensity of the 612 ± 10 cm^{-1} band of 10^{-5} M R6G molecules adsorbed on the substrate. Each point in the map was taken using 20 mW 514 nm excitation with a 1 s integration time (Reprinted with permission from [37]. Copyright (2011) American Chemical Society)

SERS mappings (as shown in Fig. 10.14f). The SERS mappings of MWNTs for ppb level are also tested (as shown in Fig. 10.14g–i). In our experiment, the detection limit reaches as low as 3 ppb which is proved by the result of SERS mapping shown in Fig. 10.14i.

This design of SNAs with uniform SERS-active properties may open a new framework for the fabrication of robust, cost-effective, large-area SERS-based sensors for trace detection of inorganic nanostructures.

10.3.2 Fluorescence Process Tailoring and Surface-Enhanced Fluorescence (SEF) Cellular Imaging

Since the early 1980s, enhancement of fluorescence from molecules near a metal nanostructure has been regarded to be associated with modification of the molecule excitation as well as the radiative and non-radiative decay rates [59, 60]. Nowadays, improved nanofabrication methods allow precise control of the nanoparticle shape and arrangement of nanoparticle ensembles thereby opening the possibility to flexibly tailor specific molecule-nanoparticle couplings [6, 7].

Herein, our group recently reported several researches of tailoring light emission properties of organic emitter by coupling to resonance-tuned silver nanocap arrays (SNAs) which support LSPR at optical frequencies.

In our research [61], Poly-[2-methoxy-5-(2′-ethyl-hexyloxy)-p-phenylene vinylene] (MEH-PPV) was chosen as the organic emitter. The PAA membrane was cut into two pieces, one being a fresh sample and the other sputtered with Ag to

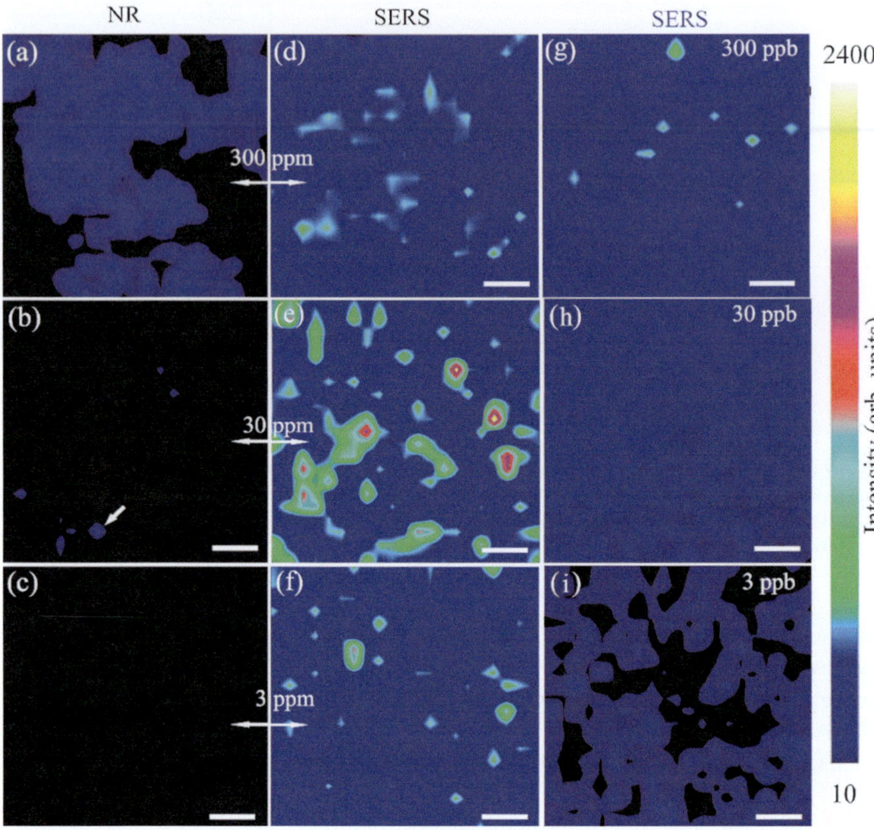

Fig. 10.14 Raman mappings of MWNTs with different concentrations adsorbed on Si chips ((**a**) 300 ppm, (**b**) 30 ppm, (**c**) 3 ppm) and on hemispherical SNAs ((**d**) 300 ppm, (**e**) 30 ppm, (**f**) 3 ppm, (**g**) 300 ppb, (**h**) 30 ppb, (**i**) 3 ppb). The scale bar is 10 μm. NR represents normal Raman signal

form the SNAs, and subsequently with MEH-PPV solution dropped onto the sample.

The PL spectra of MEH-PPV coupled two samples, acquired by excitation with the wavelength of 500 nm, are compared in Fig. 10.15a. The peak intensity of the PL spectrum of the MEH-PPV coupled SNAs is about 6 times higher than that of the MEH-PPV on the PAA membrane and the PL peaks at 575 nm are almost the same. The enhancement of the PL spectrum of the MEH-PPV coupled SNAs were confirmed by the UV-visible absorbance spectra in Fig. 10.15b, which was originated from the energy transfer effect in the SPR coupling.

The PL spectrum of MEH-PPV and the large absorption spectrum (surface plasmon resonance spectrum) of the SNAs are well overlapped (as shown in Fig. 10.15b). Similar to the donor-acceptor energy matching in Förster resonance energy transfer between two fluorophores, the critical matching of the localized resonating plasmon energy in SNAs with the electron transition energy from ground

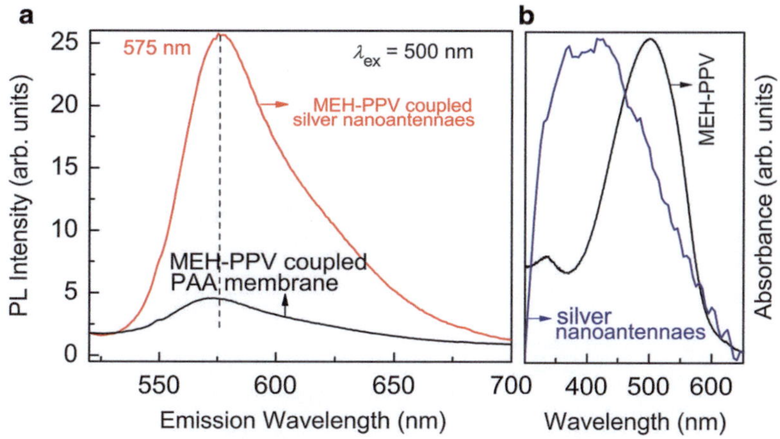

Fig. 10.15 (**a**) Comparison of PL spectra of the MEH-PPV coupled silver nanocap arrays (SNAs) and MEH-PPV coupled PAA membrane. (**b**) Normalized UV-visible absorption spectra of the MEH-PPV and SNAs (Reprinted with permission from [61]. Copyright [2009], AIP Publishing LLC)

to excited state in MEH-PPV permits the plasmon resonance energy transfer process. Then, the resulting resonance (LSPR coupling due to the hybrid junction of MEH-PPV and SNAs) could lead to energy transfer from the SNAs to the light-emitting MEH-PPV with the irradiating wavelength of 500 nm, which created more excitons in the light-emitting MEH-PPV [62].

Besides, another important factor responsible for the PL enhancement may be the enhanced local electromagnetic field of nanogaps between the SNAs in the background of light-emitting organic emitter (as shown in Fig. 10.16). As aforementioned, the local EM field enhancement can be tuned to achieve the strongest coupling between adjacent nanocaps by varying the anodic voltage of the PAA templates [12]. The maximum PL enhancement ratio was observed from the sample with anodic potential of 40 V, which can be attributed to the fact that the SNAs (under the anodic potential of 40 V) is assembled with the favorable gap configuration, a high density of both Ag nanocaps and hot gaps as well as the highly ordered arrangement.

Very similar with MEH-PPV coupled SNAs system, in another work of our group, Poly-[3-(2,5,8-trioxanonyl) thiophene] (P3TT) and poly (2,5-dioctyloxy-p-phenylene) (PPP) were chosen as the organic emitters, with the absorption peaks of ~500 and ~345 nm, respectively [63]. The conjugated polymers were spin-coated on the SNAs films and quartz substrates under the same conditions.

The PL intensity of the P3TT coupled SNAs is more than 9 times higher than that of P3TT film on quartz for excitation at 490 nm (as shown in Fig. 10.17a). The plasmon resonance energy transfer plays a much more important role in the large PL enhancement of the P3TT coupled SNAs than the increased absorption of coupled P3TT. The absorption spectrum of P3TT and that of the SNAs well overlap, which permits the plasmon resonance energy transferring from the latter

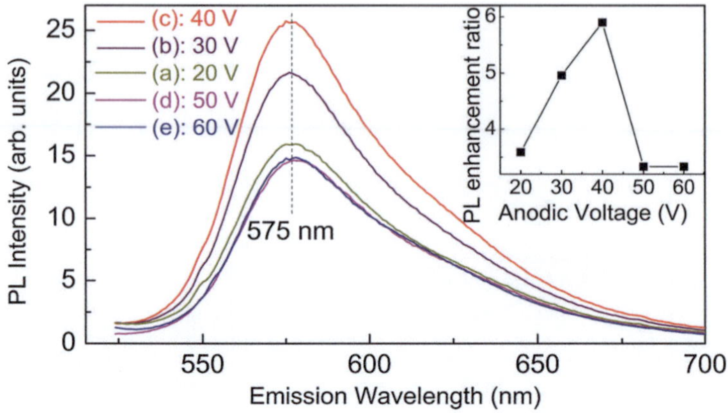

Fig. 10.16 PL spectra of MEH-PPV coupled SNAs formed under different constant d_c voltages ((a) 20 V, (b) 30 V, (c) 40 V, (d) 50 V and (e) 60 V). The sputtering times of silver are all set to be 10 min. The dependence of the PL enhancement ratio of MEH-PPV coupled SNAs to MEH-PPV coupled PAA membrane is shown in the inset (Reprinted with permission from [61]. Copyright [2009], AIP Publishing LLC)

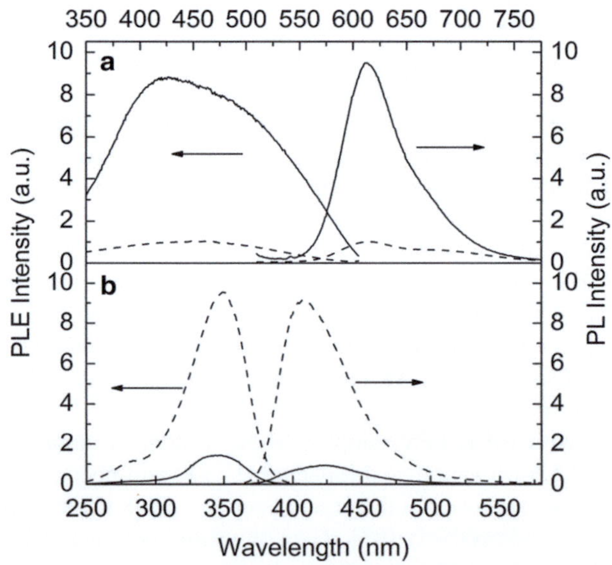

Fig. 10.17 (**a**) PL and PLE of the P3TT coupled SNAs (*solid line*) and the P3TT film (*dashed line*), the PL spectra were obtained by photoexcitation at 490 nm and the PLE spectra were measured at the monitoring emission wavelength of 620 nm. (**b**) PL and PLE of the PPP coupled SNAs (*solid line*) and the PPP film (*dashed line*), the PL spectra were obtained by photoexcitation at 340 nm and the PLE spectra were measured at the monitoring emission wavelength of 420 nm (Reprinted with permission from [63]. Copyright [2011], AIP Publishing LLC)

to the former. This energy resonance transfer process is very similar to MEH-PPV coupled SNAs system discussed above [61]. That is, if the absorption of the conjugated polymer matched that of the SNAs, the emission intensity of the conjugated polymer would be largely increased.

Totally different from P3TT, the emission from the PPP coupled SNAs is much weaker than that of the PPP film on quartz, as shown in Fig. 10.17b. By comparison, the PL spectrum of PPP well matches the absorption spectrum of the SNAs (surface plasmon resonance spectrum). Hence, the energy transfer occurs from PPP to the SNAs, resulting in the fluorescence quenching of coupled PPP [64]. And experimental evidences indicate that the energy transfer process does not directly quench the excited states in PPP, that is, the energy transfer is radiative rather than non-radiative Förster type in the PPP coupled SNAs system [65]. Therefore, if the emission from the conjugated polymer matched the absorption of the SNAs, the fluorescence quenching of the conjugated polymer would be observed.

Our results discussed above indicated that it is achievable to optimize the geometry of the SNAs in order to maximize the emission intensity of light emitters. Hence, we further applied these results to the SEF cellular imaging of biomolecules.

Cellular imaging is especially useful to non-invasive monitoring of diseases, evaluation of drug effects, assessment of the pharmacokinetic behavior of drugs, and identification of molecular biomarkers for diseases [66]. In order to accurately study the progress, a multi-color biological labeling method is needed for membrane protein imaging. Cellular fluorescence imaging meets the requirement due to the large range of fluorophores with distinctive spectral characteristics suitable for clinical practice. Moreover, it offers the specific, targeted imaging contrast needed for the study of specific cellular processes. Hence, cellular fluorescence imaging is now an irreplaceable tool in biomedical science. The advent of fluorescence microscopy has enabled routine studies of dynamic processes in living cells. These fluorescence techniques can deliver the necessary resolution to image certain cellular organelles and track proteins in for example, the nucleus, endoplasmic reticulum, and Golgi apparatus, and other biomolecules in living cells [67], so that valuable information about the dynamics of intracellular networks, signal transduction, and intercellular interactions can be obtained.

In spite of recent advances, the spatial and temporal resolution of cellular fluorescence microscopy is limited by traditional organic fluorescent dyes. When using fluorophores as dyes, the signal cannot be easily discerned [68, 69] from the background of autofluorescence, which is natural emission from biological entities [70]. Autofluorescence can be problematic in fluorescence microscopy due to several reasons. First of all, the unwanted signals may interfere with specific fluorescent signals especially when the latter are very dim. Secondly, the emission lifetime on the order of 2–4 ns is very close to that of the cell autofluorescence background, and thirdly, organic fluorophores are known to emit signals with poor stability and strong blinking [37, 71]. In many instances, the sensitivity is limited by autofluorescence from the sample rather than lack of signal. Hence, one of the main challenges is to improve the sensitivity and photostability.

In order to further improve the sensitivity, there are continuous attempts to lower the detection limits. Detection of a fluorophore is usually limited by the quantum yield, autofluorescence from the sample, and photostability of the fluorophore. And as aforementioned, there is increasing use of metallic nanostructures to modify the spectral properties of fluorophores and to overcome some of these photophysical constraints. Depending on the geometry and the distance between the metal and fluorophore, these metallic surfaces can result in fluorescence enhancement by factors of up to 10^3 [72, 73].

In our results, the silver nanocap arrays templated by the PAA membranes can obtain reliable, stable, and uniform signals spanning a wide dynamic range from the surface of SNAs [12, 37]. This uniform geometry of SNAs is necessary to make sure that different parts of the sample can have the same degree of amplification and enable the use of SEF cellular imaging in the study of intracellular processes at the single-molecule level.

Our recent work reported that this novel plasmonic substrate was successfully utilized to high-sensitivity and stable cellular fluorescence imaging [37]. - Phalloidin-fluorscein isothiocyanate (P-FITC) is used as fluorophore to label the biomolecules. Spectral analysis suggests that the SNAs can create more excitons in the light-emitting P-FITC because of plasmon resonance energy transfer from the SNAs to the nearby P-FITC. They can also act as plasmonic antennae by converting a part of the non-radiative near-field emission from the fluorophores to the far field consequently enhancing the emission. AG1522, a normal human fibroblast culture derived from foreskins of 3-day-old male, and Chinese Hamster Ovary cells are labeled by P-FITC on the cytoskeletons and the fluorescence signals from the fluorophores bound on the cell cytoskeletons on the SNAs are enhanced 8-fold compared to those on glass used in conventional imaging, as illustrated in Fig. 10.18.

It is known that fluorescence is enhanced only when the fluorophore is localized at an optimal distance close to the surface of the metal nanoparticle with a certain distance over 5 nm [74, 75]. In this case, because the membranes of the cells are about 7 nm thick [76], some fluorophores on the cell cytoskeletons that adhere to the metal substrate may be localized within this enhancement region. Hence, it can be inferred that enhancement of the emission signals from the cell cytoskeletons observed in this study is due to the coupling effect between the P-FITC and SNAs.

In addition to the intensity enhancement, the photostability is improved dramatically, as illustrated in Fig. 10.19. Fluorescence from the P-FITC labeled Chinese Hamster Ovary cells on glass is quenched in 2 min. In contrast, fluorescence from the P-FITC-labeled Chinese Hamster Ovary cells on the SNAs diminishes to half of its original intensity in 3 min, after which fluorescence is maintained for more than 12 min. The results suggest that the fluorophores near the glass are destroyed, whereas those near the SNAs are more photostable.

Apart from the geometry of plasmonic substrates, another significant factor dominating the fluorescence intensity is the distance between the metal and fluorophores. LSP works only at a very short distance range of no more than 100 nm and is influenced by the dielectric layer. In another work [77], optical

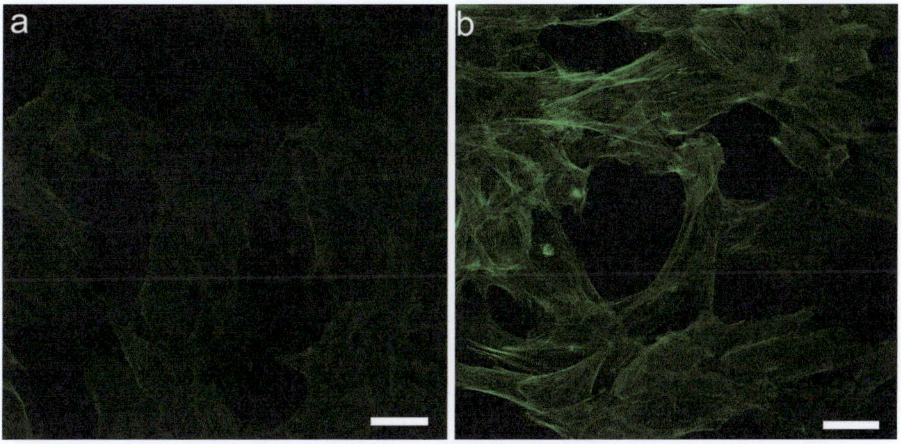

Fig. 10.18 Representative confocal images of AG1522 cells labeled with P-FITC adhered to the cytoskeleton protein f-actin on the (**a**) glass slide and (**b**) SNAs excited by the 488 nm laser line. Both images are in pseudo color and the scale bar is 20 μm (Reprinted with permission from [37]. Copyright (2011) American Chemical Society)

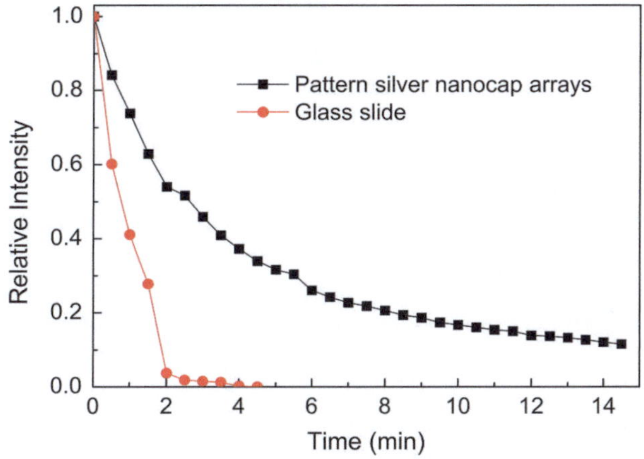

Fig. 10.19 Photostability tests on P-FITC labeled Chinese Hamster Ovary cells on the SNAs and glass slide. The excitation source is a mercury lamp operated at 450–490 nm (Reprinted with permission from [37]. Copyright (2011) American Chemical Society)

properties of R6G molecules on SNAs are tailored by precisely controlling the distance between the dyes and SNAs by coating MgF_2 film with different thicknesses. The fluorescence signal of R6G molecules depends on the distance and the largest enhancement of 10 folds is achieved when the distance between R6G molecules and SNAs is 10 nm. This result is applied in SEF cellular imaging by coupling SNAs with P-FITC staining on cytoskeleton of the rat bone mesenchymal stem cells. The strongest emission from the 5 nm thick MgF_2 film covered SNAs increases by a factor of 10.4 compared to the glass slide. As the membrane thickness of the cell is about 3–7 nm [76, 78], the distance between SNAs and dyes here should be modified from 5 to 8–12 nm, as shown in Fig. 10.20.

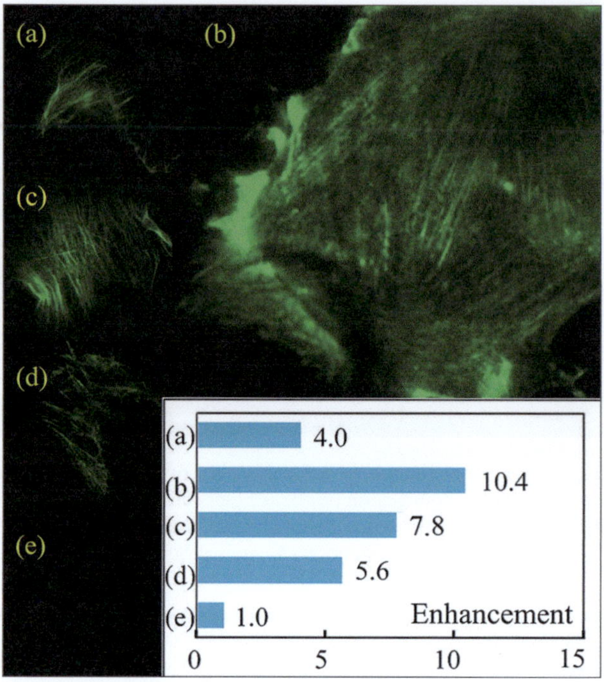

Fig. 10.20 Representative confocal images of cell cytoskeletons labelled with P-FITC on the PSNAs with different MgF_2 thicknesses: (**a**) 0 nm, (**b**) 5 nm, (**c**) 15 nm, (**d**) 35 nm and on glass slide (**e**). P-FITC is excited by the 488 nm laser line. The images are in pseudo color and the bar chart shows the fluorescence enhancement compared to that on the glass slide

We have discussed the geometry and distance dependence of the fluorophores coupled SNAs system [37, 77]. The result that the emission intensity of specific fluorophores could be dramatically increased is successfully applied to the SEF cellular imaging of biomolecules. And the results show that SNAs are good replacements for traditional glass slide. This method could be further developed in intrinsic fluorescence imaging [79] and sensor applications [80].

10.4 Summary

The use of PAA membranes as template to produce plasmonic nanostructures is very promising considering their easy fabrication, excellent reproducibility, modest cost, large area production, as well as the flexibility when spectrally tuning the SPR and controlling the ratio of the scattering to absorption cross sections. We have successfully fabricated several types of plasmonic nanostructures based on the pore surface or the bottom surface of the PAA template. Moreover, we can precisely control the assembly of nanostructures with the interparticle gap dimensions in the sub-10-nm regime, which permit achieving the strongest coupling among the adjacent nanoparticles (or closely spaced nanostructures). These novel plasmonic substrates with uniform and highly reproducible signals were successfully applied to SERS trace sensing, fluorescence process tailoring as well as SEF cellular imaging.

Acknowledgment This work was jointly supported by the National Natural Science Foundation of China under Grant No. 51271057 and the Natural Science Foundation of Jiangsu Province, China, under Grant No. BK2012757.

References

1. Kreibig U, Vollmer M (1995) Optical properties of metal clusters, vol 25. Springer, Berlin
2. Hutter E, Fendler JH (2004) Exploitation of localized surface plasmon resonance. Adv Mater 16(19):1685–1706
3. Underwood S, Mulvaney P (1994) Effect of the solution refractive index on the color of gold colloids. Langmuir 10(10):3427–3430
4. Mulvaney P (1996) Surface plasmon spectroscopy of nanosized metal particles. Langmuir 12 (3):788–800
5. Kneipp K, Wang Y, Kneipp H, Perelman LT, Itzkan I, Dasari RR, Feld MS (1997) Single molecule detection using surface-enhanced Raman scattering (SERS). Phys Rev Lett 78 (9):1667
6. Banholzer MJ, Millstone JE, Qin L, Mirkin CA (2008) Rationally designed nanostructures for surface-enhanced Raman spectroscopy. Chem Soc Rev 37(5):885–897
7. Ko H, Singamaneni S, Tsukruk VV (2008) Nanostructured surfaces and assemblies as SERS media. Small 4(10):1576–1599
8. Fleischmann M, Hendra PJ, McQuillan A (1974) Raman spectra of pyridine adsorbed at a silver electrode. Chem Phys Lett 26(2):163–166
9. Jeanmaire DL, Van Duyne RP (1977) Surface Raman spectroelectrochemistry: part I. Heterocyclic, aromatic, and aliphatic amines adsorbed on the anodized silver electrode. J Electroanal Chem Interfacial Electrochem 84(1):1–20
10. Schlegel VL, Cotton TM (1991) Silver-island films as substrates for enhanced Raman scattering: effect of deposition rate on intensity. Anal Chem 63(3):241–247
11. Lee P, Meisel D (1982) Adsorption and surface-enhanced Raman of dyes on silver and gold sols. J Phys Chem 86(17):3391–3395
12. Qiu T, Zhang W, Lang X, Zhou Y, Cui T, Chu PK (2009) Controlled assembly of highly Raman-enhancing silver nanocap arrays templated by porous anodic alumina membranes. Small 5(20):2333–2337
13. Gunnarsson L, Bjerneld E, Xu H, Petronis S, Kasemo B, Käll M (2001) Interparticle coupling effects in nanofabricated substrates for surface-enhanced Raman scattering. Appl Phys Lett 78 (6):802–804
14. Haynes CL, Van Duyne RP (2003) Plasmon-sampled surface-enhanced Raman excitation spectroscopy. J Phys Chem B 107(30):7426–7433
15. Tessier PM, Velev OD, Kalambur AT, Rabolt JF, Lenhoff AM, Kaler EW (2000) Assembly of gold nanostructured films templated by colloidal crystals and use in surface-enhanced Raman spectroscopy. J Am Chem Soc 122(39):9554–9555
16. Terekhov S, Mojzes P, Kachan S, Mukhurov N, Zhvavyi S, Panarin AY, Khodasevich I, Orlovich V, Thorel A, Grillon F (2011) A comparative study of surface-enhanced Raman scattering from silver-coated anodic aluminum oxide and porous silicon. J Raman Spectrosc 42(1):12–20
17. Wang HH, Liu CY, Wu SB, Liu NW, Peng CY, Chan TH, Hsu CF, Wang JK, Wang YL (2006) Highly raman-enhancing substrates based on silver nanoparticle arrays with tunable sub-10 nm gaps. Adv Mater 18(4):491–495
18. Schierhorn M, Lee SJ, Boettcher SW, Stucky GD, Moskovits M (2006) Metal–silica hybrid nanostructures for surface-enhanced Raman spectroscopy. Adv Mater 18(21):2829–2832

19. Lombardi I, Cavallotti P, Carraro C, Maboudian R (2007) Template assisted deposition of Ag nanoparticle arrays for surface-enhanced Raman scattering applications. Sens Actuators B 125 (2):353–356
20. Choi D, Choi Y, Hong S, Kang T, Lee LP (2010) Self-organized hexagonal-nanopore SERS array. Small 6(16):1741–1744
21. Lee SJ, Morrill AR, Moskovits M (2006) Hot spots in silver nanowire bundles for surface-enhanced Raman spectroscopy. J Am Chem Soc 128(7):2200–2201
22. Lee SJ, Guan Z, Xu H, Moskovits M (2007) Surface-enhanced Raman spectroscopy and nanogeometry: the plasmonic origin of SERS. J Phys Chem C 111(49):17985–17988
23. Keller F, Hunter M, Robinson D (1953) Structural features of oxide coatings on aluminum. J Electrochem Soc 100(9):411–419
24. Jessensky O, Müller F, Gösele U (1998) Self-organized formation of hexagonal pore arrays in anodic alumina. Appl Phys Lett 72(10):1173–1175
25. Masuda H, Yamada H, Satoh M, Asoh H, Nakao M, Tamamura T (1997) Highly ordered nanochannel-array architecture in anodic alumina. Appl Phys Lett 71(19):2770–2772
26. Masuda H, Satoh M (1996) Synthesis of highly ordered porous alumina membranes using electrochemical techniques. Jpn J Appl Phys Part 2 35:L126–L128
27. Asoh H, Nishio K, Nakao M, Tamamura T, Masuda H (2001) Conditions for fabrication of ideally ordered anodic porous alumina using pretextured Al. J Electrochem Soc 148(4):B152–B156
28. Masuda H, Fukuda K (1995) Ordered metal nanohole arrays made by a two-step replication of honeycomb structures of anodic alumina. Science 268(5216):1466–1468
29. Li A, Müller F, Birner A, Nielsch K, Gösele U (1998) Hexagonal pore arrays with a 50–420 nm interpore distance formed by self-organization in anodic alumina. J Appl Phys 84 (11):6023–6026
30. Lang X, Qiu T, Zhang W, Yin Y, Chu PK (2011) Tunable silver nanocap superlattice arrays for surface-enhanced Raman scattering. J Phys Chem C 115(49):24328–24333
31. Lang X, Qiu T, Yin Y, Kong F, Si L, Hao Q, Chu PK (2012) Silver nanovoid arrays for surface-enhanced Raman scattering. Langmuir 28(23):8799–8803
32. Lang X, Qiu T, Long K, Han D, Nan H, Chu PK (2013) Direct imprint of nanostructures in metals using porous anodic alumina stamps. Nanotechnology 24(25):255303
33. Yao J, Pan G, Xue K, Wu D, Ren B, Sun D, Tang J, Xu X, Tian Z (2000) A complementary study of surface-enhanced Raman scattering and metal nanorod arrays. Pure Appl Chem 72 (1–2):221–228
34. Yao J-L, Tang J, Wu D-Y, Sun D-M, Xue K-H, Ren B, Mao B-W, Tian Z-Q (2002) Surface enhanced Raman scattering from transition metal nano-wire array and the theoretical consideration. Surf Sci 514(1):108–116
35. Ko H, Tsukruk VV (2008) Nanoparticle-decorated nanocanals for surface-enhanced Raman scattering. Small 4(11):1980–1984
36. Sen T, Sadhu S, Patra A (2007) Surface energy transfer from rhodamine 6G to gold nanoparticles: a spectroscopic ruler. Appl Phys Lett 91(4):043104
37. Qiu T, Jiang J, Zhang W, Lang X, Yu X, Chu PK (2010) High-sensitivity and stable cellular fluorescence imaging by patterned silver nanocap arrays. ACS Appl Mater Interfaces 2 (8):2465–2470
38. Chou SY, Krauss PR, Zhang W, Guo L, Zhuang L (1997) Sub-10 nm imprint lithography and applications. J Vac Sci Technol B 15(6):2897–2904
39. Gates BD, Xu Q, Stewart M, Ryan D, Willson CG, Whitesides GM (2005) New approaches to nanofabrication: molding, printing, and other techniques. Chem Rev 105(4):1171–1196
40. Kim C, Burrows PE, Forrest SR (2000) Micropatterning of organic electronic devices by cold-welding. Science 288(5467):831–833
41. Yu J, Bulović V (2007) Micropatterning metal electrode of organic light emitting devices using rapid polydimethylsiloxane lift-off. Appl Phys Lett 91(4):043102

42. Buzzi S, Robin F, Callegari V, Löffler J (2008) Metal direct nanoimprinting for photonics. Microelectron Eng 85(2):419–424
43. Chuang S, Chen H, Kuo S, Lai Y, Lee C (2008) Using direct nanoimprinting to study extraordinary transmission in textured metal films. Opt Express 16(4):2415–2422
44. Cui B, Keimel C, Chou SY (2009) Ultrafast direct imprinting of nanostructures in metals by pulsed laser melting. Nanotechnology 21(4):19–63
45. Hsu KH, Schultz PL, Ferreira PM, Fang NX (2007) Electrochemical nanoimprinting with solid-state superionic stamps. Nano Lett 7(2):446–451
46. Kumar A, Hsu KH, Jacobs KE, Ferreira PM, Fang NX (2011) Direct metal nano-imprinting using an embossed solid electrolyte stamp. Nanotechnology 22(15):155302
47. Jiang J, Mei F, Meng WJ, Sinclair GB, Park S (2008) Direct microscale imprinting of Al at room temperature with Si inserts. Microsyst Technol 14(6):815–819
48. Lang X, Li DJ, Luo X, Zhang Y, Yin Y, Qiu T (2014) Tunable surface-enhanced Raman scattering from high-density gold semishell arrays with controllable dimensions. Chemphyschem 15(2):337–343
49. Hirai Y, Ushiro T, Kanakugi T, Matsuura T (2003) Fine gold grating fabrication on glass plate by imprint lithography. Nanofabrication Technol 5220:74–81
50. Long K, Luo X, Nan H, Du D, Zhao W, Ni Z, Qiu T (2013) Surface-enhanced Raman scattering from graphene covered gold nanocap arrays. J Appl Phys 114(18):183520, 183525
51. Moskovits M (1985) Surface-enhanced spectroscopy. Rev Mod Phys 57(3):783
52. Nie S, Emory SR (1997) Probing single molecules and single nanoparticles by surface-enhanced Raman scattering. Science 275(5303):1102–1106
53. Knoll W (1998) Interfaces and thin films as seen by bound electromagnetic waves. Annu Rev Phys Chem 49(1):569–638
54. Kubo S, Gu Z-Z, Tryk DA, Ohko Y, Sato O, Fujishima A (2002) Metal-coated colloidal crystal film as surface-enhanced Raman scattering substrate. Langmuir 18(13):5043–5046
55. Kuncicky DM, Prevo BG, Velev OD (2006) Controlled assembly of SERS substrates templated by colloidal crystal films. J Mater Chem 16(13):1207–1211
56. Yu Q, Guan P, Qin D, Golden G, Wallace PM (2008) Inverted size-dependence of surface-enhanced Raman scattering on gold nanohole and nanodisk arrays. Nano Lett 8(7):1923–1928
57. Oubre C, Nordlander P (2005) Finite-difference time-domain studies of the optical properties of nanoshell dimers. J Phys Chem B 109(20):10042–10051
58. Lang X, Qiu T, Zhang W, Ji C, Wang J, Chu PK (2010) Trace detection of multiwalled carbon nanotubes using Raman-enhancing silver nanocap arrays. J Phys D Appl Phys 43(45):455302
59. Gersten J, Nitzan A (1981) Spectroscopic properties of molecules interacting with small dielectric particles. J Chem Phys 75(3):1139–1152
60. Ruppin R (1982) Decay of an excited molecule near a small metal sphere. J Chem Phys 76 (4):1681–1684
61. Qiu T, Kong F, Yu X, Zhang W, Lang X, Chu PK (2009) Tailoring light emission properties of organic emitter by coupling to resonance-tuned silver nanoantenna arrays. Appl Phys Lett 95 (21):213104
62. Kim MS, Park DH, Cho EH, Kim KH, Park Q-H, Song H, Kim D-C, Kim J, Joo J (2009) Complex nanoparticle of light-emitting MEH-PPV with Au: enhanced luminescence. ACS Nano 3(6):1329–1334
63. Kong F, Zhang X, Lang X, Lin B, Yang Y, Qiu T (2011) Band-gap-dependent emissions from conjugated polymers coupled silver nanocap array. Appl Phys Lett 99(23):233112
64. Bhattacharyya S, Sen T, Patra A (2010) Host−guest energy transfer: semiconducting polymer nanoparticles and Au nanoparticles. J Phys Chem C 114(27):11787–11795
65. Andrew P, Barnes W (2004) Energy transfer across a metal film mediated by surface plasmon polaritons. Science 306(5698):1002–1005
66. Kovar JL, Simpson MA, Schutz-Geschwender A, Olive DM (2007) A systematic approach to the development of fluorescent contrast agents for optical imaging of mouse cancer models. Anal Biochem 367(1):1–12

67. Fernández-Suárez M, Ting AY (2008) Fluorescent probes for super-resolution imaging in living cells. Nat Rev Mol Cell Biol 9(12):929–943
68. Nagano T, Yoshimura T (2002) Bioimaging of nitric oxide. Chem Rev 102(4):1235–1270
69. Knemeyer J-P, Herten D-P, Sauer M (2003) Detection and identification of single molecules in living cells using spectrally resolved fluorescence lifetime imaging microscopy. Anal Chem 75 (9):2147–2153
70. Monici M (2005) Cell and tissue autofluorescence research and diagnostic applications. Biotechnol Annu Rev 11:227–256
71. Rosenthal SJ, Tomlinson I, Adkins EM, Schroeter S, Adams S, Swafford L, McBride J, Wang Y, DeFelice LJ, Blakely RD (2002) Targeting cell surface receptors with ligand-conjugated nanocrystals. J Am Chem Soc 124(17):4586–4594
72. Chumanov G, Sokolov K, Gregory BW, Cotton TM (1995) Colloidal metal films as a substrate for surface-enhanced spectroscopy. J Phys Chem 99(23):9466–9471
73. Sokolov K, Chumanov G, Cotton TM (1998) Enhancement of molecular fluorescence near the surface of colloidal metal films. Anal Chem 70(18):3898–3905
74. Zhang J, Matveeva E, Gryczynski I, Leonenko Z, Lakowicz JR (2005) Metal-enhanced fluoroimmunoassay on a silver film by vapor deposition. J Phys Chem B 109(16):7969–7975
75. Mertens H, Koenderink A, Polman A (2007) Plasmon-enhanced luminescence near noble-metal nanospheres: comparison of exact theory and an improved Gersten and Nitzan model. Phys Rev B 76(11):115123
76. Alberts B, Johnson A, Walter P, Lewis J, Raff M, Roberts K (2008) Molecular cell biology, 4th edn. Garland Science, New York/London
77. Hao Q, Yang F, Yin Y, Si L, Long K, Xiao Z, Qiu T, Chu PK (2013) Tunable fluorescence from patterned silver nano-island arrays for sensitive sub-cell imaging. J Phys D Appl Phys 46 (49):495302
78. Chen A, Moy VT (2000) Cross-linking of cell surface receptors enhances cooperativity of molecular adhesion. Biophys J 78(6):2814–2820
79. Akbay N, Lakowicz JR, Ray K (2012) Distance-dependent intrinsic fluorescence of proteins on aluminum nanostructures. Proc SPIE Plasmon Biol Med IX, 8234:823417
80. Doria G, Conde J, Veigas B, Giestas L, Almeida C, Assunção M, Rosa J, Baptista PV (2012) Noble metal nanoparticles for biosensing applications. Sensors 12(2):1657–1687

Chapter 11
Origin of Shifts in the Surface Plasmon Resonance Frequencies for Au and Ag Nanoparticles

Sandip Dhara

Abstract Optical properties of noble metal (Au, Ag) nanoclusters are of remarkable interest owing to the understanding fundamental issues of electronic properties in the small metallic clusters and their applications in nonlinear optics. The surface plasmon resonance (SPR), defined as resonance frequency of coherently oscillating free electrons with the exciting light, occurring in the near-UV-Visible region leads to the major applications. We discuss results, both experimental and theoretical, reporting the red or blue shift of the SPR frequency with decreasing noble metal cluster size as an effect of embedding matrix and surrounding porosity as well as in free noble metal nanoclusters without matrix. Reduction of the electron density (spillout effect) in the small nanoclusters and the interband screening of electrons in noble metals for larger nanoclusters shift the frequency of light absorbed either to the red or blue region, respectively, with decreasing cluster size. A strong dependence of the cluster size with the SPR frequency is discussed using time dependent local density approximation (TDLDA) with the consideration of porosity of surrounding medium. As the most recent report, quantum effect is also considered to understand blue shift of the SPR frequency of individual Ag nanoclusters with reducing size.

Keywords SPR • Noble metal • Spill out effect • TDLDA • Red shift • Blue shift • Nanocluster

11.1 Introduction

Optical properties of noble metal (Au, Ag) nanoclusters have attracted considerable attention in recent years mainly owing to the understanding fundamental issues related to the electronic properties in the small clusters [1] and their applications in nonlinear optics [2], optical switching [3, 4], including improved photovoltaic devices [5], cancer therapy [6] and catalysis [7] as well as surface enhanced

S. Dhara (✉)
Surface and Nanoscience Division, Materials Science Group, Indira Gandhi Centre for Atomic Research, Kalpakkam, 603 102, India
e-mail: dhara@igcar.gov.in

C.D. Geddes (ed.), *Reviews in Plasmonics 2015*,
DOI 10.1007/978-3-319-24606-2_11

spectroscopic studies at single molecule level [8–10]. In the present context, the central feature in the optical response is the surface plasmon excitation with coherent oscillation of the conduction electrons which results in a resonance band in the absorption spectra. The other interesting fact about it that for Au, Ag as well as for the other noble metals, the surface plasmon resonance (SPR) occurs in the near-UV-Visible (UV-vis) region leading to its application with visible light. A large number of experimental and theoretical results were reported for the red or blue shift of the SPR frequency with decreasing noble metal cluster size as an effect of embedding matrix and surrounding porosity [1, 11–14] as well as in free noble metal nanoclusters without matrix [15]. Reduction of the electron density (spillout effect) in the small nanoclusters and the screening of the interaction of valence (s) electrons by core (d)-electrons in noble metals for larger nanoclusters shift the frequency of light absorbed either to the red or blue region, respectively, with decreasing cluster size.

Though strong dependence of the cluster size with the SPR frequency is predicted by time dependent local density approximation (TDLDA) [1], only a few experimental evidences are available for the shift in the SPR frequency for Au clusters with different sizes embedded in alumina [12] and MgO [16, 17] matrices without any elucidation to the shift as an effect of size. As a matter of fact, few studies failed to report any shift in the SPR frequency of Au in silica [18–20]. Effect of porosity surrounding Au clusters is taken into consideration in the TDLDA calculation to support the experimental observation [1, 13]. A red shift, originating from the spillout effect with increasing polarizability in the system, was predicted by TDLDA calculation for embedded Au clusters in alumina [13]. The study in fully embedded Au nanoclusters in crystalline dielectric matrix showed first experimental evidence of red shift of the plasmon frequency with decreasing cluster size for clusters with ~157–427 number of Au atoms [21]. Experimental observations of blue shift of SPR frequency was also reported for Ag [1] and Au [12] clusters embedded in porous alumina matrix and in case of Au clusters (>5 nm) embedded in silica matrix [22]. Most recently quantum effect is also envisaged to understand blue shift of the SPR frequency of individual Ag nanoclusters with reducing size in the range of 2–20 nm [15, 23].

11.1.1 Red Shift of SPR Frequency: Spillout Effect

A red shift of the SPR frequency, measured using the UV–vis spectroscopy with decreasing cluster size was reported in the small sized Au clusters (diameter ~1.7–2.4 nm) embedded in crystalline alumina matrix [21]. The Au nanoclusters were grown in crystalline alumina matrix using 1.8 MeV Au^{++} implantation at various fluence ranges and subsequent annealing at high temperatures. The results of fully embedded Au nanoclusters in crystalline alumina matrix differ completely from other studies where clusters were grown in porous alumina matrices by co-deposition process using pulsed laser ablation technique [12] and sol–gel process [16].

Dipolar Mie resonance of a spherical metal cluster, $\omega_{SPR} = \omega_P/[2\varepsilon_m + \varepsilon_d(\omega_{SPR})]^{½}$ where Drude free-electron plasma frequency $\omega_P = (n^2/\varepsilon_0 m_e^*)^{1/2}$ with n as the electron density, ε_0 as the dielectric function of bulk metal, m_e* as the effective mass of electron and $\varepsilon_d(\omega) = (1 + \chi^d)$ is the core-electron contribution [χ^d the interband part of dielectric susceptibility (core d electrons)] to the complex dielectric function of the noble metal $\varepsilon(\omega) = 1 + \chi^s(\omega) + \chi^d(\omega)$ [χ^s the Drude-Sommerfeld part of dielectric susceptibility (valence s electrons)]. The dielectric function of matrix $\varepsilon_m \approx 3.1$ for crystalline alumina in the relevant energy range. The Mie frequency in the large-size limit can be approximately calculated by solving

$$\varepsilon(\omega) + 2\varepsilon_m = 0 \tag{11.1}$$

in the classical limit, and resulting in

$$\omega_{\mathrm{SPR}}(\infty) = \omega_{\mathrm{P}}/\left[2\varepsilon_{\mathrm{m}}(\omega_{\infty}) + \mathrm{Re}\,(\varepsilon_d(\omega_{\infty}))\right]^{½} \tag{11.2}$$

The Mie frequency in the large-size limit for embedded Au nanoclusters in crystalline alumina is approximately calculated to be ~2.3 eV (539 nm) using Eq. (11.1). Plasmon frequencies around ~2.14–2.21 eV (~579–561 nm) are observed in the UV–vis absorption study of the post-annealed samples for two different annealing conditions 1273 and 1473 K (Fig. 11.1) indicating formation of Au nanoclusters [21]. Cluster sizes were calculated to be in the range of ~1.72–2.4 nm (157–427 number of Au atoms = 32.6–45.4 atomic unit, a.u. ~0.0529 nm) using acoustic phonon confinement model in low frequency Raman spectroscopic (LFRS) studies [24]. The LFRS study for clusters in the small range (<10 nm) were performed for the determination of the size and the shape of clusters. Low-frequency Raman modes in the vibrational spectra of the materials arises due to the confined surface acoustic phonons in metallic or semiconductor nanoclusters. The Raman sensitive spheroidal motions are linked with dilation and strongly depend on the cluster material through v_t, the transverse and v_l, longitudinal sound velocities. These modes are characterized by two quantum indices l and n, where l is the angular momentum quantum number and n is the branch number. $n = 0$ represents the surface modes. The surface quadrupolar mode ($l = 2$, eigen frequency $\eta_2{}^s$) appears in both the parallel and perpendicularly polarized Raman scattering whereas, the surface spherical mode ($l = 0$, eigen frequency $\xi_0{}^s$) appears only in the parallel polarized configuration. Considering the matrix effect in the limit of elastic body approximation of small cluster (core-shell model) [25], eigen frequencies for the spheroidal modes at surface ($n = 0$; $l = 0$, 2) of Au nanocluster in alumina matrix is calculated to be $\eta^s{}_2 = 0.84$ and $\xi^s{}_0 = 0.40$. The surface quadrupolar mode frequencies corresponding to $l = 0$, and 2 are expressed by,

$$\omega_0^s = \xi_0^s v_l/Rc; \omega_2^s = \eta_2^s v_t/Rc \tag{11.3}$$

where $v_l = 3240$ m/s, $v_t = 1200$ m/s in Au and c is the velocity of light in vacuum. Average cluster radii ($<R> \approx 0.86$–1.2 nm ≈ 16.3–22.7 a.u.), calculated using

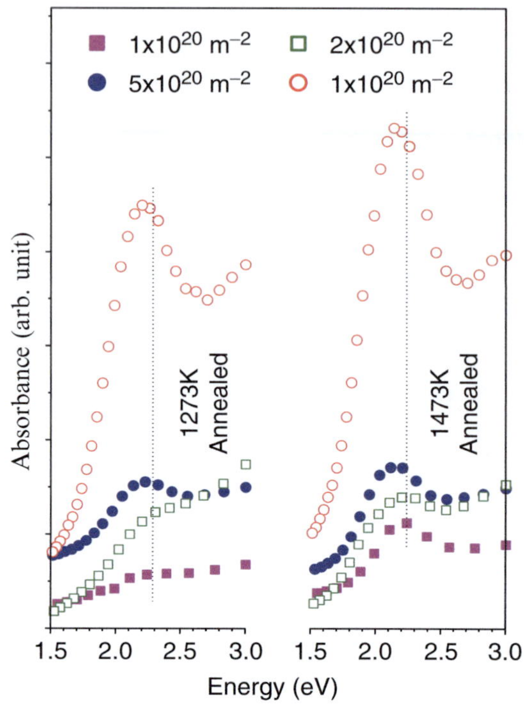

Fig. 11.1 SPR frequencies for embedded Au nanoclusters in c-Al_2O_3 annealed at 1273 and 1473 K showing redshift with decreasing fluence. *Dashed vertical line* is a guide to eye for the observed red shift of the SPR frequency with decreasing fluence (Ref. [21]; *Copyright (2004) by the Elsevier Science B.V.*)

Eq. (11.2), corresponding to $l = 0$, 2 are found to be nearly the same conforming to the indirect technique used for the calculation of very small clusters [21]. For both the annealed samples a clear red shift of the SPR frequency was observed with decreasing fluence. For these spherical clusters in the smaller size range, a red shift was observed with decreasing cluster size (Fig. 11.2a). Though the red shift effect was predicted to be quenched in case of free Au clusters [1], the large matrix-induced charge screening lead to a large electron spillout in case of embedded cluster [13]. A red shift is observed for very small clusters because of the reduction of average electron density with an increasing electron spillout effect with decreasing cluster size. The physics of the blue shift trend observed in noble metal clusters (discussed in the subsequent section) is based on the assumption that because of the localized character of the core-electron wave functions, the screening effects are less effective over a surface layer inside the metallic particle. Close to the surface, the valence electrons are then incompletely embedded inside the ionic-core background. In two-region dielectric model, this hypothesis is taken into account by assuming that the effective polarizable continuous medium responsible for the screening does not extend over the whole cluster volume, where $\chi^d(\omega) = \varepsilon_d(\omega) - 1$ vanishes for radius $>R - r$; where R is the cluster radius and r is a thickness parameter of the order of a fraction of the nearest-neighbor atomic distance (Fig. 11.2b).

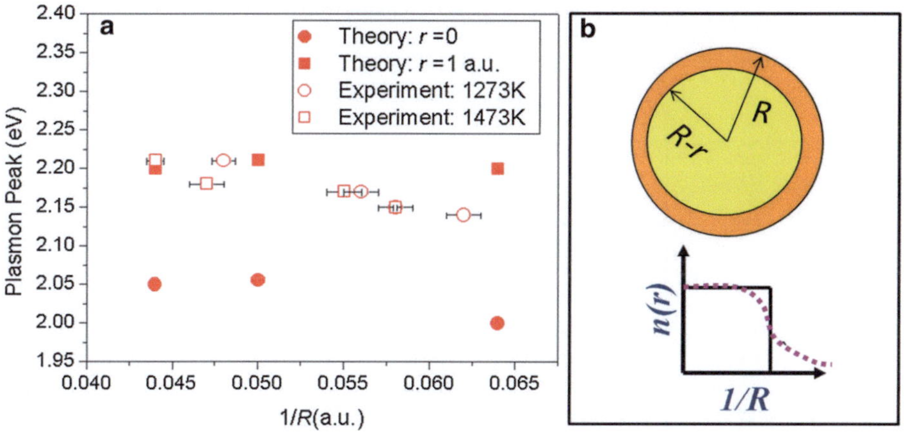

Fig. 11.2 (**a**) Size dependence of the SPR frequency for the samples annealed at 1273 and 1473 K (*unfilled symbols*) and estimated values from TDLDA calculations (*filled symbols*) (Ref. [13]) corresponding to $r=0$ and 1 a.u. are presented. Experimental values show a *red* shift with decreasing cluster size (Ref. [21]; *Copyright (2004) by the Elsevier Science B.V.*) (**b**) Schematic of two region dielectric model with nanoclusters radius R and screening length r. The reduction in the electron density as a function of r, $n(r)$ is shown to reduce with decreasing R (spillout effect)

Plasmon frequencies fall in between the values calculated using TDLDA method with $r=0$ and 1 a.u. in the specific size limit. The red shift trend with decreasing cluster size is clearly depicted for $1>r\geq 0$ as the trend is reversed for $r>1$ with increased screening effect of s and d electrons [13]. Normally spillout effect is realized in very small clusters where electron density reduces from its bulk value. Charge density for the clusters upto 170 atoms also showed deviation from its bulk value in the density functional theory (DFT) based LDA calculations applied to the spherical jellium-background model (SJBM) [26]. The SJBM at an infinite flat surface was implemented for high symmetrical potential applicable to significantly large cluster size. However, charge densities and self consistent potentials for 170 atoms vary in their values across the infinite flat surface due to the nature of electronic level filling in the spherical potential. The solution to the problem by introducing lattice structure via pseudo-potentials was limited in the presence of high lattice symmetry for specific elements. The effect of spin using local spin-density approximation was found to be important in this context. However, it was the first effort to address optical response related issues in the DFT-LDA formalism. The model was later on improved up to 440 atoms in the TDLDA calculation (Fig. 11.3) [13]. In the two-region dielectric model, in addition to the electron density beyond the radius R, the magnitude of the red shift was found to depend also on the surface electron density profile. It is analogous to SJBM model [26], involving localized pseudo-potentials. Larger red shift was anticipated with softer surface electron density, as estimated for dielectric function ε_d in the range 5–10, and was compared the results with those of the standard SJBM. The surface profiles of the mean-field potential and electron-density were found quite identical when

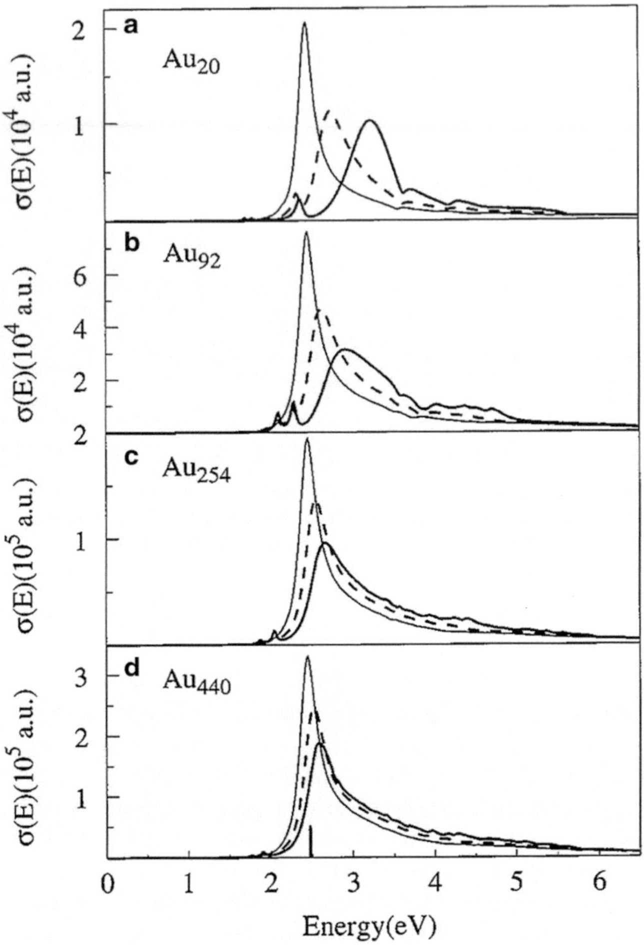

Fig. 11.3 (**a–d**) Photoabsorption spectra of free gold clusters within the two-region dielectric model, for different thicknesses of the skin region of reduced polarizability. *Thick line curves*: $d = 2$ a.u.; *dashed line curves*: $d = 1$ a.u.; *thin line curves*: $d = 0$. The *short vertical line* in (**d**) indicates the dipolar Mie resonance energy in the large-particle limit (Ref. [13]; *Copyright (1998) by the EDP Sciences, Springer-Verlag*)

$\varepsilon_m = 1$. It suggested similar magnitude of red shift of the plasmon frequency for embedded system as compared to that for free clusters. With number of atoms, $N \sim 427$ (<440) for largest cluster of radius $R \sim 1.2$ nm and 157 for the smallest cluster of $R \sim 0.86$ nm (~16.3 a.u.), in the report [21], definitely support the claim that the observed red shift of SPR frequency with decreasing size is due to electron spillout effect in the small Au clusters fully embedded in crystalline alumina matrix. Unlike co-ablated growth of Au in alumina [12], the effect of porosity leading to the blue shift of SPR frequency with decreasing cluster size was proposed to be negligible in our study as the samples were grown in a well crystalline alumina matrix. Carrier escape by interband tunneling, as observed in semiconductor heterojunction nanostructure where electron dynamics of localized Wannier–Stark states [27, 28], was reported in the presence of electric field [29, 30]. In this context dephasing of electron leading to a dynamic localization of electron in metallic nanoclusters, however was not considered as the coherence lifetime of electron (limited by scattering centers, e.g., point defects and surface states in the nanocluster) was expected to be much smaller than the period of oscillatory motion of electron for all reasonable value of applied electric field.

11.1.2 Blue Shift of SPR Frequency: Screening Effect

The optical response of free and matrix-embedded Au nanoclusters was reported in the framework of the TDLDA [13]. The characteristics of the SPR frequency with cluster size showed strong influence of the frequency-dependence of the 5*d* core-electron dielectric function in the vicinity of the interband threshold. The size evolution of the Mie-frequency in free Au nanoclusters exhibited a prominent blue shift with decreasing cluster size than that for Ag nanoclusters (Fig. 11.4). Experimental data corresponding to five different particle-size distributions were shown on composite films consisting of low-concentration (~7 %) of Au nanoclusters embedded in an amorphous alumina matrix (black squares) [12]. The experimental data were positioned between the extrapolated theoretical curves corresponding to free and fully-embedded clusters (empty triangles), suggesting that the experimental results might be explained by taking into account the matrix porosity. Neglecting volume change in these nanoclusters and with a mean porosity of about 45 % with respect to crystalline Al_2O_3 two kinds of model calculations were carried out to understand and quantify the effect of matrix-porosity in these systems. Results obtained for $\varepsilon_m = 2$, assuming a fine-grained homogeneous porous matrix, showed a value corresponding to a decrease of the matrix polarizability by a factor 2 in the visible spectral range ($\varepsilon_m \sim 3.1$) (crosses in Fig. 11.4). In spite of the quantitative agreement for large clusters, the slope of the size evolution was significantly underestimated and the predicted porosity effects might be too small for medium and small clusters. As a matter of fact $\varepsilon_m = 2$, corresponding to a large value of matrix porosity, can be assessed by calculating the dielectric function of the porous alumina matrix using the Bruggeman effective medium theory [31]. The experimentally-determined dielectric function of porous alumina films showed a decrease of ~17 % with respect to that for crystalline in the visible spectral range ($\varepsilon_m \sim 2.6$). At the cluster matrix interface the porosity is important because of the fact that the short-scale heterogeneity of the porous matrix in the vicinity of the interface will play a crucial role as the changes in the induced electron density by the oscillating external field are mostly confined to the surface. Model calculations involving a perfect vacuum "rind" were carried out for $\varepsilon_m = 1$ in the radial range $R < r < R + d_m$. The size evolutions over the entire studied size range were displayed (black triangles in Fig. 11.4). For a given mean porosity, matrix porosity

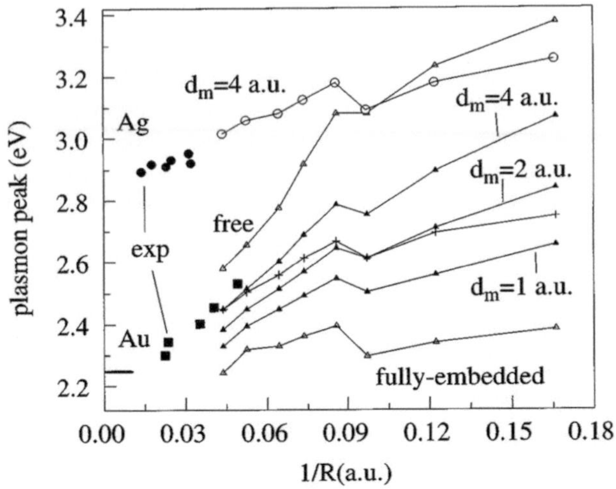

Fig. 11.4 The size evolution of the surface plasmon energy of gold clusters within different models. *Black triangles*: TDLDA results obtained with an outer vacuum "rind" at the metal/alumina matrix interface. d_m is the thickness of the outer rind. *Empty triangles*: TDLDA results for free (*upper curve*, $d_m = \infty$) and alumina matrix-embedded clusters (*lower curve*, $d_m = 0$). *Crosses*: TDLDA results with a homogeneous surrounding matrix characterized by a constant dielectric function $\varepsilon_m = 2$. In all calculations the thickness *d* of the inner skin of reduced polarizability is equal to 2 a.u. Black squares: experimental results (Ref. [12]). The short horizontal line at 2.25 eV indicates the surface plasmon energy in the large-particle limit for fully-embedded AuN clusters. The *circles* correspond to experimental (*black*) and TDLDA (*empty*) results for alumina matrix-embedded Ag_N clusters (model parameters: $d = 2$ a.u. and $d_m = 4$ a.u.) (Ref. [13]; *Copyright (1998) by the EDP Sciences, Springer-Verlag*)

effects were found stronger in this model. With the slope of the curves in good agreement with the experimental size trend, the deviation from the experimental data was minimized with a vacuum-rind thickness d_m (<< Fermi wavelength, λ_F). Assessment with recent experimental data obtained with composite films of Ag_N and alumina (black circles in Fig. 11.4) provides evidence for both the relevance of the porosity effects and suitability of the two-region dielectric model for Au and Ag. The results of TDLDA calculations on Ag_N clusters embedded in alumina, involving the values $d_m = 2$ a.u. and $d_m = 4$ a.u., were plotted (empty circles in Fig. 11.4). An agreement between theory and experiment, for both Au and Ag, in using the same model parameters, proved the correctness of the model for investigating the size effects in the location of the surface plasmon frequency. In case of matrix embedded noble metal nanoclusters, however the blue shift trend is largely reduced. Agreement with recent experimental results on size-selected Au clusters embedded in an alumina matrix was achieved by taking into account the porosity effects at the metal and matrix interface. The SPR frequencies of Ag nanoclusters, calculated in the TDLDA framework, were in good agreement (Fig. 11.5) with experimental data on free Ag_N^+ clusters [32] and on Ag_N clusters embedded in solid Ar [33]. The size effects were found weak and very sensitive to the matrix, the

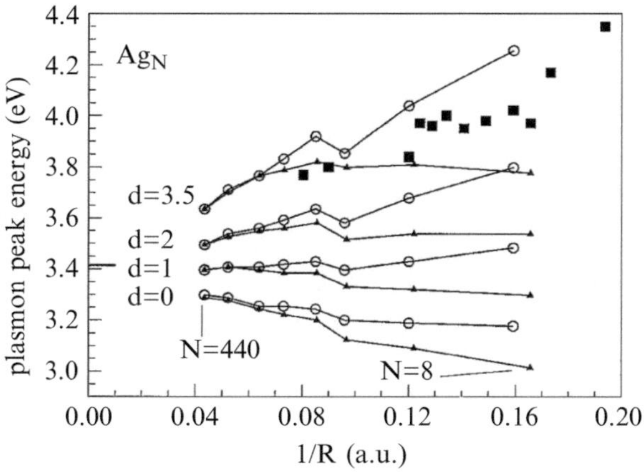

Fig. 11.5 Size evolution of the maximum of the Mie-resonance peak for free Ag_N (*triangles*) and Ag_N^+ (*circles*) clusters within the two-region dielectric model, for different values of the thickness parameter *d*. *Black squares*: experimental data on Ag_N^+ clusters (Ref. [32]). The *short horizontal line* at 3.41 eV is the Mie frequency in the large-particle limit (Ref. [1]; *Copyright (1998) by the American Physical Society*)

porosity at the interface, and the cluster charge. It is shown that the core-electron contribution to the metal dielectric function is mainly responsible for the quenching of the size effects in the optical response. In a phenomenological two-region dielectric model for the thickness of the layer of reduced polarizability, the blue shift trend is observed for $r > 1$ as the cluster radius decreases (Fig. 11.2b). Thus, depending on the experiment, a blue shift [34], or a quasi-size-independent evolution [22, 35] is exhibited.

The surface plasmon energy of Au nanoclusters formed by Ar^+ ion beam mixing of Au/silica was investigated for the size effect [22]. Core-electron contribution to the metal dielectric function was mainly responsible for the blue shift with decreasing cluster size in case of nanocluster diameter >5 nm (Fig. 11.6a). This is because of the fact that at $\omega_{SPR} > \omega_{max}$ (Fig. 11.6b) a reduction of $\mathrm{Re}[\varepsilon_d(\omega)]$, residing in the denominator of the Eq. (11.2), with increasing ω will further enhance the blue shift. The effect was depicted as quenching of size effect in the optical response [1]. The nonzero contribution of $\mathrm{Im}[\varepsilon_d(\omega)]$ in Au (Fig. 11.6b) was reported to broaden the resonance peak by coupling with the interband transition [12]. The broadening of the resonance peak for Au clusters were also reported earlier [11, 18, 19]. A blue shift of SPR frequency with reducing cluster size was also reported for Au-Ag alloy nanoclusters grown by laser ablation of an alloy target Au-Ag 1:1 atomic composition in the size range of 1.9–2.8 nm [36]. The alloy nanoculsters were embedded with a low concentration in an alumina matrix. Theoretical calculations in the frame work of the TDLDA, including an inner skin of ineffective screening and the porosity of the matrix, were in good agreement with experimental results (Fig. 11.7). In the standard Mie theory, with a single interface no size effects occur in the dipolar regime ($\lambda >> R$), except for a mere volume scaling factor.

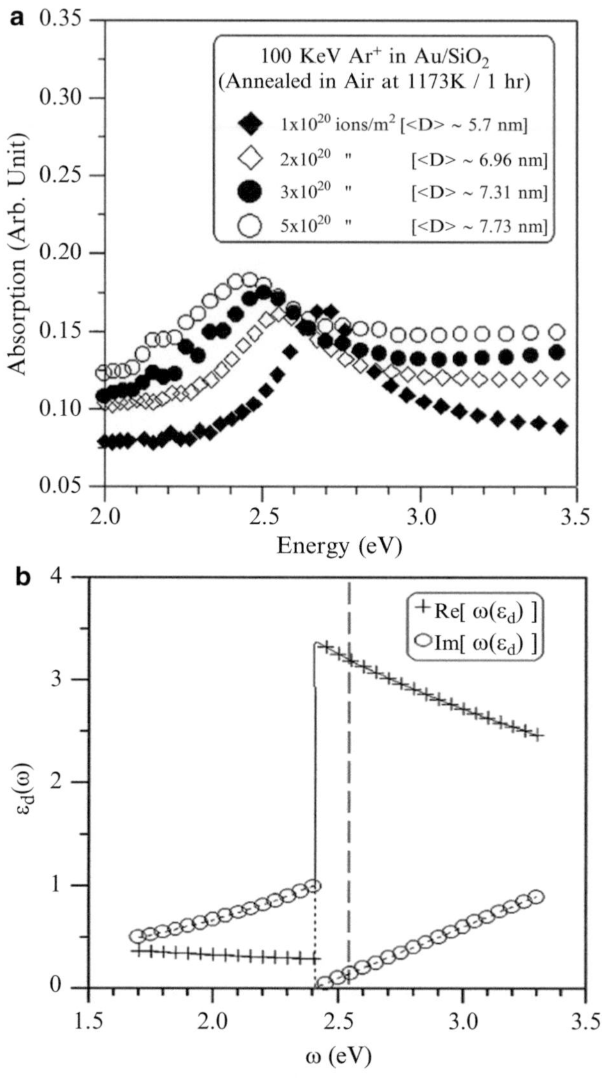

Fig. 11.6 (**a**) Surface plasmon resonance peaks for the annealed (typically for 1173 K) samples grown at various irradiation fluences showing the blue shift with decreasing fluence. The spectra are vertically shifted for clarity. (**b**) Spectral dependence of the real and imaginary components of the complex dielectric function corresponding to the core electrons for gold metal. Im $[\varepsilon_d(\omega)]$ is plotted with $\varepsilon_d(\omega) = \exp(\text{const.}(\omega\text{-}2.4)$ (near L point in the Birllouin zone: $1.7\text{ eV} \leq \omega \leq 2.4\text{ eV}$) and $\varepsilon_d(\omega) = (\omega\text{-}2.4)/\omega^2$ (near point in the Birllouin zone: $2.4\text{ eV} \leq \omega \leq 3.3\text{ eV}$) (Ref. [11]). Re$[\varepsilon_d(\omega)]$ is calculated using Kramers–Kronig relationship (Ref. [11]).for the corresponding ranges. The vertical line corresponds to the Mie frequency for large free clusters (Ref. [22]; *Copyright (2003) by the Elsevier Science B.V.*)

Finite-size effects result when the surface skins (thickness d_m) of ineffective screening are included in the embedded-cluster model. However, these results are noticeably different from those calculated within the TDLDA-based quantum model (Fig. 11.7a). This discrepancy originates from the large value of the $\lambda_F \approx 3.3r_s$ (Wigner radius, r_s) relative to d_m. The comparison with experiment has been achieved by attributing to the experimental Mie-band maximum the size corresponding to the mean diameter. A noticeable difference was observed between theory and the experiment for large Ag_n clusters. The asymptotic values correspond to a metal sphere embedded in a porous alumina matrix in the experimental

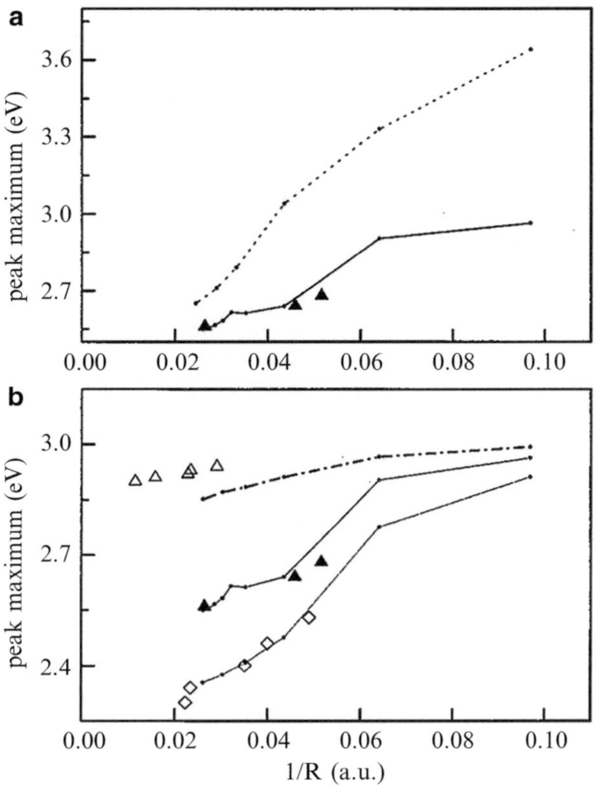

Fig. 11.7 The size evolution of the peak plasmon maximum within different models for particles embedded in an alumina matrix. (**a**) Classical results for $(Au_{0.5}Ag_{0.5})_n$ clusters (*dots*); (**a**) and (**b**) TDLDA results (*solid line*) and experimental results (*black triangles*) for $(Au_{0.5}Ag_{0.5})_n$ clusters; (**b**) TDLDA results (*dashed line*) and experimental results (*open triangles*) for Ag_n clusters; TDLDA results (*dotted line*) and experimental results (*open rhomb*) for Au_n clusters (Ref. [36]; *Copyright (2000) by the American Physical Society*)

findings. Comparing the experimental data of the pure alumina sample with the theoretically calculated value, it was found that the SPR frequency of mixed $(Au_{0.5}Ag_{0.5})_n$ clusters was slightly overestimated in the model. The qualitative analysis, however was quite applicable.

11.1.3 Blue Shift of SPR Frequency: Quantum Effect

A blue shift of SPR frequency for free Ag nanoclusters is described as quantum plasmonic property for the first time by Scholl et al. [15]. They have excited plasmons for individual particles using a scanning tunneling electron microscopic tip where fast electrons are focused on Ag nanoparticles precisely from surface to the central region. It may be noted that the interaction between electromagnetic wave corresponding to fast electrons and plasmons is equivalent to the interaction of light with plasmons. They could differentiate bulk- and surface-plasmon components and a blue shift in the frequency of surface plasmons with decreasing cluster size (Fig. 11.8). The classical electrodynamics does not predict size

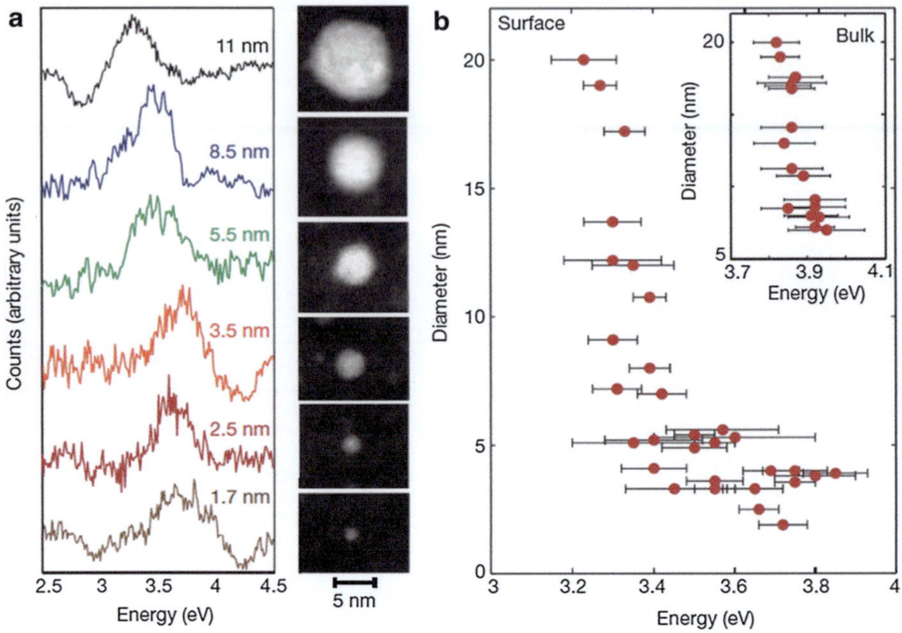

Fig. 11.8 Correlating Ag nanoparticle geometry with plasmonic EELS data. (**a**) Collection of normalized, deconvoluted EELS data from particles ranging from 11 to 1.7 nm in diameter and the corresponding STEM image of each specimen. The electron beam was directed onto the edge of the particles so that only the surface resonance is shown. (**b**) Plot of the surface plasmon resonance energy versus particle diameter, with the inset depicting bulk resonance energies. Horizontal error bars indicating 95 % confidence intervals were generated with a curve fitting and bootstrapping technique (see Methods). Vertical error bars are contained within the size of the data points (Ref. [15]; *Copyright (2012) by the Nature Publications*)

dependence of the SPR frequency in the free Ag cluster [11]. The conduction electrons respond to electromagnetic fields as a classical electron gas for nanoparticles with size >10 nm. In this size range, the contribution of each electron to the plasmons cannot be observed and the SPR frequency is uncertain because of collisions of the electrons with each other and with the atomic lattice of noble metal. However, a clear blue shift was observed for the SPR frequency of the smallest nanoparticles of ~2 nm as compared with 10 nm, along with a significant (50 %) decrease in the plasmon lifetime. The conduction electrons moving at a Fermi velocity (v_F) of ~10^6 m/s [21] take approximately 10 s to travel across a 10 nm Ag particle. Observed plasmon lifetime of ten plasmon periods of such particles is close it. However, the blue shift could not be explained using the concept of reflections of the electrons at the surface for the reduction in the plasmon lifetime [11]. The electrons appear at a discrete set of energy levels in nanoparticles smaller than 10 nm, with relatively few conduction electrons (250 electrons for

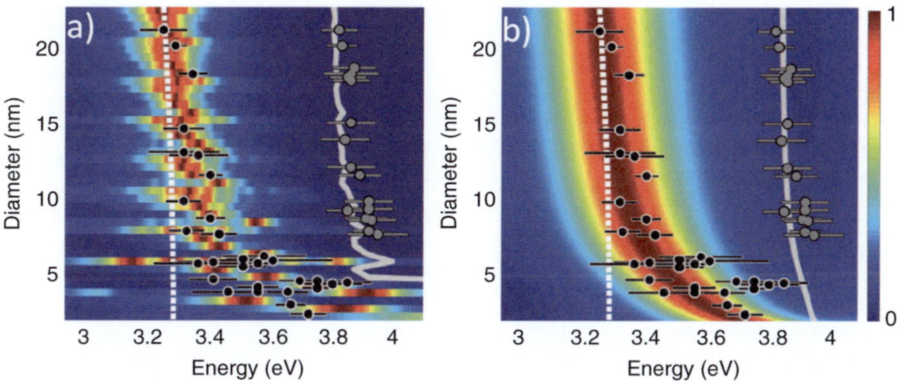

Fig. 11.9 Comparison of experimental data with quantum theory. Experimental, EELS-determined localized surface plasmon resonance energies of various Ag particle diameters are overlaid on the absorption spectra generated from the analytic quantum permittivity model (**a**) and the DFT-derived permittivity model (**b**). The experimental bulk resonance energies are also included (*grey dots*) along with the theory prediction (*grey line*). Classical Mie theory peak prediction is given by the *dashed white line*. The experimental data begin to deviate significantly from classical predictions for particle diameters smaller than 10 nm. Horizontal error bars represent 95 % confidence intervals, as calculated through curve fitting and bootstrapping techniques (Ref. [15]; *Copyright (2012) by the Nature Publications*)

2 nm nanoclusters) participating in the plasmons. These energy levels are increasingly separated from one another as the particle size is reduced and termed as quantum confinement effect. The individual electron transitions between occupied and unoccupied degenerate electron energy levels increases the uncertainty about the SPR frequency leading to the reduction of plasmon lifetime [37]. In fact, individual transitions cannot in general be resolved owing to the uncertainty produced by the collisions mentioned previously. Considering the effects of individual electron transitions, an analytical quantum mechanical model is described to understand the blue shift due to a change in particle permittivity (Fig. 11.9). The model also considers catalytic effect, and surface enhanced techniques SERS and TERS effects as manifestation of the quantum plasmon mechanical behavior of tiny noble metal nanoclusters with evanescent field are restricted only to small number of atoms corresponding to the 'hot spot' [38]. Similar observation of blue shift of SPR frequency with decreasing cluster size (Fig. 11.10), larger than that predicted in the theory, was also reported for isolated spherical Ag nanoclusters dispersed on a Si_3N_4 substrate in the diameter range 3.5–26 nm [23]. A semi-classical model corrected for an inhomogeneous electron density associated with quantum confinement, and a semi-classical nonlocal hydrodynamic description of the electron density was used to understand the observed blue shift (Fig. 11.11).

In another report, a blue shift of the quantum SPR frequency with decreasing size (Fig. 11.12) is reported for 2–10 nm Ag nanoclusters embedded in SiO_2 matrix

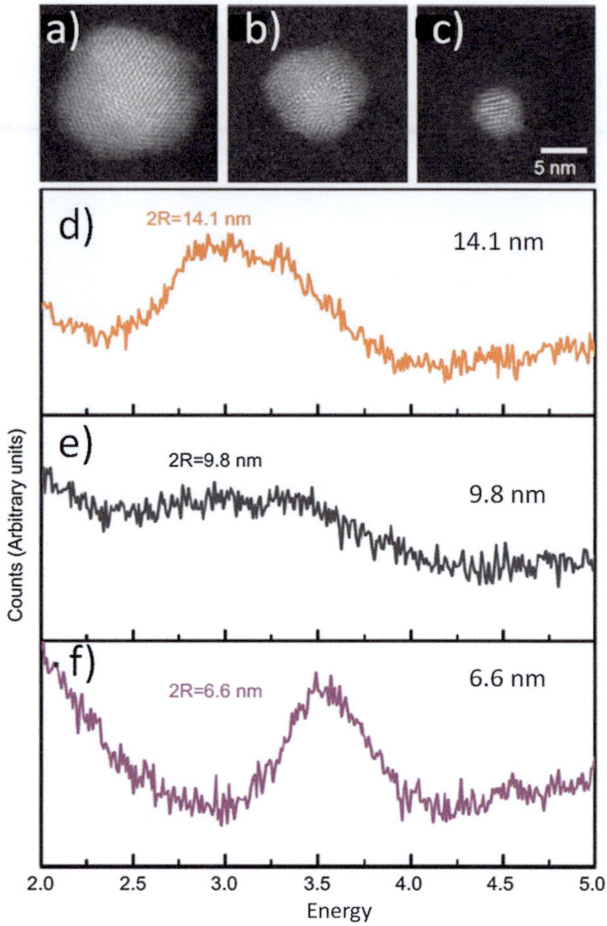

Fig. 11.10 Aberration-corrected STEM images of Ag nanoparticles with diameters (**a**) 15.5 nm, (**b**) 10 nm, and (**c**) 5.5 nm, and normalized raw EELS spectra of similar-sized Ag nanoparticles (**d–f**). The EELS measurements are acquired by directing the electron beam to the surface of the particle (Ref. [23]; *Copyright (2013) by the Science Wise Publishing and De Gruyter*)

[14]. Reduction in the size with increasing fluence, as explained by the thermal spike model, was achieved by high-energy Si^{5+} ion-irradiations in the embedded Ag nanoclusters substrate. In the noble alloy system optical properties of mixed $(Au_xAg_{1-x})_n$ clusters in the diameter range 1.5–5 nm of various relative compositions embedded in an alumina matrix were also reported [39]. Two simple phenomenological models were developed in estimating the effective dielectric function $\varepsilon_d(\omega)$ of the Au-Ag ionic background, as the optical properties of the mixture depend strongly on the topological non-uniformity in the length scales of the heterogeneities . In one model (Model 1), $\varepsilon_d(\omega)$ of the ionic mixture Au_xAg_{1-x} is assumed to be the composition-weighted average of the interband contributions of bulk Au and Ag metals. The model 1 assumes the mixture as the stacking of small homogeneous Au and Ag grains, whose optical properties are close to those of the

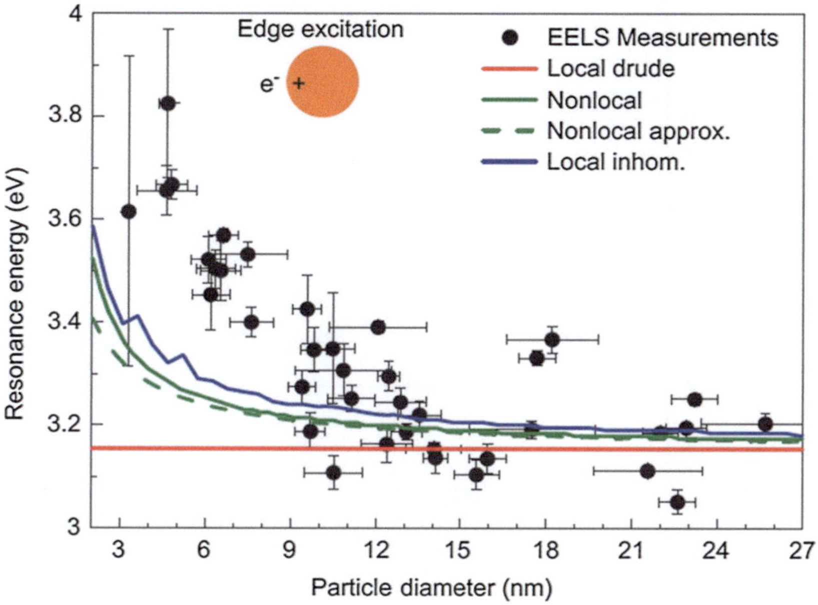

Fig. 11.11 Nanoparticle SPR energy as a function of the particle diameter. The *dots* are EELS measurements taken at the surface of the particle and analyzed using the Reflection Tail (RT) method, and the *lines* are theoretical predictions. We use parameters from Ref. [42]: $\hbar\omega_p = 8.282$ eV, $\hbar\gamma = 0.048$ eV, $n_0 = 5.9 \times 10$ 28 m^{-3} and $v_F = 1.39 \times 10^6$ m/s. From the average large-particle ($2R > 20$ nm) resonances we determine $\varepsilon_B = 1.53$ (Ref. [23]; *Copyright (2013) by the Science Wise Publishing and De Gruyter*)

corresponding bulk materials. In the second model (Model 2), relations under band structure formalism were derived from a the experimental data obtained from transmission experiments in the energy range of 2.4–4.4 eV [40], and reflection experiments in the energy range of 0.5–6.5 eV [41] on thin homogeneously alloyed Au/Ag films of various compositions. The imaginary components Im[$\varepsilon_d(x,\omega)$] are plotted for both models (Fig. 11.13). The prominent differences are (i) the interband threshold is independent of the composition in model 1, while it evolves with x in the second model; (ii) in model 1 the Im[$\varepsilon_d(x,\omega)$] exhibits a two-step pattern, while a single rising edge is exhibited in model 2, as observed for pure Au and Ag. For a given relative composition, the blue shift of SPR frequency with increasing size was reported to become more important as the Au content increases in the alloy (Fig. 11.14). The size and concentration effects in the optical properties were investigated in the light of quantum effect. These spectra were compared with model calculations, considering quantum mechanical description of s conduction electrons within the TDLDA formalism including both the porosity of the matrix and an inner skin of reduced ionic-core polarizability. Different models of alloy

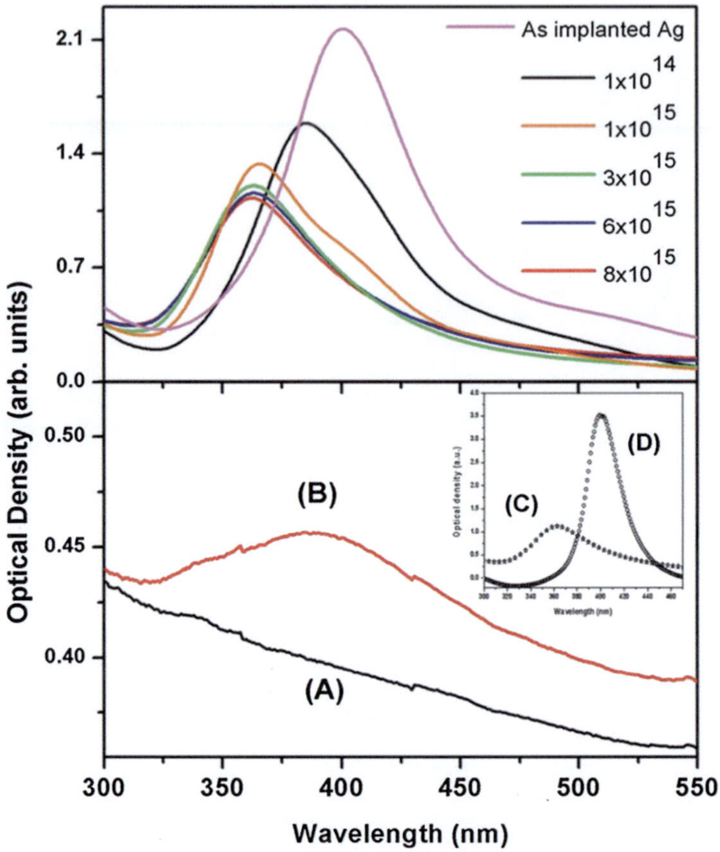

Fig. 11.12 (*Top panel*) Optical absorption spectra of Ag ion-implanted (5×10^{16} ions cm^{-2}) SiO_2 samples followed by Si ion irradiations (ion dose as indicated) are displayed. *Blue* shifts of SPR peak and systematic decrease of the resonance intensity with increase of Si ion fluence may be observed. (*Bottom panel*) Graphs A and B display the optical absorption spectra of Ag particles in SiO_2 samples before and after Si ion-irradiations (5×10^{15} ions cm^{-2}), respectively. Ag ion-implantation fluence is 1×10^{16} ions cm^{-2}. (Inset) Graph C displays the optical absorption spectra of Ag particles in SiO_2 samples after Si ion-irradiations (5×10^{15} ions cm^{-2}). Ag ion implantation fluence is 3×1016 ions cm^{-2}. Subsequent annealing (400 °C, 1 h) the SiO_2 sample in an inert gas atmosphere brings back the SPR characteristics of larger Ag nanoparticles (see graph D). Quantum nature of surface plasmon resonances in fine Ag particles vanishes, revealing ripening of the Ag particles on thermal annealing (Ref. [14]; *Copyright (2014) by the Elsevier Science B.V.*)

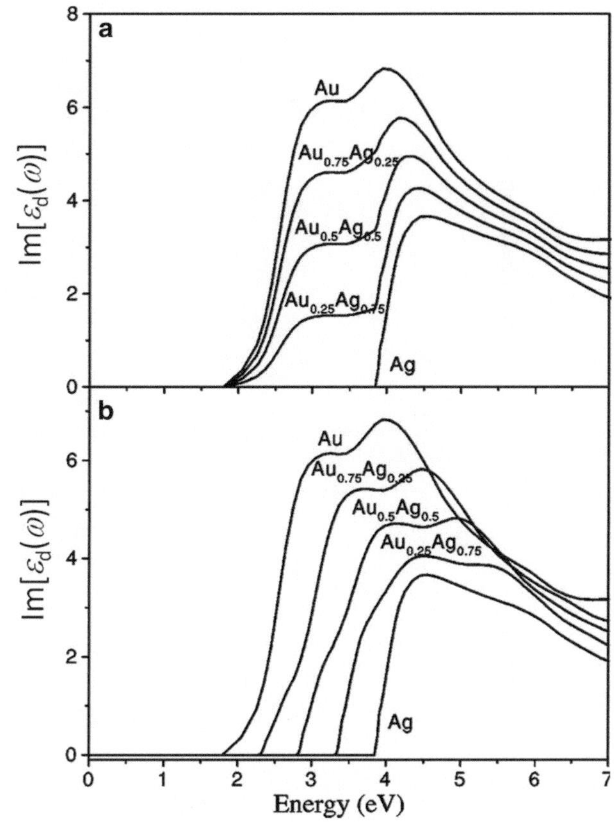

Fig. 11.13 Spectral dependence of the imaginary part of the dielectric function of the core electrons in the (Au_xAg_{1-x}) alloy, with $x = 0, 0.25, 0.5, 0.75, 1$ from top to bottom, within two different hypotheses (see text), (**a**) model 1 (**b**) and model 2 (Ref. [39]; *Copyright (2001) by the American Physical Society*)

morphology were introduced to describe the effective dielectric function of the ionic background in the noble metal alloy nanocluster. A multi-shell model showed predominant quantum effects. In classical calculations involving a multi-shell cluster of simple metals of different electronic densities, the spectra were structured with as many peaks as the number of interfaces. On the other hand, quantum TDLDA calculations within the jellium model (spillout effect) display only one main resonance peak in the spectra.

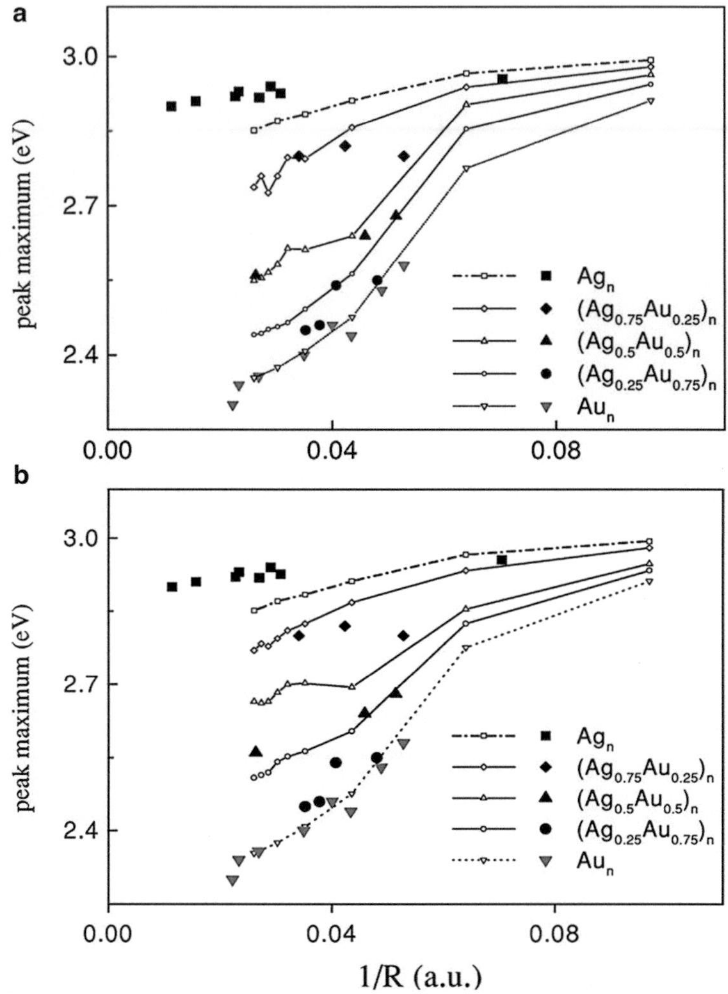

Fig. 11.14 Size evolution of the peak plasmon maximum for $(Au_xAg_{1\text{-}x})_n$ clusters ($x = 0$–1) embedded in an alumina matrix, assuming two hypotheses for the dielectric function associated with the core electrons (see text): model 1 (**a**) and model 2 (**b**). TDLDA results, open symbols connected with lines; experimental results, large black symbols. Owing to the rather tiny effects (all the values lie in the energy range 2.3–3 eV) a finer increment $\Delta\omega$, ten times smaller than the increment involved in the spectra in Fig. 6 of Ref. [39], has been used over the resonance spectral range in order to unambiguously determine the maximum locations, and thus the size evolution (Ref. [39]; *Copyright (2001) by the American Physical Society*)

References

1. Lermé J, Palpant B, Prével B, Pellarin M, Treilleux M, Vialle JL, Perez A, Broyer M (1998) Quenching of the size effects in free and matrix-embedded silver clusters. Phys Rev Lett 80:5105–5108
2. Liao HB, Xiao RF, Fu JS, Yu P, Wong GKL, Sheng P (1997) Large third-order optical nonlinearity in Au:SiO_2 composite films near the percolation threshold. Appl Phys Lett 70:1–3
3. Dhara S, Lu C-Y, Magudapathy P, Huang Y-F, Tu W-S, Chen K-H (2015) Surface plasmon polariton assisted optical switching in noble bimetallic nanoparticle system. Appl Phys Lett 106:023101
4. Dhara S, Lu C-Y, Chen K-H (2015) Plasmonic switching in Au functionalized GaN nanowires in the realm of surface plasmon polatriton propagation: a single nanowire switching device. Plasmonics 10:347–350
5. Atwater HA, Polman A (2010) Plasmonics for improved photovoltaic devices. Nature Mater 9:205–213
6. Lal S, Clare SE, Halas NJ (2008) Nanoshell-enabled photothermal cancer therapy: impending clinical impact. Acc Chem Res 41:1842–1851
7. Juluri BK, Zheng YB, Ahmed D, Jensen L, Huang TJ (2008) Effects of geometry and composition on charge-induced plasmonic shifts in gold nanoparticles. J Phys Chem C 112:7309–7317
8. Nie S, Emory SR (1997) Probing single molecules and single nanoparticles by surface-enhanced Raman scattering. Science 275:1102–1106
9. Garg P, Dhara S (2013) Single molecule detection using SERS study in PVP functionalized Ag nanoparticles. AIP Conf Proc 1512:206–207
10. Zhang R, Zhang Y, Dong ZC, Jiang S, Zhang C, Chen LG, Zhang L, Liao Y, Aizpurua J, Luo Y, Yang JL, Hou JG (2013) Chemical mapping of a single molecule by plasmon-enhanced Raman scattering. Nature 498:82–86
11. Kreibig U, Vollmer M (1995) Optical properties of metal clusters. Springer, Berlin
12. Palpant B, Prével B, Lermé J, Cottancin E, Pellarin M, Treilleux M, Perez A, Vialle JL, Broyer M (1998) Optical properties of gold clusters in the size range 2–4 nm. Phys Rev B 57:1963–1970
13. Lermé J, Palpant B, Prével B, Cottancin E, Pellarin M, Treilleux M, Vialle JL, Perez A, Broyer M (1998) Optical properties of gold metal clusters: a time-dependent local-density-approximation investigation. Eur Phys J D 4:95–108
14. Srivastava SK, Gangopadhyay P, Amirthapandian S, Sairam TN, Basu J, Panigrahi BK, Nair KGM (2014) Effects of high-energy Si ion-irradiations on optical responses of Ag metal nanoparticles in a SiO_2 matrix. Chem Phys Lett 607:100–104
15. Scholl JA, Koh AL, Dionne JA (2012) Quantum plasmon resonances of individual metallic nanoparticles. Nature 483:421–427
16. Hosoya Y, Suga T, Yanagawa T, Kurokawa Y (1997) Linear and nonlinear optical properties of sol–gel-derived Au nanometer-particle-doped alumina. J Appl Phys 81:1475–1480
17. Meldrum A, Boatner LA, White CW, Ewing RC (2000) Ion irradiation effects in nonmetals: formation of nanocrystals and novel microstructures. Mater Res Innov 3:190–204
18. Miotelo A, De Merchi G, Mattei G, Mazzoldi P, Sada C (2001) Clustering of gold atoms in ion-implanted silica after thermal annealing in different atmospheres. Phys Rev B 63:075409
19. Dai Z, Yamamoto S, Narumi K, Miyashita A, Naramoto H (1999) Gold nanoparticle fabrication in single crystal SiO_2 by MeV Au ion implantation and subsequent thermal annealing. Nucl Instr Meth Phys Res B 149:108–112
20. Ila D, Williams EK, Zimmerman RL, Poker DB, Hensley DK (2000) Radiation induced nucleation of nanoparticles in silica. Nucl Instr Meth Phys Res B 166–167:845–850
21. Dhara S, Sundaravel B, Ravindran TR, Nair KGM, David C, Panigrahi BK, Magudapathy P, Chen KH (2004) 'Spillout' effect in gold nanoclusters embedded in c-Al_2O_3(0001) matrix. Chem Phys Lett 399:354–358

22. Dhara S, Kesavamoorthy R, Magudapathy P, Premila M, Panigrahi BK, Nair KGM, Wu CT, Chen KH (2003) Quasiquenching size effects in gold nanoclusters embedded in silica matrix. Chem Phys Lett 370:254–260
23. Raza S, Stenger N, Kadkhodazadeh S, Fischer SV, Kostesha N, Jauho A-P, Burrows A, Wubs M, Mortensen NA (2013) Blueshift of the surface plasmon resonance in silver nanoparticles studied with EELS. Nanophotonics 2:131–138
24. Palpant B, Portales H, Saviot L, Lerme J, Prevel B, Pellarin M, Duval E, Perez A, Broyer M (1999) Quadrupolar vibrational mode of silver clusters from plasmon-assisted Raman scattering. Phys Rev B 60:17107–17111
25. Saviot L, Murray DB, de Lucas M (2004) Vibrations of free and embedded anisotropic elastic spheres: application to low-frequency Raman scattering of silicon nanoparticles in silica. Phys Rev B 69:113402–113404
26. Ekardt W (1984) Work function of small metal particles: self-consistent spherical jellium-background model. Phys Rev B 29:1558–1564
27. Wannier GH (1962) Dynamics of band electrons in electric and magnetic fields. Rev Mod Phys 34:645–655
28. Nenciu G (1991) Dynamics of band electrons in electric and magnetic fields: rigorous justification of the effective Hamiltonians. Rev Mod Phys 63:91–128
29. Feldmann J, Leo K, Shah J, Miller DAB, Cunningham JE, Meier T, von Plessen G, Schulze A, Thomas P, Schmitt-Rink S (1992) Optical investigation of Bloch oscillations in a semiconductor superlattice. Phys Rev B 46:7252–7255
30. Hawton M, Dignam MM (2003) Infinite-order excitonic Bloch equations for asymmetric nanostructures. Phys Rev Lett 91:267402
31. Bruggeman DAG (1935) Calculation of various physical constants of heterogeneous substances. Ann Phys (Leipzig) 24:636–679
32. Tiggesbäumker J, Köller L, Meiwes-Broer KH, Liebsch A (1993) Blue shift of the Mie plasma frequency in Ag clusters and particles. Phys Rev A 48:R1749–R1752
33. Fedrigo S, Harbich W, Buttet J (1993) Collective dipole oscillations in small silver clusters embedded in rare-gas matrices. Phys Rev B 47:10706
34. Charl'e KP, Schulze W, Winter B (1989) The size dependent shift of the surface plasmon absorption band of small spherical metal particles. Z Phys D 12:471–475
35. Genzel L, Martin TP, Kreibig U (1975) Dielectric function and plasma resonances of small metal particles. Z Phys B 21:339–346
36. Cottancin E, Lerme' J, Gaudry M, Pellarin M, Vialle J-L, Broyer M, Pre'vel B, Treilleux M, Me'linon P (2000) Size effects in the optical properties of Au_nAg_n embedded clusters. Phys Rev B 62:5179–5185
37. Thongrattanasiri S, Manjavacas A, García de Abajo FJ (2012) Quantum finite-size effects in graphene plasmons. ACS Nano 6:1766–1775
38. Bailo E, Deckert V (2008) Tip-enhanced Raman scattering. Chem Soc Rev 37:921–930
39. Gaudry M, Lerme' J, Cottancin E, Pellarin M, Vialle J-L, Broyer M, Pre'vel B, Treilleux M, Me'linon P (2001) Optical properties of $(Au_xAg_{1-x})_n$ clusters embedded in alumina: evolution with size and stoichiometry. Phys Rev B 64:085407
40. Ripken K (1972) Optical-constants of Au, Ag and their alloys in energy region from 2.4 to 4.4 eV. Z Phys 250:228–234
41. Nilsson PO (1970) Electronic structure of disordered alloys: optical and photoemission measurements on Ag-Au and Cu-Au alloys. Phys Kondens Mater 11:1–18
42. Rakić AD, Djurišić AB, Elazar JM, Majewski ML (1998) Optical properties of metallic films for vertical-cavity optoelectronic devices. Appl Opt 37:5271–5283

Chapter 12
Quantum Plasmonics: From Quantum Statistics to Quantum Interferences

Giuliana Di Martino

Abstract Plasmons are characterized by losses into the metal, here we want to investigate the effect of these losses on their quantum properties. This is a field not yet fully investigated and the work presented here will give us the possibility to understand the effect of losses on the plasmons quantum properties. This will allow us to see how plasmons can be used in the quantum information technology field, since they keep the quantum information regardless of their lossy character. Another key property yet here investigated is the bosonic character of single surface plasmon polaritons (SPPs). The quasi-particle nature of SPPs, consisting of a photon (boson) coupled to a charge density wave of electrons (fermions), makes them an unusual type of quantum excitation. Here, we will show the bosonic character of plasmons, making use of interference experiments. We describe the first direct observation of quantum interference in the Hong-Ou and Mandel quantum interference effect for single SPPs, demonstrating by this way the bosonic nature of plasmons. This study opens opportunities for controlling quantum states of light in ultra-compact nanophotonic plasmonic circuitry.

Keywords Quantum plasmonics • Plasmon losses • Hong-Ou and Mandel quantum interference • Quantum information

12.1 Introduction

Since its formulation in the early 1920s, quantum mechanics has played a fundamental role in our understanding of how Nature behaves on the microscopic scale and smaller. ‘Puzzling’ and ‘non-intuitive’ are words that were often used by scientists to describe quantum theory when it was first proposed and these sentiments are still held by researchers in the field of quantum physics today. However, compelling evidence for its correctness has been found over the last century in the form of its predictive power in a wide range of experimental settings. It has been successful in explaining many physical phenomena, such as experiments probing

G. Di Martino, PhD (✉)
NanoPhotonics Centre, Cavendish Laboratory, University of Cambridge, CB3 0HE, UK
e-mail: gd392@cam.ac.uk

C.D. Geddes (ed.), *Reviews in Plasmonics 2015*,
DOI 10.1007/978-3-319-24606-2_12

the structure of the atom during the 1930s, nuclear fusion in stars in the late 1930s, the behaviour of elementary particles since the 1940s and superconductors since the 1950s, to name only a few examples. Quantum mechanics is in essence a mathematical framework or set of rules that can be applied in order to describe a given physical system. Its use is not always necessary and depends on various factors such as the scale of the system and how well it is isolated from everything else. Since the 1970s, the ability to investigate the behaviour of single quantum systems has become possible. Scientists have developed the capability to control and manipulate these systems with numerous methods in a range of different experimental setups. The invention of the Laser [1] and its development over the years has played a crucial role in this. Techniques such as trapping single atoms in 'atom traps' [2–8] have allowed atoms to be isolated from the rest of the world and their individual behaviour to be studied by probing with laser light. Devices that enable the transfer of single electrons [9, 10] and photons (quanta of light) [11] have also been demonstrated experimentally with great success. These advances have stimulated theoretical interest in the use of quantum controlled devices for investigating and probing deeper into the intricate features of quantum physics. This interest, together with the development of quantum mechanics from a new information-theoretic point of view, has led physicists to the possibility of using quantum-based devices for carrying out information processing tasks. Technologies such as quantum computing (QC) and quantum cryptography have also emerged, offering unique advantages over their classical (non-quantum) counterparts. These advantages include a vast decrease in the running time of various mathematical algorithms and significantly higher levels of security for communication. The properties of quantum physics that at first seemed strange and rather curious to its founders are now being exploited in the design of novel schemes for quantum information protocols. The development of quantum information theory has also led to new fundamental insights into quantum physics itself, such as the roles of entanglement (quantum correlations) and measurements in the dynamics of quantum systems, both isolated and in contact with the rest of the world. Apart from the potential advantages quantum information processing (QIP) offers, there are practical reasons why microscopic scale information protocols based on quantum mechanics are needed. In 1965 Moore predicted [12] that the number of transistors per integrated circuit (for minimum component cost) would double every 2 years. This rate has now been maintained for over 40 years, with the size of the transistors becoming smaller and smaller. There will come a point, within the next decade, when quantum effects in the circuits will be unavoidable. It was first recognised by Benioff [13–15] in the early 1980s that quantum mechanical computational processes could be at least as powerful as classical computational processes. Around the same time, Feynman [16] suggested that computers based on the principles of quantum mechanics could actually overcome the essential difficulties faced when trying to simulate complex quantum systems on classical computers. In 1985, Deutsch [17] was naturally led to consider computational devices based on the framework of quantum mechanics. This work, together with Benioff's and Feynman's insights, laid the foundations of modern-day quantum computing.

The great progress of quantum cryptography is due largely to the minimal quantum resources that the protocols require. Quantum computing on the other hand requires much larger resources for performing its algorithms and numerous problems still need to be overcome, both at the experimental and theoretical level. From an experimental perspective, physically scalable qubit systems need to be identified and developed. The systems also need to allow universal sets of operations to be implemented and have the ability to initialise and read out the information being processed [18]. Unavoidable errors occur in many places in quantum systems. Effects such as decoherence (environment-induced noise) and the imperfect operation of key components in the setup act to destroy or modify the quantum information in a way that produces undesired protocol outcomes. However, it remains to be seen how these methods can be implemented in practice, given the high demand of control and isolation that even small quantum resources require. Scaling to larger resources in order to perform more computationally useful tasks also increases the quantum system's susceptibility to imperfections. There is still much work to be done before QC can be put to practical use.

It is in this framework that a new and very promising field can be introduced. Quantum plasmonics is a rapidly growing field of research that involves the study of the quantum properties of light and its interaction with matter at the nanoscale. Here, surface plasmons – electromagnetic excitations coupled to electron charge density waves on metal-dielectric interfaces or localized on metallic nanostructures – enable the confinement of light to scales far below that of conventional optics. Nanophotonic systems based on plasmonic components are currently attracting considerable attention due to the novel ways in which the electromagnetic field can be localised and controlled [19, 20]. In the classical regime, a wide range of applications are being actively pursued, including near-field nano-imaging [21], biosensing [22] and solar cells [23]. Recently, researchers have also started to investigate plasmonics in the quantum regime [24]. Devices have been proposed for a variety of applications in quantum information science such as single-photon sources [25], transistors [26] and ultra-compact quantum circuitry [27].

12.2 From Quantum Optics to Quantum Plasmonics

Photonic circuits can be much faster than their electronic counterparts, but they are difficult to miniaturize below the optical wavelength scale. Nanoscale photonic circuits based on surface plasmon polaritons (SPPs) are a promising solution to this problem because they can localize light below the diffraction limit.

Surface plasmon polaritons are electromagnetic excitations propagating at the interface between a dielectric and a conductor, evanescently confined in the direction perpendicular to the interface. These electromagnetic surface waves arise via the coupling of the electromagnetic fields to oscillations of the conductor's electron plasma. Surface plasmons propagate along the interface with electromagnetic fields, energy and charges highly localized and possessing maximum intensity at the interface. These properties depend strongly on the optical properties of both the metal (complex dielectric function, corrugations, roughness) and the dielectric (refractive index).

Recent works have shown that SPPs can maintain certain quantum properties of their exciting photon field, with the demonstration of assisted transmission of entangled photons [28, 29], sub-Poissonian statistics [25], energy-time entanglement [30], quantum superposition [31], quadrature squeezing [32], wave-particle duality [33], and single plasmon detection [34]. These results suggest that many principles of quantum optics can be transferred to the field of plasmonics, enabling novel devices, such as single photon switches, to be realized [26].

In [35] the authors described a new all-electrical SPP detection technique based on the near-field coupling between guided plasmons and a nanowire field-effect transistor. On-chip electrical detection has been demonstrated using organic photodiodes, gallium arsenide structures and germanium wires. However, none of these techniques has provided single plasmon sensitivity. By coupling a plasmon waveguide to a superconducting single-photon detector (SSPD), the authors in [34] demonstrate on-chip electrical detection of single plasmons. Plasmons preserve many key quantum mechanical properties of the photons used to excite them, including entanglement [28] and sub-Poissonian statistics [25].

Despite recent progress in using quantum optical techniques to study plasmonic systems, adapting them to realistic structures will require a much more detailed understanding of the quantum properties of SPPs when loss is present. An important goal is to extend the knowledge already existing in the field of quantum optics, and with it all its applications, to the plasmonic field. For instance, understanding how loss affects the quantum properties of SPPs may open up a route toward the realistic design and fabrication of nanophotonic plasmonic circuits for quantum information processing. With this goal in mind, we will report in the following section a study on the quantum statistics of SPPs in metallic stripe waveguides [36], assessing the realistic potential of building plasmonic waveguides for nanophotonic circuitry that operates faithfully in the quantum regime.

Another key property yet to be fully investigated is the bosonic character of single surface plasmon polaritons. This knowledge may open up new opportunities for controlling quantum states of light in ultra-compact nanophotonic plasmonic circuitry. The quasi-particle nature of SPPs, consisting of a photon (boson) coupled to a charge density wave of electrons (fermions), makes them an unusual type of quantum excitation. The bosonic character of photons was explicitly verified in the seminal work of Hong, Ou and Mandel via the observation of bunching in the output field of a 50:50 beamsplitter with identical, indistinguishable photons incident on its two input ports [37]. In the following sections, we report an investigation by Di Martino et al. [38] that confirms the bosonic nature of single SPPs in the quantum regime.

12.3 Quantum Statistics of Surface Plasmon Polaritons in Metallic Stripe Waveguides

Despite the significant progress made so far, there are many fundamental aspects of quantum plasmonic systems that remain unexplored. Plasmons are characterized by losses into the metal, therefore a crucial step is to investigate the effect of these losses on their quantum properties. This understanding is necessary for being able to show that plasmons can be used in the quantum information technology field, since they keep the quantum information regardless their lossy character [36].

Using single photons produced by parametric down conversion, Di Martino et al. [36] excited quanta of leaky SPPs in thin metallic stripe waveguides, one of the fundamental building blocks for plasmonic circuits. After having created the two heralded photon beams, light from one of the IR beams (the idler) is used as a gating beam. The detection of a photon on the idler arm signals the presence of a photon into the other arm. The other beam (the signal) is focused onto the in-coupling grating of the waveguide probed (Fig. 12.1). At the grating, the generated photons are converted into SPPs due to phase-matching conditions. These propagate along the waveguide until they reach the outcoupling grating, at which point they are converted back into light, as previously depicted in Fig. 12.2. The output is selected by an iris and injected into a multimode fiber (MMF) for analysis. Finally, the multimode fiber directs the output to a Hanbury Brown and Twiss interferometer (BS and detectors B and B′) used to measure the second-order quantum coherence function, $g^{(2)}(\tau)$.

The coherence function $g^{(2)}$ is a measure of the correlation of the intensity of a field at a time $t = 0$ and at a later time $t = \tau$ for a fixed position. By measuring $g^{(2)}(0)$ for a given field, we can determine whether or not it is in the nonclassical regime ($g^{(2)}(0) < 1$). In particular, for number states $|n\rangle$, if $g^{(2)}(0) < 0.5$ is measured in an experiment, we can be confident that the field is within the single excitation regime. Experimentally, a beamsplitter is used to symmetrically split the field into modes B and B′. In this case one can show that the definition of $g^{(2)}(\tau)$ is equivalent to [39, 40]:

$$g^{(2)}(\tau) = \frac{N_{BB'}}{N_B N_{B'}} \left(\frac{T}{\Delta t} \right)$$

where T the averaging (integration) time of the measurement, $N_{BB'}$ is the number of coincidence detections at detectors B and B′ within a coincidence time window t, and N_B and $N_{B'}$ are the number of independent detections at detectors B and B′ respectively. All detections at B′ are delayed by time t. All measurements are conditioned on the detection of a photon in mode A. Therefore, we use the conditional the previous equation, given by

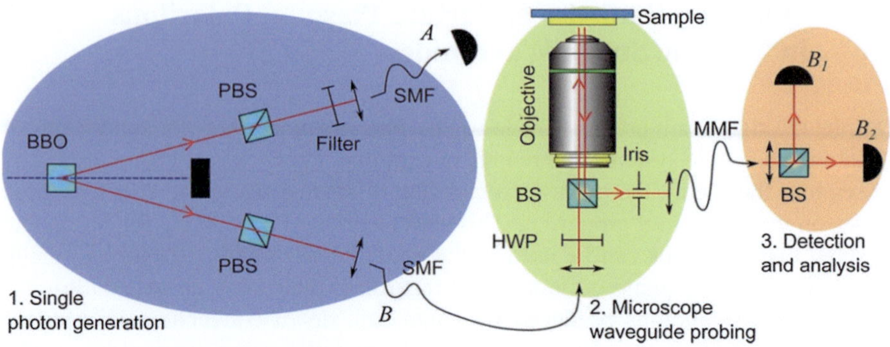

Fig. 12.1 Schematic of the experimental setup including a single-photon source stage, waveguide probing stage, and final analysis stage

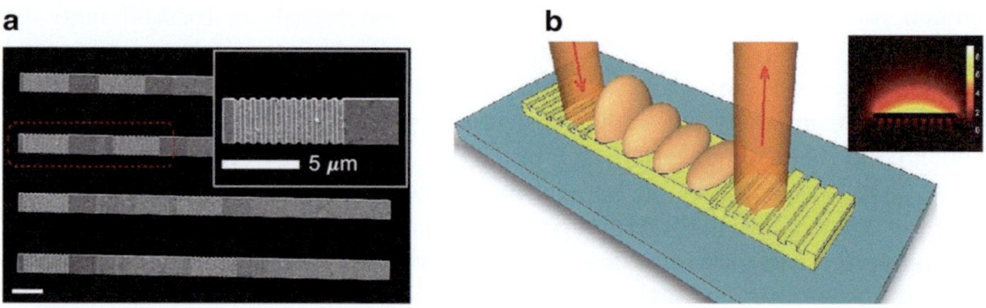

Fig. 12.2 (**a**) Scanning electron microscope image of a selection of waveguide lengths from 5 to 20 μm. All waveguides have been explored in the quantum regime, however a further time-domain analysis of the quantum statistics has been performed for the 7.5 μm waveguide highlighted by the *dashed red box*. Inset: detail of one of the in/out-coupling gratings. The scale bars denote 5 μm. (**b**) Sketch of the SPP excitation and propagation through the waveguide. Inset: fundamental SPP mode in our stripe waveguide electric field profile along the cross section of the waveguide, calculated using the Finite Element Method (FEM) for an infinitely long waveguide

$$g_c^{(2)}(\tau) = \frac{N_A N_{ABB'}}{N_{AB} N_{AB'}}$$

where $N_{ABB'}$ is the number of coincidence detections at detectors A, B, and B′, with detections at B and B′ occurring within a coincidence time window Δt centred on the detection at A. N_{AB} is the number of coincidence detections at detectors A and B within the coincidence time window Δt and similarly for $N_{AB'}$. N_A is the number of independent detections at detector A. All measurements are taken over an integration time T, which does not appear explicitly in equation. The value of $g^{(2)}{}_c\,(\tau)$ at zero time delay provides us with a measure of conditioned single-arm statistics in mode B. Additional information about the quality of the field intensity correlations can then be obtained by measuring $g^{(2)}{}_c\,(\tau)$ over a range of different time delays τ. An example of $g^{(2)}{}_c\,(\tau)$ is shown in Fig. 12.3. The rate of triples observed at zero

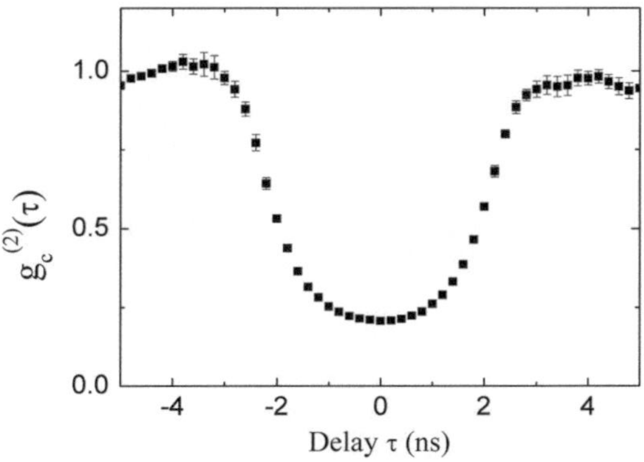

Fig. 12.3 Conditional second-order quantum coherence function, $g^{(2)}_c(\tau)$, for the down-converted light. The time window Δt used for the measurement is 2 ns, and the integration times are adjusted to obtain reasonable error bars

delay agrees very well with the value expected for solely accidental coincidences. Additionally, these accidental coincidences lead to a value of $g^{(2)}_c(0)$ higher than zero. One can show [40] that the off set on $g^{(2)}_c(0)$ due to accidental coincidences is:

$$g^{(2)}_{acc}(0) = \Delta t R_A \left(\frac{R_B}{R_{AB}} + \frac{R_{B'}}{R_{AB'}} \right)$$

The measured $g^{(2)}_c(\tau)$ for the single photon source is presented in Fig. 12.3. Using the count rates values observed at each detector, Di Martino et al. [36] calculate the value of $g^{(2)}_{acc}(0) = 0.23$ for $t = 2$ ns, which is the value experimentally observed and shown in Fig. 12.3: the nonzero value of $g^{(2)}(0)$ is solely due to accidental coincidences. Moreover, here the width and shape of the dip in $g^{(2)}_c(\tau)$ is only linked to the coincidence window used.

At this point, the effects of loss on the quantum statistics of waveguided SPPs can be characterized. The results from the single-photon source before and after the waveguides are shown in Fig. 12.4a (respectively, black and red curves). The time window Δt used for the measurement is 2 ns, and the integration times are adjusted to obtain reasonable error bars. The statistics obtained after the waveguide are identical to that of the source itself. One can see that the value of $g^{(2)}_c(0) < 0.5$ in both cases which unambiguously demonstrates that we are in the single plasmon excitation regime. Note $g^{(2)}_c(0)$ is not identically zero in either case. The finite value, however, originates solely from accidental coincidences. In Fig. 12.4b we show the unconditioned $g^{(2)}(\tau)$ as a function of time delay for the single photon source only and after a 7.5 μm length waveguide (resp. black and red). This plot shows the vital role of the detection of photons in mode A for the conditional measurements of the statistics of the out-coupled light from the waveguides. Without this 'heralding' of the photons, the statistics of the light arriving at the detectors are those of a thermal field [40]. It should be noted that the theoretically expected peak of $g^{(2)}_c(0) = 2$ for a thermal field is challenging to observe in quantum optics experiments [41], as instead of reaching the value of 2, its height above unity is effectively proportional to the ratio of the coherence time of the

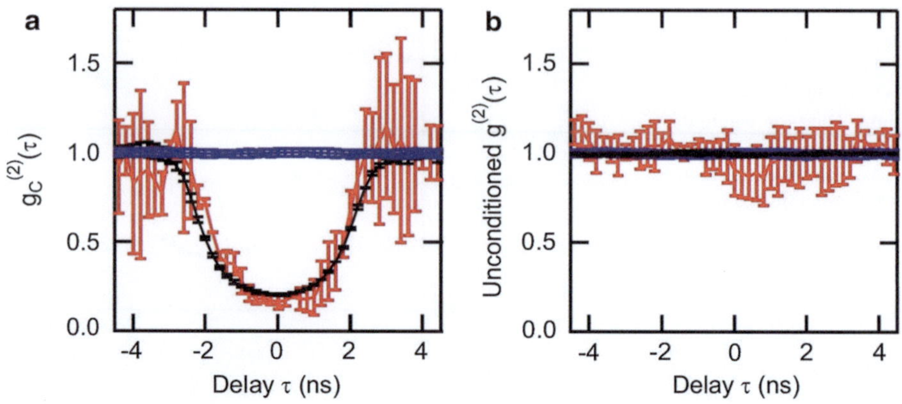

Fig. 12.4 (**a**) Conditional second-order quantum coherence function, $g^{(2)}{}_c(\tau)$, for the down-converted light in mode B before the waveguide (*black*), along with the out-coupled light when single photons are injected into a waveguide of length 7.5 μm (*red*). The classical limit is illustrated by the blue data points, corresponding to the unconditioned second-order quantum coherence, $g^{(2)}(\tau)$, for an attenuated laser injected in the waveguide. (**b**) $g^{(2)}(\tau)$ for the downconverted light in mode B (*black*) and injected into the waveguide (*red*). *Blue*: $g^{(2)}(\tau)$ for the attenuated laser injected in the waveguide

single photon source to the response time of the detection, which in our experiment is $\approx 10^{-5}$.

We now turn our attention to the effect of losses in the single excitation regime. As the SPPs propagate along the waveguide, the finite conductivity of the metal results in ohmic losses, while radiation into the substrate and surface roughness results in radiative losses. For a reasonably smooth waveguide surface and a thick gold layer, ohmic losses are the main source of damping. At the single excitation level there may be correlations between and within the different damping channels, such as excited phonons, background ion-cores, electron gas collisions and interband transition processes (involving electron-hole pairs), that cause the damping to depart significantly from the classical model. In this case we may be able to see such effects manifest themselves through a change in the behaviour of the second-order quantum coherence. Indeed, such a change is well known when light-matter scattering and absorption processes are involved [39]. Taking a simple linear loss model with uncorrelated Markovian noise, for this particular loss model, the second-order quantum coherence should remain unchanged [42]. Indeed, closer to the plasmon wavelength, the SPP character becomes more electron-like and such effects may become important. This regime remains to be investigated. To explore the quantum statistics in the presence of loss, Di Martino et al. [36] first measured the mean excitation rate over a range of waveguide lengths. To do this they measured the counts N_B at detector B for an attenuated laser at a fixed intensity and then the conditional counts N_{AB} at detectors A and B for the single photon source. In both cases, the effect of loss from the beam splitter in the analysis stage was included in the overall detection efficiency, enabling us to disregard the data

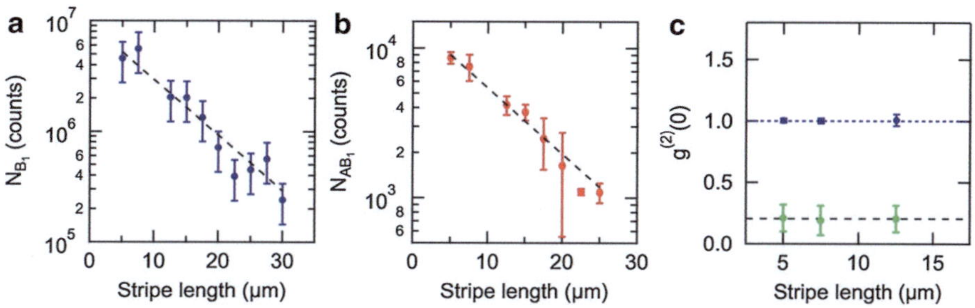

Fig. 12.5 Count rate statistics of the light out-coupled from the surface plasmon waveguides. (**a**) N_B for injected attenuated laser as the length of the waveguide is increased. *Dashed black line* is an exponential fit, yielding a propagation length $l = 8.9 \pm 1.7$ μm. (**b**) Same for N_{AB} at zero delay for injected single photons. Here, $l = 9.8 \pm 0.6$ μm. (**c**) $g^{(2)}{}_c(0)$ for single photons injected in waveguides of varying lengths (*green*) and unconditioned $g^{(2)}(0)$ for a laser injected in the waveguides (*blue*). The *black dashed line* indicates the value found for the single-photon source and the *blue dotted line* the classical limit

from detector B_0. In Fig. 12.5a Di Martino et al. [36] show N_B against waveguide length for the injected attenuated laser. Figure 12.5b presents N_{AB} at zero delay for injected single photons and as the length of the waveguide is increased. The N_{AB} trend matches N_B, providing evidence that the effect of loss on the field of the single SPPs is consistent with the classical exponential behaviour. The SPP propagation length l, defined as the length at which the intensity (mean photon number $<n_i>$) decreases to 1/e of its original value, extracted from Fig. 12.5a is $l = 8.9 \pm 1.7$ μm, a value similar to $l = 9.8 \pm 0.6$ μm obtained from Fig. 12.5b. Both values are in good agreement with each other but smaller than the propagation length expected from finite element method (FEM) calculations of 16.7 μm, due to imperfections introduced by the fabrication of the waveguides. In Fig. 12.5c, $g^{(2)}{}_c(0)$ is shown for the out-coupled light for injected single photons as the length of the waveguide is increased. The value of $g^{(2)}{}_c(0)$ for the down-converted photons only is plotted as a dashed black line for reference. One can clearly see that indeed the values remain unchanged for the lengths investigated. These results therefore provide evidence for the validity of a linear uncorrelated Markovian loss model for SPP damping at the single quanta level. This complements well and goes beyond previous studies looking into the preservation of entanglement via localized plasmons [28] and nonclassicality via long-range surface plasmons [43], where elements of plasmon loss were considered.

This result [36] implies that building longer and more complex SPP waveguide structures operating in the quantum regime is realistic and opens up the possibility for future studies of new types of functioning devices based on quantum plasmonics, assessing the realistic potential of building plasmonic waveguides for nanophotonic circuitry that operates faithfully in the quantum regime.

12.4 Quantum Interference in the Plasmonic Hong-Ou-Mandel Effect

An important property to discuss is the bosonic character of SPPs. The quasi-particle nature of SPPs, consisting of a photon (boson) coupled to a charge density wave of electrons (fermions), makes them an unusual type of quantum excitation. It is, therefore, important to understand whether SPPs are bosons, fermions, or a hybrid mixture [44]. Here, we will see the bosonic character of plasmons, making use of interferences experiments. It is important to notice that this property of SPPs open opportunities for controlling quantum states of light in ultra-compact nanophotonic plasmonic circuitry.

The bosonic character of photons was explicitly verified, in 1987, in the seminal work of Hong, Ou and Mandel via the observation of bunching in the output field of a 50:50 beamsplitter with identical, indistinguishable photons incident on its two input ports [37]. They showed the destructive interference between two single photons without a phase relationship at a beam splitter. This benchmark study directly proved the bosonic nature of the photons and allowed to probe the length of their photon wave packets with unprecedented precision.

With the current rising interest in the quantum behaviour of SPPs, our goal is to show this result with SPPs instead of free space photons [38]. Previous work using plasmonic waveguides has hinted that SPPs are bosons by observing the Hong-Ou-Mandel (HOM) effect, both indirectly using a photonic beamsplitter [45] and directly using a plasmonic beamsplitter [46]. However, the question as to whether quantum interference is involved remained for long time open due to the modest HOM interference observed in previous studies, which can also be obtained using classical light [47–51].

In the following pages, we will show how plasmonic waveguides can be used to prove that SPPs are bosons by observing the Hong-Ou-Mandel (HOM) effect. We will first understand HOM effect in the photonic case when using a conventional beamsplitter cube. After that, we will show the results obtained from the plasmonic HOM experiment.

Figure 12.6a shows the experimental setup used to perform the HOM effect investigation in the photonic case. Here, photon pairs are generated at a wavelength of 808 nm. Interference filters (IFs) with a central wavelength of 800 and 22 nm bandwidth are placed in both paths to spectrally select out the down-converted photons. The photons are then injected into single mode fibers (SMFs). After collimation of the output from the fibers, the photons pass through polarizers (Pol) in order to remove from the beams any random polarization introduced by the travel into the fibers. Their polarization is then adjusted using half-wave plates (HWPs). In order to vary the amount by which the photons from a given pair are able to interfere with each other a time delay is introduced in one path by extending it over a distance d with respect to the other using a motorized delay line. This provides a variable delay of $\Delta t = d/c$ between the single-photon wavepackets in each path. The two photons are sent to a conventional beamsplitter to make them

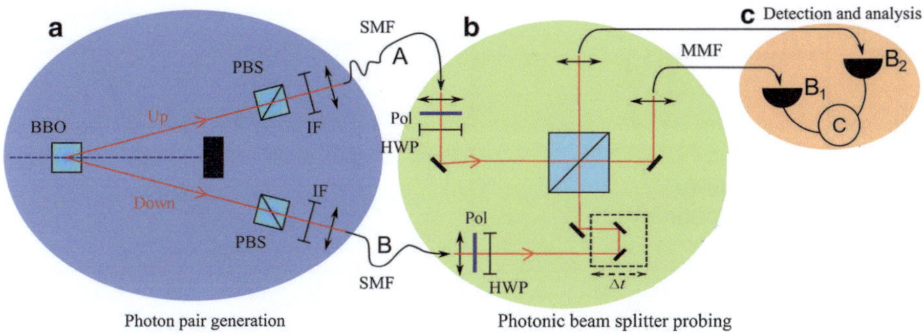

Fig. 12.6 (**a**) Photon pair generation. Photon pairs are generated via spontaneous parametric down-conversion using a 404 nm pump laser beam focused onto a Beta Barium Borate (BBO) crystal and are filtered using interference filters (IF) to narrow the bandwidth to 800 ± 11 nm. Each photon from a pair is coupled into a single-mode fiber (SMF). (**b**) Photonic beamsplitter. The photons pass through polarizers (Pol) in order to remove from the beams any random polarization. The half wave plates (HWPs) turn the polarization so that both the photons have the same polarization when they reach the beam splitter. The photons interfere at the beam splitter level and at the output they are then collected by two $10\times$ microscope objectives and coupled into multi-mode fibers (MMFs). A delay line introduces a time delay Δt in one of the photon paths. (**c**) Detection and analysis stage. The outputs of the MMFs are collected by avalanche photodiodes, B_1 and B_2, where coincident detection events within a time window t_c are collected

interfere. It is crucial to tune their polarization in order to ensure their indistinguishability at the beam splitter level. To do so, the two beams are sent through a polarizer and a HWP in order to have both the beams horizontally polarized. Multimode fibers (MMFs) collect the light coming out from the beam splitter and direct it to silicon avalanche photodiode detectors B_1 and B_2, which monitor the arrival of the photons from a given output of the splitter. All detection events are time-tagged and coincidences are evaluated within a $t_c = 2$ ns time window. In order to first confirm that the single photons generated by our source exhibit the above-described HOM effect when using a conventional beamsplitter cube, we measured the output port coincidences, as a function of the time delay between the arrival of the input port photons. At zero delay the coincidence rate drops to a minimum value, N_{min}, as expected (Fig. 12.7). This drop is quantified using the visibility, V_P, defined as the percentage drop of the coincidences from their maximum value far from the dip center, N_{max}, where the photons do not interfere, i.e. $V_P = (N_{max} - N_{min})/N_{max}$ [50, 52]. The expected coincidence rate is given by $N(\Delta t) = N_t P(\Delta t)$, where N_t is the rate of correlated pairs. $N_t(\eta_{in}\eta_{out})^2/2$ is obtained by measuring $N^{exp}(\Delta t >> \tau_c)$. Here, there is a contribution from uncorrelated (accidental) pairs, N_{acc}, which arrive at the APDs within t_c. Thus, $N^{exp}(\Delta t >> \tau_c) = Nt(\eta_{in}\eta_{out})^{2/}2 + N_{acc}$. A large proportion of N_{acc} is given by [36] $N^m{}_{acc} = t_c N_{B1} N_{B2}$, with the rest as $N^r{}_{acc} = N_{acc} - N^m{}_{acc}$, leading to $N(\Delta t) = N^r{}_{acc} + (N^{exp}(\Delta t >> t_c) - N_{acc})(1 - \mathrm{sinc}^2(\Delta t^* \Delta\omega/2))$, where $N^r{}_{acc} = N^{exp}(0) - N^m{}_{acc}$.

The photonic HOM dip obtained is shown in Fig. 12.7. It has a visibility $V_P = 0.67 \pm 0.05$. The value is limited mainly by the deviation of the beamsplitter from the ideal case of $R = T = 0.5$. We emphasize here that for a classical input

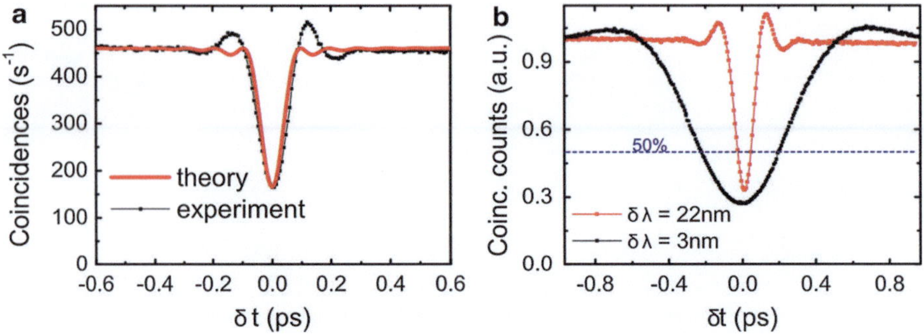

Fig. 12.7 (**a**) Comparison between the experimental photonic HOM dip, in *black*, and the expected theoretical curve, in *red*, for a top-hat shaped spectrum of Δλ = 22 nm. (**b**) Experimental photonic HOM dip with different input spectra, in *black* Δλ = 3 nm, in *red* Δλ = 22 nm

field, a drop in coincidences corresponding to a visibility of up to V = 0.5 can be observed [47, 49]. In fact, classical electromagnetic waves superposition theory provides a HOM dip with a visibility that reaches a maximum of 0.5 [51]. Thus, the visibility of V = 0.5 in HOM interference is usually considered as the border between classical and quantum physics. However, with a visibility larger than 0.5 we can confirm that the drop is due to quantum interference [49, 50]. The comparison between the experimental photonic HOM dip, in black, and the expected theoretical curve, in red, for a top-hat shaped spectrum of Δλ = 22 nm is shown in Fig. 12.7a. We can see a very good agreement between theory and experiment. The reason the theory does not match on the 'wings' at the top left and right of the curve is due to the shape of the spectrum not being exactly a top-hat shape. In addition, different HOM experiments can be performed varying the bandwidth of the IFs used, therefore varying the value of Δλ. As we can see in Fig. 12.7b, for smaller Δλ we have a wider HOM dip.

For the conduction of the plasmonic HOM experiment, the photons need to be sent to a plasmonic beam splitter. The sample is attached to a triaxis piezo stage to allow a fine positioning. In the experiment, the photons need to reach the sample be focused onto separate gratings (spot size 2 μm) at the inputs of an X-shaped plasmonic beamsplitter, shown in Fig. 12.8c, by a microscope objective (100×, NA 0.8). The same microscope objective is also devoted to the collection of the light coming from the output gratings. The plasmonic beamsplitter consists of two 2 μm wide, 70 nm thick gold stripe waveguides that cross at a right angle at their middle point, as shown in Fig. 12.9a. These asymmetric waveguides support a single low-loss, albeit leaky, SPP mode [53] and a number of short-range bound modes [54]. The waveguide X-shaped structure was defined on a glass substrate by electron beam lithography (EBL). A second EBL step is used to overlay 90 nm thick input/output gratings and the central scattering elements. The splitting operation is obtained via scattering process. The scattering element is a semi-transparent Bragg reflector, consisting of an arrangement of ridges, deposited on the top of the central part of the X-shaped splitter, as shown in Fig. 12.9a. Plasmonic Bragg

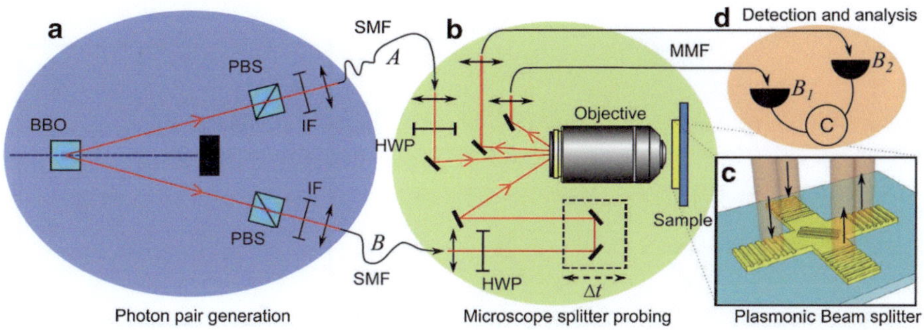

Fig. 12.8 (**a**) Photon pair generation. Photon pairs are generated via spontaneous parametric down-conversion by using a pump laser focused onto a *BBO* crystal and filtered by using interference filters (IFs). Each photon is coupled into a single-mode fiber (*SMF*). (**b**) Microscopy. The photons from the SMFs are collimated and half-wave plates (HWPs) are used to optimize *SPP* excitation. A time delay is introduced on one path. (**c**) Plasmonic beam splitter. The photons are focused onto separate spots on the input gratings by using a microscope objective. The beams at the output gratings are collected and coupled into multimode fibers (MMFs). (**d**) Detection and analysis. The outputs of the MMFs are sent to avalanche photodiodes B_1 and B_2, where coincident detection events are measured

mirrors, such as the one we use have been studied extensively in the literature, mostly as effective mirrors in the one-dimensional case of normal incidence, both on an infinite interface and on plasmonic waveguides [55]. We check the splitting ratio for a range of wavelengths for optimal splitting. As shown in Fig. 12.9d, a Bragg reflector with ridges having a period of 500 nm gives $T = 0.49 \pm 0.05$ for incident SPPs at $\lambda_0 = 808$ nm – the wavelength of the photons used in our experiment.

The plasmonic beamsplitter was then probed as depicted in Fig. 12.8. The target is to see whether the HOM effect could be observed and whether or not single SPPs interfere with each other like bosons by bunching together. When the coupling of single photons into the SPP waveguides is optimized, the count rate due to SPPs scattered by the output grating and detected by APD $B_{1,2}$ is $N_{B1,2} \sim 3.6 \cdot 10^6$ counts per hour (cph), as shown in the inset of Fig. 12.10.

A big challenge of this measurement is the acquisition procedure. We need to measure the coincidence rate for a range of Δt, by moving the motorized translation stage. Since we have low intensities, each position of the translation stage needs to be integrated for a few hours. The time-resolved correlation data give an average number of coincidences of 54.8 ± 1.4 cph far from zero time delay and 30.2 ± 2.4 cph at zero delay. However, a proportion of these counts are due to accidental coincidences from uncorrelated pairs of photons. These uncorrelated pairs are part of the field coupled into the plasmonic beamsplitter but they do not correspond to true correlated photon pairs generated by the source and, therefore, they are subtracted from the overall counts. In principle, these photons could be actively removed by better spectral filtering and reducing the coincidence time window t_c. However, this reduces the overall count rates and leads to extended data

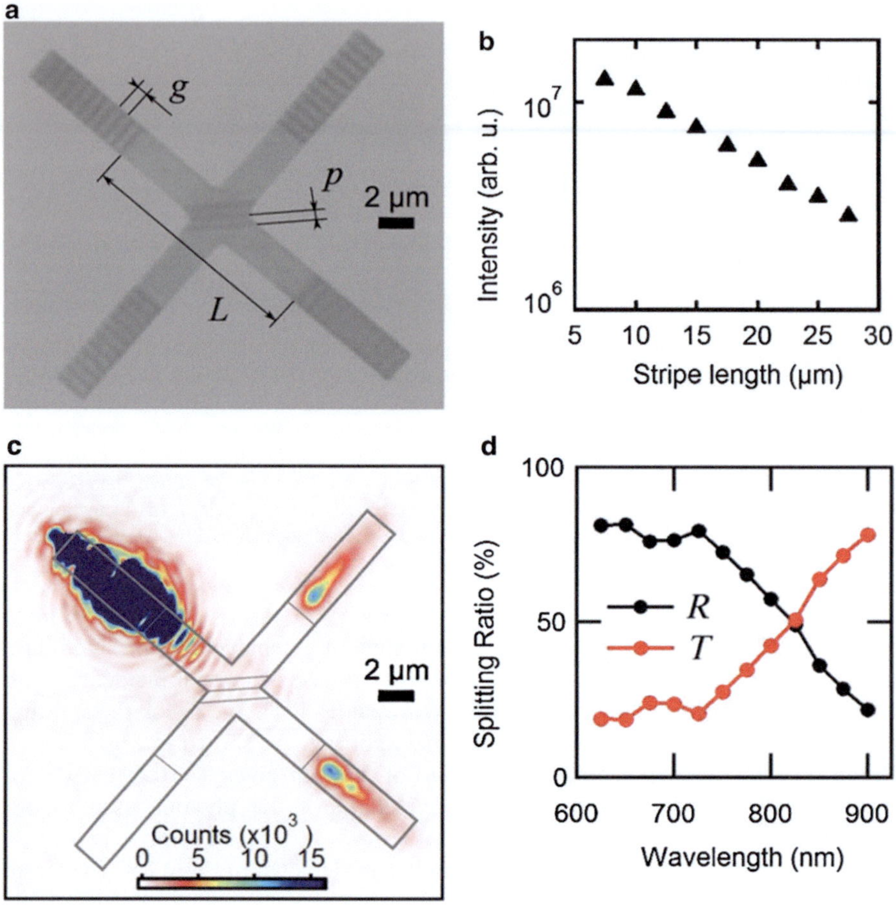

Fig. 12.9 Plasmonic beam splitter. (**a**) Optical image of the beam splitter. The in and out gratings consist of 11 ridges, each being repeated at an increment of g = 620 nm from the waveguide end. The distance between gratings is L = 12.5 μm. The Bragg reflector is made out of three ridges with a center-to-center distance of p = 500 nm. (**b**) Intensity outcoupled from a single waveguide as a function of length when excited by a laser at 808. (**c**) Optical image of the splitter when the SPPs are excited by a laser at 808 nm focused on the *top-left* grating. Light is outcoupled at the *top right* and *bottom right* with almost identical intensity. The integration time is adjusted to give a reasonable contrast for the output, leading to a saturation at the input caused by the scattered field. (**d**) Transmission T and reflection R as a function of wavelength

integration times which cause stability issues in our setup due to fluctuations in the coupling efficiency. The contribution of coincidences due to accidentals is measured by introducing an electronic time delay between the events at both detectors larger than the coincidence window, thus capturing the coincidence counts from uncorrelated pairs. After subtracting these measured accidentals the average coincidence rate far from zero delay is then 39.4 ± 0.9 cph, as shown in Fig. 12.10. The average number of coincidences far from the region of zero time delay is a weighted

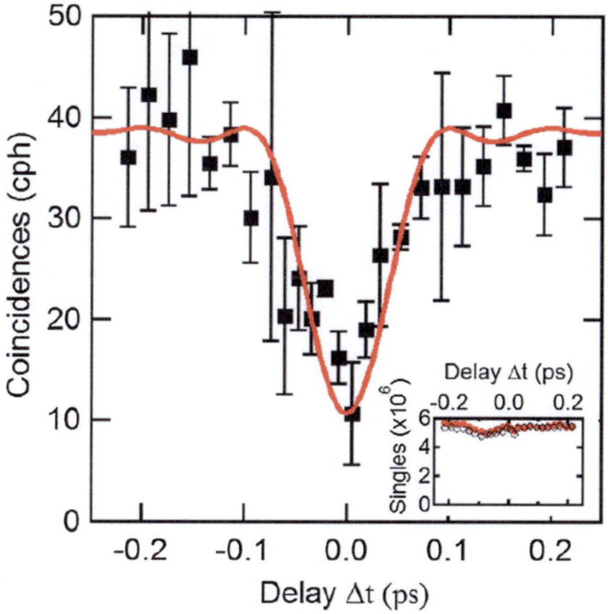

Fig. 12.10 Plasmonic Hong-Ou-Mandel dip. Black squares: coincidence rate as a function of time delay Δt. The *red curve* is a theoretical fit, $N(\Delta t)$, based on the coincidence probability $P(\Delta t)$ corrected for accidentals. From the theory fit we extract a coherence time for the single SPPs of ~0.1 ps, which is consistent with the coherence time obtained from our measured photonic dip. This confirms that during the photon-SPP conversion process the coherence properties of the single-photon wavepackets and the single SPPs are similar, and that the wavepacket has not been significantly altered by either the conversion process or propagation in the plasmonic media. The *bottom-right inset* shows the singles rates (in cph) as a function of fit: red disks for detector B_1 and *black circles* for B_2. The visibility obtained from the plasmonic dip is $V_{SPP} = 0.72 \pm 0.13$

mean over the 12 data points present in the off dip region (6 on the left and 6 on the right of the HOM dip), where the weights are the standard deviation of each data point shown in Fig. 12.10. On the other hand, at zero delay the detected coincidences drop to 10.7 ± 5.1 cph, where the reported error bar corresponds to its standard deviation. This leads to an overall visibility for the plasmonic HOM dip of $V_{SPP} = 0.72 \pm 0.13$. The observed dip confirms that single SPPs bunch together as bosons and as the visibility is larger than 0.5 in our experiment this also confirms that quantum interference is involved in the bunching process [49, 50]. The plasmonic visibility is again limited mainly by the large bandwidth of the IF's in our experiment and the resolution of the time delay used. Any loss due to radiative scattering at the beamsplitter (measured to be less than 10 %) can be included within the coefficient η_{out} and therefore does not play a role in reducing the visibility [48]. Finally, note that when the coincidence rate drops as the single SPPs are interfering, the count rate at each APD remains unchanged, as shown in the inset of Fig. 12.10. This is due to the small proportion of pairs of single excitations (as compared to the total count rate) that survive the entire process of propagation, splitting and out-coupling: in most cases at least one of the excitations from a pair will be lost in the process. Therefore, the count rate at each APD allows the efficiency of the system to be monitored to ensure the HOM dip is not caused by loss fluctuations.

In conclusion, in this work the SPPs interact via a scattering process and the HOM effect is directly observed. The SPPs showed a distinct bunching behaviour as expected for bosons, with the results clearly showing quantum interference is involved in the process. This investigation confirms the bosonic nature of single SPPs in the quantum regime and opens up new opportunities for controlling quantum states of light in ultra-compact nanophotonic plasmonic circuitry.

12.5 Conclusions

In this chapter, we gave a brief overview of the physics of SPPs at metal/insulator interface focusing our attention on the excitation of SPPs and their propagation along metal stripes. We used this knowledge in order to introduce the work done on quantum statistics of surface plasmon polaritons in metallic stripe waveguides and on quantum interference in the plasmonic Hong-Ou-Mandel effect.

We have seen how the effect of losses incurred during propagation of the single SPPs is consistent with the classical exponential behaviour and does not change the value of $g^{(2)}$, providing evidence that a linear uncorrelated Markovian loss model is valid for SPP damping at the single quanta level, implying that building longer and more complex SPP waveguide structures operating in the quantum regime is realistic and opens up the possibility for future studies of new types of functioning devices based on quantum plasmonics.

Therefore, we moved onto more complicated stripe designs, such as X shaped stripes, in order to observe nonclassical effects in the interference of two plasmons. Over the past few years there have been many works on constructing different types of scattering/interaction scenario for surface plasmons. Performing an experiment probing the interaction of two single surface plasmons with each other is an important step along the way to understanding better the behaviour of single surface plasmons at the quantum level and how one can build more complicated quantum interference networks, such as plasmonic-based quantum logic gates. A natural thing to probe in the most basic type of scatterer (a 50/50 beamsplitter) operating quantum mechanically is how it acts on two separate single surface plasmons. Here, we showed how to focus single photons onto separate gratings at the inputs of an X-shaped plasmonic beamsplitter, and convert into SPPs due to phase-matching conditions. We made them interact via a scattering process and we directly observed the plasmonic HOM effect. The SPPs show a distinct bunching behaviour as expected for bosons, with the results clearly showing quantum interference is involved in the process [38]. This investigation confirms the bosonic nature of single SPPs in the quantum regime and opens up new opportunities for controlling quantum states of light in ultra-compact nanophotonic plasmonic circuitry.

References

1. Schawlow AL, Townes H (1958) Infrared and optical masers. Phys Rev 112:1940
2. Brown LS, Gabrielse G (1986) Geonium theory: physics of a single electron or ion in a penning trap. Rev Mod Phys 58:233
3. Chu S (1998) Nobel lecture: the manipulation of neutral particles. Rev Mod Phys 70:685
4. Phillips WD (1998) Nobel lecture: laser cooling and trapping of neutral atoms. Rev Mod Phys 70:721
5. Cohen-Tannoudji CN (1998) Nobel lecture: manipulating atoms with photons. Rev Mod Phys 70:707
6. Wieman CE, Pritchard DE, Wineland DJ (1999) Atom cooling, trapping, and quantum manipulation. Rev Mod Phys 71:253
7. Metcalf HJ, van der Straten P (2001) Laser cooling and trapping. Springer, Berlin
8. Ketterle W (2002) Nobel lecture: when atoms behave as waves: Bose-Einstein condensation and the atom laser. Rev Mod Phys 74:1131
9. Kastner MA (1992) The single-electron transistor. Rev Mod Phys 64:849
10. Makhlin Y, Schon G, Shnirman A (2001) Quantum-state engineering with Josephson-junction devices. Rev Mod Phys 73:357
11. Zeilinger A, Weihs G, Jennewien T, Aspelmeyer M (2005) Happy centenary, photon. Nature 433:230–238
12. Moore GE (1965) Cramming more components onto integrated circuits. Electronics, pp. 114–117
13. Benioff PA (1980) Mechanical Hamiltonian model of computers as represented by Turing machines. J Stat Phys 22:563–591
14. Benioff PA (1982) Quantum mechanical Hamiltonian models of discrete processes that erase their own histories: applications to Turing machines. Int J Theor Phys 21:177–201
15. Benioff PA (1982) Quantum mechanical models of Turing machines that dissipate no energy. Phys Rev Lett 48:1581
16. Feynman RP (1982) Simulating physics with computers. Int J Theor Phys 21:467–488
17. Deutsch D (1985) Quantum theory, the church-Turing principle and the universal quantum computer. Proc Roy Soc Lond A 400:97–117
18. Di Vincenzo DP (1997) Topics in quantum computers. Kluwer Academic Publishers, Dordrecht
19. Gramotnev DK, Bozhevolnyi SI (2010) Plasmonics beyond the diffraction limit. Nat Photon 4:83–91
20. Giannini V, Fernández-Domínguez AI, Sonnefraud Y, Roschuk T, Fernández-García R, Maier SA (2010) Controlling light localization and light–matter interactions with nanoplasmonics. Small 6:2498–2507
21. Kawata S, Inouye Y, Verma P (2009) Plasmonics for near-field nano-imaging and superlensing. Nat Photon 3:388–394
22. Anker JN, Hall WP, Lyandres O, Shah NC, Zhao J, Van Duyne RP (2008) Biosensing with plasmonic nanosensors. Nat Mater 7:442–453
23. Atwater HA, Polman A (2010) Plasmonics for improved photovoltaic devices. Nat Mater 9:205–213
24. Tame MS, McEnery KR, Özdemir ŞK, Lee J, Maier SA, Kim MS (2013) Quantum plasmonics. Nat Phys 9:329–340
25. Akimov AV, Mukherjee A, Yu CL, Chang DE, Zibrov AS, Hemmer PR, Park H, Lukin MD (2007) Generation of single optical plasmons in metallic nanowires coupled to quantum dots. Nature 450:402–406
26. de Demler AS, Lukin EA, Chang MD, Sorensen DE (2007) A single-photon transistor using nanoscale surface plasmons. Nat Phys 3:807–812
27. de Leon NP, Lukin MD, Park H (2012) Quantum plasmonics circuits. IEEE Sel Top Quant Elec 18:1781–1789

28. Altewischer E, van Exter MP, Woerdman JP (2002) Plasmon-assisted transmission of entangled photons. Nature 418:304–306
29. García-Vidal FJ, Erni D, Cirac JI, Martín-Moreno L, Moreno E (2004) Theory of Plasmon-assisted transmission of entangled photons. Phys Rev Lett 92:236801
30. Moreno F, Erni E, Gisin D, Zbinden N, Fasel H, Robin S (2005) Energy-time entanglement preservation in plasmon-assisted light transmission. Phys Rev Lett 94:110501
31. Gisin N, Zbinden H, Fasel S, Halder M (2006) Quantum superposition and entanglement of mesoscopic plasmons. New J Phys 8:13
32. Lodahl P, Sørensen AS, Boltasseva A, Janousek J, Andersen UL, Huck A, Smolka S (2009) Demonstration of quadrature-squeezed surface plasmons in a gold waveguide. Phys Rev Lett 102:246802
33. Kolesov R, Grotz B, Balasubramanian G, Stoer RJ, Nicolet AAL, Hemmer PR, Jelezko F, Wrachtrup J (2009) Wave–particle duality of single surface plasmon polaritons. Nature 5:470
34. Heeres RW, Dorenbos SN, Koene B, Glenn GS, Kouwenhoven LP, Zwiller V (2010) On-chip single plasmon detection. Nano Lett 10:661
35. Falk AL, Koppens FHL, Yu CL, Kang K, Snapp N d L, Akimov AV, Jo M-H, Lukin MD, Park H (2009) Near-field electrical detection of optical plasmons and single plasmon sources. Nat Phys 5:475
36. Di Martino G, Sonnefraud Y, Kena Cohen S, Tame M, Ozdemir SK, Kim MS, Maier SA (2012) Quantum statistics of surface plasmon polaritons in metallic stripe waveguides. Nanoletters 12(5):2504–2508
37. Hong CK, Ou ZY, Mandel L (1987) Measurement of subpicosecond time intervals between two photons by interference. Phys Rev Lett 59:2044
38. Di Martino G, Sonnefraud Y, Tame MS, Kéna-Cohen S, Dieleman F, Özdemir ŞK, Kim MS, Maier SA (2014) Observation of quantum interference in the plasmonic Hong-Ou-Mandel effect. Phys Rev Appl 1:034004
39. Loudon R (2000) The quantum theory of light. Oxford University Press, Oxford
40. Thorn JJ, Neel MS, Donato VW, Bergreen GS, Davies RE, Beck M (2004) Observing the quantum behavior of light in an undergraduate laboratory. Am J Phys 72:1210
41. Hockel D, Koch L, Benson O (2011) Direct measurement of heralded single-photon statistics from a parametric down-conversion source. Phys Rev A 83:013802
42. Tame MS, Lee C, Lee J, Ballester D, Paternostro M, Zayats AV, Kim MS (2008) Single-photon excitation of surface plasmon polaritons. Phys Rev Lett 101:190504
43. Brongersma ML, Hartman JW, Atwater HH Plasmonics: electromagnetic energy transfer and switching in nanoparticle chain-arrays below the diffraction limit. (1999) Molecular Electronics. Symposium, Boston, 29 Nov–2 Dec)
44. Loudon R (1998) Fermion and boson beam-splitter statistics. Phys Rev A 58:4904
45. Fujii G, Segawa T, Mori S, Namekata N, Fukuda D, Inoue S (2012) Preservation of photon indistinguishability after transmission through surface-plasmon-polariton waveguide. Opt Lett 37:1535
46. Heeres RW, Kouwenhoven LP, Zwiller V (2013) Quantum interference in plasmonic circuits. Nat Nanotechnol AOP 8:719–722
47. Kim Y-S, Slattery O, Kuo PS, Tang X (2013) Two-photon interference with continuous-wave multi-mode coherent light, arXiv:1309.3017
48. Ballester D, Tame MS, Kim MS (2010) Quantum theory of surface-plasmon polariton scattering. Phys Rev A 82:012325
49. Mandel L (1999) Quantum effects in one-photon and two-photon interference. Rev Mod Phys 71:274
50. Ghosh R, Mandel L (1987) Observation of nonclassical effects in the interference of two photons. Phys Rev Lett 59:1903
51. Rarity JG, Tapster PR, Loudon R (2005) Non-classical interference between independent sources. J Opt B: Quantum Semiclassical Opt 7:S171

52. Bocquillon E, Freulon V, Berroir JM, Degiovanni P, Plaçais B, Cavanna A, Jin Y, Fève G (2013) Coherence and indistinguishability of single electrons emitted by independent sources. Science 339:1054
53. Zia R, Selker MD, Brongersma ML (2005) Leaky and bound modes of surface plasmon waveguides. Phys Rev B 71:165431
54. Lamprecht B et al (2001) Surface plasmon propagation in microscale metal stripes. Appl Phys Lett 79:51
55. Weeber JC, Gonzalez MU, Baudrion AL, Dereux A (2005) Surface plasmon routing along right angle bent metal strips. Appl Phys Lett 87:221101

Chapter 13
Lasers and Plasmonics: SPR Measurements of Metal Thin Films, Clusters and Bio-Layers

Saif Ur Rehman, Muhammad Saleem, Rizwan Raza, Ahmad Shuaib, and Zouheir SEKKAT

Abstract An extensive research has been carried out in synthesizing and characterizing single and double layer metal thin films of different thicknesses including ultrathin films. We discuss the experimental data of surface plasmon resonance (SPR) occurring at the interface between air, single and bimetallic thin layers of gold (Au) and silver (Ag) prepared on glass substrates. The bilayer configuration allowed for the measurements of the optical constants of metallic films that are ultra thin; e.g., below 10 nm of thickness since SPR modes on such thin films in a single layer configuration are shallow. In addition, using SPR technique, the optical constants of a natural biological thin layer (*Allium cepa*) is also estimated. Thickness and refractive index of metallic films were determined by matching experimental SPR curves to the theoretical ones. Thickness and roughness of the films were also measured by atomic force microscopy. The results obtained by experimental measurements are in good agreement with AFM analysis. It is estimated that the value of the real part of the dielectric constant of an onion epidermis is existing between the dielectric constants of water and air.

Keywords Surface plasmons • Thin films • Optical constants • Onion epidermis

S.U. Rehman (✉)
Department of Physics, COMSATS Institute of Information Technology, Lahore, Pakistan

Optics & Photonics Center, Rabat Design Center, Moroccan Foundation for Advanced Science, Innovation & Research (MASCIR), Ave, Med Jazouli, Al Irfane, Rabat 10100, Morocco
e-mail: saifuetian@gmail.com

M. Saleem • R. Raza
Department of Physics, COMSATS Institute of Information Technology, Lahore, Pakistan

A. Shuaib
Department of Physics, University of Engineering and Technology, Lahore 54890, Pakistan

Z. SEKKAT
Optics & Photonics Center, Rabat Design Center, Moroccan Foundation for Advanced Science, Innovation & Research (MASCIR), Ave, Med Jazouli, Al Irfane, Rabat 10100, Morocco

Faculty of Sciences, Mohammed V University of Rabat, Morocco

C.D. Geddes (ed.), *Reviews in Plasmonics 2015*,
DOI 10.1007/978-3-319-24606-2_13

13.1 Introduction

Plasmonic excitation and the associated sub-wavelength light-matter interaction has opened new and fascinating avenues for research that originates from the observations and theoretical predictions of several unique properties of surface plasmons (SPs). Plasmons are collective charge density oscillations of the nearly free electron gas in a metal [1], and surface plasmons (SPs) or surface plasmon polaritrons are surface electromagnetic waves that propagate along a metal/dielectric interface associated with high field intensity enhancement allowing for studies of surfaces and interfaces and adjacent regions [2]. Excitation of surface plasmons in Kretschmann configuration [3] is based on propagation of incident p-polarized light under total internal reflection condition through structure comprising high refractive index dielectric media, an electrically conducting metallic layer, and an external medium (gas or liquid) with low refractive index (Fig. 13.1). In a setup, the light passes through a prism reflects off the back side of the sensor chip surface and into the detector. At a given incident angle, which is called resonance angle, surface plasmon has maximal propagation length and therefore maximal absorption at the metal-dielectric interface. This results in a minimization of the intensity of the reflected light. The resonance behavior of SPs is sensitive to changes in the optical properties and thickness of the metal film, as well as of the medium which is adjacent to the metal surface allowing precise measurements of thin film properties [4].

For example, surface plasmon resonance (SPR) sensing has emerged as a powerful optical detection technique for studying label free bio-molecular interaction in real time with in a verity of diverse applications such as life science, electrochemistry, chemical vapor detection, food and environmental safety and beyond [5–7]. In this regards, bilayers of metal films of Silver (Ag) and Gold (Au) are sometimes used for biosensor applications, e.g. few nanometers of Au coating on Ag to protect Silver from oxidation. Compared to Silver, Gold is chemically stable upon exposure to atmospheric condition and it has better chemical affinity with biomolecules, and SPR modes on Silver are sharper and more sensitive to interface changes that those of Au at 632.8 nm excitation wavelength; a wavelength which is widely used in SPR sensing [8]. Furthermore, the study of ultra-thin metal films is interesting for plasmonics research. For example, long range surface plasmons are achieved when the thickness of the metal is very small [9], and plasmon excitation coupling can be tuned by using very thin metal films [10], and plasmonic coupling has been achieved in plasmonic cavities with very thin metal films [11]. It is the aim of this chapter to characterize optically very thin metal films; e.g. few nanometers metal coating. We show that SPR coupling in the bilayer configuration allows for the measurements of the optical constants of metallic films that are ultra-thin; e.g. below 10 nm of thickness. Atomic force microscopy (AFM) is used to measure the thickness and roughness of the metal films.

The use of surface plasmons in biorelated research is of great interest for biorecognition and biosensing. In this section, we use surface plasmons to study the bulk of a biomaterial; e.g. onion epidermis, and show that most of its constituants are water and air. If we look into the onion epidermis cells symmetry, each cell has an average length around 200 micron and it consists of a cell wall, cell

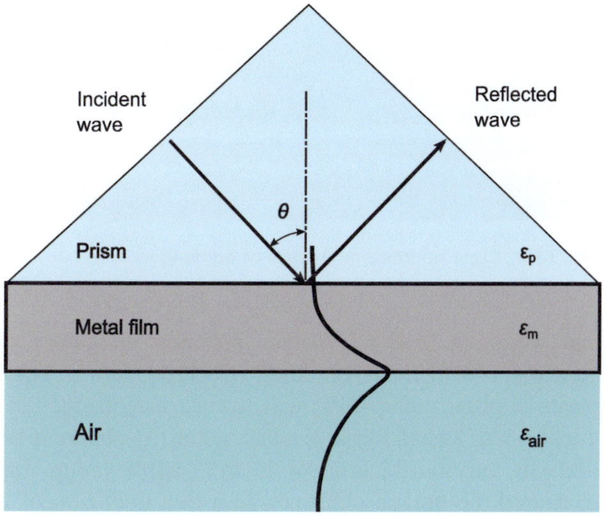

Fig. 13.1 The excitation of surface plasmons in Kretschmann geometry

membrane, cytoplasm, nucleus and a large vacuole. The vacuole is prominent and present at the centre of the cell. It is surrounded by cytoplasm. Cytoplasm comprises cytosol and is basically the substance that fills the cell. It is a jelly-like material that is about 70 % of the cell volume and is composed of water, salts and organic molecule [12]. In this part of work, we are mainly interested in determining the dielectric constant values of fresh, natural thin epidermis of onion by SPR spectroscopy to approach such a biomaterial from an optical point of view.

The study of a bio thin layer (onion epidermis) as a ligand is interesting for plasmonics research. For example, Kumar et al. [13] used inner epidermis of onion bulb scales as a natural support for immobilization of microbial cells for biosensor application. Vanstreels et al. [14] used onion epidermal peels as a model system to study the micromechanical behaviour of plant tissue in relationship to the structural parameters of the cells in the tissue. Groff et al. [15] described an experiment suitable for introductory physics where principles of wave optics are applied to probe the size of onion epidermal cells. The roughness of onion epidermis thin layer is high due to its natural cells structure as revealed in Fig. 13.2. The onion epidermal cell provides a protective layer against unwanted viruses that may harm the sensitive tissues. Because of its simple structure and transparency it is often used to introduce students to plant anatomy [16].

These properties also make it useful for demonstrations of plasmolysis [17]. The clear epidermal cells consist of a single layer and contain chloroplasts, because the onion fruiting body (bulb) is used for storing energy, not for photosynthesis.

We will go on, to discuss our experimental results. Many researchers have measured the optical constants of thin metal films by using the surface plasmon resonance curve fitting method for various thicknesses of metal thin films and have compared with other experimental techniques. In [18], the measurement of optical constants of evaporated gold films of different thicknesses was reported. The results

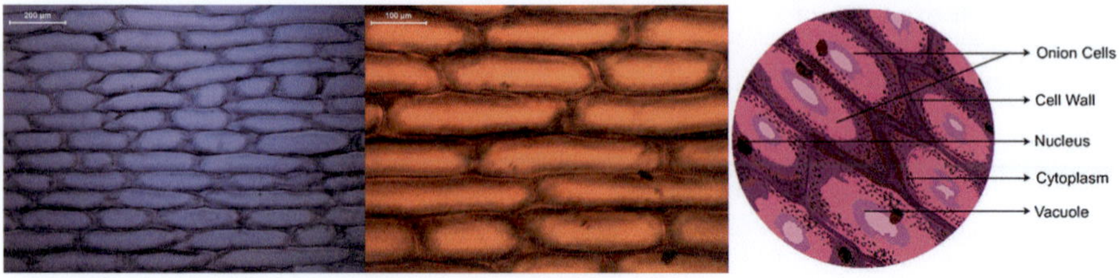

Fig. 13.2 Light microscope images of onion epidermis

that were obtained by surface plasmon resonance measurement were in good agreement with those obtained by ellipsometry. The first application of surface plasmon polaritons (SPPs) was to measure changes in the index of refraction and thickness of metal films. Kretschmann [3] obtained the refraction index of metal films of various thicknesses in the visible range. Innes and Sambles [19] also measured the optical constants of a thin gold film of about 45 nm thickness over the visible region of the spectrum. Using SPs to determine optical constants and/or thickness of metal films is important from both fundamental and industrial point of view. The method is a part of the more general issue of optical characterization of thin films and structures [20].

The thickness and optical properties of the Au layers determine the SP sensor performance. Therefore, it is important to know how accurately these parameters can be deduced from SP measurements and what factors affect the precision of the measurements. M. Yao et al. [8] discussed about the widths of the SPR curves affected by many factors. The dielectric constant and thickness of the metal film have very strong effects on the features of the resonant curve. It is widely recognized that silver (Ag) results in the sharpest resonance curve [21]. The wavelength of the incident light also has a strong effect on the resonant characteristics. A longer wavelength results in a lower resonant angle and a sharper resonant curve. Brink et al. [22] suggested that working in the near-infrared range would result in greater sensitivity. Yan et al. [23] investigated that the real metal films cannot be perfectly flat, and the influence of surface/and interfaces roughness on an SPR curve has been studied extensively. Investigations demonstrated that in the presence of surface roughness, SPs are attenuated by two processes: radiation into the vacuum, or scattering into other surface-polariton states; and the former is responsible for increase in the resonant curve angular width, the latter leads to resonant curve shift. Until now, using SPs to determine optical constants and/or thickness of thin metal films still attracts considerable attention, which is, to a great extent, due to the requirements of developing the SPR technique itself. Thin metal films are an integral part of the SPR-based devices, such as the SPR sensors, the metal-clad waveguide, and the long range SPP waveguide, and the performances of these optical devices are determined dominantly by the optical and geometrical properties of the metal films. Furthermore, parameters of the over layer (or over layers) deposited above the metal films are often required to be determined correctly, and it is evident that the metal films must first be characterized accurately. In fact, the effective optical constants of the metal films also depend on their thickness, which is related to the internal structure, owing to the variation in film density and resistivity [24–26].

In a recent research of our group, we estimated the resolution of sensing in water solution by surface plasmon resonance sensors based on Kretschmann configuration by numerical simulation for different thicknesses of combinations of silver (Ag), copper (Cu), and aluminum (Al) metallic layers (single layers) with a gold coating layer on top of these meals (bimetallic layers) in the ultraviolet, visible, and infrared regions using material dispersion taken from literature. It showed that the gold SPR biosensors have minimum resolution at 650 nm wavelength, Ag biosensors – at 500 nm, and Al biosensors – at 260 nm [27, 28]; and in the present work that describes in this chapter is consisting of two main sections. In Section I, we investigated experimentally the optical properties of ultra-thin layers of Silver and Gold, and those of single layers of Au and Ag films; as well as double layer Ag/Au and Au/Ag films of different thicknesses [29]. AFM measurements were performed for characterizing the thickness and roughness of the films. Using the measured thicknesses we study the refractive indices of the films as well as parameters of SPR curves and field enhancement of the films by numerical simulation. In Section II, we explored experimentally e.g. we compared the optical properties of bare thin films of gold (Au) and silver (Ag), and those of bio-thin layer (onion epidermis) along with Au and Ag films using surface plasmon resonance spectroscopy [30]. We show that SPR coupling in the bilayer configuration allows for the measurements of the optical constants of metallic films as well as bio thin layers (onion epidermis) when it is stacked on Au and Ag thin films. Again AFM is being used to characterizing the thickness and roughness of bare Au and Ag thin films. Using the measured thicknesses we study the dielectric constants of biological/organic thin layer by numerical simulation. The present study is focused on the modeling of ideal thin films using rigorous electromagnetic theory of wave propagation through layered media [28].

13.2 Experimental Details

For the convenience of experiment, the three-layer & four-layer Kretschmann configuration consisting of a prism-gold/silver film-air is considered. A home-constructed apparatus of SPR, based on Kretschmann geometry, was built for this study. In Section I, single layer combination (30, 45, 50 and 70 nm) and bilayer (30/30, 45/5 and 45/10 nm) thin film assemblies of (Au and Ag) were deposited on glass substrates cleaned by consecutive soaking in acetone and isopropanol for 20 min and rinsed with distilled water. In Section II, metal thin films of gold (Au) and silver (Ag) of 45 nm thickness were prepared on glass substrates and fresh, inner natural layer of onion epidermis is used as a bio layer. Generally, it would be better to deposit the metal layers directly onto the prism surface in order to achieve the lowest insertion loss and highest sensitivity. However, from the practical application point of view, it is highly desirable to implement a separate test sample into the system and keep the prism intact. Since the glass prism is a key component of the test system, the quality and mounting condition of the prism are very sensitive to the test results. Separate test samples with metal layer deposited onto a glass substrate are easy to prepare and easy to use. Glass substrate samples are cheap and disposable. It would be better to use glass substrates with the same refractive index as the prism and keep the sample substrate in close contact with the prism. To minimize the mismatch effect of the air gap between the glass substrate and the prism, matching oil with the same refractive index as the glass substrate and the prism is used to mount the test samples into the test system. The samples were evaporated by using NEXTDEP resistive thermal evaporation system (ANGSTROM ENGG.). Evaporation is one of method to deposition thin films. This method have purpose to get thin film which have good quality. In this experiment evaporation done at 1.5×10^{-6} torr vacuum pressure. The deposition rate was 0.5 Å/s at room temperature and the thickness of prepared films was measured to be between 30 and 70 nm. Later on, thickness of these thin films was also measured by Atomic Force Microscope AFM (Bruker).

The experimental arrangement is shown in Figs. 13.3 and 13.4 respectively. The setup consists of 632.8 nm wavelength, 15 mW He-Ne laser (Melles Griot) used as a light source. The power of the incident light was reduced to appropriate level for detection by a neutral density filter ND. By passing the beam splitter BS the reflected part of the laser beam is measured by photodetector PD1 to be used as reference intensity. The p-polarized light by passing the transmitted laser beam through a polarizer propagates through a right angle BK7 prism mounted on a goniometer allowing the prism to be rotated with 0.01° precision. The substrate with metal layer was put into optical contact with the prism through index matching oil. The reflection surface of the prism is placed on the rotation axis of the goniometer. The reflected beam is then focused by a lens on the photodiode PD2

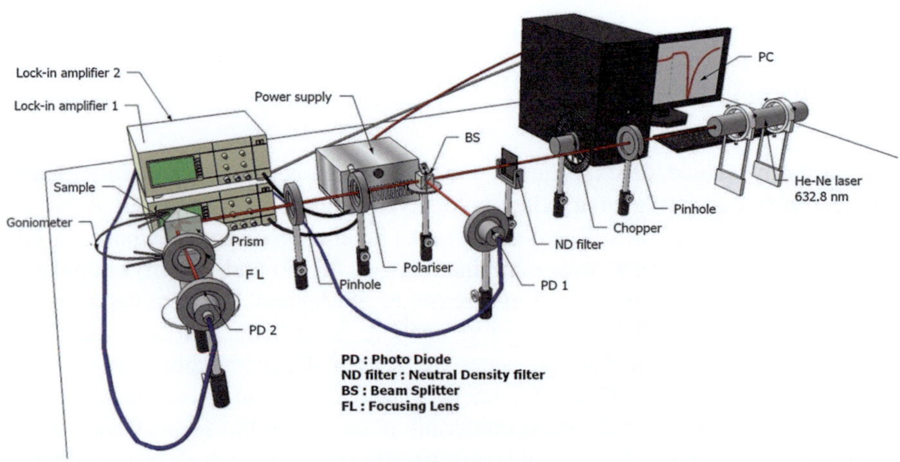

Fig. 13.3 3D schematic plan of the constructed setup

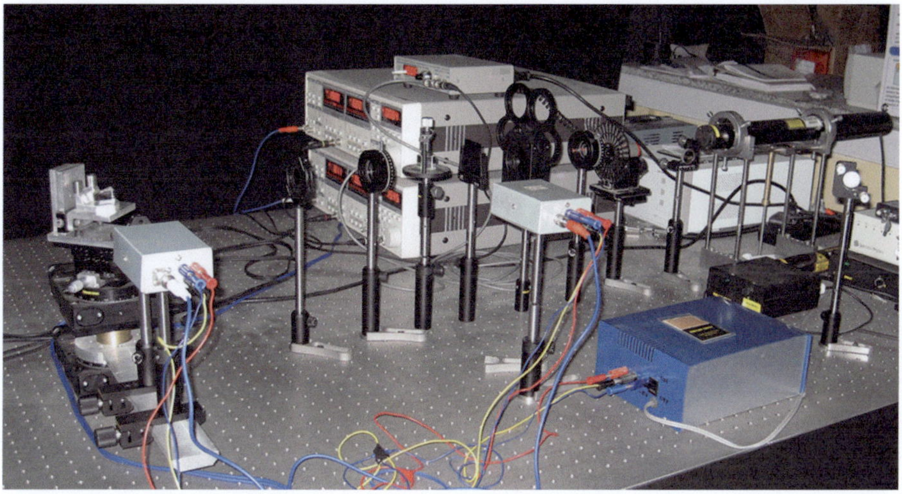

Fig. 13.4 A snapshot of constructed experimental setup

mounted on second goniometer. Then, the intensity of the light interacted with the sample and measured by PD2 is normalized by the intensity measured by PD1.

Winspall software package was also used for simulation of surface plasmon resonance phenomenon based on the Fresnel equations [31]. The thickness and dielectric constant values can be matched according to the fitting of calculated and experimental SPR curves [32]. The surfaces of the films are assumed to be planar in the software package, so the roughness of the samples is ignored. In software, the parameters of all material of multilayer system are entering into the layer system table.

13.2.1 Thin Films Deposition Technique

Thin films prepared in this research work serves as an introduction to **Thermal evaporation deposition –** In which the material to be deposited is heated to a high vapor pressure by electrical resistive heating. NEXTDEP 400 resistive thermal evaporator (ANGSTROM ENGG.) has used for preparing single and double layer metal thin films. Resistive evaporation is a commonly used vacuum deposition process in which electrical energy is used to heat a filament which in turn heats a deposition material to the point of evaporation. The process can be performed at very high levels of vacuum allowing for a long mean free path and therefore fewer tendencies to introduce film impurities. High deposition rates can be achieved and lower energy particles can reduce substrate damage.

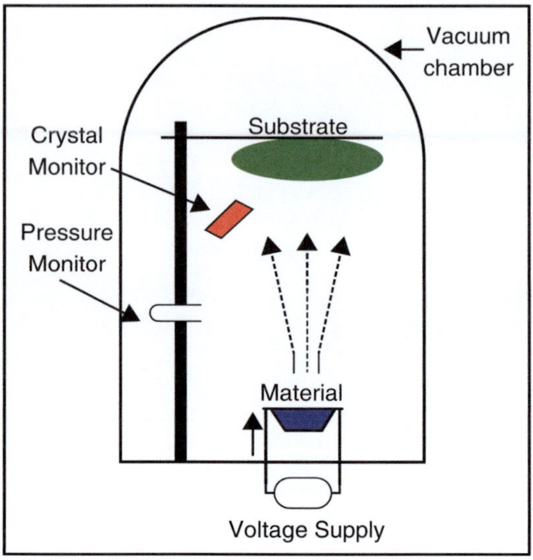

Fig. 13.5 A schematic of thermal evaporation

13.2.1.1 Introduction to Physical Vapour Deposition (PVD)

Thermal evaporation is one of the most widely used techniques of physical vapor deposition (PVD). It is a type of thin film deposition, which is a vacuum process wherein coatings of pure materials are applied over the surface of many different objects. The deposited coatings or films usually have a thickness in the range of angstroms to microns and are composed of a single material or layers of multiple materials. The coating materials can be either molecules, including nitrides and oxides, or pure atomic elements such as both metals and non-metals. The object to be coated is called the substrate, which can be many different things, including semiconductor wafers, optical components, and solar cells.

13.2.1.2 Thermal Evaporation

Thermal evaporation is a process wherein a solid material is heated inside a high vacuum chamber to a temperature which generates some vapor pressure. Inside the vacuum, even a very low vapor pressure is adequate to create a vapor cloud within the chamber. This evaporated material now consists of a vapor stream, which passes through the chamber, and strikes and sticks onto the substrate as a film or coating as shown in Fig. 13.5.

Since, in a majority of cases, the material becomes liquid by heating it to its melting temperature, it is normally placed in the bottom of the chamber, often in some form of upright crucible. The vapor then rises above from this bottom source and reaches the substrates that are held inverted in suitable fixtures at the top of the chamber, with surfaces to be coated facing down toward the rising vapor to acquire their coating.

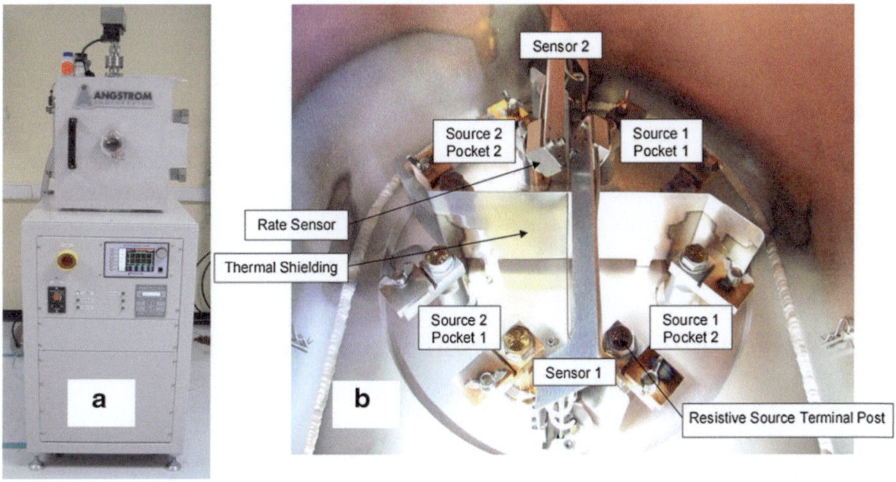

Fig. 13.6 (**a**) NEXTDEP 400 vacuum evaporator (**b**) Interior outlook of evaporator chamber

Various measures may have to be taken in order to ensure film adhesion and control different film properties as desired. The evaporation system design allows process engineers to adjust a number of parameters to obtain desired results for variables such as grain structure, uniformity, thickness, stress, adhesion strength, optical or electrical properties, and much more. Angstrom Engineering has developed thin film deposition systems based on this technique (Fig. 13.6) which can deposit a wide range of materials including metals, organic, and inorganic polymers. The process can be controlled using quartz rate sensor, temperature, or optical monitoring systems to ensure consistent high-quality results.

NEXTDEP 400 resistive thermal evaporator can be operated at high level (1.5×10^{-6} torr) vacuum pressure for achieving high quality films. This vacuum system has four thermal resistive evaporation sources and one substrate shutter with which deposition is achieved the interior outlook of evaporator chamber is revealed in Fig. 13.6b. To access the evaporation sources, substrate holder, or deposition rate sensors, the chamber must be open.

13.2.2 *Thin Films Preparation*

Regarding thin films preparation, there are two main sections to describe here in this chapter.

Section I In Section I the following combinations of single and double layer films are prepared on glass substrate.

Single Layers. Ag: 30 nm, 45 nm, 50 nm and 70 nm
Au: 30 nm, 45 nm, 50 nm and 70 nm

Double Layers. Ag + Au: 30 + 30 nm, 45 + 5 nm and 45 + 10 nm

Au + Ag: 30 + 30 nm, 45 + 5 nm and 45 + 10 nm

Figure 13.7a, b, and c illustrates different thickness combinations of metal thin films prepared on glass substrate.

Section II In this section, single layer films of bare Au and Ag of 45 nm are prepared on glass substrate by thermal evaporation. Later on we peeled off the fresh onion inner thin epidermis and stacked it on metal thin films as shown in Fig. 13.8.

13.2.3 Thin Films Characterization (Morphology)

13.2.3.1 Atomic Force Microscopy (AFM)

Introduction

Atomic force microscopy (AFM) is a technique to obtain images and other information from a wide variety of samples, at extremely high (nanometer) resolution. AFM works by scanning a very sharp (end radius ca. 10 nm) probe along the sample surface, carefully maintaining the force between the probe and surface at a set, low level. Usually, the probe is formed by silicon or silicon nitride cantilever with a sharp integrated tip and the vertical bending (deflection) of the cantilever due to forces acting on the tip is detected by a laser focused on the back of the cantilever. The laser is reflected by the cantilever onto a distant photo detector. The movement of the laser spot on the photo detector gives a greatly exaggerated measurement of the movement of the probe. This set-up is known as an optical lever. The probe is moved over the sample by a scanner, typically a piezoelectric element, which can make extremely precise movements. The combination of the sharp tip, the very sensitive optical lever, and the highly precise movements by the scanner, combined with the careful control of probe-sample forces allow the extremely high resolution of AFM.

Principle of Operation

The basic principle of AFM is that a probe is maintained in close contact with the sample surface by a feedback mechanism as it scans over the surface, and the movement of the probe to stay at the same probe-sample distance is taken to be the sample topography. A variety of probes have been used but the most commonly used are micro-fabricated silicon (Si) or silicon nitride (Si_3N_4) cantilevers with integrated tips. The tips can be very sharp.

Typically, probe radius varies from 5 to 20 nm. The bending of the cantilever normal to the sample surface is usually monitored by an optical lever, although

Fig. 13.7 (**a**) Ag single layer thin films of different thicknesses (**b**) Au single layer thin films of different thicknesses (**c**) Ag/Au and Au/Ag double layer thin films of different thicknesses

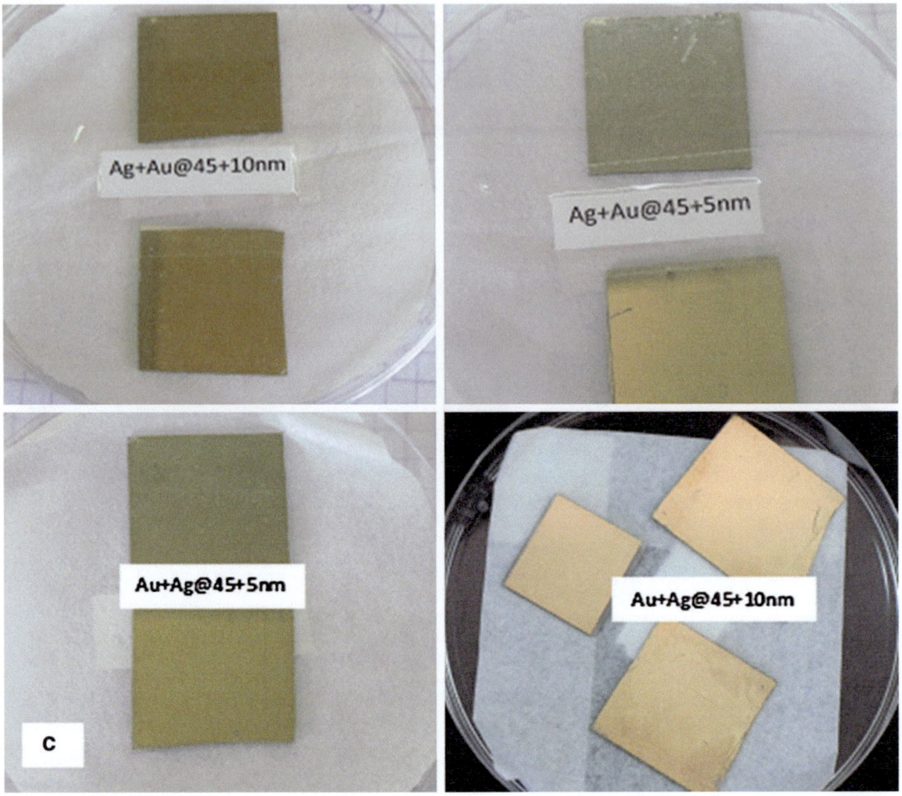

Fig. 13.7 (continued)

other methods have been investigated. This system magnifies the normal bending of the cantilever greatly, and is sensitive to Angstrom-level movements. There are a variety of modes of scanning but in the simplest mode, the probes gently touch the sample as it moves over the surface. The movement of the probe over the surface is controlled by a scanner. This is normally made from a piezoelectric material, which can move the probe very precisely in the x, y, and z axes, although other types of actuator are also used. The signal from the photo detector passes through a feedback circuit, and into the z-movement part of the scanner, in order to maintain the probe-sample distance at a set value (Fig. 13.9). Because the cantilever acts as a spring, this fixed cantilever deflection means a fixed probe-sample force is maintained. The amount by which the scanner has to move in the z-axis to maintain the cantilever deflection is taken to be equivalent to the sample topography.

A great advantage of AFM compared to for example TEM or SEM, is that it is simple to operate in almost any environment, such as aqueous solutions, but also other solvents, in air, vacuum, or other gases. Depending on the details of the experiment, the resolution can be very high. Z (height) resolution is extremely high and can be sub-Angstrom, whereas lateral resolution could be of the order of 1 nm.

Fig. 13.8 (**a**) Au and Ag single layer thin film attached with bio-thin layer (onion epidermis) (**b**) Ag single layer thin film attached with bio-thin layer (onion epidermis)

To obtain such high resolution work, the state of the sample, the levels of acoustic and electronic noise and experimental conditions must all be highly optimized.

In a present research, thickness of thin films has measured by Atomic Force Microscope AFM. Dimension Icon AFM system (Bruker) delivers uncompromised performance, robustness, and flexibility to perform nearly every measurement at scales previously obtained by extensively customized systems. Utilizing an open-access platform, large- or multiple-sample holders, and numerous ease-of-use features. With more features than any other AFM, the new NanoScope-based control system sets the standard for power and flexibility. At the same time, the software's simplified work flow and ergonomic system design make the Icon ideal for multi-user environments. The Dimension Icon AFM system and overall instrument view are shown below in Fig. 13.10.

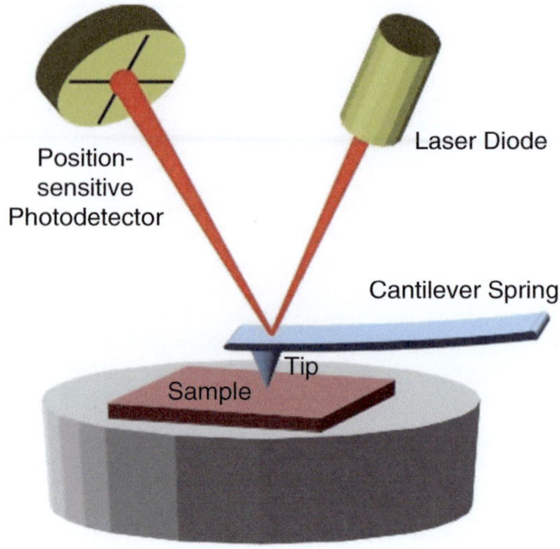

Fig. 13.9 Block diagram of atomic force microscope using beam deflection detection

13.3 Results and Discussion (Section I)

13.3.1 Single Layer Films

We deposited three samples of Au and Ag thin films with thicknesses ranging from 30 to 60 nm on glass substrates. We performed SPR experiments to find out the mode position and shape versus the thickness. Experimental data was obtained in case of Au and Ag thin films. Computer simulation has been extensively used in the bimetallic layer SPR and comparison with monolayer SPR sensors. Data fitting was used throughout the simulation. It is a software package for simulation of surface plasmon resonance performance based on the Fresnel equations [33]. The thickness, refractive index can be entered, the dielectric constant is calculated according to the Fresnel's equation and the SPR curve is plotted [32]. However, the assumption that the surface of the SPR sensor chip is planar results in that the roughness of every layer on the surface of the SPR sensor chip is ignored in the software package. Therefore, in the simulation of the bimetallic layer system, all the layers on the surfaces of the SPR sensor chips are assumed to be planar.

Simulation was carried out by assuming that the total thickness of metal layers on the surface of sample are between 30 and 70 nm in case of single and double layer. The measured reflectivity versus incident angle for the Au and Ag samples and corresponding fitted SPR curves are shown in Fig. 13.11.

The thicknesses and surface roughness of the films were also measured by atomic force microscope (AFM), Dimension Icon AFM (Bruker) in tapping mode for confirming the validity of the SPR technique. Figure 13.12 shows the AFM image of Ag(2) single layer sample.

The thicknesses obtained by the two approaches and corresponding estimated refractive indices of single metal layers, the roughness of the samples are

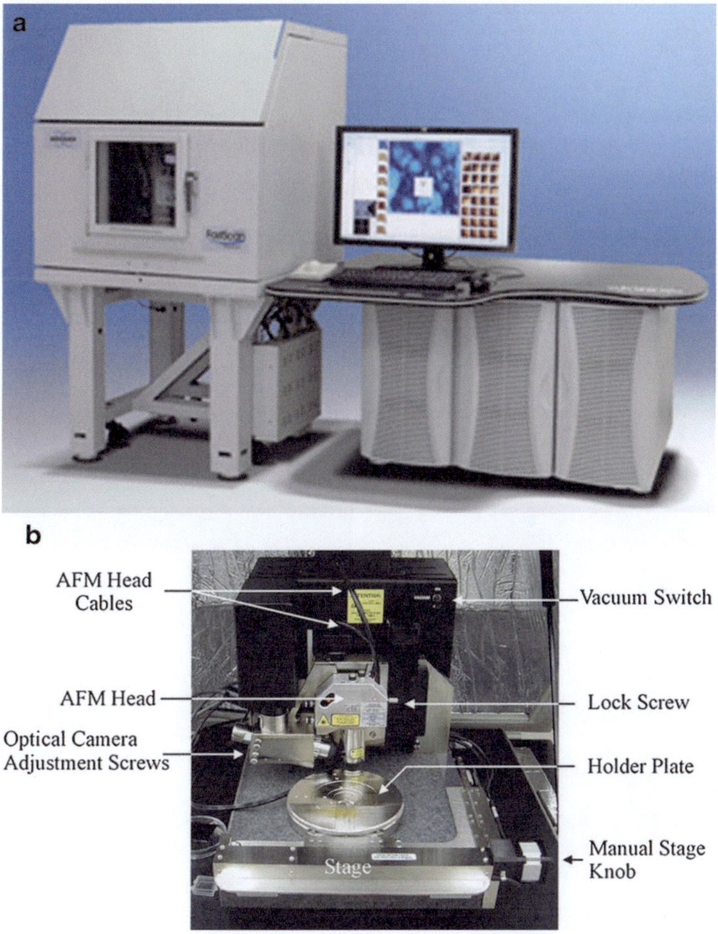

Fig. 13.10 (**a**) Dimension icon AFM system (**b**) overall instrument view

summarized in Table 13.1. The thickness difference between two methods is less than 0.8 nm for Au and less than 6.5 nm for Ag samples, and the roughness of Ag single layer is higher compared to Au samples; a feature which could be caused by the chemical instability and oxidization of silver thin films.

The refractive index values for Au(3) film of 0.198 – i3.182 and for Ag(1) film of 0.119 – i4.010 are in agreement with those given in the literature [34] at $\lambda = 632.8$ nm wavelength.

As we mentioned above, during data fitting, the roughness of layers of the SPR structure is not considered in the simulations. From Table 13.1, it can be seen that roughness almost does not depend on thickness of the single layer films. The real part of refractive index decreases in both gold and silver single layer cases with increasing thickness.

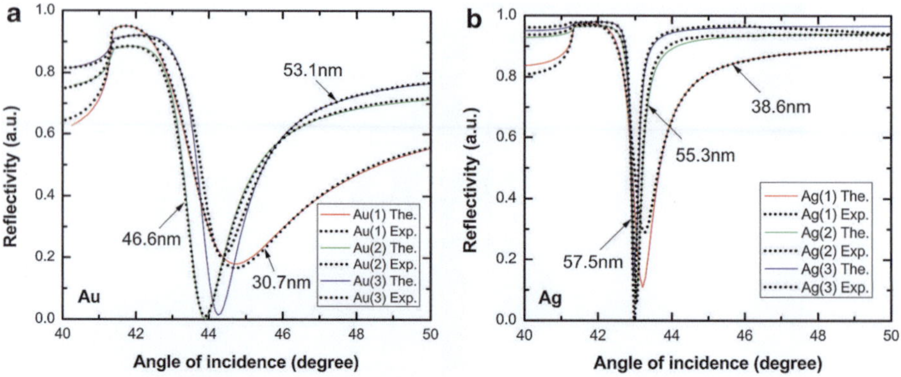

Fig. 13.11 Measured SPR data (*dots*) and their fitted curves (*solid line*) for the Au(a) and Ag (b) samples

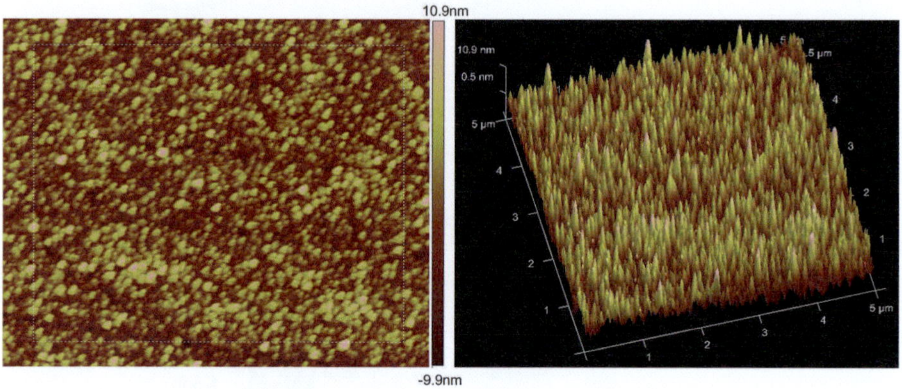

Fig. 13.12 The AFM micrograph of Ag(2) sample

We examined the following parameters of SPR curve for each sample presented in Table 13.2, resonance angle, full width at half maximum (FWHM), height of curve and α parameter which equals to the relation of the height to the FWHM of the curve; and calculated values of field enhancement based on the estimated refractive indices. It can be seen that Au(2) with 46.7 nm thickness and Ag (3) with 57.5 nm thickness samples indicate deeper SP dip, narrowest FWHM values, and maximal values of α as compared to other samples. Moreover, the field enhancement of these samples is highest.

Ag SPR curves are narrower compared to Au ones that is indicating in the Table 13.2. Au samples have longer SP wavelengths compared to Ag samples.

13.3.2 Double Layer Films

Silver and gold perform differently in plasmon resonance when they are used as coating materials on glass slides. Silver produces a narrower SPR curve compared

Table 13.1 The characterization of Au and Ag single layer thin films

Sample	AFM thickness, nm	Real part (n)	Imaginary part (k)	SPR thickness, nm	AFM roughness, nm
Au(1)	30.76	0.2492	3.3041	30.00	1.7–2.2
Au(2)	46.69	0.2151	3.4352	46.57	1.6–2.2
Au(3)	53.11	0.1985	3.1821	53.00	1.8–2.9
Ag(1)	38.60	0.1194	4.0097	39.00	2.2–2.8
Ag(2)	55.30	0.1007	4.0360	57.55	2.3–2.9
Ag(3)	57.50	0.0606	4.0809	64.00	2.6–3.3

Table 13.2 Parameters of SPR of Au and Ag single layer structures

Sample	Resonance angle, degree	SP wavelength, nm	FWHM, degree	Height of curve	α, $degree^{-1}$	Field enhancement
Au(1)	44.75	446	2.70	0.39	0.14	13.5
Au(2)	43.88	439	1.42	0.70	0.49	21.5
Au(3)	44.37	443	1.61	0.57	0.35	19.6
Ag(1)	43.26	434	0.90	0.61	0.68	46.0
Ag(2)	43.05	432	0.28	0.88	3.14	63.8
Ag(3)	43.00	432	0.20	0.93	4.65	114.9

to gold, as we have discussed in 'Single layer films' section. But it is chemically unstable and tends to be oxidized. Gold is chemically stable. One way to take advantages of the both metals is to manage deposition with a silver layer at the bottom and a thin gold layer on the top to reduce oxidation. We use this approach to characterize ultrathin films of Au and Ag by depositing of 5, 10 and 30 nm thick layers of each metal over second metal.

We deposited three sets of Au/Ag and three sets of Ag/Au (double layer) thin films of different thicknesses onto glass substrates. In case of double layer samples, the thicknesses estimated by SPR technique were again matching with the thicknesses measured by AFM. Table 13.3 shows the thickness of each metal estimated by SPR fitting and it is in good agreement with the combined double layer structure deliberated by AFM. It means in case of double layer samples, we can characterize metal clusters or ultra small thickness (less than 10 nm) by SPR technique.

In case of double layer thin films, the real part of refractive indices in both cases of Au/Ag and of Ag/Au having thicknesses of 30/30, 45/5 and 45/10 nm show unstable behaviour. This instability might be caused by surface roughness of sample and oxidation of silver films. Moreover, in case of double layer thin films, by increasing gold thickness more than 10 nm as a top layer it might broad the SPR response curve (see Fig. 13.13). It can be clearly seen from Fig. 13.13 that the reflectivity curves are different for the two configurations; e.g. Au/Ag and Ag/Au; a feature which puts in evidence the evanescent character of the SPR mode. Even

Table 13.3 The characterization of Au/Ag and Ag/Au double layer thin films

Sample	AFM thickness, nm	Real part (n)	Imaginary part (k)	SPR thickness, nm	AFM roughness, nm
Au/Ag (1)	61.64	0.1423 0.4739	3.9746 3.2723	31.00 30.00	0.8–1.1
Au/Ag (2)	56.00	0.1900 0.2317	3.6643 3.0340	46.00 09.00	1.6–2.1
Au/Ag (3)	59.55	0.1996 0.1976	3.5131 2.5566	46.00 13.00	2.6–3.3
Ag/Au (1)	62.15	0.1560 0.2611	4.3338 3.3940	31.00 31.00	1.1–1.5
Ag/Au (2)	51.40	0.1314 0.4412	4.0770 3.2769	46.00 05.00	1.7–2.2
Ag/Au (3)	55.97	0.1359 0.3479	4.2483 3.2371	47.00 10.00	1.6–2.0

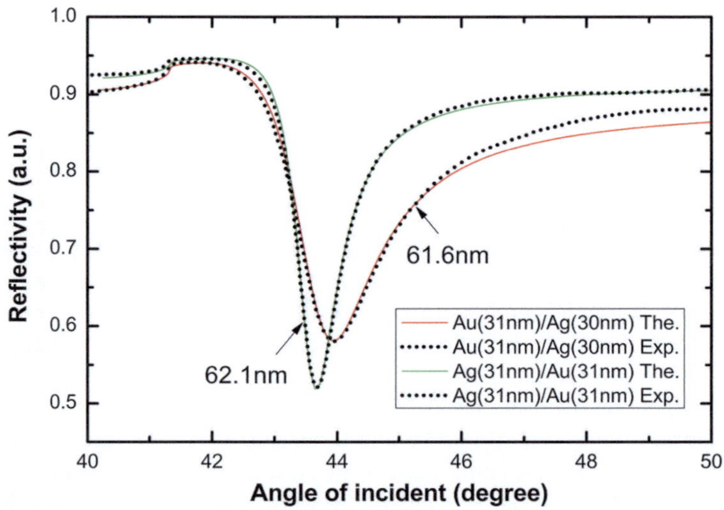

Fig. 13.13 Measured SPR data (*dots*) and corresponding fitted curves (*solid line*) of Au/Ag (31/30 nm) and Ag/Au (31/31 nm) double layer samples

though the total thickness and individual thicknesses of the metal layers are the same, the wave feels more the metal that is in direct contact with the glass. The measured reflectivity versus incident angle for Au/Ag(1) and Ag/Au(1) double layer samples and corresponding fitted SPR curves are shown in Fig. 13.13.

The parameters of SPR curve of double layer films are presented in Table 13.4. Au/Ag(2) with 56.0 nm overall thickness and Ag/Au(2) with 51.4 nm overall thickness samples indicate deeper SP dip, narrowest FWHM values, and maximal

Table 13.4 Parameters of SPR by Au/Ag and Ag/Au double layer structures

Sample	Resonance angle, degree	SP wavelength, nm	FWHM, degree	Height of curve	α, $degree^{-1}$	Field enhancement
Au/Ag(1)	43.95	439	1.60	0.30	0.19	4.5
Au/Ag(2)	43.97	439	1.07	0.60	0.56	19.5
Au/Ag(3)	44.68	445	1.27	0.63	0.50	15.6
Ag/Au(1)	43.68	437	0.81	0.38	0.47	9.4
Ag/Au(2)	43.26	434	0.66	0.73	1.11	22.7
Ag/Au(3)	43.38	435	0.69	0.53	0.77	14.0

values of α as compared to other samples. The field enhancement of these samples is highest.

It is shown in Table 13.3 that the refractive index values of silver are in agreement with those given in the literature [34] at $\lambda = 632.8$ nm wavelength when it is behaving as a bottom/base layer and gold is on top. This is confirming that gold layer on top of silver may protect silver layer to be oxidised.

SPR structures coated with silver produce narrower and sharper SPR response curves, due to the real part of refractive index which is slightly higher than that of gold. It can be seen in Table 13.4 that SPR structure with 45 nm thick silver and 5 and 10 nm thick gold layers on top produce narrower SPR response curves than 45 nm single gold layer. On contrary, we have also examined experimentally 45 nm thick gold layer coated by 5 and 10 nm silver layers. Due to very thin silver layer on top it produces narrow SPR curves compared to single gold layer. It is noted that Au/Ag configurations have longer SP wavelengths compared to Ag/Au configurations due to influence of metal under layer to SP propagation.

The efficiency of detection of the presence of small amount of the absorbing dielectric, such as fluorophores or fluorophore-labeled molecules, by employing the surface-plasmon field-enhanced fluorescence is proportional to the intensity of generated SP field [35]. The ratio of the field intensity I at the analyte-metal interface to the incident wave intensity I_0 is referred to as field enhancement. As can be seen from Table 13.2, the maximum enhancement of Au-based structure is about 22× for 47 nm thickness; about 115× for Ag-based structure of 57.5 nm thickness (see Table 13.4). The maximum field enhancement of bimetallic structures does not exceed 23× due to higher overall thickness (see Table 13.4).

13.4 Summary (Section I)

We have measured the thickness and optical constants of metal thin films and ultra-thin metal clusters (sub 10 nm layers) by SPR spectroscopy. We have also measured the variation of the refractive indices of gold and silver films of single and double-layer structures for a set of film thicknesses ranging from 30 to 60 nm by using the SPR measurements at 632.8 nm wavelength. The thicknesses of the metal films estimated by SPR technique are in good agreement with those measured by AFM. Thus, we showed that the SPR technique is confirmed to be a simple method for the measurement of the optical constants at ultra-thin film thicknesses (5 nm in case of bimetallic layers). The narrowest resonant curves with deeper surface plasmon dip are observed for both gold and silver 45 nm thin films. In case of bimetallic films, silver/gold layers produced narrower SPR response curves compared to single gold layer as well as 5 and 10 nm gold as a top layer, protected silver layer to be oxidized.

13.5 Results and Discussion (Section II)

We prepared single layer thin films of bare Au and Ag of 45 nm thickness on glass substrates by thermal evaporation. Later on we peeled off the fresh onion inner thin epidermis and stacked it on metal thin films as shown in Fig. 13.14. SPR experiments were performed using single layer (metal) and double layer (metal/onion epidermis) assemblies and Winspall curve fitting methodology was carried out to determine the thickness and optical constants of bare Au and Ag thin films and dielectric constant values of onion epidermis. We assume that the total thickness of all single and double layer thin films on the surface of prism is between 45 and 70 nm.

The measured reflectivity versus incident angle for single and double layer samples and corresponding fitted SPR curves at 632.8 nm wavelength are illustrated in Fig. 13.15a and b. Resonance angles and shift of single layer (bare Au and Ag), and of double layer (Au/onion epidermis and Ag/onion epidermis) thin films were studied from SPR curves. Resonance angles of bare Au 45 nm and Au/onion epidermis thin layers were found to be 43.90° and 44.18° respectively. So, the shift in resonance angle was deliberated as 0.28°. On the other hand, resonance angles of bare Ag 45 nm and Ag/onion epidermis thin layers were found 43.62° and 44.38° so the shift in resonance angle was observed 0.38°. The thickness and dielectric constant values of single layer (bare Au and Ag), and double layer (Au/onion epidermis and Ag/onion epidermis) thin films obtained by SPR method and atomic force microscopy (AFM) are set out in Table 13.5.

The thicknesses and surface roughness of the films were also measured by atomic force microscope (AFM), Dimension Icon AFM (Bruker), in tapping mode for confirming the validity of the SPR technique. Figure 13.16 spectacles the AFM image of bare Ag 45 nm sample. The surface roughness of onion epidermis is high due to natural asymmetrical structure of the epidermis cells. That is already observed in Fig. 13.2, the average diameter of one onion cell is around 200–250 micron and it varies from cell to cell. The surface of epidermis

Fig. 13.14 Ag and Au films of 45 nm with fresh onion epidermis layer

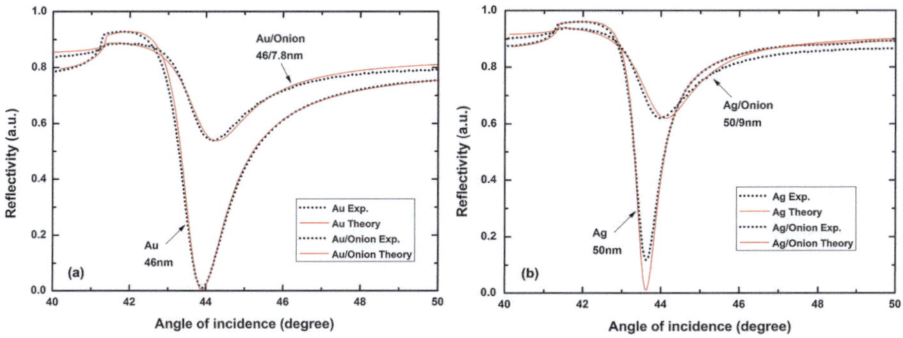

Fig. 13.15 (**a**) Measured SPR data (*dots*) and their fitted curves (*solid line*) for bare Au and Au/onion epidermis (**b**) bare Ag and Ag/onion epidermis samples excited at 632.8 nm wavelength

Table 13.5 The characterization of single and double layer thin film assemblies

Sample, (thin films)	AFM thickness, nm	Real(ε)	Imag.(ε)	SPR thickness, nm	AFM roughness, nm
Au	46	−11.80	1.567	45.0	1.5–2.1
Ag	50	−12.80	0.835	50.7	2.3–2.8
Au/Onion epidermis	46	−11.89 1.2	1.59 0.515	62.2 7.8	1.5–2.1
Ag/Onion epidermis	50	−12.80 1.2	0.83 0.60	63 9	2.3–2.8

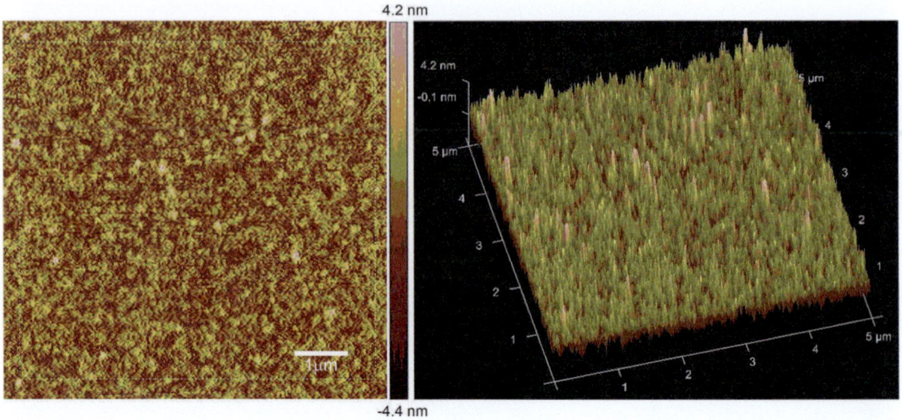

Fig. 13.16 The AFM micrograph of Ag thin layer sample

layer was not smooth that's why AFM could not be helpful to measure the layer thickness.

The thickness difference found between two methods is 1.6 nm in case of Au single layer sample and 0.1 nm in case of Ag single layer sample. The roughness of the Ag thin film is higher compared to the Au thin film. It might be originated by the chemical instability and oxidization of silver film. The dielectric constant values for bare Au and Ag thin films are in agreement with those given in the literature [36]. SPR thickness of onion epidermis loaded on Au film comprises 7.8 nm while SPR thickness on Ag film is found to be 9 nm. The dielectric constant values of onion epidermis have also been determined. The real part of dielectric constant, *Real(ε)* of onion epidermis is estimated to be 1.2 and imaginary part of dielectric constant, *Imag.(ε)* of onion epidermis is 0.515 when it is stacked onto Au thin film surface. While the *Real(ε)* of onion epidermis is found to be 1.2 and *Imag.(ε)* of onion epidermis is 0.600 when it is loaded onto Ag thin film surface.

To describe the general behavior of onion epidermis, water and air in terms of optical constants, we assume that these components (water and air) are individually stacked onto the surface of bare Au and Ag 45 nm thin films. The standard dielectric constant values of water and air have been taken from [37]. Since, the values of real part of the dielectric constants, *Real(ε)* of water is 1.744, air is 1.000 and of onion epidermis is found to be 1.20. In the light of above discussion, optical constant value of onion epidermis is laying between water and air which is confirmed by [12]. On the other hand, the comparison of the imaginary part of the dielectric constants, *Imag.(ε)* of water is 0.00 and of air is also 0.00 while *Imag.(ε)* of onion epidermis has some values that are estimated to be 0.515 and 0.600 respectively as shown above in Table 13.5. It might be possible due to the surface roughness of natural cells symmetry of onion epidermis. As the surface of onion epidermis layer is not smooth light scattering takes place and we observed shallow dips and non zero imaginary part of dielectric constants.

13.6 Summary (Section II)

Surface plasmon resonance technique is found to be simple and excellent to investigate a natural transparent, thin layer of a biological material (Allium Cepa) the inner epidermis of onion bulb. The dielectric constant values of this thin bio-film e.g., (onion epidermis) were measured. We found that the values of the real part of this dielectric constant is between water and air, a feature which suggests that thin onion epidermis is mostly composed of water and air.

Acknowledgements I acknowledge Prof. Shinji Hayashi for his fruitful guidelines and scientific input to improve the quality of the manuscript. I also acknowledge Dr. Dmitry Nesterenko and Dr. Anouar Rahmouni for their help during the experimental design and data analysis.

References

1. Goyer C, Labbe P (2011) Analytical tools for protein-carbohydrate interaction studies. In: Spinelli N, Renaudet O (eds) Synthesis and biological applications of glycoconjugates. Bentham Science University of Grenoble, France, pp 255–266
2. Kooyman RPH (2008) Physics of surface plasmon resonance. In: Tudos AJ, Schasfoort RBM (eds) Handbook of surface plasmon resonance. Royal Society of Chemistry, Cambridge, pp 15–34
3. Kretschmann E (1971) Determination of optical constants of metals through the stimulation of surface plasma oscillations. Z Phys 241:313–324
4. Raether H (1988) Surface plasmons on smooth and rough surfaces and on gratings. Springer, Berlin
5. Ou S, Kwok K-C (2004) Ferulic acid: pharmaceutical functions, preparation and application in foods. J Sci Food Agric 84:1261–1269
6. Zhai PM, Guo J, Xiang J, Zhou FM (2007) Electrochemical surface plasmon resonance spectroscopy at bilayered silver/gold films. J Phys Chem C 111:981–986
7. Shan XN, Foley KJ, Zhang PM (2010) Measuring surface charge density and particle height using surface plasmon resonance technique. J Anal Chem 82:234–240
8. Yao M, Tan OK, Tjin SC, Wolfe JC (2008) Effects of intermediate dielectric films on multilayer surface plasmon resonance behavior. Acta Biomater 4:2016–2027
9. Sarid D (1981) Long-range surface plasma waves on very thin metal films. Phys Rev Lett 47:1927
10. Balci S, Kocabas C, Ates S, Karademir E, Salihoglu O, Aydinli A (2012) Tuning surface plasmon-exciton coupling via thickness dependent plasmon damping. Phys Rev B 86:235402
11. Balci S, Kocabas C, Aydinli A (2011) Critical coupling in plasmonic resonator arrays. Opt Lett 36:2770–2772
12. Luby-Phelps K (2000) Cytoarchitecture and physical properties of cytoplasm: volume, viscosity, diffusion intracellular surface area. Int Rev Cytol 192:10–11
13. Kumar J, D'Souza SF (2011) Immobilization of microbial cells on inner epidermis of onion bulb scale for biosensor application. Biosens Bioelectron 26:4399–4404
14. Vanstreels E, Alamara MC, Verlindena BE (2005) Micromechanical behaviour of onion epidermal tissue. Postharvest Biol Technol 37:163–173
15. Groff JR (2012) Estimating the size of onion epidermal cells from diffraction patterns. Phys Teach 50:420

16. Plant Anatomy at the University of Hamburg (2014) Biologie.uni-hamburg.de. http://en.wikipedia.org/wiki/Onion_epidermal_cell. Accessed 17 June 2014
17. Smith M (2014) Plasmolysis. Microscopy-uk.org.uk. Accessed 10 July 2014
18. Lee W-J, Kim J-E, Park HY, Park S, Kim M-s, Kim JT, Ju JJ (2008) Optical constants of evaporated gold films measured by surface plasmon resonance at telecommunication wavelengths. J Appl Phys 103:0737131–0737135
19. Innes RA, Sambles JR (1987) Optical characterization of gold using surface plasmon-polaritons. J Phys F 17:277
20. Heavens OS (1960) Optical properties of thin films. Rep Prog Phys 23:1–65
21. Gent J-v, Lambeck PV, Kreuwel HJM, Gerritsma GJ, Ernst JR et al (1990) Optimization of a chemo-optical surface plasmon resonance based sensor. Appl Opt 29:2843–2849
22. Brink G, Sigl H, Sackmann E (1995) Near-infrared surface plasmon resonance in silicon-based sensor: new opportunities in sensitive detection of biomolecules from aqueous solutions by applying microstep for discriminating specific and non-specific binding. Sens Actuators B Chem 25:756–761
23. Yan H, Hong-An Y, Song-Quan L, Yen-Fing D (2013) The determination of the thickness and the optical dispersion property of gold film using spectroscopy of a surface plasmon in the frequency domain. Chin Phys B 22:027301
24. Yano M, Fukui M, Haragichi M, Shintani Y (1990) In situ and real-time observation of optical constants of metal films during growth. Surf Sci 227:129–137
25. Gadenne P, Vuye G (1977) In situ determination of the optical and electrical properties of thin films during their deposition. J Phys E 10:733–736
26. Reale C (1970) Optical constants of vacuum deposited thin metal films in the near infrared. Infrared Phys 10:173–181
27. Nesterenko DV, Rehman SU, Sekkat Z (2012) Surface plasmon sensing with different metals in single and double layer configurations. Appl Opt 51:6673–6682
28. Nesterenko DV, Sekkat Z (2013) Resolution estimation of the Au, Ag, Cu, and Al single- and double-layer surface plasmon sensors in the ultraviolet, visible, and infrared regions. Plasmonics 8:1585–1595
29. Rehman SU, Rahmouni A, Mahfoud T, Nesterenko DV, Sekkat Z (2014) Determination of the optical thickness of sub 10-nm thin metal films by SPR experiments. Plasmonics 9:381–387
30. Rehman SU, Mumtaz H, Hayashi S, Shaukat SF, Sekkat Z (2014) Estimation of optical constants of a bio-thin layer (onion epidermis), using SPR spectroscopy. J Opt 16:125014
31. Worm J (2009) WINSPALL, version 3.02. Available http://archive.is/6yLw9. Accessed 15 July 2013
32. Cardona M (1971) Fresnel reflection and surface plasmon. Am J Phys 39:1277
33. Pan M (2009) Using multiple layers and surface roughness control for improving the sensitivity of SPR sensors. M.Phil dissertation, University of Birmingham, Birmingham, United Kingdom
34. Palik ED (1985) Handbook of optical constants of solids. Academic, Orlando
35. Ekgasit S, Yu F, Knoll W (2005) Fluorescence intensity in surface plasmon field-enhanced fluorescence spectroscopy. Sens Actuators B Chem 104:294–301
36. Palik ED (1998) Handbook of optical constants of solids. Academic, New York, p 999
37. Hale GM, Querry MR (1973) Optical constants of water in the 200-nm to 200-μm wavelength region. Appl Opt 12:555–563

Chapter 14
Plasmon Assisted Luminescence in Rare Earth Doped Glasses

M. Reza Dousti and Raja J. Amjad

Abstract Incorporation of metallic nanoparticles in rare earth doped oxide glasses has been introduced as an interesting method to enhance their optical properties. Controlling the size and shape of metallic nanoparticles is a challenging issue which depends strongly and ambiguously on concentration, time and temperature of heat-treatments. In this chapter, we firstly revisit the importance of materials science in general and the rare-earth doped glasses in particular. The performance of rare earth ions in glassy matrices, the probable interactions and the related theories are discussed. Moreover, the incorporation of metallic nanoparticles and their effect on modification of optical properties of rare earth doped glasses is reviewed. Finally, several examples of enhancement and quenching of stokes and anti-stokes luminescence of rare earth ions doped in different glasses are summarized. The incorporation of the metallic nanoparticles is a promising method to improve the optical properties of different oxide glasses for diverse applications such as amplifiers, solid state lasers, sensors, etc. However, to understand the complete role of metallic nanoparticles and control their size distribution and shape, further research is necessary.

Keywords Rare earth doped glasses • Metallic nanoparticles • Amorphous nanocomposites • Optical materials

The necessity to develop noble materials for various applications – from decorative objects, optoelectronic devices, and military facilities, to medical interferences such as drug-deliverers, artificial bones and sturdy dental ceramics – has motivated the materials engineers and scientists to demand for new materials, with low-cost, high efficiency, long durability and recyclability. Glass and glass-ceramic technology is classified to be among the most important branch in the materials science since these composites could provide a wide range of optical, chemical, thermal,

M.R. Dousti (✉)
Instituto de Física de São Carlos, Universidade de São Paulo, 13560-970 São Carlos, SP, Brazil
e-mail: mrdousti@ifsc.usp.br; mrdphysics@gmail.com

R.J. Amjad
Department of Physics, COMSATS Institute of Information Technology, Lahore 54000, Pakistan
e-mail: rajajunaid25@gmail.com; mjunaidamjad@ciitlahore.edu.pk

C.D. Geddes (ed.), *Reviews in Plasmonics 2015*,
DOI 10.1007/978-3-319-24606-2_14

electrical and magnetic properties due to the large diversity in their structure and compositional elements. Glasses, by definition, are non-crystalline solids which show glass transition temperature and can form the glass-ceramics by the controlled heat-treating procedures. Glasses have emerged in various fields and show significant contribution in development of new functional materials. Glass-ceramics show typically better mechanical, thermal and optical performance than their mother counterpart, due to the partial presence of nano- or micro-crystalline structures. The glasses and glass ceramics can be used as solders, laser hosts, memory planes, switching panels, optical fibers, sensors, lenses, optical limiters, waveguides, army devices, medical devices and dental materials etc. The research on the glassy materials shows a rapid growth as evaluated by a recent analysis of the major scientific databases. The glasses are widely used as optical materials as they usually show a wide transparency in the ultraviolet (UV) to near-infrared (NIR) region and non-linear optical absorption, which could be combined with their other promising properties to obtain novel materials. When doped with rare earth (RE) ions, the glasses can emit in a wide range of spectral energy; from ultraviolet to visible and to near-infrared and far-infrared regions. The intensity, peak wavelength, excited state lifetime and quantum efficiency of those emissions are determinant characteristics of the RE-doped materials which could be engineered by selection of the host glass, concentration of RE ions, the co-doping species, heat-treatments etc. For example, the selection of the glass hosts with low phonon energies, can reduce the nonradiative losses, and increases the quantum efficiency, while upconversion emissions could occur to generate a high energy photon, through the excitation of the RE ions by two incoming photons. The upconversion (anti-Stokes) and normal (Stokes) emissions in the RE-doped glasses are mainly favored by the various energy transfer mechanisms among the neighboring ions. Such energy transfer processes are listed in this review paper and discussed in details. Moreover, the Judd-Ofelt formalism as the most important theoretical approach to evaluate the radiative properties of RE ions is also revisited. The radiative properties such as emission cross-section of most of the RE ions in the current glasses is not very effective to obtain a high quantum yield device, and in most of the cases, increasing the concentration of RE ions results in the concentration quenching effect. Therefore, improving the emission efficiency of the RE ions is the current issue and several proposals have been demonstrated to optimize the latter properties. For example, the incorporation of Yb^{3+} ions or metallic species as the sensitizers, or addition of shells or OH-removers which could decrease the quenching centers.

On the other hand, the incorporation of metallic particles (such as nanoparticles, clusters, atoms, dimers, ions, etc) has attracted a large interest due to the remarkable interaction of the excitation beam with the electronic structure of the metal. Meanwhile, in the last decade, nanoscience and nanotechnology has been merged as a new filed of science which attempt to synthesize the particles having dimensions in the order of few nanometers to a hundred of nanometers and to embed and confine such particles in several functional materials. In the nanosized metallic particle, the interaction of the light with metal results in the so-called surface plasmon resonance which is the collective oscillations of the electrons of the metal, thereby, it modifies the optical, electrical and structural properties of the containing host, especially the dielectric glassy media which will be discussed in this work. The glass containing metallic nanoparticles has gained a large interest thanks to their potentiality to develop waveguides, optical limiters, micro-lenses, non-linear devices, solid state batteries, electrochemical sensors etc. On the other hand, it has been also proposed that the optical properties (emission intensity) of the RE doped glasses could be altered by embedding the metallic nanoparticles. The latter application is the main aim of this chapter. Beside a great number of reports on the enhancement and quenching of the emissions of the RE ions doped glasses by incorporation of metallic species, the origins of both enhanced or quenched luminescence is still not fully understood and needs further attention. Several works have shown that the metallic nanoparticles could improve the emission intensity of RE ions (e.g. Eu^{3+}, Er^{3+}, Nd^{3+}, Tb^{3+} etc) doped glasses. The size of particles observed in those reports could vary from a few angstrom to tens of nanometer, while in many cases the plasmon resonance absorption band are not observed. On the other hand, the quenching in the luminescence of some RE ions (mainly Dy^{3+}, and Sm^{3+} etc) has been observed in presence of metallic nanoparticles, while there are several reports which attributed the enhancement of the luminescence to other metallic species such as ions, dimers, clusters, etc, rather than the nanoparticles. However, up to this stage, the contribution of each species in luminescence intensification is not clearly understand. Therefore, in this work, we would like to review the existing literature, up to our knowledge and time, on the history of the glass science and technology, the optical properties of rare earth doped glasses and their relevant theories, which would be followed by an introduction on the importance of the nanoscience with an emphasizing on the optical properties of the nanoparticles in an dielectric media. We will also discuss the important characteristics of the metallic nanoparticles, such as size and shape of them, which could alter their optical properties. The Mie theory, and related theories to understand the optical properties of nanoparticles, their formation and growth mechanisms are also discussed in this review paper. Later, the most utilized spectroscopic and imaging techniques to characterize the nanoparticles will be quickly overviewed. At the end, the effect of metallic species (mainly nanoparticles) on the absorption and luminescence characteristics of the RE doped glasses will be discussed according to the existing literature.

14.1 Glasses and Glass Ceramics

Nowadays, materials science is a crucial area of research due to the limited range of organic materials on earth to supply fuels and power for living, factories and technology. Therefore, lot of efforts have been put on the design of new materials to alternate inorganic sources. Solar cells are good example of those substances. Polymers, ceramics, glasses and new light sources based on lasers, amplifiers in networking, Li-ion batteries, etc are some of the new materials which emerged to assist the "green energy" agenda. Currently, glasses exist all around the world and on our daily life, starting from drinking cups to the dressing mirrors, from electric lamps to the communication fibers, from window glass to wine bottle and many attractive decorative jams. There are magnificent collections and museum of glasses which inspire the human mind by those timeless and limitless colorful and shaped objects of art. The first glass on the earth, indeed, has been made by nature. The ashes of overflowed volcanoes get cooled down slowly and made natural glasses, containing aluminum silicon, sodium, potassium, calcium and iron.

But, what is a glass? Glass is an amorphous solid (non-crystalline) material which has no long- or short-range order. It shows a glass transition, which is the reversible transition from a rigid and relatively brittle state into a molten or rubber-like state. Glasses are usually fragile and preferably transparent in visible region. Glasses can be formed by cooling a melt rapidly to glass transition temperature such that the quenching time is insufficient to form any crystalline phase. In discipline, the glass is referred to all non-crystalline solids (with an amorphous structure), exhibiting a glass transition when heated towards the liquid state. Therefore, glasses can be classified to various sorts of materials, such as ionic melts, metallic alloys, aqueous solutions, molecular liquids, and polymers.

On the other hand, glasses can be transformed to glass-ceramics (polycrystalline materials) by controlled crystallization process. Glass-ceramics possess versatile properties of both glasses and ceramics. The structure of a glass-ceramic includes an amorphous phase and one or more crystalline phases (usually between 30 % and 90 % of whole bulk). Glass-ceramics are among interesting materials since they can be fabricated as easy as glasses, while having advantageous properties of ceramics, such as low porosity, high toughness, strength, low or even negative thermal expansion, high temperature stability, good chemical durability, resorbability, ion conductivity, superconductivity, isolation capabilities, low dielectric constant and loss, better optical and emission properties when doped with rare earth ions, high resistivity and break down voltage. These properties can be engineered by appropriate base glass composition and controlled heat treatment/crystallization process. Glass-ceramics are usually obtained in two steps: Firstly, a glass is formed by the melt-quenching or sol-gel techniques. Then, it is cooled down and is again reheated at an appropriate temperature for a certain time. As a consequence, the glass partly crystallizes. In most cases nucleating agents are added to the base composition of the glass-ceramic. These nucleation agents can speed-up and control the crystallization process. Unlike sintered ceramics, glass-ceramics have no pores, because of no pressing and sintering procedures. They can be mass-produced and their nano-structure can be designed to combine a variety of properties [1].

The history of glasses made by man has started since 4000 B.C. in Mesopotamia, western Asia. These glasses were glazed due to application of copper compounds. Colorful glasses have been prepared between fourteen and sixteen century B.C. in Egypt. However, the art of glass transferred to Syria, Cyprus and Palestine at eleventh century B.C. All nations have used same technique to prepare the glasses; the glasses were melted, drown out and winded around a clay core which was kept by an iron rod. Around 4000 B.C., Greek and Macedonia become centers of glass makers. They used new approaches such as sandwiching techniques, adding gold layers and colorful mosaics. At third and second century B.C., Roman Empire was emerged which also continued glass blowing up to a new revolution which occurred in Greek Empire and Muslims countries, especially in Syria and Palestine at seventh century A.D. On the other side of the earth, China (second century B.C.) and India (fifth century B.C.) also were heading in blowing the glasses containing barium and lead. In the ninth century, Baghdad was on the headline of art of glass blowing. Persia also became famous in preparing the glasses and grinding the glassy surfaces. In the thirteenth century, once again, Syria became the center of glasses due to conquest of Persia by Chengis Khan, and fantasy mosque lamps in Damascus were built up. However, the war of Timur in Damascus contributed to immigration of many glass makers from Syria to Samarkand. In this time, large tax reliefs and social recognition from European governments pushed the glass makers to develop the techniques of glass blowing. Addition of calcia to make shinier and brilliant glasses (Crystallo-glasses) and "ice glasses" methods were two biggest achievements of that time. After immigration of glass workers from Italy, Germany became a center of glass blowing. Patronage of Church helped the regime to prepare new glasses such as "waldglas" and "potash-lime". In the nineteenth century, Europe, United States, Bohemia and many other countries developed the techniques of glass blowing. In that time, crystals became very popular. Next, fluorescent glasses, Opaline, HF-etched glasses and cobalt blue glasses were prepared [2].

In the twentieth century, by developments of chemistry and physics of inorganic materials, glasses were also grown, not only for decorative objects, and daily life facilities, but for scientific purposes. Silica glass is the first production of this century which was mixed with other oxides such as CaO, PbO and CoO. By understanding the amorphous nature of glasses, study of the local structure and bonding in glasses were acknowledged as important factors. On the other hand, glass blowing became industrialized where the sheets of few millimeters in thickness, couple of meters in width with desirable length were seen using developed machines. The cutting, polishing and coating methods led to develop glass pieces potential for microscope, telescope, glass-fibers (cables), electronic bulbs and beverage bottles [2]. Nuclear wastes are also cleared using borosilicate glasses and calcium phosphate and silicate were emerged in repairing the bones and bio-techniques [3, 4].

More recently, zinc phosphate and lead borate glasses were used as solders [5–7]. Phosphate and fluro-phosphate glasses were applied as laser hosts [8]. Memory panels and switching materials, which turn from resistive to conductors at a certain voltage, were prepared by chalcogenide glasses [9]. Superior transparency of halide glasses made them, recently, a good candidate to be substituted for silica fibers in telecommunication technology [10, 11]. Modern glasses are versatile, from soda-lime silicate, borosilicate and alumina-silicate to borofluorite, phosphate, germanate, antimony and tellurite glasses. They show different physical and optical properties due to their specific structures which make them excellent candidates for applications in different branches of science and technology. In a scientific approach, tellurite glasses attracted large interest due to its significant optical, thermal and physical properties [12]. Moreover, oxyfluoride glasses have attracted a large attention for optical properties because they provide comparatively low phonon energies and high chemical and mechanical stabilities [13].

Providing diverse applications from solar cells, optical amplifiers, solid state lasers, army tools, and medical devices, the necessity of research on glasses is increased to understand their structure, optophysical properties and functionality. An exponential increase in number of publications in the field of glasses in general and tellurite glass (as an example) in particular is an evident of such needs. Figure 14.1 presents the number of publication on "glasses" and "tellurite glasses" indexed by SCOPUS database from 1930 up to the end of June 2014 [14]. The increased number of publications on the glass technology was also highlighted by Mauro and Zanotto [15]. Analyzing the data given by SCOPUS, they concluded that the China and United States are the frontiers in glasses in terms of number of publications, while Journal of Non-Crystalline Solids is the most significant journal which publishes "glass-related" papers.

Glasses are excellent candidates to host the so called rare earth ions (lanthanides), since they possess large refractive index, wide transparency window, high RE solubility, etc. Rare earth ions show exceptional optical properties such as long lifetimes, sharp absorption intensities, and excellent coherence properties due to their $4f^N$ to $4f^N$ optical transitions and other significant properties like large oscillator strengths, broad absorption and emission bands and short lifetimes due to their $4f^N$ to $4f^{N-1}5d$ transitions [16]. Special optical properties of the RE ions and their photonic applications in addition to nonradiative (NR) energy transfer (ET) processes led to a wide study on REs. In principle, ET processes may favor particular applications (such as operation of anti-Stokes emitters) but it may be detrimental as in the case of RE based lasers because interactions among the active ions contribute for the increase of the laser threshold. In particular the study of ET processes in glasses having frequency gap in the visible region deserves large attention because when doped with RE ions some glasses may present efficient visible luminescence. Visible and infrared (IR) emissions of RE ions in glasses and glass-ceramics are known to be applicable in solid state lasers and broadband communications, respectively [17].

Many efforts have been done to enhance the intensity and gain of emissions in RE-doped glasses. First, it was reported that increment of concentration of REs in the system could intensify the Stokes and/or anti-Stokes luminescence intensities. However, a quench is observed often, after the introduction of 1–2 mol% of the RE ions [18, 19]. This happens due to further energy transfer among the RE ions, which results to increase the lifetime. Another approach was demonstrated to increase the absorption cross section of REs by introduction of second dopant, commonly trivalent ytterbium ions. The tripositive ytterbium ion shows large absorption cross section, therefore the large concentration of Yb^{3+} in vicinity of Er^{3+} provide larger absorption and emission gain, through energy transfer from $\{^2F_{5/2}: Yb^{3+}\}$ to $\{^4F_{7/2}:^4F_{9/2}:Er^{3+}\}$ [20]. In order to overcome such deficiencies, it has been demonstrated that metallic nanoparticles (NPs) may enhance the RE luminescence doped glasses and improve their nonlinear optical properties [21, 22]. In majority of the

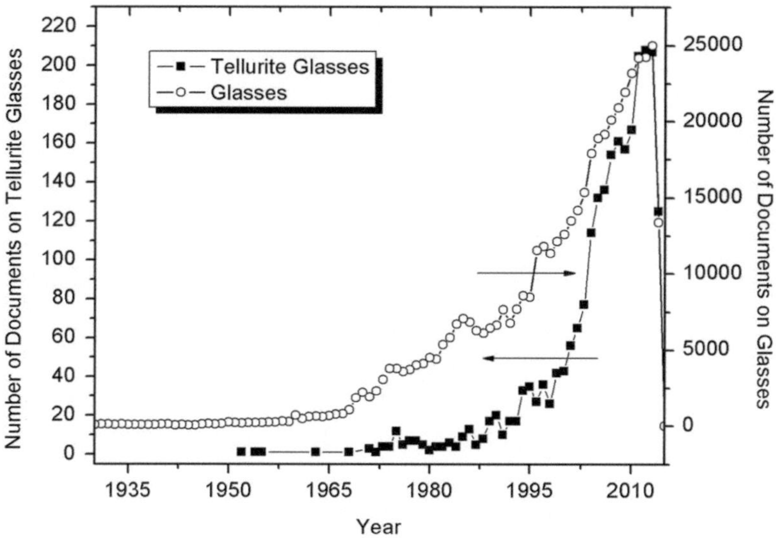

Fig. 14.1 Number of Scopus-index publications on glasses and tellurite glasses (Data taken from Scopus on 15 July 2014 [14])

studies, the presence of NPs contributes to enhance the material's luminescence efficiency either due to ET from the NPs to the RE ions or by influence of the large local field on the RE ions positioned in the vicinity of the metal. Indeed, the presence of nanostructures in the glass alters the luminescence efficiency. It is expected that RE doped glassy systems may be optically and thermally optimized by appropriate doping with metallic NPs.

14.2 Trivalent Rare Earth Ions Doped Glasses

Rare earth ions with their exclusive unfilled 4f shells are promising dopants in glasses to obtain optical fibers and solid state lasers [17]. At the atomic number of 57, 5s and 5p shells are full and 4f is unfilled. By increasing the atomic number in lanthanide group, the radius of 4f shell decreases gradually [23]. The outer 5s and 5p shells shield the 4f electrons and sensitize their environment so that several laser transitions are available in RE ions. For instance, there are 11 electrons in trivalent erbium ions (Er^{3+}:[Xe]$4f^{11}$), and 12 in Tm^{3+} ions. The schematic energy level of Er^{3+} ion is shown in Fig. 14.2. Russell-Saunders symbols ($^{2S+1}L_J$) are used to label the energy levels, where L, S and J are the orbital angular, spin angular and total angular momentums, respectively.

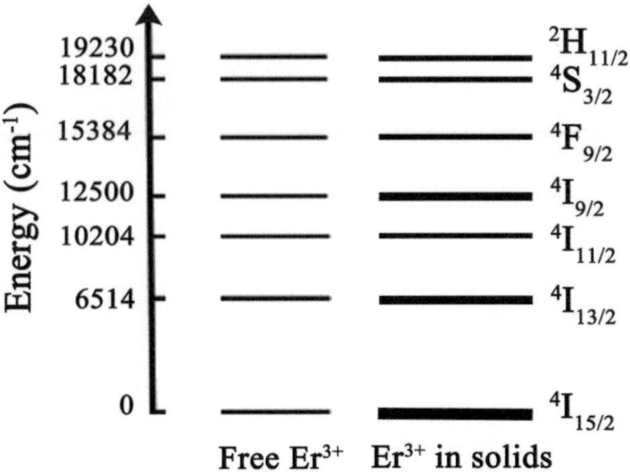

Fig. 14.2 Energy level of free-Er^{3+} ions and Er^{3+}-doped in a solid medium

Energy levels of rare earth ions can be explained by following Hamiltonian operator [24]:

$$H = H_{free-ion} + H_{''Crytal-field''} \tag{14.1}$$

where

$$H_{free-ion} = -\frac{\hbar^2}{2}\sum\nolimits_{i=1}^{N} \nabla_i^2 = \sum\nolimits_{i=1}^{N} \frac{Z^* e^2}{r_i} + \sum\nolimits_{i<j}^{N} \frac{e^2}{r_{ij}} + \sum\nolimits_{i=1}^{N} \zeta(r_i)\vec{s}_i.\vec{l}_i + negligible\ terms. \tag{14.2}$$

Here, N is the number of electrons in 4f, Z^* is the effective nuclear charge, including the inner electrons and nuclei, $\zeta(r_i)$, s_i and l_i are the spin-orbit coupling efficiency, and spin and orbit angular momentum, respectively. The terms in this equation (from left to right) define the kinetic energy, Coulomb interaction, mutual Coulomb repulsion and spin-orbit interaction of the 4f electrons. The last two terms are responsible for the broadening of energy level structure of RE ions in a host, which lift the degeneracy of the $4f^N$ electron configuration. A non-spherical symmetric crystal field in solids splits the energy levels of ion, which is frequently called as "Stark splitting". Due to shielding by 5s and 5p electrons, the crystal field Hamiltonian is 100 times weaker than electrostatic and spin-orbit interactions, in 4f electrons [24]. It is worth to mention that electric-dipole intra $4f^N$ transitions are forbidden due to matching parity of all levels, however they became allowed as a result of mixture into the $4f^N$ configuration of a small amount of excited opposite parity configuration.

In the presence of magnetic field, and considering the ion-ion interaction, two more terms will be added into the Eq. 14.1

$$H = H_{f-ion} + H_{CF} + V_{EM} + V_{ion-ion} \tag{14.3}$$

where V_{EM} is the Hamiltonian of interaction of light by ion, and the last term represent the interaction of two neighboring ions. V_{EM} is responsible for absorption transitions, when frequency of incoming magnetic field is in resonance or near-resonance with transition between different energy levels of RE ions.

The efficient emission intensity of REs embedded in glasses is limited to particular concentration so that further introduction of REs results in clustering of dopants or inefficient energy transfers. The formation of such clusters increases the nonradiative transfer rates between REs; therefore "concentration quenching" phenomena may occur [19, 25]. Several methods have been proposed to avoid the quench phenomena by increasing the efficiency of RE emissions. Enhancements through the effect of co-dopants on local symmetry of RE, adjusting the local interaction with suitable crystal structure, reduction of quench centers, semiconductor induced ET and introduction of noble metallic NPs are some of appropriate proposals [26].

14.2.1 Radiative Properties and Judd-Ofelt Theory

Judd [27] and Ofelt [28], independently and simultaneously formulated the theory of absorption and emission of lanthanides. Interestingly, approaches, assumptions and results of both theories were one and the same; however there are some differences in definitions. Judd defines the theory as the optical absorption while Ofelt referred to crystal spectra of RE ions [29]. Judd-Ofelt theory defines the properties of the radiative transitions in REs. Using this theory, radiative transition probabilities, branching ratios and intrinsic lifetime of an excited state to its lower states can be determined by assessing the three intensity parameters, Ω_i ($i = 2, 4$ and 6). The calculated oscillator strength (f_{cal}) of an electric-dipole absorption from the ground state $/(S,L)J>$ to the excited state $/(S',L')J'>$ depends on the Judd-Ofelt intensity parameters by

$$f_{cal}\left[(S,L)J;\left(S',L'\right)J'\right] = \frac{8\pi^2 mc}{3h\lambda e^2(2J+1)}\left[\frac{\left(n^2+2\right)^2}{9n}S_{ed} + n^3 S_{md}\right] \tag{14.4}$$

where α is the absorption coefficient (cm^{-1}), N is the number of active ions (mol. L^{-1}), e, m, c, h and λ have their common definitions in physics. Here, S_{ed} and S_{md} are electric dipole and magnetic dipole linestrenghts, respectively.

$$S_{ed} = e^2 \sum\nolimits_{t=2,4,6} \Omega_t \left|\left\langle (S,L)J\left\|U^{(t)}\right\|\left(S',L'\right)J'\right\rangle\right|^2 \tag{14.5}$$

$$S_{md} = \frac{e^2\hbar^2}{4m^2c^2}\left|\left\langle (S,L)J\|J\vec{L}+2\vec{S}\|\left(S',L'\right)J'\right\rangle\right|^2 \tag{14.6}$$

Thus, three intensity parameters can be evaluated by a least square fitting to equalize the calculated and experimental (f_{exp}) oscillator strengths,

$$f_{exp} = \frac{4.318\times10^{-9}}{N}\int\alpha(\omega)d\omega \tag{14.7}$$

The reduced matrix elements, $\|U^{(t)}\|^2$ ($t=2$, 4 and 6) [30, 31] are consistent from host to host, and Ω_t ($t=2$, 4 and 6) are known as Judd-Ofelt intensity parameters that can be estimated by a least-square fitting method of experimental oscillator strengths on calculated ones. Magnetic dipole term is resulted from orbit-spin coupling and its different probabilities, considering the selection rules, are discussed and given in Ref. [32]. The transition probabilities A, branching ratio β and radiative lifetime τ can be calculated by

$$A_{J\to J'} = \frac{64\pi^4}{3h(2J+1)\lambda^3}\left[\chi S_{ed}+n^3S_{md}\right] \tag{14.8}$$

$$\beta = \frac{A_{J\to J'}}{\sum_{J'}A_{J\to J'}} \quad \text{and} \quad \tau = \frac{1}{\sum_{J'}A_{J\to J'}} \tag{14.9}$$

The Judd-Ofelt intensity parameters of Er^{3+} ions in different glassy hosts are tabulated in Table 14.1. The lifetime of excited states of RE ions is an important parameter that defines the possibility of achieving the population inversion and efficiency of pumping in amplifiers and laser applications. For a particular excited sate, transition rate is given as inverse of the lifetime. Transitions from a state include both radiative and NR decays. Radiative transitions are due to absorption and emission of a photon, while nonradiative transitions correspond to the interaction of ions with lattice quantized network; the phonons and energy transfers etc. Nonradiative decay rates from excited states play a significant role to choose the suitable host for different applications. For example, multi-phonon (MP) relaxation rate in borate and phosphate glasses in $^4I_{11/2}(Er^{3+})$ excited states is large which reduces the radiative emissions and quantum efficiency of $^4I_{13/2}(Er^{3+})$ excited state. On the other hand, tellurite glass possesses lower phonon energy which is favorable to enhance the lifetime of $^4I_{11/2}$ level in Er^{3+}; therefore, the efficiency of the tellurite glass is large enough to develop the erbium doped fiber amplifiers (EDFA), pumping at 980 nm, as well as solid state upconverters.

The RE-doped glasses have many technological applications. They can be used as solid state lasers, sensors, optical fibers, amplifiers and etc. It is just at 1961 that Snitzer [36] reported the first fiber prepared by Nd^{3+} ions in solid state flash lamp pumped laser, which operates at 1061 and 1062 nm. In 1969, Koester and Snitzer reported on the near single-Nd^{3+}-doped fiber laser. Kao and Hockham [37] developed the theory of propagation in core-clad fibers and studied the structured optical

Table 14.1 Judd-Ofelt parameters of the Er^{3+} ion in various glasses ($\times 10^{-20}$ cm^2) [33]

Glass type	Ω_2	Ω_4	Ω_6	Ω_4/Ω_6
Silicate	5.59	1.42	0.87	1.63
Phosphate	4.67	1.37	0.77	1.78
Tellurite	5.34	1.75	0.94	1.86
Germanate	5.72	0.91	0.32	2.84
Fluoride	2.91	1.27	1.11	1.14
Borate [34]	4.11	1.45	1.42	1.02
Phospho-tellurite [35]	4.25	1.50	0.43	2.44

fiber. In 1972, Sandoe et al. [38] reported the first phosphate glass containing Er^{3+} ions which emits at 1530–1560 nm regions. In addition, Stone and Burrus [39] studied the 800 nm emission of Nd^{3+} fibers (CW laser). Mears et al. presented the first tunable and Q-switched fiber laser which operates in two regions; 1528–1542 nm and 1544–1555 nm. First Transatlantic fiber optic TAT-8 cables were developed in 1990 and later Svendsen developed the optical network WDM system and he installed this system for the first time on long-haul routes in Norway on 1997 [40].

Broad and flat stimulated emission cross-section in communication band and large amplifications gain in L-band of Er^{3+}-doped tellurite glass, introduced them as promising materials for broadband applications. The cross-section of this broadband in heavy metal oxide glasses such as bismuth and tellurite glasses (having refractive index >2 and low phonon energy) is larger than phosphate, silicate and fluoride glasses (see Table 14.2). Solubility of RE ions is another factor to select the suitable host matrix. The solubility of a dopant directly depends on the strength of the structural bonding. For instance, in silicate glass four oxygen atoms are tightly bounded to silicon atom by the strong covalent linkages. Therefore, the incorporation of RE ions in silicate glass is weak, and uniform distribution is difficult. Due to this fact, remarkable amount of modifiers (usually alkalis) are required to break the covalent bonds, weakening the network structure and to form the non-bridging oxygens (NBO) [41]. Phosphate glasses are also based on tetrahedral structure. However their covalency is five. The double bond between phosphorus and oxygen increases the number of NBOs. Therefore phosphate glasses show better spectroscopic properties and emissions than silicates when doped with REs [42]. On the other hand, tellurite glass shows a 2-dimensional system, where the Te-O linkage is significantly weaker than Si-O bond in silicate glass. Thus, it is much easier to break the atomic network of tellurites. Moreover, the atomic/ionic diameter of Te is larger than Si, therefore the network is not tightly closed. The open and weak network in tellurite glass facilitates the incorporation of RE ions and formation of uniform doped glasses is easier than silicate and phosphate glasses [43].

Table 14.2 Full-width at half maximum, emission cross-section and lifetime of 1.5 μm broadband of Er^{3+} ion in different hosts (AFP stands for alumina fluorophosphates) (All data taken from [44])

Glass	FWHM (nm)	σ_e ($\times 10^{-20}$ cm^2)	FWHM $\times$ σ_e	Lifetime (ms)
AFP	53	0.60	3.18	7.6–8.4
Silicate	40	0.55	2.2	5–8
Phosphate	37	0.64	2.37	6–10
Tellurite	65	0.75	4.88	2.5–4
Bismuth based	79	0.70	5.54	1.6–2.7

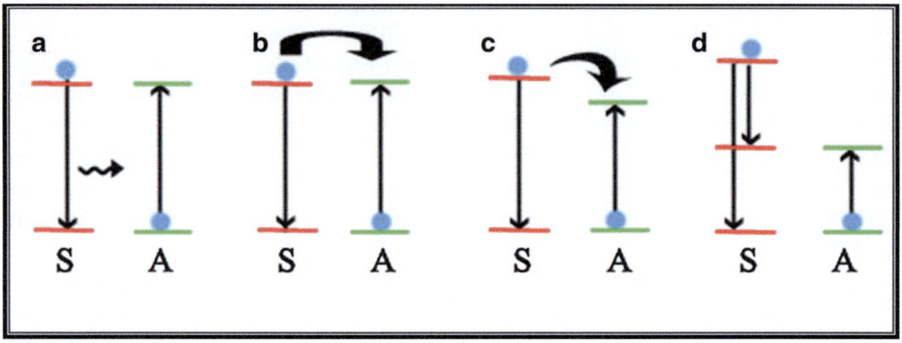

Fig. 14.3 Different energy transfer processes from a sensitizer (S) to an activator (A) in its ground state. Resonant radiative transfer (**a**), resonant energy transfer (**b**), Energy transfer assisted by phonons (**c**), and example of quenching of the fluorescence of *S* by energy transfer to *A* (**d**)

14.2.2 Energy Transfers and Cooperative Process

The neighboring active ions in short distances can interact with each other in two different approaches, (i) by summing up the photon energy by energy transfer (ET), and (ii) cooperative effects due to emission, absorption or sensitization. Given in Fig. 14.3 is the first proposal of the energy transfer processes, where activator ion is in its ground state and sensitizer ion is in an excited state.

Auzel introduced the situations when activator ion is excited, considering the exchanging of energy due to difference between ions energy, not only absolute energy [45]. He also presented and discussed different 2-step absorption, 2-photon excitation, cooperative luminescence, second harmonic generation (SHG), and cooperative sensitization. Cooperative upconversion (CUC) includes both cooperative sensitization and cooperative luminescence mechanisms. In high concentration of rare earth ions, the interaction between two electric dipole moments of ions is effected by short distance neighboring. CUC depends greatly on pumping intensity and concentration of ions. At low laser intensity, the CUC is not efficient. Cross-relaxation process is the reverse mechanism of CUC as shown in Fig. 14.4. The transition of two electrons in high and low energy levels leads to populate the

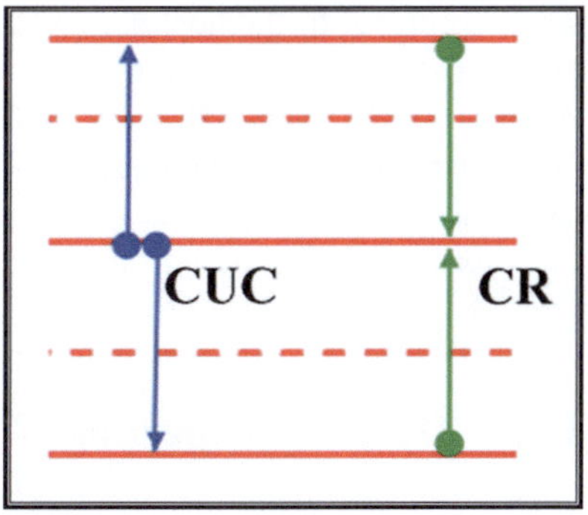

Fig. 14.4 Cooperative upconversion and cross-relaxation process in a schematic energy levels diagram of rare earth ions

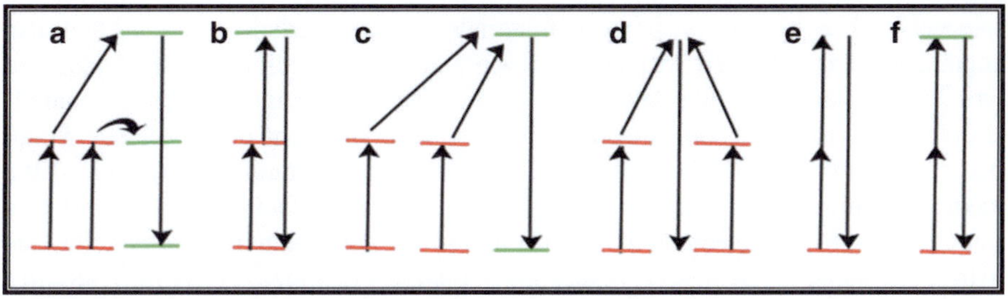

Fig. 14.5 Schematic energy transfer mechanism for different two photon upconversion processes. APTE effect in YF_3:Yb:Er (**a**), 2-steps absorption in SrF_2:Er (**b**), cooperative sensitization in YF_3: Yb:Tb (**c**), cooperative luminescence in Yb:PO_4 (**d**), Second harmonic generation in KDP (**e**), and 2-photon absorption excitation in CaF_2:Eu^{2+} (**f**)

middle energy state. Second-order cross-relaxation of excited levels is negligible due to low concentration of ions, in this excited state.

14.2.3 Non-linear and Upconversion Processes

Excited state absorption or 2-step absorption is shown in Fig. 14.5. An ion in its metastable level can absorb a second photon and be excited to higher energy levels. Normally, nonradiative decays through multi-phonon relaxations will help to populate lower metastable levels where the absorption of second or third photons may excite the ion, gradually.

The so-called "energy transfer" up-conversion is the general form of Dexter energy transfer [46] when the activator is in a metastable excited state. In a Dexter ET process, two ions, or two molecules or two parts of a molecule mutually swap their electrons. Increasing the distance between two particles result in an exponential decrease in the rate of process. This exchange mechanism is also named as "short-range" energy transfer. In addition, the interaction of activator and sensitizer should be weaker than the vibronic interaction of two parties. The latest circumstance, suggest the coupling of single-ion level to the host network. By and large, this situation is more probable in high-concentration of REs in glass or crystal, where the splitting of pairs' level is as small as 0.5 cm^{-1} [47]. Besides, the transfer probability of such ET processes must be faster than radiative and nonradiative transitions from the metastable level. Therefore, two or three photons lead to generation of only one metastable state ion. The excited state absorption depends strongly on the pump intensity.

14.3 Optical Properties of Metallic Nanoparticles

Starting from early 1980, the field of nanoscience has been rapidly extended. It is a branch of science which deals with phenomena of 1–100 nm scales. Many basic sciences such as chemistry, physics, materials science, medicine, biology and also engineering are subjected with nanoscience and nanotechnology applications. The term, nanotechnology is the knowledge of produce, control and operate such small-sized particles to create noble materials, applicable for different area of science. Small particles firstly were used to color the metallic or glassy decorative subjects during the mediaeval times [48]. Lycurgus cup is an ancient Roman artifact – which relates to the fourth century – are appeared red in transmission and pale green in reflected light, due to incorporation of gold or silver or an alloy of both (average sizes of ~70 nm) [48]. In a scientific effort, initially Faraday (1857) studied the colloidal gold to determine the origin of its red color in compare to yellowish bulk gold [49]. However, it is just in recent decade that nanotechnology emerged as a basic and important science to develop scanning probe microscope and scanning tunneling microscope (STM) [50, 51] to study the isolated nanoclusters. In 1980, the first clusters of alkali metals with hundred atoms were produced and studied [52]. Present nanoscience techniques aimed to study the catalysts and to optimize the reaction rate, which reduce the iteration of experiments by trial and errors. More recently, nano-biology and nanomedicine used the nanoscience to overcome the cancer cells [53] and solid tumors [54]. Data storages with high capacity are developed by tunneling of the magnetization of a cluster through a magnetic anisotropy barrier [55].

The shape and the size of NPs affect their optical properties, such as surface plasmon resonance (SPR) absorption band position and shape [56]. By and large, formation of non-spherical metallic NPs results in red-shift of SPR peak and changes its Gaussian shape. Practically, any desirable shape of NPs can be created, such as nano-rods, nano-wires, nano-cages, aggregates, nano-prisms, nano-spheres, nano-sheets and plates [57–59] which show different absorption spectrum and optical properties. Complicated structures like nano-prisms and aggregates showed broad spectrum peaks due to large non-degenerate resonance modes. The non-degeneracy increases with lowering the symmetry of system [60]. Besides, the size of NPs, refractive index of environment medium and other proximal NPs are important factors to determine the SPR band. Therefore, the chief challenge is to optimize the optical properties of nanostructure by adjusting the frequency, FWHM, and shape of the SPR absorption band as shape, size, dielectric function, electron density and effective mass of nanoparticles changes.

14.3.1 Interaction of Light with Nanoparticles

The study of optical properties of metallic nanoparticles interacting with light has rapidly grown and become known as plasmonics [61, 62]. Nanoparticles undertake diverse response under the excitation by light due to scattering or absorption of photons. The photons with efficient energy can either excite the electrons by an internal electronic transition (optical absorption) or ionize the particle by emitting an electron from NP at high energy levels (photoemission spectroscopy). The absorption takes place due to SPR in the particles with sizes d, where $d << \lambda$, and λ is the wavelength of excitation light. The electrons of the metal are subjected into the electric field of incoming light which relocates them with respect to positive charge of nuclei. On the other hand, static Columbic force between positive and negative charges resist to this shift. Therefore, electrons of metal begin to oscillate. The resonant phenomena, typically is determined with a peak in visible region of absorption spectrum. Mie used the classical concept of dielectrics to develop a theory based on this behavior [63]. If the particle is not spherical, the absorption spectrum will be complicated (Fig. 14.6).

Gustav Mie developed the theory on the interaction of NPs with light [63]. Strong absorption band accompanies temperature increment of the particle. Large enhancement of the electric field in particle and its adjacent environment can be observed by near resonance excitations. Optical properties of metallic NPs strongly depend on the dielectric function of the host matrix (environment). The optical properties of a metal can be described by a simple Drude-Lorentz-Sommerfeld model [64]. The effect of an external field $E = E_0 exp(-i\omega t)$ on an electron with mass, m and charge, $-e$, can be written as

$$m\frac{d^2\vec{r}}{dt^2} + m\Gamma\frac{d\vec{r}}{dt} = -e\vec{E} \tag{14.10}$$

where Γ is the damping constant which is metal-dependant parameter. The polarization is $\vec{P} = -ne\vec{r}$, where n is the density of the electrons. Dielectric function, thereby, is given as

$$\varepsilon = 1 + \frac{4\pi P}{E} \tag{14.11}$$

in cgs system. Dielectric function is related to frequency as follow

$$\varepsilon = 1 - \frac{\omega_P^2}{\omega^2 + i\Gamma\omega} \tag{14.12}$$

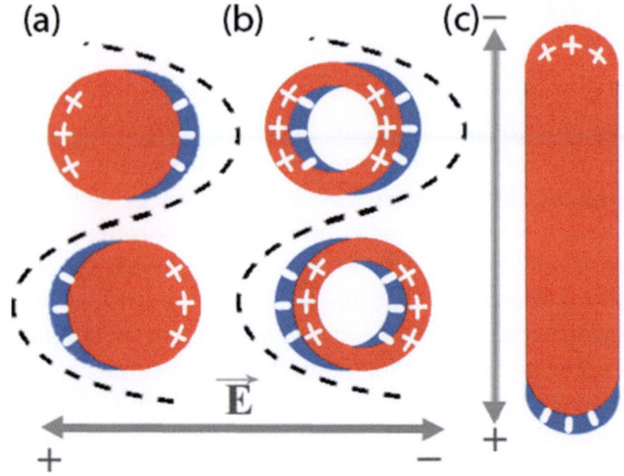

Fig. 14.6 Schematic interaction of light with metallic particles (surface plasmon resonance) for spherical nanoparticles (**a**), hallow nano-spheres (**b**) and nano-rods (**c**) [61]. In addition, this figure exhibits the increase number of resonant modes with decreasing the symmetry

where ω_P is the frequency of plasma,

$$\omega_P^2 = \frac{4\pi ne^2}{m} \tag{14.13}$$

Dielectric function is intrinsically an imaginary function, $\varepsilon = \varepsilon_1 + i\varepsilon_2$, where

$$\varepsilon_1 \approx 1 - \left(\frac{\omega_P}{\omega}\right)^2 \quad \text{and} \quad \varepsilon_2 \approx 1 - \left(\frac{\omega_P}{\omega}\right)^2 \frac{\Gamma}{w} \tag{14.14}$$

for damping constant smaller than frequency, $\Gamma << \omega$. The absorption coefficient is given by $= 2\omega k/c$, where $n + ik = \varepsilon$. In the case of clusters, the nuclei are static and electrons are assumed to move freely in an external field. For the particle sizes smaller than wavelength of light, at any time, the electric field is uniform around the cluster. For a cluster embedded in a medium, the electric field inside the cluster is

$$\frac{E}{E_0} = \frac{3\varepsilon_m}{\varepsilon + 2\varepsilon_m} \tag{14.15}$$

where E_0, ε and ε_m are the static electric field, dielectric constant of the sphere cluster and dielectric constant of medium, respectively. Mie frequency is defined as

$$\omega_M^2 = \frac{\omega_P^2}{1 + 2\varepsilon_M} \tag{14.16}$$

Having the volume of the particle V_p, the extinction cross section for a particle is

$$\sigma_{ext} = 9\varepsilon_m^{3/2}\left(\frac{\omega}{c}\right)V_p\frac{\varepsilon_2}{(\varepsilon_1 + 2\varepsilon_m)^2 + \varepsilon_2^2} \tag{14.17}$$

According to Mie [63], in a medium with refractive index, n, containing the NPs smaller than excitation wavelength, λ, the absorption coefficients can be determined by free electron approximation as

$$\alpha = \frac{9\pi p n^3 c \lambda^2}{\sigma_{dc}\left[\left(\lambda_m^2 - \lambda^2\right)^2 + \lambda^2 \frac{\lambda_m^4}{\lambda_a^2}\right]} \tag{14.18}$$

where p, σ_{dc}, c and λ_m are the volume fraction of the metal spheres in the glass matrix, the dc electrical conductivity, the velocity of the light, and the wavelength at maximum absorption, respectively. Reduced length scales are defined as

$$\lambda_a = \frac{2\lambda_c^2 \sigma_{dc}}{c} \quad \text{and} \quad \lambda_c = \frac{(2\pi c)^2 m}{4\pi N_c \varepsilon_m^2} \tag{14.19}$$

Here, m, N_c and ε_m are the electron mass, the number of electrons per unit volume and the complex form of dielectric function of metallic NPs. The wavelength at which the maximum absorption takes place is given as

$$\lambda_m = \lambda_c\left(\varepsilon_0 + 2n^2\right)^{1/2} \tag{14.20}$$

where ε_0 is the frequency dependent part of ε_m. FWHM of this band can be found as

$$FWHM = \frac{\lambda_m^2}{\lambda_a} = \frac{(\varepsilon_0 + 2n^2)c}{2\sigma_{dc}} \tag{14.21}$$

and the dc conductivity is given by

$$\sigma_{dc} = \frac{N_c e^2 L}{2 m u_F} \tag{14.22}$$

where e, L and u_F are electron charge, diameter of NP and Fermi velocity.

$$u_F = \left(\frac{2E_F}{m}\right)^{1/2} \tag{14.23}$$

14.3.2 Preparation and Observation of Metallic Nanoparticles

There are several physical and chemical techniques to synthesize the NPs. One of the important physical techniques is to evaporate the metal from a source and deposit the gas on a substrate [65]. This method is more efficient since controlling the shape and size of the NPs are more facile. One of the important chemical techniques is to reduce the metal ions or metal salts with a suitable reducing agent. This method is known as most suitable technique to produce the spherical NPs [57, 58, 66, 67]. For example, Xu et al. [68] embedded the silver NPs into the sodium-silicate glass through an ion exchange method and subsequent heat-treatments. There are also many other approaches such as electrochemistry techniques which are beyond the scope of this chapter [69].

Synthesis of the NPs includes two processes [70]: (i) formation of small seeds (nucleation) and (ii) growth process. The capping material plays an important role to determine the shape and size of NPs during the growth. If the capping material is very weak, the growth will continue to produce big crystallized particles. Contrary, if it is too strong, it may reduce or prevent the growth. Therefore, nanomaterial, capping material or reduction agent and medium are important factors to increase the efficiency of synthesize. Moreover, the concentration of agent is another factor to determine the concentration of initial seeds of metallic particles.

The clustering of the NPs can be discussed by following suggested models [71]:

(1) Coagulation process; in which the pairwise particles will be destroyed through the collisions and larger particles will be formed. In a simple model [72], by considering the identical probability of collision (K_D) for different particles, one can describe the coagulation process by

$$\frac{\partial Z_n}{\partial t} = -\frac{1}{2} K_D Z^2 \tag{14.24}$$

where, Z_n is the concentration of particles of kind n. The solution for given time evaluation equation is

$$Z(t) = \frac{Z_0}{1 + 0.5 K_D Z_0} \tag{14.25}$$

And the required time for coagulation is

$$t_c = \frac{2}{K_D Z_0} \tag{14.26}$$

For the coagulation of two particles with radius R_1 and R_2 and respective diffusion velocity of $u(R_1)$ and $u(R_2)$, the probability constant is:

$$K_D = \pi (R_1 + R_2)^2 [u(R_2) - u(R_1)]. \tag{14.27}$$

(2) Ostwald ripening process; in which the large particle will be grown through feeding by smaller particles. Therefore, the smaller particles start to vanish [73]. The matter in its thermodynamic balance consists of surrounding gaseous cloud. The bigger the particle is, the lower number of particles in cloud exists. The

diffusion from cloud to matter and vice versa is consistent with growth and disaggregation of particles, respectively.

(3) Coalescence of particles; in this process, the particles grow due to a strong chemical or physical bonding which is consistent with coagulation in first step and the consequent Ostwald ripening effect. The critical size of particle to grow or dissolve is [74]

$$r^{*} = \frac{2V\gamma}{3k_B T ln(S_r)} \tag{14.28}$$

where V, γ and S are the molecular volume, surface free energy per unit surface area and saturation ratio. Therefore, for a given $S > 1$ and temperature (T), the particle with radius $r < r^*$ starts to grow, while the particles with $r > r^*$ will dissolve.

Properties of NPs are characterized by their extremely small size which needs appropriate apparatuses to be probed. The first instrument to observe the NPs is usually UV-Vis absorption spectroscope since NPs show strong absorption peak at near-UV to near-infrared region. Up to now, many instruments were developed to characterize/observe the structure of NPs such as X-Ray diffraction (XRD), atomic force microscopy (AFM), scanning electron microscopy (SEM), transmission electron microscopy (TEM), X-Ray absorption spectroscopy (XAS) and its fine extended device (EXAFS), dynamic light scattering, energy dispersion X-ray (EDX), IR and Raman [75–77].

XRD technique can be used to determine the crystalline phases, average inter-particle distances, and atomic structure of the nanoclusters. The size of particles, defects and strains of nanocrystals can be defined by measuring the width of diffraction lines. The broadening of the diffraction linewidth is directly related to reduction of nanocrystal size.

AFM, STM, and chemical force microscopy are usually called in a group named: the scanning probe microscopy (SPM). It can be applied on versatile organic and inorganic materials in their gassy or liquid phases. AFM captures the image of specimen through the difference between the atomic forces in short-range and long-range which are repulsive and attractive, respectively [78]. Scanning force microscopy (SFM) also is a useful tip to measure the electrostatic and magnetic interactions between the molecules. SEM is another commanding technique to observe the surface of any type of materials with a resolution of 1 nm [77]. The principle of SEM is based on the interaction of incoming electron with the sample, which will be captured by its backscattered electron. Therefore, both structural and topological properties can be defined through two captured beams.

TEM is also another powerful instrument to characterize the chemical composition and spatial structure [76]. Recently, high-resolution-TEM is developed which allow the imaging of crystals as small as 1 Å. TEM is usually used to determine the shape, size, crystallinity and inter-particle interactions in nanomaterials.

14.3.3 Surface Enhanced Raman and Fluorescence Spectroscopy (SERS, SEFS)

Raman spectroscopy is a powerful technique to study the vibrational modes of molecules through an elastic scattering of a monochromatic excitation source in the routine range of 200–4000 cm^{-1}. Molecular vibrations are either infrared (IR) and/or Raman active. Therefore, Raman technique is mainly used to complement the IR measurements. The critical issue in Raman spectroscopy is the small cross section which requires lot of molecules to achieve valuable results. The efforts to develop the detection ability by Raman spectroscopy have been yield to introduce the Surface-Enhanced Raman Spectroscopy (SERS) [79], where the enhancement factor up to $10^{10} \sim 10^{11}$ times may aim to detect single molecules [80–82]. The enhancement results from the enhanced localized electromagnetic field in the surface of metallic NP and dielectric environment. Furthermore, topological change of NPs in dielectrics adds promoted electric field by lightening rod effects (LRE) in non-spherical (elliptical, cube, pyramid shapes and so on) surface of NPs which enhances the intensity of scattered Raman beam up to 10^{14} times [82]. Messinger, Wang and Kerker developed the theory of the SERS [83, 84]. Fleishmann et al. [85] reported the first SERS using roughened metal electrodes.

The enormous localized electric field between two NPs, firstly, enhances the excitation light, and as the result, the Raman modes of probing molecular will enhance by a factor of η^2, where η is the enhancement factor. Moreover, the emitted Raman signals experience further enhancement by same SPR effect. Therefore, output Raman signal is enhanced by a factor of η^4 [86]. The maximum enhancement happens near the plasmon frequency (ω_P) [87]. Even if, the plasmon resonance may not be efficient in special cases, LRE aims to confine the large electric field in the sharp edges of metal surface or at curved surfaces.

The SERS effect is reported for the first time in a glassy system by Dousti et al. [88]. The incorporation of silver NPs as small as 12 nm into the zinc tellurite glass increases the Raman signals as large as eight times. However, controlling the resonance frequency, shape and size of the NPs is still a controversial subject which effectively determines the enhancement factor. For instance, the addition of concentration of NPs can result in a red shift of SPR band and evolution of enhancement factors of Raman and photoluminescence (PL) signals [89].

On the other hand, the presence of metallic NPs in vicinity of luminescent centers can alter the radiative emission properties of such dopants, so called surface enhanced fluorescence spectroscopy (SEFS). Large electromagnetic field can be localized in vicinity of the metallic NPs. Therefore, they can trap the photons (photon catchers). One of the advantages of such system is fluorescence enhancements due to controlling the light and enhancing the local field around the fluorophores. Such enhancement by SPR was firstly reported by Malta et al. [21]. The application of SEFS on RE-doped glasses containing noble metallic NPs shows significant results which are promising to develop the optical amplifiers, solid state lasers, color displays and versatile nanophotonics devices [90–92].

The enhancements of emissions in the hosts containing metallic NPs with average size of D and separated by the distance d, are mainly attributed to enlarged local electric field. The enhancement factor η, of the system can be evaluated by

$$\eta = \frac{E_{loc}}{E_i} = \frac{(D+d)}{d} \tag{14.29}$$

where E_{loc} and E_i are the amplified local field (effective field, E_{eff}) and the incident field, respectively [62]. The effective field is interpreted by Malta et al. [21] as

$$\vec{E}_{eff} = (\varepsilon_0 + 2)\left[1 + \frac{q\omega_p^2}{3\varepsilon_0\left\{(1-q)\left(\omega_p^2/3\varepsilon_o\right) - \omega^2 + i\gamma\omega\right\}}\right]\frac{\vec{E}_0}{3} \tag{14.30}$$

where ε_0 is the dielectric constant in the presence of external electric field E_0, and γ is the damping coefficient of plasmon resonance.

Radiative decays of any fluorophore (e.g. RE ions) can be modified by plasmonics approach. Various shapes, sizes, corresponding local field of metallic NPs and distances between the fluorophore and NP can result in different enhancement order of emissions and quantum efficiency. Modification of radiative decay is usually called as "radiative decay engineering" [62]. In the next section, we review some of the available reports on the optical properties of the rare earth doped oxide glasses containing metallic nanoparticles.

14.4 Rare Earth Doped Glasses Embedded with Metallic NPs

Glasses are superior candidates to embed the RE ions and metallic NPs due to their high transparency, mechanical strength, and simple preparation in any size and shape and high energy stretching vibration which extinct the interaction of ligand with metals. Therefore, metallic glasses composites containing nanoclusters or NPs are introduced as potential substrate materials (hosts) for significant local field enhancement to increase the luminescence of RE ions [90, 93–95]. Moreover, the introduction of metallic NPs may modifies the thermal and structural properties as well as chemical durability of oxide glasses [96, 97], which are beyond the scope of this chapter. Thermal features of glasses can be also modified by incorporation of metallic NPs and may result in considerable increase in thermal diffusivity and thermal conductivity [98]. This is particularly important considering that cooling is one of the most challenging technical issues to be overcome in the areas of microelectronics and solid-state lighting. It has been reported that the thermal diffusivity of materials doped with metallic NPs depends on the size and concentration of NPs, even though the subject is still a matter of controversy in the literature. The large nonlinear absorption of silver nanoclusters and NPs are recently reported in the silicate, borate and oxyfluoride glasses, which are promising for applications in optical limiting and object's contrast enhancement due to the non-saturated and saturated nonlinear absorptions [99–101].

The formation of NPs in the glass commonly carries out by thermodynamic reduction of NPs in the presence of an oxidant agent. Som and Karmakar showed the feasible reduction process of silver [90] and gold [95] in antimony glass, as follows

$$Sb^{5+}/Sb^{3+},\ E^0 = 0.649\ V \tag{14.31a}$$

$$Au^{3+}/Au^0,\ E^0 = 1.498\ V \tag{14.31b}$$

$$Ag^{+}/Au^0,\ E^0 = 0.7996\ V \tag{14.31c}$$

$$3Sb^{3+} + 2Au^{3+} \rightarrow 3Sb^{5+} + 2Au^0, \Delta E^0 = 1.05V \tag{14.31d}$$

$$Sb^{3+} + 2Ag^{+} \rightarrow Sb^{5+} + 2Ag^0, \Delta E^0 = 1.02V \tag{14.31e}$$

They suggested that the reduction of gold is faster than the silver ions. This idea led to preparation of $gold_{core}@silver_{shell}$ nanoparticles [102]. Wu et al. [103] investigated the reduction of silver ions to neutral NPs in a bismuth-borate glass system, and the possible reduction is proposed as

$$Bi^{3+}/Bi^0,\ E^0 = 0.3172\ V \tag{14.32a}$$

$$Ag^{+}/Ag^0,\ E^0 = 0.7996\ V \tag{14.32b}$$

$$Bi + 3Ag^{+} \rightarrow Bi^{3+} + 3Ag^0, \Delta E^0 = 2.0816V \tag{14.32c}$$

Dousti et al. [104] illustrated the reduction of silver ions to silver NPs in Er^{3+}-doped zinc tellurite glass. The reduction of Ag^+ particles to Ag^0 NPs can be discussed by the reduction potential (E^0) of redox system elements. The E^0 values of each component in this system are:

$$Te^{6+}/Te^{4+},\ E^0 = 1.02\ V \tag{14.33a}$$

$$Ag^{+}/Ag^0,\ E^0 = 0.7996\ V \tag{14.33b}$$

$$Er^{3+}/Er^0,\ E^0 = -2.331\ V \tag{14.33c}$$

$$Er^{3+}/Er^{2+},\ E^0 = -3.0\ V \tag{14.33d}$$

Probable reduction process and their total potentials are as following:

$$Te^{4+} + 2Er^{3+} \rightarrow Te^{6+} + 2Er^{2+} \qquad \Delta E^o = -7.02 \tag{14.34a}$$

$$3\ Te^{4+} + 2\ Er^{3+} \rightarrow 3\ Te^{6+} + 2\ Er^0 \quad \Delta E^o = -7.722 \tag{14.34b}$$

$$Te^{4+} + 2\ Ag^{+} \rightarrow Te^{6+} + 2\ Ag^0 \quad \Delta E^o = +0.5792 \tag{14.34c}$$

Therefore, from the thermodynamic point of view, only the last redox reaction (14.34c) is feasible. The reduction of Ag^+ ions to Ag^0 neutral particles and growth of silver NPs are also reported by addition of other reducing agents such as SnO [105].

In the past decade, there were group of authors who contributed in the studies of effect of the metallic NPs on optical properties of RE-doped glasses, glass-ceramics and thin films. The effect of asymmetric silver NPs reduced in an antimony glass system is reported by Som and Karmakar [90], where intense SPR peak were probed in UV-Vis-IR absorption spectra. They concluded that the decrease in NP-NP distances by increasing the concentration of Ag NPs enhances the localized electric field and broadens the plasmon peak by a red-shift up to 1100 nm. The formation of NPs with different shapes and size ranging between 4 and 31 nm was discussed as an Ostwald's ripening process.

Upconversion of only Ag NPs (without any RE ion in system) is also reported by few authors [90, 106]. Som and Karmakar [90] presented the upconversion of Ag NPs on antimony glass under 798 nm excitation wavelength where the SPR band was observed around 598 nm in UV-Vis absorption spectrum. The upconversion emission wavelengths were located at ~536 and 654 nm. Dousti et al. [106] also observed same behavior of Ag NPs in tellurite glass while broad emission band centered at 500 nm originated due to 786 nm excitation wavelength. Luminescence of metallic NPs are also observed by different research groups [107]. Such radiative emissions from metallic NPs can either contribute to further enhance or quench of the neighboring luminescent ions through energy transfer mechanism.

14.4.1 Eu^{3+}-Doped

The introduction of metallic NPs into the glassy hosts was first reported by Malta et al. [21]. A large enhancement in the order of 5.6 for luminescence of Eu^{3+} in borate glass (emission at 612 nm and under 312 nm excitation wavelength) has been observed due to the presence of small silver particles by a concentration around 7.5 wt%. The absorption peak of small silver particles in this study showed a sharp peak at 312 nm. In 1999, Hayakawa et al. [108] reported on the enhanced luminescence of Eu^{3+} ions doped in silica glass in presence of Ag NPs having sizes of about 4.3 nm (surface plasmon band was observed at 394 nm). The enhancement of $^5D_0 \rightarrow {}^7F_J$ ($J = 0$, 1, 2, 3 and 4) emissions of Eu^{3+} ions doped borosilicate glass (derived by a sol-gel method) in vicinity of polymer-protected gold NPs is also reported by Hayakawa et al. [109], where 6 times enhancement is observed under a long UV excitation light.

The luminescence of Eu^{3+}-doped lead-tellurite glass embedding gold NPs were characterized by Almeida et al. [110]. The sample containing 0.5 wt% Au NPs and annealed for 41 h (average particle diameter ~4 nm) show largest enhancements in Eu^{3+} spectra under 405 nm excitation. Since, the electric dipole transitions such as $^5D_0 \rightarrow {}^7F_{4,2}(Eu^{3+})$ are sensitive to polarizability and environment around the RE, rather than those magnetic dipoles ($^5D_0 \rightarrow {}^7F_{1,3}$), the presence of metallic NPs alert them progressively. They concluded that the presence of small NPs cannot result in to the energy transfer process, hence the main contribution for such intensification is attributed to increased local field around the Eu^{3+} ions. Such enhancements are also observed in Eu^{3+}-doped GeO_2-Bi_2O_3 glasses containing Au NPs by the same research group [111], where SPR band of NPs observed at 500 nm and contributes to 1000 % and 500 % enhancement of $^5D_0 \rightarrow {}^7F_{2,4}$ and $^5D_0 \rightarrow {}^7F_{1,3}$ emissions, respectively.

The luminescence enhancements of Eu^{3+}-doped zinc-tellurite glass and lead-tellurite glass are also reported. Two-times enhancement is observed for Eu^{3+}-doped ZnO-TeO2 [112] and PbO-TeO2 [113] glasses after 12 and 9 h of heat-treatments which lead to formation of silver NPs with an average size of 14 and 10 nm, respectively. Such enhancements are attributed to intensified local field in distances between NPs and RE ions, induced by surface plasmons. In both reports, the silver NPs are grown along (200) crystallographic direction (JCPDS no. 030931), as revealed by TEM images (lattice constant of about 2–2.05 Å).

Jimenez at al. also reported on the enhanced UV-excited luminescence of Eu^{3+} ions in silver/tin-doped glass [114]. However, such enhancements are attributed to Ag^+ ions, and not Ag NPs. Moreover, the quenched PL is caused by Ag NPs, by providing "the paths for the nonradiative loss of excitation energy in europium ions through coupling with plasmon resonance modes". Eu^{3+}-doped aluminosilicate glasses containing different Ag species are reported by Li et al. [115]. The observed broadbands in UV-Vis absorption, photoluminescence excitation and emission spectra of the glasses suggest the presence of Ag ions and molecular-like Ag species. However, after 30 and 120 min of heat-treatments, the silver NPs are formed in the glasses and are discussed in terms of following redox reaction; $Eu^{2+} + Ag^+ \rightarrow Eu^{3+} + Ag^0$. The surface plasmon band of the silver NPs in this glass is observed at 440 nm by taking the difference between absorption spectra of samples with and without NPs. Although the excitation lines of Eu^{3+} ions are suppressed in the spectra of Ag NPs-doped samples, the luminescence emissions in the visible region enhances under 350 nm excitation wavelength. The authors concluded that the observed enhancement can be purely associated to the energy transfer from silver aggregates to Eu^{3+} ions and not an enlarged local field by SPR of Ag NPs (Fig. 14.7).

Jiao et al. [116] investigated the effect of concentration of Eu^{3+} ions on the formation and growth of silver NPs. They concluded that increasing the Eu_2O_3 content led to increase the Eu^{2+} ions and increasing the concentration of Eu^{3+} ions and Ag NPs as; $Eu^{2+} + Ag^+ \rightarrow Eu^{3+} + Ag^0$. In this regard, the enhancement and quenching of Eu^{3+} emissions are observed under 280 and 340 nm excitation wavelengths, respectively. These results are in good agreement with Riano et al. [117] reporting on the intense surface plasmon band of Ag NPs in presence of Eu^{3+} ions, rather than Pr^{3+} ions. However, they observed quenching of the luminescence of Eu^{3+} ions, an indicative of energy transfer from Eu^{3+} ions to Ag NPs. Wei et al. [118] also worked on the preparation of Ag NPs-embedded Eu^{3+}-doped oxyfluoride glasses. The enhancement in emissions of Eu^{3+} ions are attributed to presence of silver NPs, small-molecular like silver and isolated Ag^+ ions under 464, 350 and 270 nm excitation wavelengths, respectively.

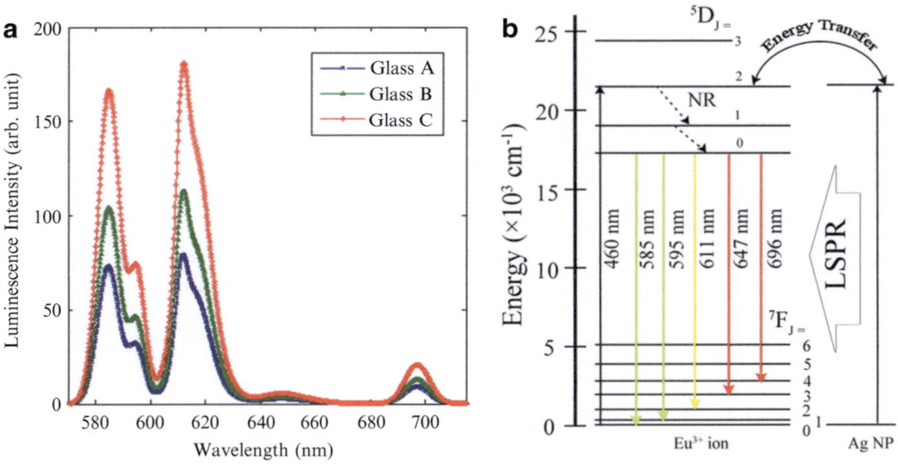

Fig. 14.7 (**a**) Luminescence spectra of Eu^{3+} ions in tellurite glasses embedding (A) 0 mol%, (B) 0.5 mol% and (C) 1 mol% silver nanoparticles (NPs) under 460 nm excitation wavelength. (**b**) Schematic partial energy level diagram of Eu^{3+} ions in vicinity of silver NPs showing localized surface plasmon resonance (LSPR), nonradiative (NR) and radiative decays (Figures are adapted from [113])

Kumar et al. reported on the enhancement of the luminescence of Eu^{3+}-doped titanosilicate glass by introduction of silver NPs [119]. The silver NPs with average particle size of 14.9 nm and TiO_2 polycrystalline are observed in SAED and XRD pattern of the glass samples. A broad absorption band in the 300–400 nm region is assigned to SPR band of NPs. The photoluminescence emissions ($\lambda_{exc} = 393$ nm) and excitations ($\lambda_{emi} = 612$ nm) showed enhancements after incorporation of silver NPs. The asymmetry ratio (AS, the ratio of integrated emissions bands of Eu^{3+} ions) are given as $(\int {}^5D_0 \rightarrow {}^7F_2\, d\lambda)/(\int {}^5D_0 \rightarrow {}^7F_1\, d\lambda)$, which varies by the addition of silver NPs due to modification of (i) ligand filed and (ii) refractive index around the RE ions. AS factor increases from 2.627 to 3.615 for singly-doped and co-doped samples, respectively.

Culea et al. [120] also reported on the effect of Ag_2O and Ag NPs on the spectroscopic and structural properties of lead-tellurite glasses. Enhancement of red-emissions of Eu^{3+} ions by Ag NPs is also reported in other medium [121] under a green light excitation which is known to be promising materials for solar cell and nano biotechnology. Moreover, white light emission is observed in Eu^{3+}-doped oxyfluoride glass containing molecular-like (ML) silver, where no proof was observed to attribute the enhancement of photoluminescence to plasmonic silver NPs [122].

14.4.2 Er^{3+}-Doped

In 2002, Strohhofer et al. [123] reported on the enhanced emission of Er^{3+} in a borosilicate glass by an ion-exchange process. They observed 70 and 220 times enhancements in broadband line under 488 and 360 nm excitation wavelength. They concluded that such enhancement can be attributed to the silver ions/atoms defects and an ET to Er^{3+} ions. Chiasera et al. [124] reported on the silver-sodium exchange process in soda-lime silicate glass containing Er^{3+} ions. The plasmon band of silver was observed in blue region, and its intensity increased by further heat-treatments. The silver exchange has no effect on the broadband emission of erbium; however, it increased the lifetime of this metastable state, which was in disagreement with some older reports [123, 125]. The increase in lifetime of $^4I_{13/2}$ was attributed to silver-induced radiation trapping [124], while the decrements is featured by silver-induced defects in glassy hosts [123, 125]. Using the gold NPs in an Er^{3+}-doped SiO_2 thin film, prepared by sol-gel method, the optical absorption and emission at 1.54 μm were characterized by Fukushima et al. [126]. One hundred times enhancement emissions of sample with 1 mol% Au NPs were attributed to strong filed originated from confined surface plasmon, while the excitation wavelength were selected to line in the Au plasmon band region, located at 520 nm. Lin et al. [127] fabricated the Au NPs-doped erbium optical fiber in a germane-silicate glass which showed plasmon resonance band around 498.2 nm, and the net gain of broadband emission at 1535.6 and 1551.2 nm experienced enhances under 980 nm excitation wavelength. The quenches are observed under 488 nm excitation wavelength and are attributed to absorption of the incident energy by Au NPs. They discussed the observed loss by an ET from Au NPs to lattice, and not to Er^{3+} ions. The effect of the temperature on XRD pattern, absorption and emission spectra and Judd-Ofelt parameters in a Er^{3+}-Au NPs co-doped alumina-silicate glass is investigated by Watekar et al. [128]. A blue shift in plasmon absorption band of gold NPs is observed by increasing the annealing temperature. The radiative lifetime and integrated emission cross-section of infrared emissions of Er^{3+} ion decreased after introduction of Au NPs. Such quenches in non-resonance excitation process is attributed to ET from Er^{3+} ions ro Au NPs. Singh et al. [129] studied the effect of annealing time interval on the size of Ag NPs in Er^{3+}-doped tellurite glasses. They exposed the samples containing Ag NPs under the temperature below the glass transition temperature for different periods of time. They observed increased size of NPs and enhancements in up-conversion emissions in visible range (green and red lines) by increasing the annealing time.

The introduction of Au NPs in Er^{3+}-doped antimony glass raise up to preparation of dichroic nanocomposite [95], showing different colors in transmittance and reflecting surfaces. The enhancements in the order of 3.4 and 7.5 times in green (536 nm) and red (645 nm) bands through an upconversion luminescence were attributed to the enlarged local field by asymmetric Ag NPs. Diffraction peaks of Au NPs in XRD pattern was in good conformity with SAED results for NPs with sizes about 11–30 nm. Once more, broadening of plasmon peak (612–664 nm) was observed due to presence of non-spherical metallic NPs. They conclude that optimized upconversion intensity of Er^{3+}-doped antimony glass occurs with 0.03 wt% of Au. However, further increase of Au concentration results in quenched luminescence, showing the ET from Er^{3+} ion to Au NPs and reabsorption due to SPR of gold NPs [95].

Rivera et al. [130] showed that by exciting the Er^{3+} ions doped in tellurite glass containing silver NPs upon 980 nm laser, a blue shift occurs in the peaks of broadband emission (~1.55 μm). The modification of Stark energy levels (blue shift) was attributed to oscillator strengths of NPs which results in the ET from NP to Er^{3+} ions. The small SPR peaks were revealed in an Er^{3+}-free Ag NP-doped tellurite glass centered at 479 and 498 nm for 3 and 6 h annealed samples, respectively. The peaks in XRD pattern of tellurite glass containing silver NPs ($2\theta = 44.8 \pm 0.4°$, where $d' = 2.0231 \pm 0.0169$ Å) belongs to the (hkl – 200) diffraction planes of Ag crystals (JCPDS Card File No. 4–0783.). Slight increase in FWHM and intensity of broadband emission of Er^{3+} ion (~1.55 μm) were observed by increasing the annealing time interval. The lifetime of 1.55 μm decreases by introduction of Ag NPs in their system comparing to Er^{3+}-single-doped tellurite glass. In another study, Rivera et al. [131] showed that the lifetime of $^4I_{13/2}$ level increases due to presence of heat-treated gold NPs. Therefore, the upconversion luminescence ($^2H_{11/2} \rightarrow {}^4I_{13/2}$, 805 nm) upon the 980 nm excitation enhances drastically (~75 times) due to presence of annealed Au NPs up to 7.5 h. The large enhancement is attributed to LSPR of Au NPs which is located at 800 nm, as evidenced in UV-Vis-IR absorption spectrum, which modifies the local electric field through an electric coupling by Er^{3+} ions. The XRD peaks of Au NPs are revealed at $2\theta = 38.3 \pm 0.4°$ and $44.6 \pm 0.3°$, respectively corresponding to (111) and (200) diffraction planes of gold [131].

de Campos et al. [132] investigated the Er^{3+}-doped bismuth-tungsten-tellurite glasses containing silver NPs and heat-treated for 1, 24, 48 and 72 h. The maximum enhancement in upconversion luminescence were observed for sample with 24 h of heat-treatments, due to presence of NPs with average size of 35 nm. There was no plasmon peak reported in this study, however, they conclude that a tail on the absorption spectrum in the blue region belongs to SPR which is not clearly observable due to small amount of silver NPs.

Amjad et al. prepared the Er^{3+}-doped magnesium phosphate [133, 134] and magnesium-tellurite [135] glasses embedding silver NPs. Introduction of silver NPs with average size of 37 nm enhanced the upconversion intensities of Er^{3+} ions under 797 nm excitation wavelength by a factor of 2.04 and 1.99 for 540 and 634 nm emission bands, respectively. SPR band of Ag NPs was observed in an Er^{3+}-free phosphate glass to be centered at 528 nm [133]. The enhancement is mainly attributed to enhanced local field and partly discussed in terms of ET from Ag NPs to Er^{3+} ions. HR-TEM imaging revealed the cubic closed pack structure of the silver NPs as the lattice constant were measured to be 2.13 Å which is attributed to d_{200} crystallography characterize of silver ($d_{200} = 2.05$Å, JCPDS No.030931).

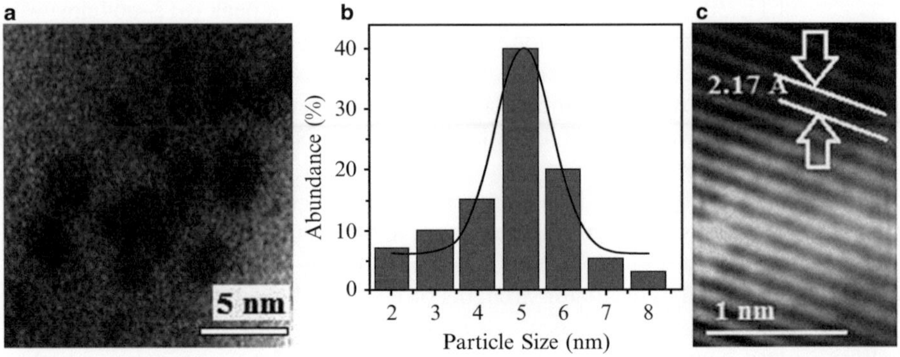

Fig. 14.8 (**a**) TEM, (**b**) size abundance, and (**c**) HR-TEM microscopic images of silver NPs embedded phosphate glass having average size of about 5 nm and grown along (200) crystallographic plane (Micrograph is taken from [134])

The effect of heat-treatment on Ag NPs-Er^{3+}-co-doped phosphate glass is also investigated [134]. The enhancement in visible bands up to 2.2 times was reported due to annealing the glasses up to 40 h at 300 °C. Comparing to their previous report [133], the plasmon peak shows a blue-shift (observed at 488 nm) and the size of NPs are clearly smaller, about 5 nm in average. However, HR-TEM image confirms the presence of Ag NPs with $d_{200} = 2.17$ Å. In the case of Er^{3+}-doped magnesium tellurite glasses, the addition of 0.5 mol% Ag NPs resulted in more efficient enhances in upconversion emission up to 3.33 times for the red emission after 24 h of annealing [135]. The silver NPs with average size of 12 nm were observed with SPR band located at ~534 nm. However, it is reported in their work that luminescence intensities in whole visible range quenched due to over-heat-treatments. The ET from Er^{3+} ions to Ag NPs and reabsorption by SPR were mentioned as main explanations for such quenches (Fig. 14.8).

The effect of silver NPs on the optical and structural properties of Er^{3+}-doped zinc tellurite glass is investigated by Dousti et al. [96, 106, 136]. The absorption bands of Er^{3+} ions are located at 445, 488, 522, 654, 800, 976 and 1526 nm and ascribed to the electric transitions from the $^4I_{15/2}$ ground state to $^4F_{3/2}/^4F_{5/2}$, $^4F_{7/2}$, $^2H_{11/2}$, $^4F_{9/2}$, $^4I_{9/2}$, $^4I_{11/2}$ and $^4I_{13/2}$ excited states, respectively. The introduction of Ag NPs enhanced the upconversion emissions of green and red bands centered at 520, 550 and 640 nm by 4–6 times. The enhancement is attributed to the presence of silver NPs with average size ~10 nm and with SPR localized at 522 nm. By introduction of 0.5 mol% of Ag NPs with an average size of 12 nm, 3.5-fold enhancement was observed for green emission ($^2H_{11/2} \rightarrow {}^4I_{15/2}$) due to formation and growth of NPs after 8 h of annealing (Fig. 14.9) [136]. After 2 h annealing, two peaks in UV-Vis-IR spectra were attributed to SPR, centered at 550 and 580 nm. After 8 h annealing, they observed three SPR peaks which are ascribed to different modes of oscillating particles. In the case of 1 mol% of Ag NPs in the same glassy system, enhances up to 6.5 folds were observed for upconversion emissions after 4 h annealing at temperatures above the T_g. Average size of manipulated Ag NPs

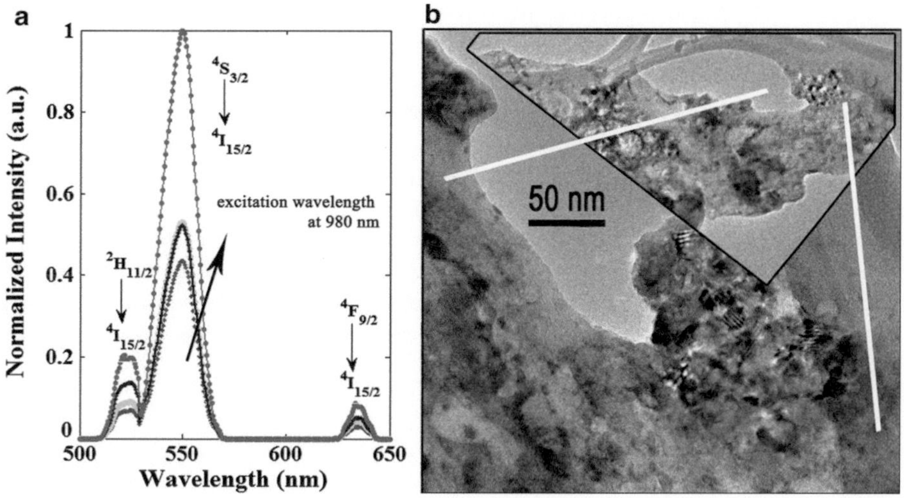

Fig. 14.9 (**a**) Upconversion emission of Er^{3+}-doped tellurite glasses is enhanced in presence of silver nanoparticles and heat-treatment up to 20 h (Adopted from [104]) (**b**) TEM image of silver nanoparticles with average size of about 12 nm which are formed by heat-treatment up to 8 h (Micrograph is taken from [136])

were 14 nm, with two SPR bands at 560 and 594 nm, an indicative of formation of non-spherical NPs.

Qi et al. [137] showed that presence of silver NPs enhances the 1.53 μm emission of Er^{3+}-doped TeO_2-Bi_2O-TiO_2 glasses under 980 nm LED excitation. SPR band is observed at 537 nm and size of NPs was found to be ~13 nm. Er^{3+}-doped lead-tellurite glass containing silver ions (using Ag_2O) and silver nanoparticles (using AgNPs) are also studied by Culea et al. [138]. The structural and optical properties of these sets of glasses are investigated using different techniques. However, there is no evidence of existence of Ag NPs using XRD, UV-Vis absorption and photoluminescence spectroscopic techniques.

The influence of gold NPs on the upconversion emission and Judd-Ofelt parameters of Er^{3+}-doped tellurite glass is investigated by Awang et al. [139, 140] and Sazali et al. [141]. However, the correlation of upconversion and Judd-Ofelt parameters is still not clearly understood. Refractive index, density, the quality factor, and thermal stability of this glassy system are also increased by addition of Au content. Upon heat-treatments at various temperatures, the glasses showed further enhancements of upconversion luminescence due to formation of non-spherical Au NPs [142]. The effect of the heat-treatment duration on the same glass shows improvement of green emission [143]. Surface plasmon band of silver NPs in the bismuth glass is observed at 555 nm [144]. In this glass, the infrared emission of Er^{3+} ions centered at 1554 nm experienced an enhancement in the order of 7.2 times due to local filed enhancement by SPR. Judd-Ofelt intensities parameters of these glasses are also increased by addition of Ag NPs. The enhancement of upconversion luminescence of Er^{3+} ions in zinc boro-tellurite glass is also reported [145]. Surface plasmon band is observed at 630 nm for silver NPs with average size of 4.5 nm as captured by TEM imaging for 0.1 mol% of silver NPs doped tellurite glass. In another study, the luminescence of Er^{3+} ions enhanced by 4 times due to introduction of silver NPs which are embedded through an heat-treatment of borate glass, where SPR band is observed at 410 nm [146]. They showed that extra heat-treatments results in decrease of the size of NPs, as observed by a blue-shift (380 $\leftarrow$ 410 nm) in SPR band of Ag NPs.

Similar to above-mentioned reports, the effect of noble metallic NPs on the optical properties of Er^{3+}-doped in various glasses are investigated wide and large [147–150]. The effect of gold and silver NPs on the upconversion emissions of Er^{3+}/Yb^{3+} co-doped glasses are also reported by different authors, and evident enhancements are observed [151–154].

14.4.3 Nd^{3+}-Doped

Nd^{3+}-doped antimony glass embedding $gold_{core}$@$silver_{shell}$ NPs are studied by different spectroscopic techniques [155]. XRD and SAED showed the formation of core-shell NPs by diffraction patterns along the (111) and (200) crystal planes. TEM imaging confirmed the presence of NPs by average size of about 22–107 nm. The plasmon peaks of NPs are observed by UV-Vis-IR absorption spectroscopy in the range of 532–675 nm, which showed a red-shift by increasing the concentration of $gold_{core}$ NPs. Five-fold intensity enhancement of upconversion emissions of Nd^{3+} ion is observed under 805 nm excitation wavelength. The enhancements of two-major bands centered at 540 nm ($^4G_{7/2} \rightarrow {}^4I_{9/2}$; green) and 649 nm ($^4G_{7/2} \rightarrow {}^4I_{13/2}$; deep-red) are attributed to local field effect induced by plasmonic core-shell metallic NPs.

Frequency upconversion emissions in Nd^{3+}-doped lead-germanate glass containing silver NPs are enhanced under 805 nm excitation wavelength [156]. The absorption band of silver NPs are not observed due to the low concentration of this specie, however, TEM showed the nanoparticles with varying size from 2 to 50 nm. Enhancement of upconversion emissions of Nd^{3+}-doped lead-tellurite glass under 800 nm excitation wavelength is also reported by Dousti [157]. Sixteen-fold enhancement is attributed to large local field in vicinity of silver NPs having an average size of 18 nm. Different interactions and growth process is also described and silver crystalline peaks are observed in XRD patterns of glassy system at $2\theta = 44°$ (Fig. 14.10).

14.4.4 Sm^{3+}-Doped

The effect of noble metallic NPs in antimony glass and glass-ceramic containing Sm^{3+} ions were reported by Som and Karmakar [155, 158–160], and enhanced

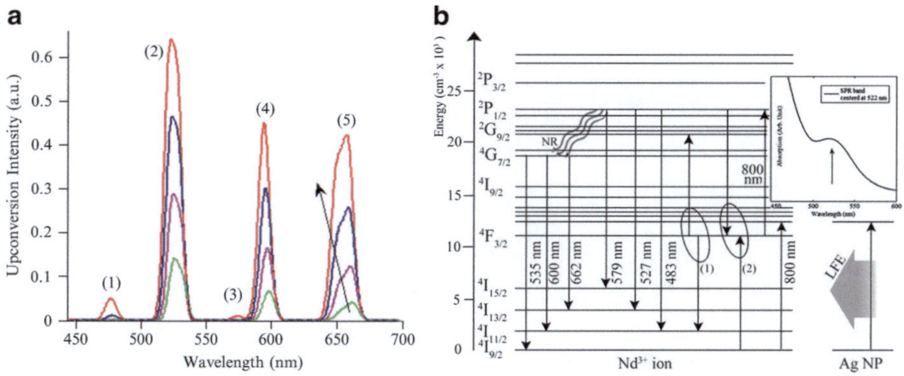

Fig. 14.10 (**a**) Enhancement in upconversion emission of Nd^{3+}-doped tellurite glass by increasing the concentration of silver nanoparticles from 0 to 0.1, 0.5 and 1 mol%. (**b**) SPR band of silver NPs in tellurite glass is centered at 522 nm [157]

luminescence in samarium emissions were observed. The red upconversion luminescence (centered at 636 nm) of Sm^{3+} ions is studied in presence of core-shell bimetallic nanoparticles (Au-Ag NPs) in an antimony glass system. Two-fold enhancement is observed under the excitation at 949 nm and the surface plasmon band is observed in the range of 554–681 nm for various concentrations of metal [102]. Similar results are observed by addition of silver NPs in Sm^{3+}-doped silicate glass, where Ag ions are reduced to Ag neutral particles by antimony oxide as the oxidant agent [161].

Li et al. [162] also investigated the effect of silver NPs on optical properties of Sm^{3+}-doped silicate glasses. Different silver species are formed in the silicate glass by an Ag^{+}-Na^{+} ion exchange process. Although enhancement of luminescence under 270/250 and 355 nm is observed due to energy transfer from Ag^{+} and Ag^{+}-Ag^{+} to Sm^{3+} ions, respectively, the presence of NPs quenches the luminescence under 401 excitation wavelength. They concluded that the competitive absorptions by Sm^{3+} ions and Ag NPs (SPR ~ 420 nm) suppress the luminescence of Sm^{3+} ions.

Jimenez and Sendova [163] studied the effect of silver species (Ag NPs and non-plasmonic clusters) on the luminescence intensity of Sm^{3+}-doped aluminophosphate glass as a function of holding heat-treatment time. The SPR band is not observed up to 50 min of heat-treatments, however the luminescence enhances gradually. The SPR band emerges and intensifies by further heat-treatments up to 120 min, while the luminescence intensity of Sm^{3+} ions quenches progressively. They concluded that the enhancements and quench in the luminescence are associated to the presence of non-plasmonic clusters and NPs, respectively.

In a recent study, the effect of the heat-treatment on the upconversion luminescence of Sm^{3+}-doped borosilicate glasses containing silver NPs are examined [161]. The silver ions are reduced by Sb^{3+} ions as oxidation agents and NPs are grown gradually by increasing the time of heat-treatments up to 20 h. It is stated that

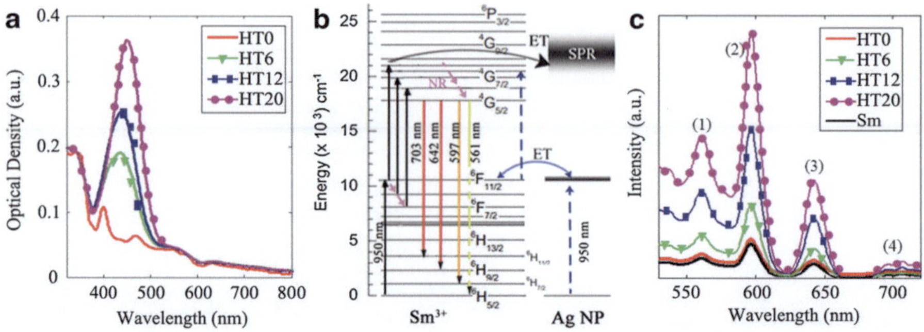

Fig. 14.11 (**a**) UV-Vis-NIR absorption spectra of Sm^{3+}-AgNPs doped borosilicate glass without (HT0) and after 6 (HT6), 12(HT12) and 20 h (HT20) heat-treatments. (**b**) Schematic partial energy level diagram of Sm^{3+} ion in vicinity of silver NP where two-photon absorption mechanism results in upconversion emission of this ion. Probable energy transfer (ET) and nonradiative (NR) decay mechanisms are also illustrated. (**c**) Emission spectra of Sm^{3+}-doped glasses without (Sm) and after incorporation of Ag NPs (Data are adopted from [161])

the further heat-treatments result in a translucent glassy component which is not favorable for optical applications. The surface plasmon band of silver NPs in this glass is observed at ~436 nm which red-shifts to 450 nm by increasing the heat-treatments, indicative of a growth in the size of NPs from 8 to 14 nm (Fig. 14.11).

The effect of annealing temperature on the surface plasmon band position of silver NPs embed in Sm^{3+}-doped lithium borate and sodium borate glasses are also investigated [164]. The SPR band shows a disordered shift in the range of 425–445 nm with varying the annealing temperature from 430 to 510 °C. However, the plasmonic effect of silver is too diminutive and causes a small improvement of emissions from Sm^{3+} ions.

The enhancement in the luminescence of Sm^{3+}-doped tellurite glass by introduction of silver NPs is also given in Ref. [165]. The author showed that under 406 nm excitation wavelength, the emission line at 645 nm enhances up to 130 % by addition of concentration of Ag NPs up to 1 mol%. The increased luminescence of the Sm^{3+} ions are attributed to localized surface plasmon resonance of silver NPs.

The effect of the silver NPs on the Sm^{3+}-doped different media are also available. For instance, Kaur et al. [166] reported on the enhanced luminescence of Sm-complex (PVA) where the Ag NPs were formed by laser irradiation at 355 nm. The SPR band is observed around 402–405 nm and emissions in the visible range are enhanced for both 355 and 400 nm excitation wavelengths. The lifetime of 595 nm emission of Sm^{3+} ions under 355 nm is increased in presence of Ag NPs. Although, there are not many reports on the effect of noble metallic NPs on the Sm^{3+}-doped glasses and glass-ceramics, different authors contributed to study the influence of other NPs (such as CdS NPs) on the luminescence of Sm^{3+} ions doped glasses [167].

14.4.5 *Dy^{3+}-Doped*

The influence of Ag NPs on the luminescence decay of Dy^{3+}-doped aluminophosphate is reported by Jimenez [168]. Although the incorporation of metallic NPs in this system quenched the luminescence of Dy^{3+} ions under 450 nm excitation wavelength, the new concept of "plasmonic diluents" highlighted this work as a worthy publication to think over. The mechanism behind the plasmonic diluents is similar to lowering the effective concentrations of ions, where lower absorptions takes place in the system. On the other word, the resonance excitation of the system results in an energy transfer from ion to particle (silver NPs) which is a detrimental factor for subsequent PL processes, and results in luminescence quenching. Moreover, Jimenez showed that the increasing the volume fraction of the silver NPs by increasing the heat-treatment durations, prolongs the fast and slow decay times of Dy^{3+} ions.

On the other hand, in another study, the upconversion emissions of Dy^{3+}-doped tellurite glasses are enhanced by heat-treated silver NPs [169]. Four-time enhancement is observed for visible emissions under 800 nm excitation wavelength. Silver NPs having an average size of 18 nm are observed and the enhancement in photoluminescence feature is described as the modification of local field due to difference between dielectric constant of medium and metallic particle (Fig. 14.12).

There are not many reports on the influence of noble metallic NPs on the optical properties of Dy^{3+}-doped glasses or glass-ceramics. However, the influence of Cu NPs on the luminescence of Dy^{3+}-doped barium-phosphate glasses is studied [170]. Cu^{2+} and Cu^{+} ions are reduced to Cu^{0} NPs as $Cu^{2+} + Sn^{2+} \rightarrow Cu^{0} + Sn^{4+}$ and $2Cu^{+} + Sn^{2+} \rightarrow 2Cu^{0} + Sn^{4+}$, respectively. The glasses are annealed for 30, 60 and 120 min, but all the observed photoluminescence emissions are quenched under 350 and 450 nm excitation wavelength. The quench is attributed to nonradiative loss of excitation energy in Dy^{3+} ions with an energy transfer from ion to NP. The effect of gold is also reported on the optical properties of Dy^{3+}- and Eu^{3+}-doped silica nanoparticles [171].

14.4.6 *Tm^{3+}-Doped*

Assumpcao et al. [172] studied the upconversion emission of Tm^{3+}-doped zinc tellurite glasses containing silver NPs. In the latter work, they investigated the infrared-to-visible and infrared-to-infrared upconversion process under 1050 nm excitation wavelength and the observed enhancements of luminescence is attributed to the increased local field by silver NPs after heat-treatments. Assumpcao et al. [173] also studied the frequency upconversion emissions from Tm^{3+}-Yb^{3+} co-doped germanate glasses embedding silver NPs. They showed that infrared (980 nm)-to-visible (480 nm) upconversion emission in the current system is due to the energy transfer from Yb^{3+} ions to Tm^{3+} ions. They concluded that the

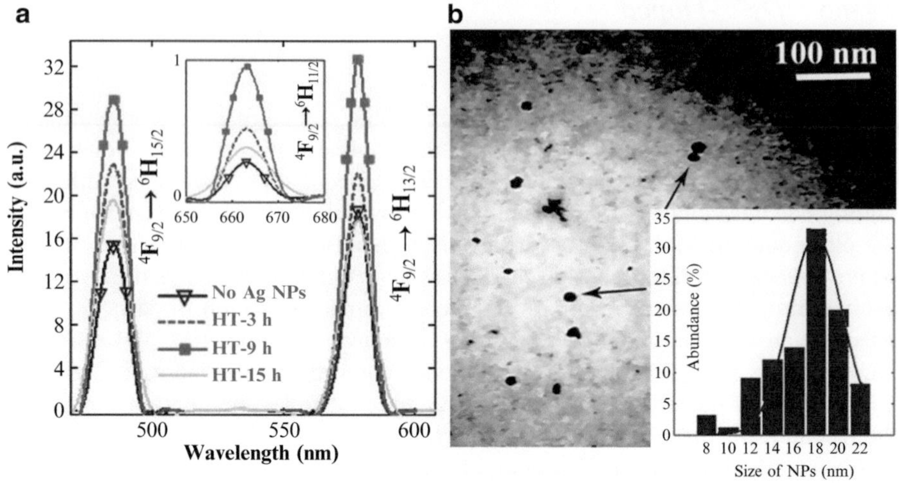

Fig. 14.12 (**a**) Upconversion emissions of Dy^{3+}-doped tellurite glasses are enhanced and quenched after 9 and 15 h of heat-treatments (HT), respectively, under 800 nm excitation wavelength. (**b**) TEM image of silver NPs embedded tellurite glass after 9 h heat-treatments with an average size of about 18 nm [169]

absorption of two, three, and four photons result in upconversion emissions at 800, 652–477 and 542 and 455 nm, respectively. Presence of silver NPs is confirmed by TEM technique and NPs with average size of about 10 nm (isolated) and 80 nm (aggregated) are captured. Intensity of the upconversion emissions at vicinity of SPR band are enhanced up to 30 %. Moreover, Assumpcao et al. [174] showed that the SPR band for this glassy system can be observed by annealing the samples at higher temperature ($T > 480$ °C) for 6 h. As reported, the intensity of SPR absorption band and upconversion emissions increases by increasing the heat-treatment temperature up to 540 °C.

The influence of silver NPs on upconversion emission of Tm^{3+}-Yb^{3+} co-doped zinc tellurite glass is also reported [175]. However, there is no SPR band observed for this glass system up to 72 h of heat-treatment at 325 °C. Upconversion emissions in this system are associated to the energy transfer from Yb^{3+} ions to Tm^{3+} ions, with only 2 and 3-photons absorption mechanism for bands at infrared and visible regions, respectively. Silver NPs enhances the upconversion emissions in order of 300 % under 980 nm excitation wavelength. Kassab et al. [176] reported that the large Judd-Ofelt intensity parameters of Tm^{3+} ions ($\Omega_2 = 15.65 \times 10^{-20}$ cm^2) in $Tm^{3+}-Yb^{3+}$-co-doped zinc-tellurite glass containing silver NPs can nominate them as optically stimulated quantum electronic devices and optically operated fibers.

Singly Tm^{3+}-doped PbO-GeO_2 oxide glasses containing silver NPs are also investigated by the latter group [177]. Upconversion emissions are enhanced in presence of silver NPs under 1050 excitation wavelength for heat-treatment up to 24 h. Further heat-treatments (up to 72 h) resulted in a quenching in luminescence spectra. Moreover, they concluded that the continuous heat-treatments stimulate the NPs to aggregate while non-continues heat-treatments with a step of 12 h prevents the aggregation of NPs. In another study, Qi et al. [178] observed significant enhancement in emissions of Tm^{3+} ions due to participation of Ag NPs in the bismuth germanate glasses.

Triply doped Tm^{3+}-Er^{3+}-Yb^{3+} lead germanate oxide glasses containing silver NPs is also investigated [179]. The SPR band is observed at 400–500 nm region and the broad FWHM of this band is attributed to presence of NPs with different sizes and shapes. For excitation at 980 nm, enhancements of about 60 % are observed for five emission bands. Two bands are associated to Tm^{3+} ions which are centered at 477 and 652 nm, and three bands located at 530, 550 and 660 nm are originated form Er^{3+} ions. Energy transfers among these species are discussed in latter reference. Figure 14.13 shows a schematic partials energy level diagram of this tri-doped glass. The same doping system embedded in a tellurite glass containing silver NPs is also reported by the same group [180] where efficient mid-infrared emissions from Er^{3+} (1.55 μm) and Tm^{3+} ions (1.86 μm) are observed. Moreover, efficient energy transfer from Yb^{3+} ions to Er^{3+} and Tm^{3+} ions are concluded by time-resolved luminescence investigations. The decay lifetime of $^4F_3 \rightarrow {}^3H_6$ (Tm^{3+}) shortened from 3.4 to 2.9 ms by addition of silver NPs and lifetime of $^4I_{13/2} \rightarrow {}^4I_{15/2}$ (Er^{3+}) decreased from 2.4 to 1.7 ms. $Tm^{3+}/Yb^{3+}/Er^{3+}$ tri-doped oxyfluorogermanate glasses containing silver NPs are also reported [181]. The authors showed that the introduction of silver NPs reduces the glass thermal stability as well as glass transition and crystallization temperatures. A broad absorption band around 400–500 nm is attributed to surface plasmon band of silver which is extended up to 800 nm for further annealing time intervals. Silver NPs are grown from 4 to 10 nm by increasing the annealing time from 34 to 51 h. The intensity of all the observed upconversion emissions (emissions at 476, 524, 546 and 658 nm, and excitation at 980 nm) increases by increasing the annealing time up to 34 h, while it quenches for further heat-treatments. All the emissions originate from a two-photon absorption process. The effect of Ag_2O concentration on the upconversion emission of the $Tm^{3+}/Yb^{3+}/Er^{3+}$ doped oxyfluorogermanate glasses is also studied by the same research group [13]. They showed that by increasing the molar concentration of Ag_2O up to 1.5 mol%, the upconversion luminescence of both Tm^{3+} and Er^{3+} ions increases, whereas further introduction of Ag NPs result in the quenching of emissions. The presence of non-spherical NPs in this study is emerged by appearance of two SPR bands; transverse and longitudinal modes located at 344 and 425 nm, respectively.

14.4.7 Tb^{3+}-Doped

A photoluminescence enhancement of about 1.6 times is observed in Tb^{3+}-doped silicate glass due to incorporation of silver NPs [182]. Under excitation at 488 nm, the emissions of Tb^{3+} ions at visible region (537, 578 and 612 nm) are enhanced

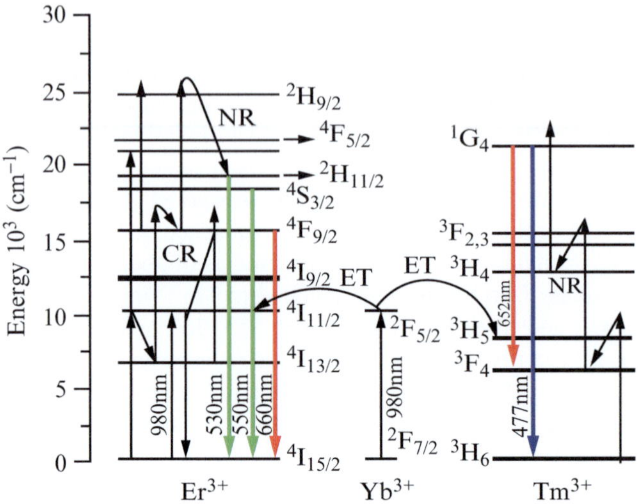

Fig. 14.13 Schematic partial energy level diagram of Tm^{3+}-Er^{3+}-Yb^{3+}tri-doped glasses showing upconversion emissions, energy transfer (ET), cross-relaxation (CR) mechanisms and nonradiative decays (NR) (Adapted from [179])

thanks to the increased local field by metallic silver NPs. Tb^{3+}-doped silicate glass containing silver NPs are also investigated by Piasecki et al [183, 184]. SPR band is observed at 480 nm and the luminescence lines are enhanced under 325 nm excitation wavlenegth. Maximum enhancement of 1.8 times is obtained for sample containing 0.5 mol% of silver after 3 h of heat-treatments.

The SPR band of silver NPs embedded Tb^{3+}-doped silicate glass is located at 420 nm and its intensity increases by increasing the concentration of silver NPs [185]. The emissions of Tb^{3+} ions in the range of 400–700 nm are enhanced for the sample containing 3 mol% of $AgNO_3$ and quenched in the glass sample embedding 5 mol% of silver NPs. The enhancement of about 35 % in emission intensity of Tb^{3+} ions (543 nm) in a silica system is also observed as the lifetime of this fluorescence reduces. Such fluorescence enhancement and increased radiative decay rates are attributed to the increased local filed induced by silver NPs [186].

Verma et al. [187] investigated the effect of silver NPs on the fluorescence of Tb^{3+}-doped aluminosilicate glass. Silver NPs are prepared by laser ablation in distilled water, and embedded in the glass following a sol-gel technique. Surface plasmon resonance band of silver is observed at 404 nm, and their radius is estimated to be ~6.48 nm by measuring the FWHM of SPR band. The lifetime of the $^5D_4$ level of terbium is increased from 310 to 420 μs in the presence of silver NPs. The emission intensity of rare earth ions enhances up to 100 % which is a result of the energy transfer from the excited silver NPs to Tb^{3+} ions.

Sodium-lead-zinc-tellurite glass doped with Tb^{3+}/Eu^{3+} ions and silver NPs showed a plasmon peak centered at 490 nm after a long heat-treatment [188]. The large enhancement in luminescence of Eu^{3+} ions (centered at 590 and 614 nm) is attributed to simultaneous contributions of (i) energy transfer among Eu^{3+}-Tb^{3+} ions and (ii) intensified local field in vicinity of Ag NPs and Eu^{3+} ions. They observed that the integrated emission of Eu^{3+} ions increases by increasing the annealing time of sample up to 62 h.

In addition, Tb^{3+}-Yb^{3+} doped aluminosilicate glasses containing silver NPs are investigated by Pan et al. [189]. The SPR peak is observed at 420 nm which shows no shift by further annealing. However, the presence of silver NPs with size of about 3–7 nm is confirmed by TEM. Upconversion luminescence of Tb^{3+} ions is enhanced after 5 h of heat-treatments, but longer treatments quenched the intensity of PL spectrum. Moreover, the normal luminescence under 488 nm excitation wavelength is quenched due to the quantum cutting effect.

14.4.8 Pr^{3+}-Doped

Kassab et al. [94] reported that the luminescence characteristics of Pr^{3+} in lead-tellurite glass enhances due to presence of silver NPs with average size around 3.5 nm, which are formed by annealing at 350 °C for 7 h. Rai et al. also studied the influence of silver NPs on the optical properties of Pr^{3+}-doped zinc-tellurite glasses [190]. Upconversion emission of Pr^{3+} ions centered at ~482 and ~692 nm are observed under the excitation with a nanosecond laser operating at 590 nm. An enhancement of about 120 % is achieved for the integrated intensities of those emissions after heat-treating the sample containing silver NPs up to 40 h. Enhancement in luminescence of Pr^{3+}-doped PbO-GeO_2 glasses are also observed in presence of isolated silver particles and aggregated silver NPs with an average diameter of 2 nm and less than 100 nm, respectively [191]. They showed that the amplitude of surface plasmon band of silver NPs is centered at 464 nm in this glass and increases by increasing the heat-treatment durations. Both energy transfer mechanism from nanoparticle to ion and modified local filed around the Pr^{3+} ions are discussed as the factors of enhancement, however the exact contribution of each phenomenon was not concluded. Enhancement in emissions from Pr^{3+} ions doped zinc-tellurite glass containing silver NPs is also reported [192]. The influence of large local field on the Pr^{3+} ions is discussed as the major factor for enhancements of visible emission under both 470 nm (Stokes) and 586 (anti-Stokes) excitation wavelengths.

Furthermore, the effect of silver NPs on the luminescence of Pr^{3+}-doped borate glass is investigated [117]. Both excitation and luminescence spectra showed enhancement in the emission line centered at 558 nm and excitation at 442 nm, respectively. The increased intensity of luminescence is attributed to "the resonance between the energy band with the levels of the Pr^{3+} ions". As mentioned before (see 4.1), the formation of silver NPs in this system is slower than its counterpart Eu^{3+}-doped glass, and the size of NPs are 4 times large in the case of Eu^{3+}-doped samples.

14.4.9 Ho^{3+}-Doped

The enhancement in the upconversion emissions of Ho^{3+}/Yb^{3+} co-doped tellurite glass is also observed [193] by incorporation of silver NPs which are formed by addition of $AgNO_3$ to the glassy host, melted at 900 °C for 30 min and annealed above its glass transition temperature. Silver NPs with average diameter size of about 3–12 nm show a SPR absorption band centered at 560 nm and contribute to ~2.5 times enhancement in intensity of emissions at 546 and 657 nm under 980 nm excitation wavelength.

Although there are not many reports on the effect of noble metallic NPs on the luminescence of Ho^{3+}-doped glasses, some research group investigated the effect of nanocrystals on such systems. For instance, Zhang et al. [194] reported on the enhanced 2.0 μm emission of Ho^{3+}-Tm^{3+}-co-doped glass ceramics containing BaF_2 nanocrystals. Efficient energy transfer from Tm^{3+} to Ho^{3+} ions is attributed to low phonon energy environment and reduced ionic distance of lanthanide species after incorporation into nanocrystal embedding glass ceramics.

14.5 Summary

Improving the optical and thermal properties of glasses, glass-ceramics and crystals is of supreme importance in order to develop the smart optical devices, such as solid state lasers, amplifiers, undersea cameras, telecommunication devices, color displays and so on. The aim of this chapter was to revisit the current achievements on the state of art in understanding the role of metallic NPs on the optical properties of some rare earth-doped oxide glasses. Until now, various studies have been done to enhance the properties of glasses containing RE ions. Incorporation of larger concentrations of RE ions, introduction of second dopant, thermal treatments, different synthesizing methods and varying the glass host matrix are among the common techniques in order to modify the environment of the RE ions, which significantly can alter its optical properties. Room temperature optical properties of RE-doped glasses containing metallic NPs synthesized by a melt quench technique have been studied by many researchers. Analytical techniques such as UV-Vis absorption and photoluminescence spectroscopy have been used to optically characterize the glasses and the data showed that Stokes and anti-Stokes luminescence of most of the proposed glassy systems has improved by addition of metallic NPs.

In this chapter, initially, a short introduction on the importance of material science, glass and glass-ceramic technology was explained in the first section. It was followed by a chronological study on the history of glasses during the last centuries. This section was continued by a review on spectroscopic properties of rare earth ions in materials in general and glasses in particular. The developed theories and models on optical properties of rare earth ions and interactions between light and ions and ion-ion interactions were explained. The energy level diagram of electronic configuration of rare earth ions and probable mechanisms, such as energy transfer and relaxation processes were summarized. Section 14.3 was devoted to study the interaction of light with metallic NPs in the dielectric media. The effect of shape and size of NPs on the optical behavior of the system are given. The custom methods to prepare and observe the NPs and widespread applications of plasmonic phenomenon were asserted. Finally, recent results on the influence of metallic NPs on the glasses containing RE ions are listed in Sect. 14.4. Based on the results, the potentiality and applicability of these studies are concluded in current section and the new achievements and importance of the research is ascertained. However, there are many characteristic analyses which are not provided in the literature. Different preparation methods and environmental situations may alter the functionality and potentiality of the samples and several suggestions are listed for further research as follows:

(i). Structural properties of the studied glass samples can further be investigated through Raman spectroscopy, NMR, and ESR.
(ii). Thermal properties can be analyzed to correlate the nucleation of NPs with thermal characteristics of glasses, especially at the heat-treatment temperatures.
(iii). Trivalent rare earth ion possesses various excited states. The optical investigation while doped with different co-dopants can be interesting where the system can provide white-light emissions through the second and third non-linear processes, and energy transfers.
(iv). Various preparation techniques are suggested by different authors to synthesize a glass. The usage of different melting and annealing temperature, and pouring in different environments may also modify the structural and consequently the thermal and optical properties of glasses.
(v). Ceramics show completely different structural, thermal and optical properties with respect to amorphous glass system. Therefore, the preparation of ceramics and glass-ceramics with the same compositions can be also recommended to observe the behavior of a noble system. One may arise for preparation of thin films by same composition.

As the final point, it is of utmost importance to recall that the controlling the size and shape of the NPs is an imperative factor to establish an optically favorable glassy system with enhanced optical properties. As revealed in this chapter, the incorporation of metallic NPs can enhance the optical properties of the RE-doped oxide glasses. Therefore, further investigations are suggested to provide a suitable glassy system and to correlate the controlling parameters with size, shape and performance of the NPs in such optical materials.

Acknowledgments The experimental data given in this work has been collected by the authors during their doctoral degree and later while working as a post-doctoral fellowship or a professorship. The authors would like to thanks many people who has contributed in discussing the results, providing funds and grants, giving feedbacks and commenting on the work to improve it in both terms of scientific and publication subjects. Especial thanks to Prof. Dr. Md. Rahim Sahar for his supports, and to Prof. Dr. Andrea de Camargo for her valuable comments. Moreover, last but not least, we would like to acknowledge the kind invitation form Prof. Chris. D. Geddes for giving us such a great opportunity to contribute in this Book.

References

1. Zanotto ED (2010) A bright future for glass-ceramics. Am Ceram Soc Bull 89:19–27
2. Tait H (ed) (1991) Five thousand years of glass. The British Museum Press, London
3. Plodinec MJ (2000) Borosilicate glasses for nuclear waste immobilisation. Glass Technol 41:186–192
4. Brian Starling L, Stephan JE (2002) Calcium phosphate microcarriers and microspheres. US Patent 6210715 B1, 6,358,532 B2
5. Clinton JM, Coffeen WW (1984) Low melting glasses in the system B_2O_3-ZnO-CaO-P_2O_5. Am Ceram Soc Bull 63:1401–1404
6. Daimer J, Paschke H (1990) Glass sealant containing lead borate glass and fillers of mullite and cordierite. US Patent 5145803 A, 5,145,803
7. Koudelka L, Mošner P (2001) Study of the structure and properties of Pb–Zn borophosphate glasses. J Non Cryst Solids 293–295:635–641
8. Stokowski SE, Saroyan RA, Weber MJ (1981) Nd-doped laser glass spectroscopic and physical properties. Lawrence Livermore National Laboratory, University of California, Livermore
9. Stocker HJ (1969) Bulk and thin film switching and memory effects in semiconducting chalcogenide glasses. Appl Phys Lett 15:55
10. Poulain M (1983) Halide glasses. J Non Cryst Solids 56:1–14
11. Baldwin CM, Almeida RM, Mackenzie JD (1981) Halide glasses. J Non Cryst Solids 43:309–344
12. El-mallawany RAH (2002) Tellurite glasses handbook: physical properties and data. CRC Press, Boca Raton
13. Hu Y, Qiu J, Song Z, Zhou D (2014) Ag_2O dependent up-conversion luminescence properties in $Tm^{3+}/Er^{3+}/Yb^{3+}$ Co-doped oxyfluorogermanate glasses. J Appl Phys 115:083512
14. www.scopus.com. Accessed 15 July 2014
15. Mauro JC, Zanotto ED (2014) Two centuries of glass research: historical trends, current status, and grand challenges for the future. Int J Appl Glass Sci 15:313–327
16. Thiel CW, Sun Y, Cone RL (2002) Photonic materials and devices progress in relating rare-earth Ion 4f and 5d energy levels to host bands in optical materials for hole burning, quantum information and phosphors. J Mod Opt 49:2399–2411
17. Jha A, Richards B, Jose G, Teddy-Fernandez T, Joshi P, Jiang X, Lousteau J (2012) Rare-earth Ion doped TeO_2 and GeO_2 glasses as laser materials. Prog Mater Sci 57:1426–1491
18. Pan Z, Sekar G, Akrobetu R, Mu R, Morgan SH (2012) Visible to near-infrared down-conversion luminescence in Tb^{3+} and Yb^{3+} Co-doped lithium–lanthanum–aluminosilicate oxyfluoride glass and glass-ceramics. J Non Cryst Solids 358:1814–1817
19. Dai S, Yu C, Zhou G, Zhang J, Wang G, Hu L (2006) Concentration quenching in erbium-doped tellurite glasses. J Lumin 117:39–45
20. Pokhrel M, Kumar GA, Balaji S, Debnath R, Sardar DK (2012) Optical characterization of Er^{3+} and Yb^{3+} Co-doped barium fluorotellurite glass. J Lumin 132:1910–1916
21. Malta OL, Santa-Cruz PA, De Sá GF, Auzel F (1985) Fluorescence enhancement induced by the presence of small silver particles in Eu^{3+} doped materials. J Lumin 33:261–272
22. De Araújo CB, Oliveira TR, Falcão-Filho EL, Silva DM, Kassab LRP (2013) Nonlinear optical properties of PbO–GeO_2 films containing gold nanoparticles. J Lumin 133:180–183
23. Hufner S (1978) Optical spectra of rare earth compounds. Academic, New York

24. Becker PC, Olsson NA, Simpson JR (1999) Erbium-doped fiber amplifiers. Academic, San Diego
25. Maheshvaran K, Marimuthu K (2012) Concentration dependent Eu^{3+} doped boro-tellurite glasses – structural and optical investigations. J Lumin 132:2259–2267
26. Zheng H, Gao D, Fu Z, Wang E, Lei Y, Tuan Y, Cui M (2011) Fluorescence enhancement of Ln^{3+} doped nanoparticles. J Lumin 131:423–428
27. Judd BR (1962) Optical absorption intensities of rare-earth ions. Phys Rev 127:750–761
28. Ofelt GS (1962) Intensities of crystal spectra of rare-earth ions. J Chem Phys 37:511–520
29. Walsh BM (2006) Judd-Ofelt theory: principles and practices. In: Di Bartolo, Baldassare, Forte, Ottavio (eds) Advances in spectroscopy for lasers and sensing. Springer, Netherlands, pp 403–433
30. Carnall WT, Fields PR, Rajnak K, Ranjak K (1968) Electronic energy levels in the trivalent lanthanide aquo ions. I. Pr^{3+}, Nd^{3+}, Pm^{3+}, Sm^{3+}, Dy^{3+}, Ho^{3+}, Er^{3+}, and Tm^{3+}. J Chem Phys 49:4424
31. Weber MJ, Myers JD, Blackburn DH (1981) Optical properties of Nd $^{3+}$ in tellurite and phosphotellurite glasses. J Appl Phys 52:2944–2946
32. Rahman HU (1972) Optical intensities of trivalent erbium in various host lattices. J Phys C Solid State Phys 5:306–315
33. Yang J, Dai S, Dai N, Wen L, Hu L, Jiang Z (2004) Investigation on nonradiative decay of 4I13/2-4I15/2 transition of Er^{3+} -doped oxide glasses. J Lumin 106:9–14
34. Chen Y, Huang Y, Huang M, Chen R, Luo Z (2004) Spectroscopic properties of Er^{3+} ions in bismuth borate glasses. Opt Mater 25:271–278
35. Nandi P, Jose G (2006) Spectroscopic properties of Er^{3+} doped phospho-tellurite glasses. Phys B Condens Matter 381:66–72
36. Snitzer E (1961) Optical maser action of Nd^{3+} in a barium crown glass. Phys Rev Lett 7(12):444–446
37. Kao KC (1966) Dielectric-fibre surface waveguides for optical frequencies. Proc Inst Electr Eng London 113:1151
38. Sandoe JN, Sarkies PH, Parke S (1972) Variation of Er^{3+} cross-section for stimulated emission with glass composition. J Phys D Appl Phys 5:1788
39. Stone J, Burrus CA (1973) Neodymium-doped silica lasers in end-pumped fiber geometry. Appl Phys Lett 23:388–389
40. Svendsen SM (1997) First installation of WDM, optical ADM and optical in-line amplifiers on long-haul cable links in Norway. Telektronikk 3:115
41. Snitzer E (1973) Lasers and glass technology. Ceram Bull 52:516–525
42. Izumitani T, Toratani H, Kuroda H (1982) Radiative and nonradiative properties of neodimium doped silicate and phosphate glasses. J Non Cryst Solids 47:87–99
43. Wang JS, Vogel EM, Snitzer E (1994) Tellurite glass: a news candidate for fiber devices. Opt Mater 3:187–203
44. Li-Yan Z, Li-Li H (2003) Evaluation of broadband spectral properties of erbium-doped aluminium flourophosphate glass. Chin Phys Lett 20:1836–1837
45. Auzel F (1990) Upconversion processes in coupled Ion systems. J Lumin 45:341–345
46. Dexter DL (1953) A theory of sensitized luminescence in solids. J Chem Phys 21:836–851
47. Kaplyanskii AA, MacFarlane RM (eds) (1987) Spectroscopy of solids containing rare-earth ion. North-Holland, Amsterdam
48. Barber DJ, Freestone IC (1990) An investigation of the origin of the colour the Lycurgus Cup by analytical electron microscopy. Archaeometry 32:33–45
49. Faraday M (1857) The Bakerian lecture : experimental relations of gold (and other metals) to light. Philos Trans R Soc London 147:145–181
50. Binning G, Rohrer H, Gerber C, Weibel E (1982) Surface studies by scanning tunneling microscopy. Phys Rev Lett 49:57–61
51. Binnig G, Rohrer H (1982) Vakuum Tunnel Mikroskop. Helv Phys Acta 55:128

52. Knight WD, Clemenger K, de Heer WA, Saunders WA, Chou MY, Cohen ML (1984) Electronic shell structure and abundances of sodium clusters. Phys Rev Lett 52:2141–2143
53. Brigger I, Dubernet C, Couvreur P (2002) Nanoparticles in cancer therapy and diagnosis. Adv Drug Deliv Rev 54:631–651
54. Hirsch LR, Stafford RJ, Bankson JA, Sershen SR, Rivera B, Price RE, Hazle JD, Halas NJ, West JL (2003) Nanoshell-mediated near-infrared thermal therapy of tumors under magnetic resonance guidance. Proc Natl Aced Soc USA 100:13549–13554
55. Storhoff JJ, Mirkin CA (1999) Programmed materials synthesis with DNA. Chem Rev 99:1849–1862
56. Lee K-C, Lin S-J, Lin C-H, Tsai C-S, Lu Y-J (2008) Size effect of Ag nanoparticles on surface plasmon resonance. Surf Coat Technol 202:5339–5342
57. Schwartzberg AM, Grant CD, Buuren TV, Zhang JZ (2007) Reduction of $HAuCl_4$ by Na_2S revisited : the case for Au nanoparticle aggregates and against Au_2S/AuCore/shell particles. J Phys Chem 111:8892–8901
58. Gou L, Murphy CJ (2005) Fine-tuning the shape of gold nanorods. Chem Mater 17:3668–3672
59. Sun Y, Mayers B, Xia Y (2003) Transformation of silver nanospheres into nanobelts and triangular nanoplates through a thermal process. Nano Lett 3:675–679
60. Gonzalez AL, Noguez C (2007) Influence of morphology on the optical properties of metal nanoparticles. J Comput Theor Nanosci 4:231–238
61. Zhang JZ, Noguez C (2008) Plasmonic optical properties and applications of metal nanostructures. Plasmonics 3:127–150
62. Prasad P (2004) Nanophotonics. Wiley, Hoboken
63. Mie G (1908) Beitrage Zur Optik Truber Medien, Speziell Kolloidaler Metallosungen. Ann Phys 3:377–445
64. Wooten F (1972) Optical properties of solids. Academic, New York
65. Shanmukh S, Jones L, Driskell J, Zhao Y, Dluhy R, Tripp RA (2006) Rapid and sensitive detection of respiratory virus molecular signatures using a silver nanorod array SERS substrate. Nano Lett 6:2630–2636
66. Schwartzberg AM, Olson TY, Talley CE, Zhang JZ (2006) Synthesis, characterization, and tunable optical properties of hollow gold nanospheres. J Phys Chem B 110:19935–19944
67. Sun Y, Mayers BT, Xia Y (2002) Template-engaged replacement reaction: a one-step approach to the large-scale synthesis of metal nanostructures with hollow interiors. Nano Lett 2:481–485
68. Xu K, Heo J (2013) Luminescence enhancement of CdS quantum dots in glass by Ag^+ ion exchange. J Am Ceram Soc 96:1138–1142
69. Zoval JV, Stiger RM, Biernacki PR, Penner RM (1996) Electrochemical deposition of silver nanocrystallites on the atomically smooth graphite basal plane. J Phys Chem 100:837–844
70. Henglein A (1993) Chemical and optical properties of small metal particles in aqueous solution. Isr J Chem 33:77–88
71. Quinten M (2011) Optical properties of nanoparticle systems: mie and beyond. Wiley-VCH Verlag & Co. KGaA, Weinheim
72. Smoluchowski MV (1916) Drei Vorträge Über Diffusion, Brownsche Molekularbewegung Und Koagulation von Kolloidteilchen. Phys Z 17:557–599
73. Ostwald W (1900) On the supposed isomerism of red and yellow mercury oxide and the surface tension of solid substances. Z Phys Chem 34:495–503
74. Yu CH, Tam K, Tsang ESC (2009) Chemical methods for preparation of nanoparticles in solution. In: Blackman JA (ed) Handbook of metal physics, 1st edn. Elsevier, Oxford
75. Liu G, Xu S, Qian Y (2000) Nanofabrication of self-assembled monolayers using scanning probe lithography. Acc Chem Res 33:457–466
76. Wang ZL (ed) (2000) Characterization of nanophase materials. Wiley, New York

77. Wang ZL (2000) Transmission electron microscopy of shape-controlled nanocrystals and their assemblies. J Phys Chem B 104:1153–1175
78. Binning G, Quate CF, Gerber C (1986) Atomic force microscope. Phys Rev Lett 56:930–933
79. Fleischmann, M, Hendra PJ, McQuillan AJ (1974) Raman spectra of pyridine adsorbed at a silver electrode. Chem Phys Lett 26:163–166
80. Kneipp K, Wang Y, Kneipp H, Perelman LT, Itzakan I, Dasari RR, Feld MS (1997) Single molecule detection using surface-enhanced raman scattering (SERS). Phys Rev Lett 78:1667–1670
81. Blackie EJ, Le Ru EC, Meyer M, Etchehoin PG (2007) Surface enhanced raman scattering enhancement factors: a comprehensive study. J Phys Chem C 111:13794–13803
82. Nie S, Emory SR (1997) Probing single molecules and single nanoparticles by surface-enhanced Raman scattering. Science 275:1102–1106
83. Messinger BJ, Raben KUV, Chang RK, Barber PW (1981) Local fields at the surface of noble-metal microspheres. Phys Rev B 24:649–657
84. Wang DS, Kerker M (1981) Enhanced raman scattering by molecules adsorbed at the surface of colloidal spheroids. Phys Rev B 24:1777–1790
85. Fleischmann M, Hendra PJ, McQuillan AJ (1974) Raman spectra of pyridine adsorbed at a silver electrode. Chem Phys Lett 26:163–166
86. Moskovits M (2006) Surface-enhanced Raman spectroscopy: a brief perspective. In: Kneipp KM, Kneipp H (eds) Surface-enhanced Raman scattering – physics and applications. Springer, Berlin/New York, pp 1–18
87. Campion A, Kambhampati P (1998) Surface-enhanced raman scattering. Chem Soc Rev 27:241–250
88. Dousti MR, Sahar MR, Amjad RJ, Ghoshal SK, Awang A (2013) Surface enhanced raman scattering and up-conversion emission by silver nanoparticles in erbium–zinc–tellurite glass. J Lumin 143:368–373
89. Amjad RJ, Sahar MR, Dousti MR, Ghoshal SK, Jamaludin MNA (2013) Surface enhanced Raman scattering and plasmon enhanced fluorescence in zinc-tellurite glass. Opt Express 21:14282–14290
90. Som T, Karmakar B (2009) Nanosilver enhanced upconversion fluorescence of erbium ions in Er^{3+}: Ag-antimony glass nanocomposites. J Appl Phys 105:013102–013108
91. Amjad RJ, Sahar MR, Ghoshal SK, Dousti MR, Samavati AR, Riaz S, Tahir BA (2013) Spectroscopic investigation of rare-earth doped phosphate glasses containing silver nanoparticles. Acta Phys Pol A 123:746–749
92. Karmakar B, Som T, Singh SP, Nath M (2010) Nanometal-glass hybrid nanocomposites: synthesis, properties and applications. Trans Indian Ceram Soc 69:171–186
93. Speranza G, Minati L, Chiasera A, Ferrari M, Righini GC, Ischia G (2009) Quantum confinement and matrix effects in silver-exchanged soda lime glasses. J Phys Chem C 113:4445–4450
94. Kassab LR, De Araújo CB, Kobayashi RA, Pinto RDA, Silva DM (2007) Influence of silver nanoparticles in the luminescence efficiency of Pr^{3+} -doped tellurite glasses. J Appl Phys 102:103515
95. Som T, Karmakar B (2009) Enhancement of Er^{3+} upconverted luminescence in Er^{3+}: Au-antimony glass dichroic nanocomposites containing hexagonal Au nanoparticles. J Opt Soc Am B 26:B21–B27
96. Dousti MR, Sahar MR, Ghoshal SK, Amjad SK, Samavati AR (2013) Effect of AgCl on spectroscopic properties of erbium doped zinc tellurite glass. J Mol Struct 1035:6–12
97. Dousti MR, Ghassemi P, Sahar MR, Mahraz ZSA (2014) Chemical durability and thermal astability of Er^{3+}-doped zinc tellurite glass containing silver nanaparticles. Chalcogenide Lett 11:111–119
98. Silva AP, Carmo AP, Anjos V, Bell MJV, Kassab LRP, Pinto RDA (2011) Temperature coefficient of optical path of tellurite glasses doped with gold nanoparticles. Opt Mater (Amst) 34:239–243

99. Adamiv VT, Bolesta IM, Burak YV, Gamernyk RV, Karbovnyk ID, Kolych II, Kovalchuk MG, Kushnir OO, Periv MV, Teslyuk IM (2014) Nonlinear optical properties of silver nanoparticles prepared in Ag doped borate glasses. Phys B Condens Matter 449:31–35
100. Mai HH, Kaydashev VE, Tikhomirov VK, Janssens E, Shestakov MV, Meledina M, Turner S, Van Tendeloo G, Moshchalkov VV, Lievens P (2014) Nonlinear optical properties of Ag nanoclusters and nanoparticles dispersed in a glass host. J Phys Chem C 118:15995–16002
101. Ganeev RA, Ryasnyansky AI, Stepanov AL, Usmanov T (2004) Characterization of nonlinear optical parameters of copper- and silver-doped silica glasses at $\lambda = 1064$ nm. Phys Status Solidi 241:935–944
102. Som T, Karmakar B (2009) Core-shell Au-Ag nanoparticles in dielectric nanocomposites with plasmon-enhanced fluorescence: a new paradigm in antimony glasses. Nano Res 2:607–616
103. Wu Y, Xu T, Shen X, Dai S, Nie Q, Wang X, Song B, Zhang W, Lin C (2011) Effect of silver nanoparticles on spectroscopic properties of Er^{3+} -doped bismuth glass. In: Proceedings of the 6th IEEE Conference on Industrial Electronics and Applications, Beijing, China, pp 1464–1467
104. Dousti MR, Sahar MR, Amjad RJ, Ghoshal SK, Khorramnazari A, Dordizadeh Basirabad A, Samavati A (2012) Enhanced frequency upconversion in Er^{3+}-doped sodium lead tellurite glass containing silver nanoparticles. Eur Phys J D 66:237
105. Rao GV, Shashikala HD (2014) Optical, dielectric and mechanical properties of silver nanoparticle embedded calcium phosphate glass. J Non Cryst Solids 402:204–209
106. Dousti MR, Sahar MR, Ghoshal SK, Amjad RJ, Arifin R (2012) Up-conversion enhancement in Er^{3+}-Ag Co-doped zinc tellurite glass: effect of heat treatment. J Non Cryst Solids 358:2939–2942
107. Lin A, Son DH, Ahn IH, Song GH, Han W-T (2007) Visible to infrared photoluminescence from gold nanoparticles embedded in Germano-silicate glass fiber. Opt Express 15:6374–6379
108. Hayakawa T, Selvan ST, Nogami M (1999) Enhanced fluorescence from Eu^{3+} owing to surface plasma oscillation of silver particles in glass. J Non-Cryst Solids 259:16–22
109. Hayakawa T, Furuhashi K, Nogami M (2004) Enhancement of 5D0-7FJ emissions of Eu^{3+} ions in the vicinity of polymer-protected Au. J Phys Chem B 108:11301–11307
110. De Almeida R, da Silva DM, Kassab LRP, de Araujo CB (2008) Eu^{3+} luminescence in tellurite glasses with gold nanostructures. Opt Commun 281:108–112
111. Kassab LRP, da Silva DS, de Almeida R, de Araújo CB (2009) Photoluminescence enhancement by gold nanoparticles in Eu[sup $^{3+}$ doped GeO_2–Bi_2O_3 glasses. Appl Phys Lett 94:101912
112. Amjad RJ, Dousti MR, Sahar MR, Shaukat SF, Ghoshal SK, Sazali ES, Nawaz F (2014) Silver nanoparticles enhanced luminescence of Eu^{3+}-doped tellurite glass. J Lumin 154:316–321
113. Dousti MR, Sahar MR, Rohani MS, Samavati A, Mahraz ZSA, Amjad RJ, Awang A, Arifin R (2014) Nano-silver enhanced luminescence of Eu^{3+}-doped lead tellurite glass. J Mol Struct 1065–1066:39–42
114. Jiménez JA, Lysenko S, Liu H (2008) Enhanced UV-excited luminescence of europium ions in silver/tin-doped glass. J Lumin 128:831–833
115. Li L, Yang Y, Zhou D, Yang Z, Xu X, Qiu J (2013) Investigation of the interaction between different types of Ag species and europium ions in $Ag^+ - Na^+$ ion-exchange glass. Opt Mater Express 3:806
116. Jiao Q, Qiu J, Zhou D, Xu X (2014) Contribution of Eu ions on the precipitation of silver nanoparticles in Ag-Eu Co-doped borate glasses. Mater Res Bull 51:315–319
117. Riano LP, de Araujo CB, Malta OL, Santa Cruz P, Couto dos Santos MA (2004) Growth of metallic Ag nanoparticles in fluoroborate glasses doped with rare-earth ions and their optical characterization. Proc SPIE 5622:551–555

118. Wei R, Li J, Gao J, Guo H (2012) Enhancement of Eu^{3+} luminescence by Ag species (Ag NPs, ML- Ag, Ag^{+}) in oxyfluoride glasses. J Am Ceram Soc 95:3380–3382
119. Kumar KVA, Revathy KP, Prathibha V, Sunil T, Biju PR, Unnikrishnan NV (2013) Structural and luminescence enhancement properties of Eu^{3+}/Ag nanocrystallites doped SiO_2-TiO_2 matrices. J Rare Earths 31:441–448
120. Culea E, Vida-Simiti I, Borodi G, Culea EN, Stefan R, Pascuta P (2014) Structural and spectroscopic effects of Ag–Eu^{3+} codoping of TeO_2–PbO glass ceramics. J Mater Sci 49:4620–4628
121. Buch Z, Kumar V, Mamgain H, Chawla S (2013) Silver nanoprism enhanced fluorescence in YVO4: Eu^{3+} nanoparticles. Chem Commun 49:9485–9487
122. Guo H, Wang X, Chen J, Li F (2010) Ultraviolet light induced white light emission in Ag and Eu^{3+} Co-doped oxyfluoride glasses. Opt Express 18:18900–18905
123. Strohhöfer C, Polman A (2002) Silver as a sensitizer for erbium. Appl Phys Lett 81:1414
124. Chiasera A, Ferrari M, Mattarelli M, Montagna M, Pelli S, Portales H, Zheng J, Righini GC (2005) Assessment of spectroscopic properties of erbium ions in a soda-lime silicate glass after silver–sodium exchange. Opt Mater 27:1743–1747
125. Portales H, Mattarelli M, Montagna M, Chiasera A, Ferraris M, Martucci A, Mazzoldi P, Pelli S, Righini GC (2005) Investigation of the role of silver on spectroscopic features of Er^{3+}-activated Ag-exchanged silicate and phosphate glass. J Non Cryst Solids 351:1738–1742
126. Fukushima M, Managaki N, Fujii M, Yanagi H, Hayashi S (2005) Enhancement of 1.54-Mm emission from Er-doped Sol–gel SiO2 films by Au nanoparticles doping. J Appl Phys 98:024316
127. Lin A, Boo S, Moon DS, Jeong HJ, Chung Y, Han W-T (2007) Luminescence enhancement by Au nanoparticles in Er^{3+}-doped Germano-silicate optical fiber. Opt Express 15:8603–8608
128. Watekar PR, Ju S, Han W-T (2008) Optical properties of the alumino-silicate glass doped with Er-ions/Au particles. Colloids Surf A Physicochem Eng Asp 313–314:492–496
129. Singh SK, Giri NK, Rai DK, Rai SB (2010) Enhanced upconversion emission in Er^{3+}-doped tellurite glass containing silver nanoparticles. Solid State Sci 12:1480–1483
130. Rivera VAG, Osorio SPA, Ledemi Y, Manzani D, Messaddeq Y, Nunes LAO, Marega E (2010) Localized surface plasmon resonance interaction with Er^{3+}-doped tellurite glass. Opt Express 18:25321–25328
131. Rivera VAG, Ledemi Y, Osorio SPA, Manzani D, Messaddeq Y, Nunes LAO, Marega E (2012) Efficient plasmonic coupling between Er^{3+}:(Ag/Au) in tellurite glasses. J Non Cryst Solids 358:399–405
132. De Campos VPP, Kassab LRP, de Assumpção TAA, da Silva DS, de Araújo CB (2012) Infrared-to-visible upconversion emission in Er^{3+} doped TeO_2-WO_3-Bi_2O_3 glasses with silver nanoparticles. J Appl Phys 112:063519
133. Amjad RJ, Sahar MR, Ghoshal SK, Dousti MR, Riaz S, Tahir BA (2012) Enhanced infrared to visible upconversion emission in Er^{3+} doped phosphate glass: role of silver nanoparticles. J Lumin 132:2714–2718
134. Amjad RJ, Sahar MR, Ghoshal SK, Dousti MR, Riaz S, Samavati AR, Jamaludin MNA, Naseem S (2013) Plasmon enhanced upconversion fluorescence in Er^{3+}: Ag phosphate glass: effect of heat treatment. Chin Phys Lett 30:027301
135. Amjad RJ, Sahar MR, Ghoshal SK, Dousti MR, Riaz S, Samavati AR, Arifin R, Naseem S (2013) Annealing time dependent up-conversion luminescence enhancement in magnesium–tellurite glass. J Lumin 136:145–149
136. Dousti MR, Sahar MR, Ghoshal SK, Amjad RJ, Arifin R (2013) Plasmonic enhanced luminescence in Er^{3+}: Ag Co-doped tellurite glass. J Mol Struct 1033:79–83
137. Qi Y, Zhou Y, Wu L, Yang F, Peng S, Zheng S, Yin D, Wang X (2014) Annealing time dependent 1.53 μm fluorescence enhancement in Er^{3+}-doped tellurite glasses containing silver NPs. Mater Lett 125:56–58

138. Culea E, Vida-Simiti I, Borodi G, Nicolae Culea E, Stefan R, Pascuta P (2014) Effects of Er^{3+}: Ag codoping on structural and spectroscopic properties of lead tellurite glass ceramics. Ceram Int 40:11001–11007
139. Awang A, Ghoshal SK, Sahar MR, Dousti MR, Amjad RJ, Nawaz F (2013) Enhanced spectroscopic properties and Judd–Ofelt parameters of Er-doped tellurite glass: effect of gold nanoparticles. Curr Appl Phys 13:1813–1818
140. Ghoshal SK, Awag A, Saar MR, Amjad RJ, Dousti MR (2013) Spectroscopic and structural properties of TeO_2-ZnO-Na_2O-Er_2O_3-Au glasses. Chalcogenide Lett 10:411–420
141. Sazali ES, Sahar MR, Ghoshal SK, Arifin R, Rohani MS, Awang A (2014) Optical properties of gold nanoparticle embedded Er^{3+} doped lead–tellurite glasses. J Alloys Compd 607:85–90
142. Awang A, Ghoshal SK, Sahar MR, Arifin R, Nawaz F (2014) Non-spherical gold nanoparticles mediated surface plasmon resonance in Er^{3+} doped zinc–sodium tellurite glasses: role of heat treatment. J Lumin 149:138–143
143. Awang A, Ghoshal SK, Sahar MR, Dousti MR, Nawaz F (2014) Growth of Au nanoparticles stimulate spectroscopic properties of Er^{3+} doped TeO_2-ZnO-Na_2O glasses. Adv Mater Res 895:254–259
144. Qi J, Xu T, Wu Y, Shen X, Dai S, Xu Y (2013) Ag nanoparticles enhanced near-IR emission from Er^{3+} ions doped glasses. Opt Mater 35:2502–2506
145. Mahraz ZSA, Sahar MR, Ghoshal SK, Dousti MR, Amjad RJ (2013) Silver nanoparticles enhanced luminescence of Er^{3+} ions in boro-tellurite glasses. Mater Lett 112:136–138
146. Obadina VO (2013) Investigation of silver nanostructures and their influence on the fluorescence spectrum of erbium-doped glasses. Opt Photonics J 03:45–50
147. Dousti MR, Amjad RJ, Mahraz ZAS (2014) Enhanced greena Nd red upconversion emissions in Er^{3+}doepd boro-tellurite glass containing gold nanoparticles. J Mol Struct 1079:347–352
148. Pan Z, Ueda A, Aga R, Burger A, Mu R, Morgan SH (2010) Spectroscopic studies of Er^{3+} doped Ge-Ga-S glass containing silver nanoparticles. J Non Cryst Solids 356:1097–1101
149. Speranza G, Bhaktha SNB, Chiappini A, Chiasera A, Ferrari M, Goyes C, Jestin Y, Mattarelli M, Minati L, Montagna M et al (2006) Nanocomposite Er–Ag silicate glasses. J Opt A Pure Appl Opt 8:S450–S454
150. Da Silva DM, Kassab LRP, Lüthi SR, de Araújo CB, Gomes ASL, Bell MJV (2007) Frequency upconversion in Er^{3+} doped PbO–GeO_2 glasses containing metallic nanoparticles. Appl Phys Lett 90:081913
151. Kassab LRP, Bomfim FA, Martinelli JR, Wetter NU, Neto JJ, Araújo CB (2009) Energy transfer and frequency upconversion in Yb^{3+}–Er^{3+}-doped PbO-GeO_2 glass containing silver nanoparticles. Appl Phys B 94:239–242
152. Qi Y, Zhou Y, Wu L, Yang F, Peng S, Zheng S, Yin D (2014) Silver nanoparticles enhanced 1.53 μm band fluorescence of Er^{3+}/Yb^{3+} codoped tellurite glasses. J Lumin 153:401–407
153. De Araujo CB, Silvério da Silva D, Alves de Assumpção TA, Kassab LRP, Mariano da Silva D (2013) Enhanced optical properties of germanate and tellurite glasses containing metal or semiconductor nanoparticles. Sci World J 2013:385193
154. Ma Y, Lin J, Chen J, Feng Z, Wei H, Mao J (2011) Plasmon-enhanced luminescence of YAG: Yb, Er nanopowders by Ag nanoparticles embedded in silicate glass. Mater Lett 65:282–284
155. Som T, Karmakar B (2011) Synthesis and enhanced photoluminescence in novel Aucore Au–Agshell nanoparticles embedded Nd^{3+}-doped antimony oxide glass hybrid nanocomposites. J Quant Spectrosc Radiat Transf 112:2469–2479
156. Da Silva DS, de Assumpção TAA, Kassab LRP, de Araújo CB (2014) Frequency upconversion in Nd^{3+} doped PbO–GeO2 glasses containing silver nanoparticles. J Alloys Compd 586:S516–S519
157. Dousti MR (2013) Efficient infrared-to-visible upconversion emission in Nd^{3+}-doped PbO-TeO_2 glass containing silver nanoparticles. J Appl Phys 114:113105
158. Som T, Karmakar B (2010) Surface plasmon resonance and enhanced fluorescence application of single-step synthesized elliptical nano gold-embedded antimony glass dichroic nanocomposites. Plasmonics 5:149–159

159. Som T, Karmakar B (2010) Enhanced frequency upconversion of Sm^{3+} ions by elliptical Au nanoparticles in dichroic Sm^{3+}: Au-antimony glass nanocomposites. Spectrochim Acta A Mol Biomol Spectrosc 75:640–646
160. Som T, Karmakar B (2009) Surface plasmon resonance in nano-gold antimony glass–ceramic dichroic nanocomposites: one-step synthesis and enhanced fluorescence application. Appl Surf Sci 255:9447–9452
161. Dousti MR (2014) Plasmonic effect of silver nanoparticles on the upconversion emissions of Sm^{3+}-doped sodium-borosilicate glass. Measurement 56:117–120
162. Li L, Yang Y, Zhou D, Yang Z, Xu X, Qiu J (2013) Investigation of the role of silver species on spectroscopic features of Sm^{3+}-activated sodium–aluminosilicate glasses via $Ag^{+} - Na^{+}$ ion exchange. J Appl Phys 113:193103
163. Jiménez JA, Sendova M (2012) In situ isothermal monitoring of the enhancement and quenching of Sm^{3+} photoluminescence in Ag Co-doped glass. Solid State Commun 152:1786–1790
164. Dyrba M, Miclea P-T, Schweizer S (2010) Surface plasmons for fluorescence enhancement in Sm-doped borate glasses. Radiat Meas 45:314–316
165. Fauzia Abdullah NA, Sahar MR, Hamzah K, Ghoshal SK (2014) Luminescence enhancement of samarium-doped tellurite glass containing silver nanoparticles. Adv Mater Res 895:260–264
166. Kaur G, Verma RK, Rai DK, Rai SB (2012) Plasmon-enhanced luminescence of Sm complex using silver nanoparticles in polyvinyl alcohol. J Lumin 132:1683–1687
167. Rai S, Bokatial L, Dihingia PJ (2011) Effect of CdS nanoparticles on fluorescence from Sm^{3+} doped SiO_2 glass. J Lumin 131:978–983
168. Jiménez JA (2013) Influence of Ag nanoparticles on the luminescence dynamics of Dy^{3+} ions in glass: the "plasmonic diluent" effect. Phys Chem Chem Phys 15:17587–17594
169. Dousti MR, Hosseinian SR (2014) Enhanced upconversion emission of Dy^{3+}-doped tellurite glass by heat-treated silver nanoparticles. J Lumin 154:218–223
170. Jiménez JA, Hockenbury JB (2013) Spectroscopic properties of CuO, SnO, and Dy_2O_3 Co-doped phosphate glass: from luminescent material to plasmonic nanocomposite. J Mater Sci 48:6921–6928
171. Petit L, Griffin J, Carlie N, Jubera V, García M, Hernández FE, Richardson K (2007) Luminescence properties of Eu^{3+} or Dy^{3+}/Au Co-doped SiO_2 nanoparticles. Mater Lett 61:2879–2882
172. De Assumpção TAA, da Silva DM, Camilo ME, Kassab LRP, Gomes ASL, de Araújo CB, Wetter NU (2012) Frequency upconversion properties of Tm^{3+} doped TeO_2–ZnO glasses containing silver nanoparticles. J Alloys Compd 536:S504–S506
173. Assumpção TAA, da Silva DM, Kassab LRP, de Araújo CB (2009) Frequency upconversion luminescence from Yb^{3+}–Tm^{3+} codoped PbO–GeO_2 glasses containing silver nanoparticles. J Appl Phys 106:063522
174. De Assumpção TAA, Da Silva DMM, Kassab LRP, Martinelli JRR, De Araújo CB (2010) Influence of the temperature on the nucleation of silver nanoparticles in Tm^{3+}/Yb^{3+} codoped PbO–GeO_2 glasses. J Non Cryst Solids 356:2465–2467
175. Kassab LRP, Freitas LF, Assumpção TAA, Silva DM, Araújo CB (2011) Frequency upconversion properties of Ag: TeO_2–ZnO nanocomposites codoped with Yb^{3+} and Tm^{3+} ions. Appl Phys B 104:1029–1034
176. Kassab LPR, Ferreira Freitas L, Ozga K, Brik MG, Wojciechowski A (2010) ZnO–TeO_2–Yb/Tm glasses with silver nanoparticles as laser operated quantum electronic devices. Opt Laser Technol 42:1340–1343
177. Assumpção TAA, Kassab LRP, Gomes ASL, Araújo CB, Wetter NU (2010) Influence of the heat treatment on the nucleation of silver nanoparticles in Tm^{3+} doped PbO-GeO_2 glasses. Appl Phys B 103:165–169

178. Qi J, Xu Y, Huang F, Chen L, Han Y, Xue B, Zhang S, Xu T, Dai S (2014) Photoluminescence of Ag nanoparticles and Tm^{3+} ions in the bismuth germanate glasses for the blue light-excited W-LED. J Am Ceram Soc 97:1471–1474
179. Camilo ME, Assumpção TAA, da Silva DM, da Silva DS, Kassab LRP, de Araújo CB (2013) Influence of silver nanoparticles on the infrared-to-visible frequency upconversion in Tm^{3+}/ Er^{3+}/Yb^{3+} doped GeO_2-PbO glass. J Appl Phys 113:153507
180. Silva GH, Holgado DPA, Anjos V, Bell MJV, Kassab LRP, Amâncio CT, Moncorgè R (2014) Effect of Ag nanoparticles on the radiative properties of tellurite glasses doped with Er^{3+}, Yb^{3+} and Tm^{3+} ions. Opt Mater 2:6–11
181. Hu Y, Qiu J, Song Z, Yang Z, Yang Y, Zhou D, Jiao Q, Ma C (2014) Spectroscopic properties of Tm^{3+}/Er^{3+}/Yb^{3+} Co-doped oxyfluorogermanate glasses containing silver nanoparticles. J Lumin 145:512–517
182. Bi G, Wang L (2012) Photoluminescence enhancement induced from silver nanoparticles in Tb^{3+} -doped glass ceramics. Chin Opt Lett 10:092401
183. Piasecki P, Piasecki A, Pan Z, Ueda A, Aga R Jr, Mu R, Morgan SH (2010) Plasmon enhanced luminescence of Tb^{3+} doped Li_2O-LaF_3-Al_2O -SiO_2 glass containing Ag nanoparticles. Proc SPIE 7757:77572M
184. Piasecki P (2010) Formation of Ag nanoparticles and enhancement of Tb^{3+} luminescence in Tb and Ag Co-doped lithium-lanthanum-aluminosilicate glass. J Nanophotonics 4:043522
185. Li L, Yang Y, Zhou D, Xu X, Qiu J (2014) The influence of Ag species on spectroscopic features of Tb^{3+}-activated sodium–aluminosilicate glasses via Ag + −Na + ion exchange. J Non Cryst Solids 385:95–99
186. Zhang D, Hu X, Ji R, Zhan S, Gao J, Yan Z, Liu E, Fan J, Hou X (2012) Influence of Ag nanoparticles on luminescent performance of SiO2: Tb^{3+} nanomaterials. J Non Cryst Solids 358:2788–2792
187. Verma RK, Kumar K, Rai SB (2010) Pulsed laser ablation synthesis of silver nanoparticles and their use in fluorescence enhancement of Tb^{3+}-doped aluminosilicate glass. Solid State Commun 150:1947–1950
188. Kassab LRP, De Almeida R, da Silva DM, De Assumpão TAA, De Araújo CB, de Assumpção TAA, de Araújo CB (2009) Enhanced luminescence of Tb^{3+}/Eu^{3+} doped tellurium oxide glass containing silver nanostructures. J Appl Phys 105:103505
189. Pan Z, Sekar G, Akrobetu R, Mu R, Morgan SH (2011) Cooperative infrared to visible upconversion and visible to near-infrared quantum cutting in Tb and Yb Co-doped glass containing Ag nanoparticles. SPIE Proceedings 8096:809633–809633–7
190. Rai VK, Menezes LDS, de Araújo CB, Kassab LRP, Davinson M (2008) Surface-plasmon-enhanced frequency upconversion in Pr^{3+} doped tellurium-oxide glasses containing silver nanoparticles. J Appl Phys 103:093526
191. Naranjo LP, de Araújo CB, Malta OL, Cruz PAS, Kassab LRP (2005) Enhancement of Pr^{3+} luminescence in PbO–GeO_2 glasses containing silver nanoparticles. Appl Phys Lett 87:241914
192. Lakshminarayana G, Qiu J (2009) Enhancement of Pr^{3+} luminescence in TeO_2–ZnO–Nb_2O_5–MoO_3 glasses containing silver nanoparticles. J Alloys Compd 478:630–635
193. Qi Y, Zhou Y, Wu L, Yang F, Peng S, Zheng S, Yin D (2014) Enhanced upconversion emissions in Ho^{3+}/Yb^{3+} codoped tellurite glasses containing silver NPs. J Non Cryst Solids 402:21–27
194. Zhang WJ, Zhang QY, Chen QJ, Qian Q, Yang ZM, Qiu JR, Huang P, Wang YS (2009) Enhanced 2.0 microm emission and gain coefficient of transparent glass ceramic containing BaF_2: Ho^{3+}, Tm^{3+} nanocrystals. Opt Express 17:20952–20958

Chapter 15
Surface Enhanced Fluorescence by Plasmonic Nanostructures

Jun Dong, Hairong Zheng, Zhenglong Zhang, Wei Gao, Jihong Liu, and Enjie He

Abstract The optically generated collective electron density waves on metal-dielectric boundaries known as surface plasmons have been of great scientific interest since their discovery. Being electromagnetic waves on noble nanostructure's surface, surface plasmons resonance can strongly enhance the electromagnetic field. These strong electromagnetic fields near the metal surfaces have been used in various applications like surface enhanced spectroscopy, plasmonic lithography, plasmonic trapping of particles and plsmonic catalysis etc. Resonant coupling of localized surface plasmons to fluorescent emitters can strongly modify the emitted intensity, the angular distribution and the polarization of the emitted radiation and even the speed of radiative decay, which is so-called surface/metal enhanced fluorescence (MEF/SEF). In this chapter, we illustrate current progress in design of metallic nanostructures for efficient fluorescence signal amplification that utilizes propagating and localized surface plasmons, and also some critical parameters in SEF, such as spacer, wavelength dependence effect are also discussed.

This work was supported by the National Science Foundation of China (11574190 and 11304247), Shaanxi Provincial Research Plan for Young Scientific and Technological New Stars (Program no.2015KJXX-40),and the New Star Team of Xi'an University of Posts & Telecommunications.

J. Dong (✉) • W. Gao • J. Liu
School of Electronic Engineering, Xi'an University of Posts and Telecommunications, Xi'an 710121, China
e-mail: dongjun@xupt.edu.cn

H. Zheng (✉)
School of Physics and Information Technology, Shaanxi Normal University, Xi'an 710062, China
e-mail: hrzheng@snnu.edu.cn

Z. Zhang
Leibniz Institute of Photonic Technology, Albert-Einstein-Str. 9, 07745 Jena, Germany

E. He
Department of Opto-electronic Technology, Anhui Science and Technology University, Fengyang 233100, China

C.D. Geddes (ed.), *Reviews in Plasmonics 2015*,
DOI 10.1007/978-3-319-24606-2_15

Keywords Surface plasmon plartion • Enhanced fluorescence • Metallic nanostructure

15.1 Introduction

Since Prof. Ritchie firstly predicted that the existence of self-sustained collective oscillations at surface of metallic nanostructure by treatment of energy losses characteristic of fast electrons passing through thin metal films in 1957 [1, 2]. Many researchers are devoted to study the properties of metallic nanostructure from theoretical and experimental aspect [3–5]. And it shows that the band intensity and wavelength position of SPR strongly depend on geometrical parameter of the nanostructures, such as the size, shape et. al, as well as the such as the composition of the nanostructure, and also the dielectric properties of the surrounding environment [6–8]. The corresponding properties enabled researchers to tailor design surface topographies on the nanoscale which revealed new aspects of their underlying science, and in-doing so, inspired and encouraged new groups to join the ever-growing field of plasmonics [9, 10]. With the help of excitation of Surface Plasmon (SP), the electron-photon interaction and fluorophore photonic state can be modulated. Therefore, the study on the coupling effect among plasmonic metallic nanostructure and surface enhanced fluorescence (SEF) have been one of most active and important topic, particularly since the firstly classical observation of the effect of media on the emission properties of a molecular excited electronic state by Drexhage [11]. As a result, manipulation of optical nanostructure substrate with plasmonic properties is considered as the mostly promising direction in the SEF research area, and it has been widely used in the diagnosis, biological detection and so on [12–14].

As we know that, in order to pursue a more convenient and controllable avenue is still of great interest for the application of SEF, deeply understanding of coupling and inter-conversion mechanism among free electrons, surface plasmon, photons and fluorophores in SEF effect is still should be highlighted. In this regard, many methods have been developed for fabrication of metal substrate with various surface configurations recently, which including the polish of the surface, vapor deposition method, self-assembly of metal colloids method, chemical reduction method and so on [9, 15, 16]. In this chapter, we present the latest developments of this fast growing field boosted, in particular, by the ability to design nanometric plasmonic structures and objects. We address the case of surfaces as well as that of nanoparticles with respect to their enhancement properties. The whole chapter is divided into four parts. The first part introduces the concepts and principles of plasmonics and SEF. The three other parts are devoted to SEF for fluorophores coupled to propagating surface plasmons existed on periodical metallic nanostucture substrate (Sect. 15.3.1), and to SEF for fluorophores coupled to localized enhancement of the EM field existed non-periodical metallic nanostucture substrate (Sect. 15.3.2), and to SEF for the distance and wavelength dependence factor (Sect. 15.3.3).

15.2 SEF Principles

15.2.1 Principles of Fluorescence

Luminescence is the emission of light from any substance that has absorbed energy, and do not care that the excited source, so it is a generally concept. And fluorescence is the property of some luminescent center, such Re-earth ions and organic molecules, to absorb a particular wavelength photon and to subsequently emit a longer wavelength photon after a brief interval. The processes which occur between the absorption and the emission of light are usually illustrated by a Jablonski diagram, as shown in Fig. 15.1.

Figure 15.1 shows a simplified version of the diagram which illustrates the electronic transition processes of some molecular which occur in the excited states. The fluorescence process can be divided with three important processes. Firstly, excitation of a susceptible molecule by a photon to a higher vibrational energy level, that is, electronically excited state S_1 and S_2, which happens in femtoseconds. Secondly, while vibrational relaxation of excited state electrons to the lowest energy level is much slower and can be measured in picoseconds. And the final process, emission of a longer wavelength photon from the excited singlet states to the ground state, occurs in the relatively long time period of nanoseconds or longer. There are two important parameters are used in consider the properties of emission processes, that is, fluorescence lifetime τ and the quantum efficiency Q. Fluorescence lifetime τ can be defined as the average relaxation time to the electronic ground state, and quantum efficiency Q can be defined as the ratio of the number of photons emitted over the number of excited photons absorbed. As we know, the excited electron transition from the upper energy level to ground level, often by two methods, one is emit photon, and the other way is relaxation of energy into phonon [15, 17, 18].

With the condition of weak excitation, that is, far from saturation of the excited state, [9, 15–19] the fluorescence emission rate γ_{em}^{0} can be regarded as the excitation

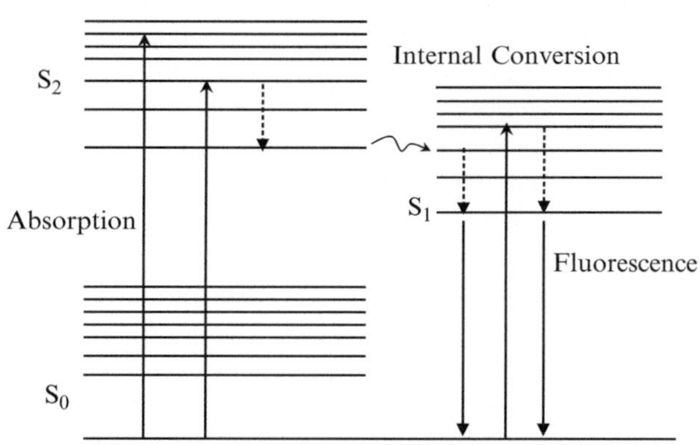

Fig. 15.1 Simplified Jablonski diagram illustrating the molecular process involved during a fluorescence excitation and emission cycle

from ground state to excited state and the subsequent relaxation back to the ground state via emission of a fluorescence photon, i.e

$$\gamma_{em}^0 = \gamma_{exc}^0 Q_i^0 \quad (15.1)$$

where γ_{exc}^0 is the excitation rate and Q_i^0 is the quantum yield. The superscripts '0' specify that the molecule is in free space and does not couple to the local environment. The subscript 'i' indicates that the quantum yield is defined by the intrinsic properties of the molecule.

As indicated before, Q_i^0 is the probability of relaxing from excited to ground state by emission of a fluorescence photon. In terms of the radiative decay rate γ_r^0 and the non-radiative decay rate γ_{nr}^0, so the intrinsic quantum yield with more general definition for Q_i^0 can be defined as follows:

$$Q_i^0 = \frac{\gamma_r^0}{\gamma_r^0 + \gamma_{nr}^0} \quad (15.2)$$

If the local environment of the fluorophore has been changed, the excitation and decay rates will be changed corresponding. Then the Eqs. (15.1) and (15.2) get modified as

$$\gamma_{em} = \gamma_{exc} Q \quad (15.3)$$

and

$$Q = \frac{\gamma_r}{\gamma_r + \gamma_{nr}} = \frac{\gamma_r}{\gamma_r + \gamma_{nr}^0 + \gamma_{abs} + \lambda_m} \quad (15.4)$$

Here, γ_{abs} accounts for dissipation to heat in the environment and γ_m accounts for coupling to non-radiative electromagnetic modes, such as the emitted energy is dissipated into heat through the interaction between the electron and lattice. The total decay rate $\gamma = \gamma_r + \gamma_{nr}$ defines the lifetime $\tau = 1/\gamma$ of the excited state. In a homogeneous non-absorbing media, such as a fluorophore ions in water, the definition 1 of Q is equivalent to the classical one. In general, the fluorescence emission is not only a function of the fluorophore properties but also of external parameters accounting for the local environment of the emitter.

For a conventional fluorescence technique measurement, pursuing the brighter and more stable signals of the fluorophore performed at the minimization of the internal and environmentally conditioned non-radiative processes, increasing higher spontaneous-emission rate mainly depending on its nature properties [15]. Though the fluorescence is known as one of the best choice optical method for the detection of chemical and biological species, typical disadvantage for the conventional fluorescence technique is the relatively low signal-to-noise (background) ratio which restrict its applicationsto important areas of medical diagnostics, food control, and security, particularly realization the purpose of the detection of a wide range of adsorbate molecules down to the single molecule detection limit and boost the emission efficiency of Re-earth ions. So the exploring a proper solution to overcome the disadvantage mentioned before which has attracted much attentions because of their potential application in many fields.

15.2.2 Interaction of Fluorophores with Surface Plasmons

Since the pioneering work of Purcell [20], many researches have been demonstrated that the lifetime of an excited atomic state is critically dependent on the inner properties of the atom and its environment both from experimental and theoretical aspect [17–19]. Due to the interaction of the fluorophore with its environment nearby, the fluorescence processes including both the excitation and the emission processes, can be modulated through the modification of its local EM field. As a result, pushing fluorescence detection to the limit of sensitivity, in order to overcome the disadvantage of the convention fluorescence technique, can be performed by controlling the local electromagnetic (EM) environment of the fluorophores.

Fluorophore labels are coupled with the tightly confined field of surface plasmons-collective oscillation of charge density and associated electromagnetic field on a surface of metallic films and nanostructures. With the help of the coupled effect of light with localized surface plasmons (LSPs-supported by single nanoparticles or nanoperodical typed metallic nanostructure) and surface plasmon polaritons (SPPs-traveling along periodical typed metal nanostructure), which can provide strong confinement of electromagnetic field intensity. These confined EM fields can interact with fluorophores at their absorption λ_{ab} and emission λ_{em} wavelengths which alter respective transitions between the ground state and higher excited states (see Fig. 15.2). It is reported that the internal conversion process depends mainly on the electronic configuration of the fluorophore through the wave-functions overlap of the fluorophore [9]. As a result, the internal conversion process is insensitive to the modification of the local EM environment to a first approximation. The effect of enhanced fluorescence which is critical dependence on the local EM field is defined as surface enhanced fluorescence. It is important to understand the energy transfer between the fluorophore and the metallic surface through near-field component, which will help us to study the contributed effect of the modification of the relaxation rate of a fluorophore towards the radiative emission rate.

Based on the EM mechanism, the cross section of SEF from the metal nanostructure can be defined as $\sigma_{SEF}(\lambda_L, \lambda, d_{av})$, where λ_L and λ are the excitation and emission wavelengths, respectively, d_{av} is the average distance between a molecule and metal surface. The total enhanced effect was induced by coupling of plasmon resonance with incident light and with fluorescence light, then the following equation could be obtained [21]

Fig. 15.2 Schematic of confined field of SPP and LSP modes coupled with a fluorophore excited with external EM field

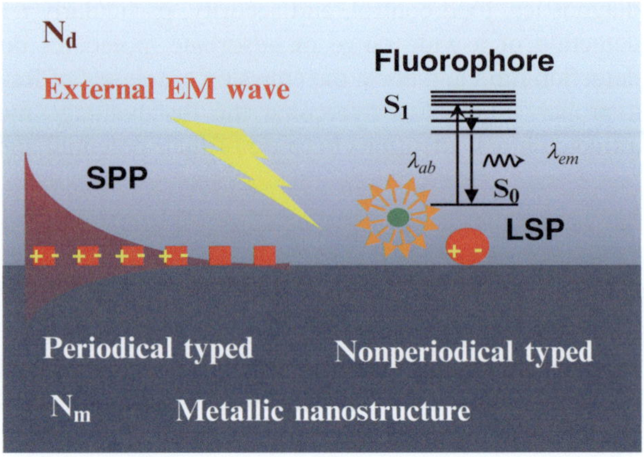

$$\begin{aligned} |M_{EM}(\lambda_L, \lambda, d_{av})|^2 &= \left|\frac{E_{loc}(\lambda_L, d_{av})}{E_{in}(\lambda_L)}\right|^2 \times \left|\frac{E_{loc}(\lambda, d_{av})}{E_{in}(\lambda)}\right|^2 \\ &= |M_1(\lambda_L, d_{av})|^2 |M_2(\lambda, d_{av})|^2 \end{aligned} \tag{15.5}$$

While $|M_{EM}|^2$ is total EM enhancement factors, $|M_1|^2$ and $|M_2|^2$ are the EM enhancement factor, corresponding to the coupling interaction between plasmon resonance with the incident light and with Raman scattering plus fluorescence light, respectively.

In the case of $\sigma_{SEF}(\lambda_L, \lambda)$, the decay rate of fluorescence is also enhanced by a factor of $|M_d|^2$ through energy transfer from a molecule to an metal surface. Thus, $1/|M_d|^2$ is multiplied with fluorescence scattering cross section $\sigma_{FL}(\lambda_L, \lambda)$ of a molecule in free space. Note that the use of $\sigma_{FL}(\lambda_L, \lambda)$ is the first approximation for evaluation of SEF and assumes that a quantum yield of fluorophore is unity located in free space [22]. Thus

$$\sigma_{SEF}(\lambda_L, \lambda, d_{av}) = |M_{EM}(\lambda_L, \lambda, d_{av})|^2 \times \frac{\sigma_{FL}(\lambda_L, \lambda)}{|M_d(\lambda, d_{av})|^2} \tag{15.6}$$

From Eq. (15.4), the fluorescence enhanced effect shows a competition of several processes including a increase of the excitation rate by the local field enhancement, an enhancement of radiative decay rate by the surface plasmon-coupled emission (SPCE) and quenching that reduces the efficiency caused by the non-radiative energy transferring from the fluorophore ions to the metal surfaces, all of which will be greatly dependent on the spacing distance between the fluorophore ions and the metal [23]. Since the balance between dissipation and coupling to radiation depends strongly on distance, the introduction of various typed spacer to isolated them plays a crucial role in the optimization of plasmon-enhanced luminescence.

From the point of view above, we know that SEF is a multiprocess system, and the enhanced effect depended on many critical factors. In order to simplify the interaction of fluorophore in the vicinity of the metal surface, we will review the latest recent progress of surface enhanced fluorescence and related research works from the geometrical shape of metal nanostructure, the selective of excitation condition and fluorophore and the tip enhanced fluorescence.

15.3 SEF from Various Geometrical Metallic Plasmonic Nanostructure

Plasmons lead to large enhancements of the local electromagnetic field in the proximity of metal plasmonic nanostructure surface, an essential aspect of surface enhanced spectroscopy. An investigation of the relationship between the plasmonics properties with geometry of nanostructure enables the rational design and fabrication of composite plasmonic nanostructures tailored for specific applications such as SEF, where one seeks to increase the near-field electromagnetic field across a wide frequency range, where location of SPR dip and a high sensing sensitivity to the changes of the dielectric environment nearby are desirable. In this sections, we will review the recently progress of enhanced fluorescence effect from the periodical and nonperiodical typed plasmonic nanostructure.

15.3.1 Fluorescence Enhancement from Periodical Metallic Plasmonic Nanostructure

SPPs are transverse magnetic polarized electromagnetic surface modes associated with delocalized collective electron oscillation propagating along the interface between a metal and a dielectric materials. With the highly efficiently coupled with incident light, SPR has been applied for label-free biosensing and as an excitation field for sensitive fluorescence detection. So a conventional enhanced substrate, where a flat metal coated glass surface, has been developed for enhanced fluorescence bioassay and fluorescence imaging of cells. In general, due to the existence of momentum mismatch between the wave vector of the SPPs and the incident EM wave, the excitation of SPPs on a flat continual metal surface is difficult to realize [24]. So the flat metal surface morphology limits the plasmonic coupling and fluorescence enhancement efficiency. It is showed that the mismatch can be overcome by introduction of an external periodical metal surface such as gratings, nanohole arrays, nanocaps arrays, and so on. As a result, the precise fabrication of nanostructure surfaces with a large area to provide high plasmonic coupling efficiency and high enhancement factors remains a challenge therefore for application platforms in biosensing and imaging. Here we will review the application of the various shaped periodical distributed nanostructure in SEF (Fig. 15.3).

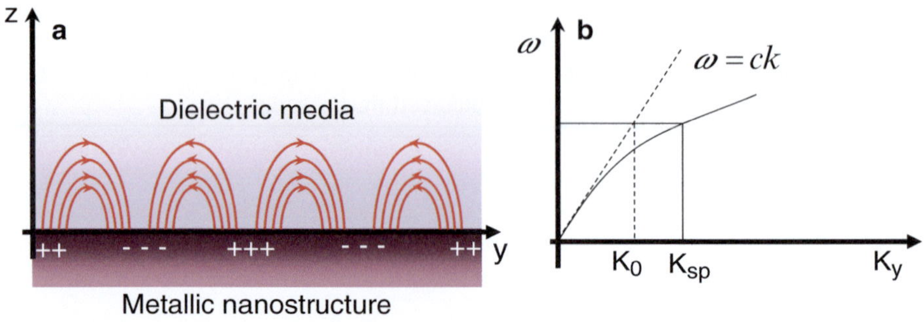

Fig. 15.3 SPs at the interface between a metal and a dielectric material have a combined electromagnetic wave and surface charge character (**a**), and the dispersion curve for a SP mode (**b**)

15.3.1.1 Surface Enhanced Fluorescence from Nanograting Substrate

Anomalies in the intensity distribution of light spectrum diffracted from a grating were first observed by Wood' in 1902 [25], and latterly interpreted in terms of an interaction between the incoming photon and a surface plasmon resonance in the grating surface [26]. So patterning the metallic surface with a periodic wavelength scale corrugation provides an excellent candidate, which can break the translational invariance of the thin film and allows one to relax the momentum conservation restriction. For the excitation of grating-coupled SPR, the wave vector of the incident EM wave is increased by the grating vector as the wave vector of the light in air is smaller than that of the SPs. The excitation condition is defined as

$$K_{sp} = K_{px} + G = \frac{2\pi}{\lambda}\sqrt{\varepsilon_m(\omega)}\sin\theta + \frac{2\pi}{\Lambda}m \quad (15.7)$$

Where Λ is the diffraction grating pitch, λ is the wavelength of incident EM wave, m is an integer, is the diffraction order, and $\varepsilon_m(\omega)$ is the wavelength-dependent dielectric constant of metal given by the classical Drude free-electron model.

The mismatch in wave vector between the in-plane momentum $K_x = K\sin\theta$ of impinging photons and can also be overcome by using diffraction effects at a grating pattern on the metal surface. As a result, the moderated diffraction on periodically corrugated metallic surfaces provides an alternative methods for simultaneous SPP-enhanced excitation at λ_{ab} and excitation of SPP-driven emission of fluorescence light at λ_{em}. By fabricating series of tailored silver plasmonic grating nanostructure, Prof. K. Tawa systematically investigated the relationship between the grating parameters (depth and duty ratio) and the fluorescence enhanced effect [27, 28], and found that with the condition of depth equal to 20 nm and duty ratio equal to 0.43, the fabricated grating substrate achieve optimal fluorescence enhancement. They also observed the distinguish transfect GFP cell binding with an antibody-Cy5 conjugation on the grating-typed substrate, which has a great potential for cell-imaging application.

It is found that the fluorescence enhancement mechanisms from grating arise from both the SPP enhanced absorption and SPCE. Prof. Sun also adapted silver sinusoidal nanograting as the enhanced substrate to boost the Rh6G dye molecules, and obtained maximal fluorescence enhancement factor is more than 30-fold [29, 30]. By increasing the thickness of the metal film in the grating-based SPCE system, the enhanced spectra was demonstrated significantly affect by both depth and gold thickness. Higher angular sensitivity will be achieved at short wavelengths [31]. The wavelength resolution ability of SPCE makes it potential for the applications such as wavelength-ratiometric measurements and fluorescence sensing [32].

Recently, rare-earth (RE) doped upconversion nanoparticles (UCNP) have attracted remarkable attentions due to their potential applications in ion sensing and biomedical systems. To enhance the upconversion efficiency, surface plasmon has been employed but the reported enhancements vary widely and the exact enhancement mechanisms are not clearly understood. [33] By deposition of $NaYF_4$:Yb^{3+}, Er^{3+} nanocrystals on silver nanograting with polyelectrolyte-mediated layer-by-layer deposition technique. The green and red photoluminescence intensity of UCNPs on nanograting was up to 16 and 39-fold higher than the reference sample deposited on flat silver film were experimentally obtained. By analyze the absorption, emission (Purcell effect), and Förster energy transfer effect towards the fluorescence enhancement, the author found that the absorption enhancement plays the dominant role in general, the enhancement of Förster energy transfer is important only under weak excitation conditions, and the Purcell effect is unimportant [34] (Fig. 15.4).

15.3.1.2 Surface Enhanced Fluorescence from Nanohole Arrays Substrate

Enhanced extraordinary optical transmission (EOT) effect was firstly observed in the metallic films perforated with a tiny hole. Compared with the expectation from classic diffraction theory, the transmitted of light can be greater than the percentage area occupied by the holes [35]. As a result, the periodic arrays of subwavelength holes in optically metallic films act as an excellent antenna to couple the incident light into SPs at a given frequency. Both the position and the width of the transmission peak can be tuned by adjusting the period and the symmetry.

For a square array of period a_0, the position of transmission peaks $\lambda_{\max}$ for normal incidence can be identified approximately from the dispersion relation [36], and they are given by:

$$\lambda_{\max}\sqrt{i^2 + j^2} \approx a_0\sqrt{\frac{\varepsilon_m \varepsilon_d}{\varepsilon_m + \varepsilon_d}} \tag{15.8}$$

where indices i and j are the scattering orders from the array [37].

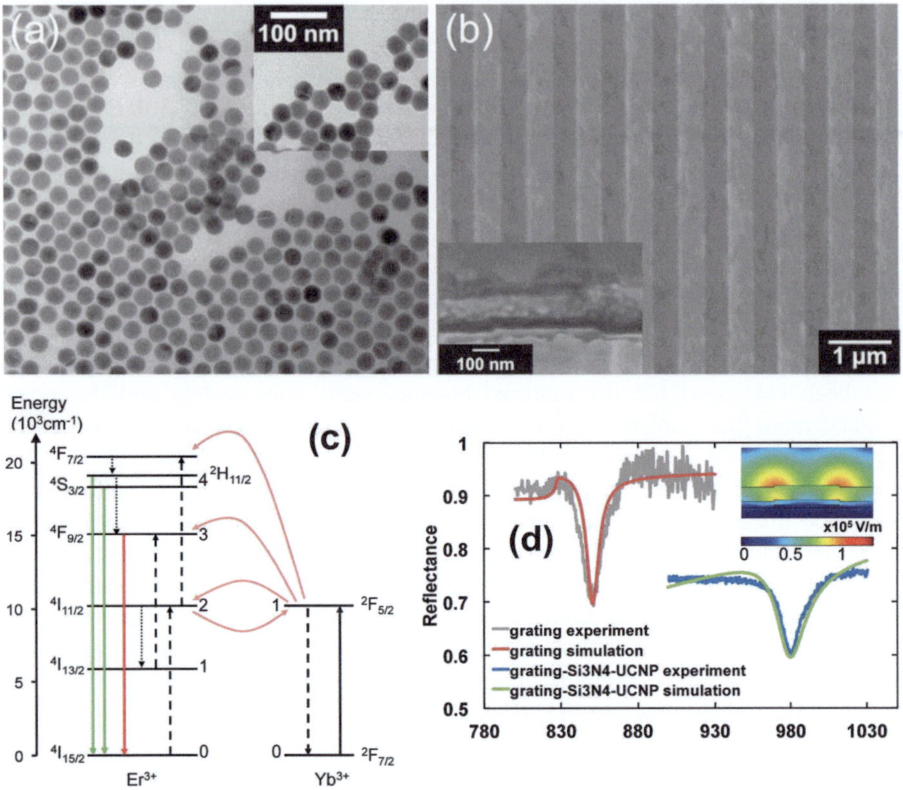

Fig. 15.4 (**a**) TEM of as-synthesized β-phase $NaYF_4$:Yb^{3+}, Er^{3+} UCNPs. The mean lateral size was 32 nm. The inset shows the TEM of PMAO-coated UCNPs showing a uniform 2 nm thick coating. The scale bar applies to both TEM images. (**b**) SEM of silver nanograting with period of 830 nm and line width of 410 nm. (**c**) Energy levels of Yb^{3+} and Er^{3+} ions relevant to the energy-transfer upconversion process. (**d**) Experimental reflectance spectra of as-fabricated nanograting (*gray*) and nanograting-Si_3N_4-UCNP structure (*blue*). The simulated reflectance spectra for the two structures are also plotted with *red* and *green* lines. The inset shows the field profile at the resonance wavelength for the nanograting-Si_3N_4-UCNP sample under normally incident plane wave excitation with a power density of 1.2 MW/m^2 adapted with permission from [34]. Copyright (2014) American Chemical Society

For a free-standing hole array, incident light coupled with the SPs in order to enhance the fields associated with the evanescent waves, thus producing a way to increase the transmittance. As a result, continuous metallic film perforated with arrays of nanoholes that can both supports the excitation of SPP and LSP. And the enhanced EM field which will help to increase fluorescence efficiency of the fluorophore located in vicinity of substrate. In order to obtain the largest enhanced fluorescence signal, the LSPR resonance peak should be about 40–120 meV higher in energy than a dye's emission peak, and the plasmonic spectral shape should coincide with the dye's excitation spectrum [38]. The plasmon coupling conditions,

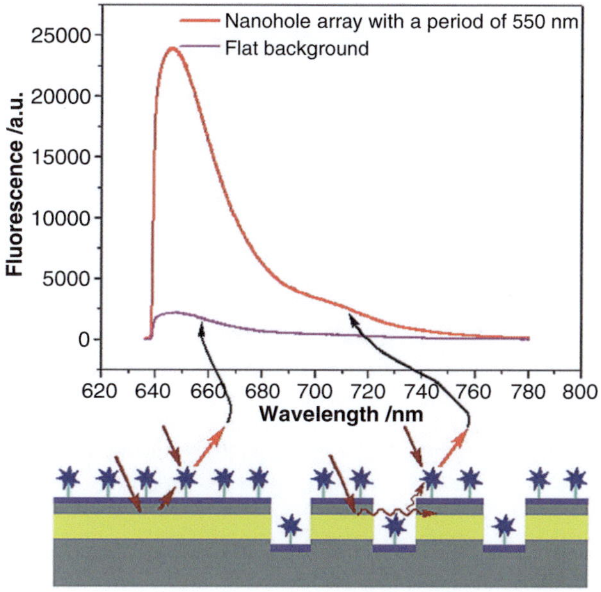

Fig. 15.5 Schematic illustration of surface modification and fluorescence labeling on SiO_2-coated silver nanohole arrays (adapted with permission from [40]. Copyright (2010) American Chemical Society)

such as angle of incidence, metal composition, laser frequency and excitation/emission properties of fluorophores, are play an important role in plasmon enhanced fluorescence on metallic nanohole arrays. By tuning the transmission and localization of laser excited conditions using various geometric parameters of the arrays in the Au film, Prof. Brolo obtains a significant enhanced fluorescence effect relative to that obtained from unpatterned identical films coated onto glass slides. The results suggest that Au films decorated with nanoholes arrays at resonance condition can improve both the sensitivity and the limit of detection of fluorescence-based analysis [39]. In order to avoid the fluorescence quench effect, Prof. Zhu designed a nanoarray nanostructure decorated with 30 nm SiO_2 dielectric film as spacer to separate the fluorophore from substrate, as shown in Fig. 15.5 [40]. The detected fluorescence is strongly dependent on the nanohole radius. When an array period of 550 nm and a nanohole radius of 100 nm were optimized to match the most efficient fluorescence excitation and emission of a boron-dipyrromethene fluorophore (BODIPY 630/650) on the array region, the maximum fluorescence enhancement of 11-fold on a silver nanohole array region in reflection mode was obtained experimentally.

Combining template-stripping with atomic layer deposition (ALD) method, Prof. Sang-Hyun Oh fabricate periodic nanohole arrays in smooth and optically thick Ag films, which are subsequently encapsulated with an ALD grown Oxide layer, such as silica shell or alumina, to prevent fluorophore quenching [41, 42]. Selecting Rh6G-Quasar670™ FRET pair as acceptor molecule, by tuning the plasmon wavelength with the maximal overlap of the donor's emission and

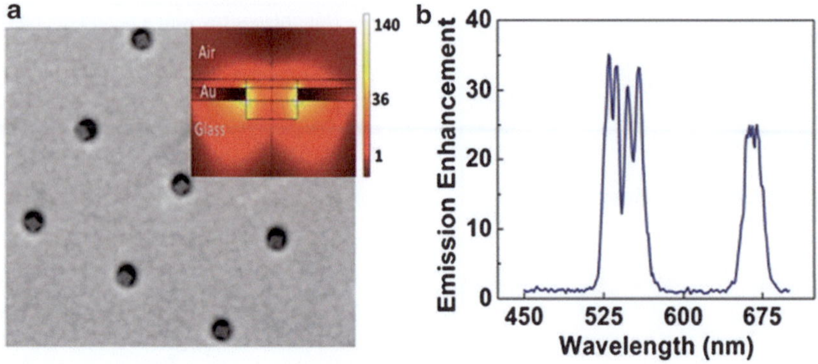

Fig. 15.6 (**a**) SEM images of the fabricated nanohole arrays, and (**b**) Upconversion luminescence spectra of $NaYF_4$:Yb^{3+}, Er^{3+} UCNPs in a Au nanohole array under 980 nm excitation (adapted with permission from [47]. Copyright (2013) American Chemical Society)

acceptor excitation, the mostly efficiency enhancement effect was achieved on the nanohole arrays [43].

Due to the front side illumination provides a stronger coupling between the incident electromagnetic field and the nanostructure as compared to back side illumination [44], the investigation of the relationship between the incident angle and the fluorescence enhancement on gold nanohole arrays has been reported [45]. Furthermore, Bai's group systematically studied the plasmonic responses of four plasmonic metals (Au, Ag, Cu, Al) and three composite plasmonic metals (Ag/Au, Cu/Au, Al/Au) nanohole arrays in the plasmon enhanced fluorescence spectroscopy biosensing setup.[46] By taking advantage of the enhanced electromagnetic field generated in metallic nanohole arrays at resonance, the dramatic plasmonic enhancement of UC luminescence with 35-fold in Yb^{3+} and Er^{3+} co-doped $NaYF_4$ nanocrystals was observed, as shown in Fig. 15.6. The increased UC emission come from the combined effects of enhancement in the excitation of sensitizers and increase in radiative decay rate of emitters, which is further supported by the measured decrease in the rise and decay times of the UC nanoparticles in nanohole arrays [47].

15.3.1.3 Plasmon Enhanced Fluorescence from Nanoparticle Arrays Substrate

It is well known that the plasmonic properties determining the final enhancement or quenching of the fluorescence depend not only on the characteristics of the NP (its nature, shape, and dimension) but also on the position and orientation of the fluorophore with respect to the NP as well as on the interparticle distance if a multi-NP system is used. Prof. Benedetta Mennucci studies the effect of solvent properties and plasmon couplings on the fluorescence intensity of fluorophore

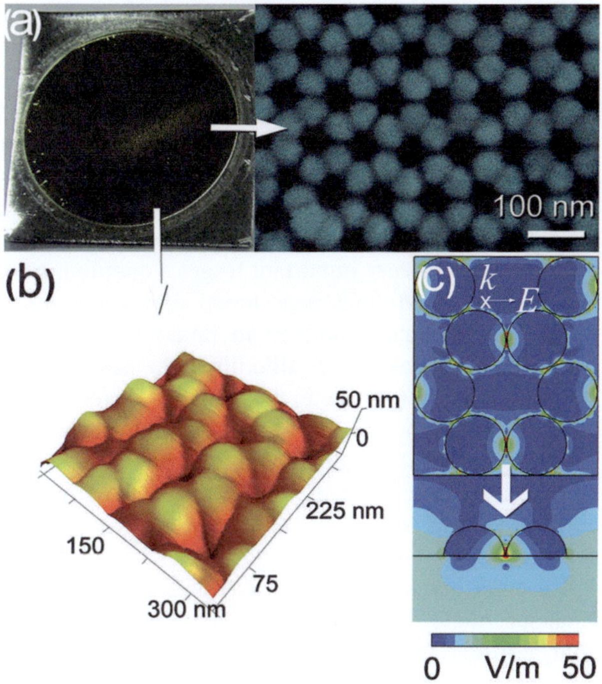

Fig. 15.7 (**a**) SEM image and (**b**) 3D AFM image acquired from the PSNAs. The inset in (**a**) shows large-area PSNAs with the size exceeding 3 cm^2. (**c**) Simulated electromagnetic field distribution map of the PSNAs with parameters D = 50 nm and G = 1 nm. The inset in (**c**) shows the k-vector and polarization of the incident light in the simulation (adapted with permission from [50]. Copyright (2010) American Chemical Society)

positioned within a four-metal NP square array. and also the effects of the metal nature, period constant of NP arrays, fluorophore/particle distance, fluorophore orientation, and addition of a solvent (dimethylformamide) on the SEF intensity of the dye and on the various quantities that determine such enhancement (radiative and nonradiative decay times, absorption) are systematic analyzed theoretically [48]. Prof. Lu developed a chemistry-based method to prepare silver nanoparticle (SNP) arrays by means of a combination of nanoimprint lithography (NIL) and electrochemical deposition (ECD). And he demonstrated that these SNP arrays show strong fluorescence enhancement because of coupling between the emission of the fluorophores and the SPR band of the metallic nanostructures. In addition, the intensity of SEF can be tuned by varying the pattern features of the SNP arrays [49].

Though the deposition of metal nanoparticles onto the substrate often produce some local "hot spot" randomly, which will helpful to obtain a high fluorescence signal, it is hard to obtain a reliable, stable, and uniform signal with a large area due to particle aggregation. Prof. Qiu reported a convenient method to fabricate highly ordered hemispherical silver nanocap arrays templated by porous anodic alumina membranes suitable for robust and cost-efficient enhanced fluorescence substrates, as shown in Fig. 15.7 [50]. Compared to those on glass used in conventional imaging, eightfold enhancement factor was obtained. Spectral analysis suggests that the PSNAs can create more excitons in the light-emitting P-FITC because of plasmon resonance energy transfer from the silver nanocaps to the nearby P-FITC.

15.3.1.4 Plasmon Enhanced Fluorescence from Nanorod Arrays Substrate

LSPR excitation are associated with a strongly enhanced electromagnetic (EM) field in the vicinity of the particles, which is the key mechanism in enhanced EM field and exploited in enhanced spectroscopy such as surface enhanced Raman scattering (SERS)[51] and enhanced fluorescence. The shape and arrangement of the metal nanostructures play a major role on the LSPR properties. Therefore, their precise tuning is very important to get a coupling of the molecular resonance and the enhanced EM field near metal nanostructures [52]. Compared with local enhanced EM field generated in randomly distributed metal nanostructures or nanoscale roughness of metallic films, the near-field hybrid coupling among particle pairs can lead to shifts in the resonant wavelength and anomalously large fields for a array of nanostructure. By using an electrodeposition method with an AAO template, the highly regular nanorod arrays were fabricated. The author reports that considerable enhancement in fluorescence of up to a factor of two orders of magnitude was observed as compared to the emission on the controlled substrate [53].

Recently, Abdulhalim et al. has performed a comparative study of SEF on the Glancing angle deposition (GLAD) thin films of different materials like silver, gold, copper, and silicon. However, further research and experiments are still required to improve the enhancement factor for sensitive studies as well as understanding the effect of film morphology on the SEF phenomenon. The effect of silver nanorods arrays morphology on the SEF has been investigated. And the author found that with increasing the nanorod length, a significant rise in SEF enhancement factor up to 32-fold was observed with respect to the conventional silver film. The variation of SEF efficiency was qualitatively explained using lightning-rod and surface plasmon resonance effects [54].

15.3.2 Non-Periodical Metallic Plasmonic Nanostructure

Spatial correlations and distribution of EM field and intensity have been extensively investigated in the context of EM wave propagation in nonlinear disordered systems and are indicative of the nature of wave transport in random media [55]. In disordered metallic nanostructure, the enhancement of local EM intensity correlations due to the localized surface plasmons and the lightning-rod effect which governed by structural inhomogeneities, has been widely studied, as reviewed by Prof. V. M. Shalaev [56, 57]. Due to the resonant dipolar excitations in disorder nanostructures localized in subwavelength sized regions and exhibited strong frequency and polarization dependence of their spatial location, the near-field intensity distribution across a semicontinuous metal film, such as rough, fractal-like and so on, near the "percolation threshold" is extremely inhomogeneous with giant local field maxima (hot spots). So the spatial location of light-induced dipole excitations is determined by both the local topography and the large-scale geometrical structure [58]. For the giant local EM field excited in the nonperiodical structure, the spectrum properties of molecules located in the vicinity of "hot spots" existed in the plasmonic nanostructure can be moderated. As a result, the investigation of enhanced fluorescence effect of many typed nonperodical nanostructure, such as silver island film (SiF), fractal-like, and et. al. on fluorescence have been widely performed.

15.3.2.1 Plasmon Enhanced Fluorescence from Metallic Silver Island Substrate

Fluorescence enhancement on substrates with metallic islands film was subject to research since the 1980s [59]. This approach offers the advantage of a relatively simple preparation method and provides moderate enhancement factors through the combined effect of Plasmon enhanced fluorescence excitation rate and increased quantum yield. With the help of thermal vapor deposition method, different thicknesses of copper films from 1 to 5 nm were deposited onto glass slides has been used to study the enhanced fluorescence effect. The experimental observations show that fluorescence intensity of fluorophores located in close proximity to the Cu films critical dependence on the Cu thickness, and reaches a maximum 2.5-fold at 3 nm. The surface plasmons from Cu film can radiate and therefore enhance a fluorophore's spectral properties, dedicating to the fluorescence enhancement [60]. Furthermore, a bimetallic island film (Ag/Cu) was fabricated to study the enhanced fluorescence. This simple procedure yielded a silver overlayer with features ranging from random silver nanoparticles to silver thin films on top of the copper thin films. Compared with free-space emission of fluorophore, the largest enhancement in PFE was obtained with the silver island deposited for 50 s onto copper film. The angle of SPCF emission was shifted to smaller angles was also observed, which is predicted by the Fresnel calculations. So the Ag/Cu bimetallic combination has enormous implications for SFE-based bioassays than that using fluorophores emitting over a broad range of wavelengths on a single assay platform [61].

15.3.2.2 Surface Enhanced Fluorescence from Metallic Fractal-Like Substrate

The nonlinear electrical and optical properties of disorder metal nanostructured composites, including thin fractal like films, colloidal aggregates, etc., have been attracted much attention in recent years. Nanostructured composite materials are potentially of great practical importance as media with intensity-dependent dielectric functions. Due to the strong fluctuations of the local fields, and these fluctuations of the optical response could be tuned by controlling the volume fraction and morphology of constituents [57, 62]. Nonperiodical nanostructures have different surface plasmonic modes that can interact and form collective surface plasmons [63]. With the selection of proper EM frequencies, the delocalized surface Plasmon excitations were created extending over the whole substrate. And also the localization of plasmon excitations in small areas with giant EM field named as "hotspot". Due to localization of optical excitations in fractal, with the strong morphology-dependent resonances of nanostructure, dedicating to the EM field concentration, nonperiodical metal nanostructures with fractals-like type have the potential to strongly amplify a range of effects associated with EM radiation, such as SFE. As we know, the resonant excitation of plasmons in metallic nanostructures can generate intense electromagnetic field on the surfaces of metals, which in turn provide dramatic increases in the detected spectroscopic signals for molecules located in the close vicinity of metal surfaces. It is has been shown that fluorescence enhancement effect fluorophore emission was observed on "hotspots" in silver fractal-like nanostructure [64]. The combination of extended fractals' area and strong interactions of the excited-state fluorophores with the collective surface plasmons of the fractal-like structure are contributed to the fluorescence enhancement on silver fractals nanostructure. Prof. Tanya Shtoyko study the enhanced fluorescence effect from the silver fractal-like nanostructures fabricated by electrochemical reduction of silver on the surface of glass slides [65]. A significant enhancement (more than 100-fold) of the signal was detected on fractals compared to bare glass. It is demonstrated that such fractal-like structures can assist in improving the signals from assays used in medical diagnostics, especially those for analytes with molecular weight under 100 kDa (Fig. 15.8).

The fluorescence enhancement effect is very sensitive to the topography and dielectric property of metal substrate. Prof. Zheng investigated enhanced fluorescence effect from the metallic substrates with complex structures, which are made of silver fractal-like structure and nanoparticles, are prepared through electrochemical reduction followed by physical deposition [66]. The surface enhanced fluorescence of Rh6G monolayer molecules deposited on the prepared complex substrates are investigated with laser spectroscopic technique. The experimental results show that the fractal-like structure decorated with silver nanoparticles presents stronger fluorescence enhancement, as high as 12.0-fold, comparing with silver nanoparticles or pure silver fractal-like structure. The better enhancement effect from the fractal-like sample was attributed to the fact that localized plasmons created more "hot spots" in the formed three dimensional nanostructure.

Furthermore, the experimental observation of surface enhanced spectroscopy, including SERRS and SEF, from Rh6G monolayer deposited on the electrochemically fabricated fractal-like silver substrate is presented [67]. Based on the AFM characterization of fractal nanostructure substrate, we numerically study the EM field distribution of fractal substrate with finite element method (FEM), it shows that giant local EM enhancement is formed on the prepared substrate, which boosts Raman and fluorescence signals simultaneously. When the fractal-like silver nanostructure composing constituent particles are illuminated by incident light, oscillation of dipole moments are induced which interact strongly via dipolar interactions,

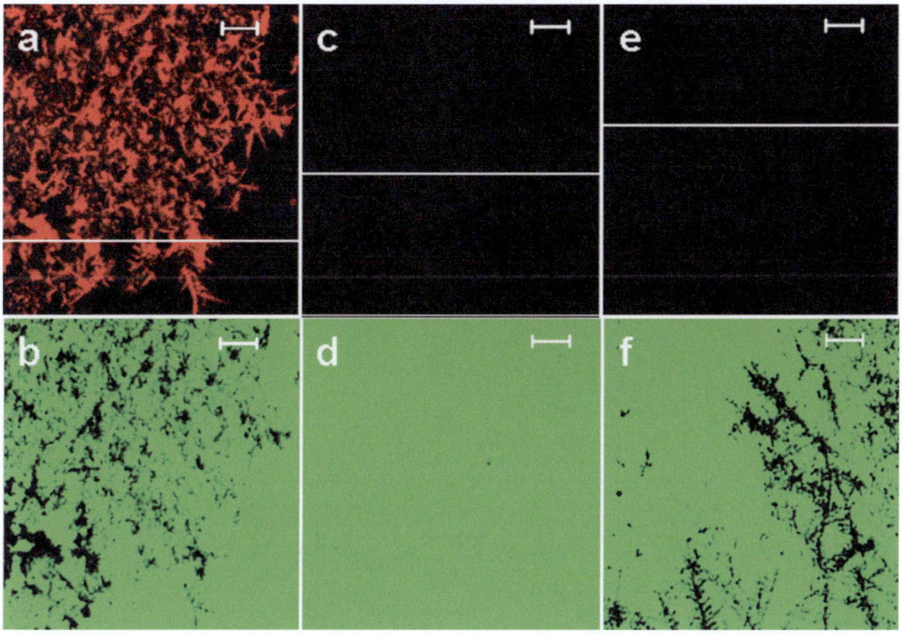

Fig. 15.8 High-magnification laser scanning confocal microscope images; the images of the immunoassay performed on a glass slide with fractal-like structures (*left column*), the immunoassay performed on a clean glass slide (*middle column*), and a nonspecific binding performed on a glass slide with fractal-like structures (*right column*): (**a**) the fluorescence images of the immunoassay prepared on the glass slide with fractal-like structures, (**c**) the immunoassay prepared on a clean glass slide, and (**e**) a nonspecific binding performed on a glass slide with fractals; (**b**) the images in transmission mode of the immunoassay prepared on the glass slide with fractals, (**d**) on a clean glass slide, and (**f**) on a glass slide with fractal-like structures with nonspecific protein binding. The bars on all images represent 5 μm (adapted with permission from [65]. Copyright (2010) American Chemical Society)

resulting in the formation of collective optical mode. Since the enhanced optical process is proportional to the local EM field in the silver fractal-like surfaces, the resulting enhancement associated with the "hotspots" can be extremely large. Specifically, because the local enhancement factor $\propto |E|^4$ for SERRS and $\propto |E|^2$ for SEF, it can reach giant local enhancement factor, making Raman and fluorescence spectroscopy of adsorbed fluorophore located in vicinity of fractal-like structure could be experimentally observed. The result is helpful for exploring the possibility of studying the properties of the molecules with both vibrational and electronic transitions simultaneously (Fig. 15.9).

As reported in Ref. [68], the diffusion-limited aggregation model is widely used to explain the growth of silver dendrites. The evolution of the silver aggregates to a thermodynamically stable flower-like dendrite nanostructure attributed to strong anisotropic growth. Based on the galvanic displacement reaction method, with properly controlled experimental condition, such as reaction time and solution

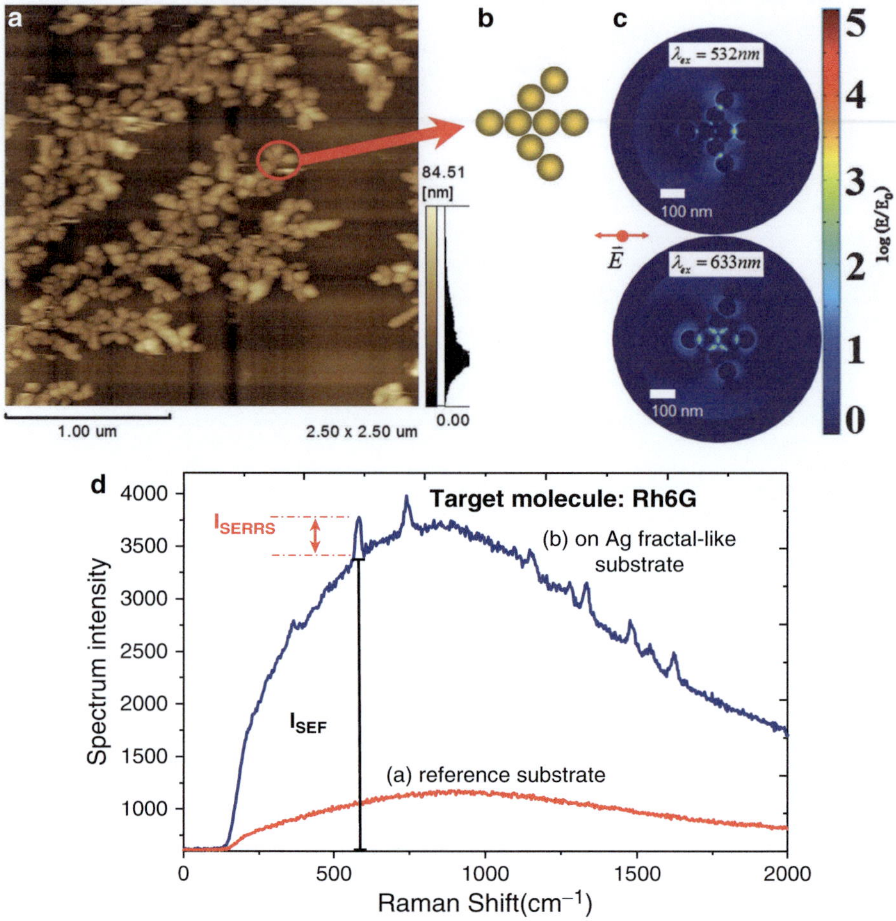

Fig. 15.9 (**a**) AFM image of silver fractal nanostructure, (**b**) the simulated model of fractal nanostructure marked with *red line circle*, (**c**) the electrical field with 532 nm (*up*) and 633 nm (*down*), and (**d**) enhanced spectra of Rh6G molecules on (**a**) reference glass substrate, and (**b**) silver fractal-like substrate (adapted with permission from [67])

concentration, a heterogeneous silver flower-like dendrites nanostructures with well-crystallized separate hexagonal plates was fabricated, which exhibits very good fluorescence enhancement capabilities [69]. According to the distance dependent nature of the SEF effects, Prof. Li study the distance effect on the enhanced fluorescence by introduction of stimulus-responsive PAA/PDDA multilayer film as an interlayer, the distance between the fluorophores and the silver dendrites nanostructures could be tuned by immersing the substrates into solutions of different ionic strength or pH [70].

15.3.2.3 Surface Enhanced Fluorescence from Deposited Metallic Nanoparticle Substrate

Noble-metal (such as Cu, Ag, and Au) nanoparticles have been the subject of extensive studies due to their extraordinary geometrical-dependent optical properties. Specifically, these nanoparticles exhibit strong UV-visible absorption band that is not present in the spectrum of the bulk metal. This absorption band results when the incident photon frequency is resonant with the collective electrons and is known as LSPR, which is strongly depends on the nanoparticle size, shape, and the surrounding medium. With the help of LSPR, the local enhanced EM field can be formed. As a result, by introduction of nanoparticles with properties geometrical shape, the emission of fluorophore can be moderated. However, Due to non-radiative decay rat is critically depends on the fluorophore–metal particles distance, SPR do not always enhance the emission properties of the fluorophore [71]. So the investigation of the coupling effect between the resonance of the metal particle and the emission of the fluorophore is extremely importance.

In order to prepare the enhanced fluorescence substrate with uniform distribution of metal nanoparticles, the self-assembly method are often adapted in the sample fabrication process. For example, when the silanized glass substrate coated with 3-aminopropyltrimethoxysilane (APS) immersed into a colloidal solution, the silver or gold nanoparticles can be immobilized on the surface, and the density of silver nanostructures could be controlled by both the amount of time the substrate was immersed in the colloidal solution and the concentration of the colloidal solution. Prof. E. M. Goldys [72] has been developed for the deposition of uniform silver-coated nanoparticles on glass substrate and demonstrate that the Au core-Ag shell nanostructures on glass surfaces are promising substrates for fluorescence enhancement with outstanding macroscopic. The origin of this enhancement for appropriate size nanoparticles is attributed to the effect of an increased excitation rate from the local field enhanced by the interaction of incident light with the nanoparticles and/or higher quantum yield from an increase of the intrinsic decay rate of the fluorophore. By employing sequential surface modification method, Prof. C. D. Geddes employed polycarbonate (PC) films as substrate to deposit silver nanoparticles on it [73]. It was found that the silver nanoparticles were deposited more homogeneously onto PC films etched by UV-radiation as compared to PC films etched with sodium hydroxide. And Enhanced emission of fluorescence from FITC was observed from the silver-deposited side of the PC film as compared to the non-silvered side. By comparing the fluorescence enhancement of dyes on different shaped silver nanoparticle self-assembled films, such as spherical, rod-shaped, nanotetrahedral and prismatic silver NP SA films, with a proper thickness polyelectrolyte spacing layer, the localized plasmon controlled fluorescence has been systemically investigated [74]. It is found that the strategy of LSPR band coupling with the excitation band of dyes is more efficient than the strategy of the LSPR band overlapping the emission band.

Due to the Plasmon enhanced fluorescence is a property of the near-field interaction of an excited state fluorophore with the enhanced local EM field generated on the metal nanoparticle, the coupled metal NPs could produce larger enhancement when the fluorophore is located in the proper position [75]. Silver nanopartilce arrays were prepared by means of combining nanoimprint lithography and electrochemical deposition methods. Due to the different coupling efficiency between the emission of the fluorophores and surface plasmon resonance band of the metallic nanostructures, the fluorescence intensity can be tuned by engineering the feature size of the SNP arrays. It is found that the fabrication of effective SEF substrate with high local EM field or "hot spots" will produce better fluorescence enhancement. Prof. Zheng observed the simultaneous enhanced Raman and fluorescence spectrum from the gold dimers and trimers nanostructure [76]. Particularly the dimensional of enhanced substrates play an important role for enhancing the fluorescence emission. It is shown that two-dimensional (2D) silver nanoparticle (AgMy) sheets can enhance the fluorescence only when the excitation wavelength overlaps with the plasmon resonance wavelength [77]. Compared to 2D nanostructure, the larger enhanced local EM field with further deposition of SNPs on the top of monolayer SNPs structure, more "hot spots" are produced and larger enhanced local EM field were formed in the 3D nanostructured substrate. When the dye molecules located nearby the 3D nanostructure, giant enhanced fluorescence emission will be observed. The result suggests that a metallic substrate with 3D nanostructures can produce better fluorescence enhancement, which is important for studying the mechanism and expanding the potential applications of enhanced fluorescence effect [78] (Fig. 15.10).

It is reported that the luminescence enhanced effect from Ln-doped NCs situated near noble metal nanostructures is generally attributed to the local field enhancement associated with the excitation of SPR in the metal nanostructures [79, 80], which under favorable circumstances can increase the sample's absorption and emission cross sections. As in SERS, the plasmonic metal NPs act as antennas which concentrate (and therefore increase) the strength of the optical fields exciting the transitions in the luminescent NCs (and also the upconversion process). Recently, the enhancement of lanthanide-based UC and downshifting (DS) luminescence were experimental observed simultaneously on metal nanodisk – insulator – metal (MIM) nanostructure fabricated via a nanotransfer printing method [81]. The optimized Ag nanodisk array and conversion layer thickness of the structure matched the MIM resonance mode, which resulted in 174- and 29-fold enhancement of UC and DS luminescence, respectively. The proposed MIM structure is a promising way to harness the entire solar spectrum by converting both ultraviolet and near-infrared to visible light concurrently through resonant-mode excitation (Fig. 15.11).

The robust DS and UC luminescence enhancement in lanthanide ions residing on silica hybrid nanostructures ($LaF_3:Yb^{3+}$, $Er^{3+}@SiO_2$) decorated with varying quantities of Ag NPs was also studied recently [82]. By using an appropriate

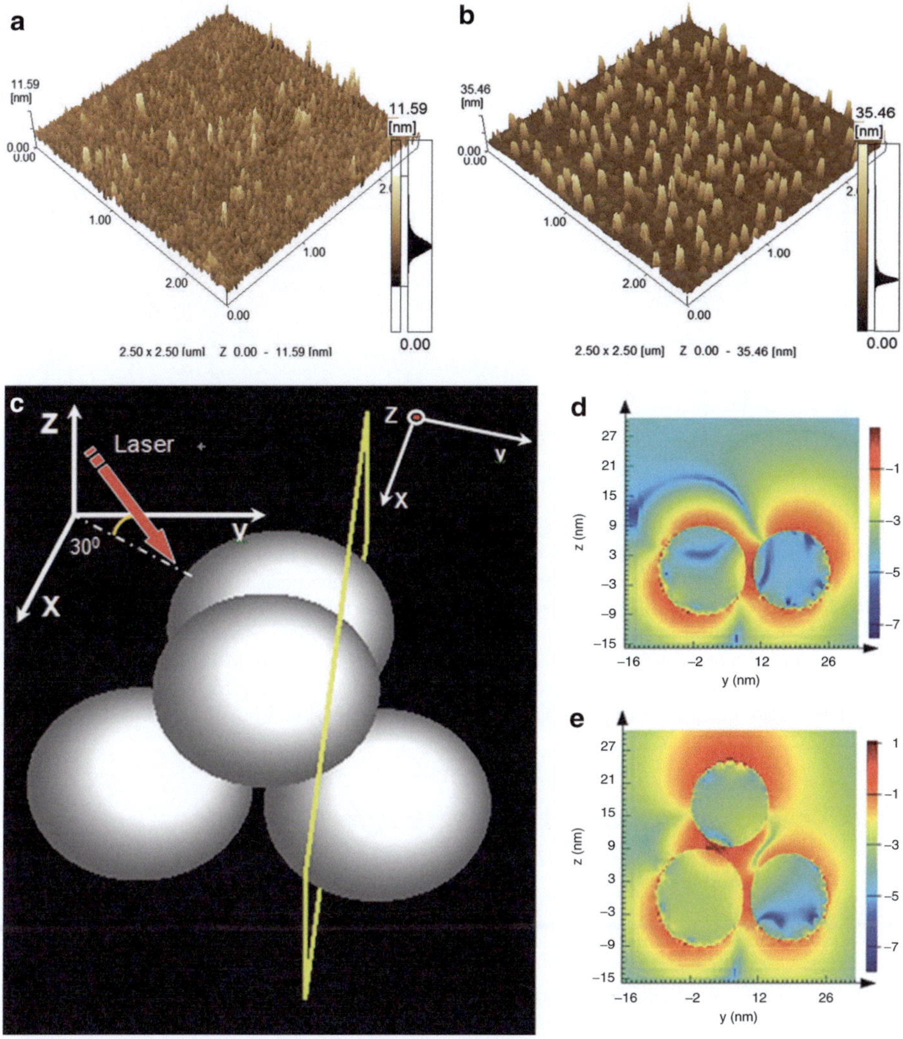

Fig. 15.10 AFM image of prepared substrates (**a**) 2D nanostructure with SNP monolayer and (**b**) 3D nanostructure with double-layer SNPs, and the simulation of EM field distribution in 2D (**d**), 3D (**e**) nanostructure with FDTD method (Lumerical FDTD solutions 7.0.0), and (**c**) is the 3D model (adapted with permission from [78])

model for the fluorescence enhancement and the simultaneous quenching effects induced by the presence of the plasmonic metal NPs, the luminescence enhancement as a function of Ag nanoparticle (NP) loading with the rates of luminescence buildup and decay are observed.

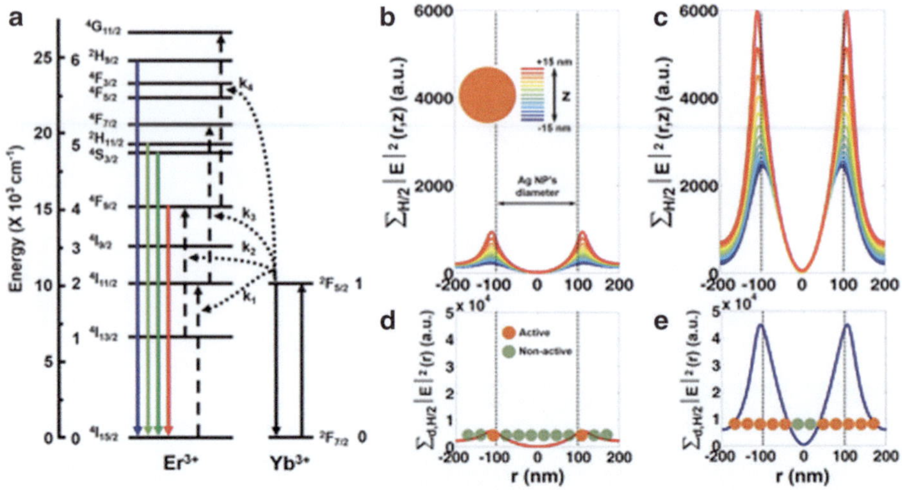

Fig. 15.11 (**a**) Energy diagram of UC material ($NaYF_4$:Yb^{3+},Er^{3+}) and comparison of the plasmonic effects of the MI and MIM structures simulated using the FDTD method. Distribution of the squared intensity of electromagnetic field in the *x*,*y* plane measured at different *z* positions (color-coded), $\sum_{H/2}|E|^2$ (*r*,*z*), for the (**b**) MI and (**c**) MIM structures, respectively. *Vertical dotted lines* guide the position of the top of the Ag nanoparticle (NP). Integrated $\sum_{H/2}|E|^2$ (*r*,*z*) over the *z* space occupied by the UCNPs, $\sum_{d,H/2}|E|^2$ (*r*), for the (**d**) MI and (**e**) MIM structures, respectively. *Color-filled circles* guide the array of the UCNPs (sized to scale) (*blue* for active and brown for nonactive under the plasmonic effect) (adapted with permission from [81]. Copyright (2015) American Chemical Society)

15.3.3 The Spacer and Wavelength Effect Towards the Fluorescence Enhancement

The enhanced and quenched effect of fluorescence are critically effect by the separation distance between fluorophore and plasmnoic nanostructure and spectra characteristics (absorption and emission of the fluorescence materials and LSPR wavelength of the plasmon nanostructures) [84]. Tuning of the spacing between the fluorophore and metallic substrate is also expected to change the coupling between LSPR and excited state fluorophore. As a result, many researches have been done to precisely modulate the distance with high resolution, such as adapted SiO_2 [83], TiO_2 [85], Al_2O_3 [86], LB monolayer [87] and DNA molecules [88] as spacers. Prof. Stuart Lindsay and Yan Liu Using stoichiometrically controlled 1:1 functionalization of gold nanoparticles with fluorescent dye molecules (Cy3 and Cy5 molecules) in which the dye molecule is held away from the particle surface by a rigid DNA spacer allows precise determination of the distance-dependent effect of the metal nanoparticles on fluorescence intensity [88].

Prof. Lukas Novotny investigates the fluorescence rate of a single molecule as a function of its distance to a laser-irradiated gold nanoparticle from experimental and theoretical aspect. By varying the distance between molecule and particle, they

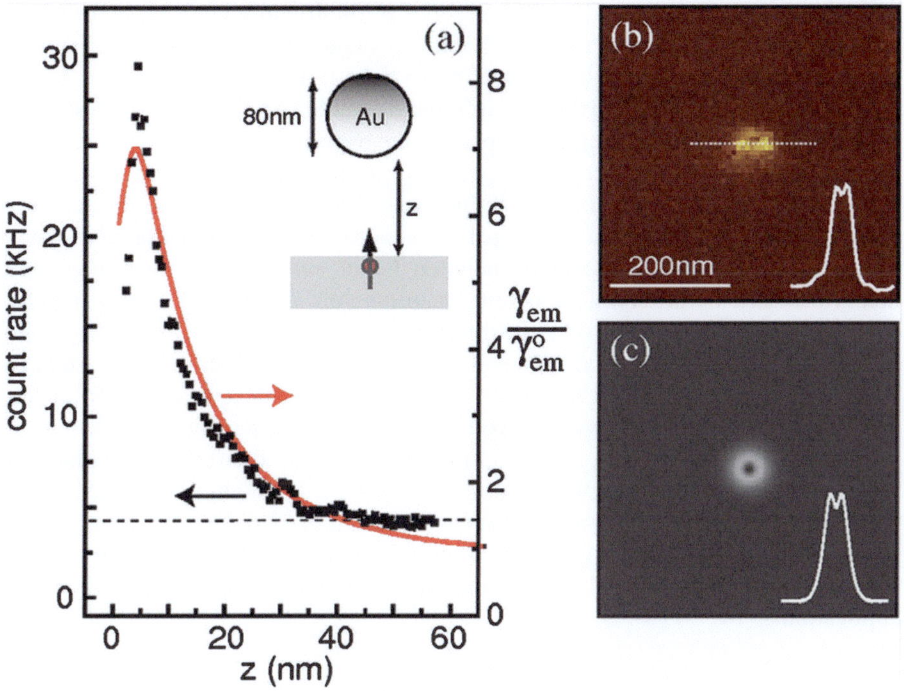

Fig. 15.12 (**a**) Fluorescence rate as a function of particle-surface distance z for a vertically oriented molecule (*solid curve*: theory, *dots*: experiment). The *horizontal dashed line* indicates the background level. (**b**) Fluorescence rate image of a single molecule acquired for z ≈ 2 nm. The dip in the center indicates fluorescence quenching. (**c**) Corresponding theoretical image (adapted with permission from [89])

realized that the continuous transition from fluorescence enhancement to fluorescence quenching, as shown in Fig. 15.12. It is proved that the local field enhancement leads to an increased excitation rate whereas nonradiative energy transfer to the particle leads to a decrease of the quantum yield (quenching) [89].

Besides to the spacer effect, the spectra dependence of plasmon enhancement for fluorescence is also important. Prof. C. D. Geddes study the metal-enhanced fluorescence (MEF) from 6-propionyl-2-dimethylaminonaphthalene (Prodan) in different solvents when placed in close proximity to silver island films (SIFs), surface-deposited nanoparticles [90]. They observe that MEF is wavelength dependent, that is, the enhanced emission of Prodan on SIFs in different solvents changes from 1.5- to 3-fold as compared to a glass control sample containing no silver nanoparticles. Our findings strongly suggest that MEF of Prodan in different solvents is correlated with the scattering portion of the extinction spectrum for metallic particles, i.e., the fluorophore couples and radiates through that scattering mode, which further confirms our laboratories current interpretation of the MEF effect. Several works have been reported that the investigation of the distance and

Plasmon Wavelength dependent Fluorescence of fluorophore via nanorod or nanoparticle substrate [91, 92]. Prof. Udo Bach invesitgate the fluorescence quenching of four different dye molecules, which absorb light at different wavelengths across the visible spectrum and into the near infrared, is studied using a rigid silica shell as a spacer for the first time [92]. A comprehensive experimental determination of the distance dependence from complete quenching to no coupling is carried out by a systematic variation of the silica shell thickness. The plasmonic field enhancement in the vicinity of the 13 nm gold nanoparticles is calculated as a function of distance and excitation wavelength and is included in all calculations. Electrodynamic theory predicts the observed quenching quantitatively in terms of energy transfer from the molecular emitter to the gold nanoparticle.

Recently, Lin's group report a novel plasmon-enhanced fluorescence system in which the distance between UCNPs and nanoantennae (gold nanorods, AuNRs) was precisely tuned by using layer-by-layer assembled polyelectrolyte multilayers as spacers [93]. By modulating the aspect ratio of AuNRs, localized surface plasmon resonance (LSPR) wavelength at 980 nm was obtained, matching the native excitation of UCNPs resulting in maximum enhancement of 22.6-fold with 8 nm spacer thickness. These findings provide a unique platform for exploring hybrid nanostructures composed of UCNPs and plasmonic nanostructures in bioimaging applications.

15.4 Conclusion

Placing luminescent center (Ln-doped NCs or fluorescent molecules) near plasmonic nanostructure (with resonances close to the absorption and/or emission frequencies of the fluorohphore), by tailoring the electromagnetic environment and mastering the local environment, offers additional channels for luminescence quenching (and enhancement) over and above what occurs with bulk metal surfaces. SPP and LSPR excited in periodical and nonperiodical metal nanostructures provide also very intense electromagnetic enhancement (for instance in ‘hot spots’). So the spectrum emission from the fluorophore in the vicinity of nanostructured metallic surface can be modified by controlling the geometrical shape, composition of the metallic nanostructure and dielectric properties of the surrounding media. This chapter described for emission control and enhancement of the fluorophores are readily applicable to other sources of fluorophore luminescence (such as chemiluminescence, electroluminescence and phospholuminescence) and their applications in point-of-care medical diagnostics, food control, and safety.

References

1. Ritchie RH (1957) Plasma losses by fast electrons in thin films. Phys Rev 106:874
2. Pitarke JM, Silkin VM, Chulkov EV, Echenique PM (2007) Theory of surface plasmons and surface-plasmon polaritons. Rep Prog Phys 70:1–87
3. Ringe E, Zhang J, Langille MR, Sohn K, Cobley C, Au L, Xia Y, Mirkin CA, Huang J, Marks LD, Van Duyne RP (2010) Effect of size, shape, composition, and support film on localized surface plasmon resonance frequency: a single particle approach applied to silver bipyramids and gold nanocubes. Mater Res Soc Symp Proc 1208:O10–02
4. Ringe E, Langille MR, Sohn K, Zhang J, Huang J, Mirkin CA, Van Duyne RP, Marks LD (2012) Plasmon length: a universal parameter to describe size effects in gold nanoparticles. J Phys Chem Lett 3:1479–1483
5. Jain PK, El-Sayed MA (2007) Surface plasmon resonance sensitivity of metal nanostructures: physical basis and universal scaling in metal nanoshells. J Phys Chem C 111:17451–17454
6. Brandl DW, Mirin NA, Nordlander P (2006) Plasmon modes of nanosphere trimers and quadrumers. J Phys Chem B 110:12302–12310
7. Ghosh SK, Pal T (2007) Interparticle coupling effect on the surface plasmon resonance of gold nanoparticles: from theory to applications. Chem Rev 107:4797–4862
8. Sun MT, Zhang ZL, Chen L, Sheng SX, Xu HX (2014) Plasmonic gradient effects on high vacuum Tip-enhanced raman spectroscopy. Adv Optical Mater 2:74–80
9. Fort E, Gresillon S (2008) Surface enhanced fluorescence. J Phys D Appl Phys 41:013001 (1–31)
10. Xu HX, Wang XH, Persson MH, Johansson P (2004) Unified treatment of fluorescence and raman scattering processes near metal surfaces. Phys Rev Lett 93:243002(1–4)
11. Drexhage KH (1974) Interaction of light with monomolecular dye layers. Prog Opt 12:163–232
12. Sun MT, Zhang ZL, Zheng HR, Xu HX (2012) In-situ plasmon-driven chemical reactions revealed by high vacuum tip-enhanced Raman spectroscopy. Sci Rep 2:647
13. Lantman EMV, Deckert-Gaudig T, Mank AJG, Deckert V, Weckhuysen BM (2012) Catalytic processes monitored at the nanoscale with tip-enhanced Raman spectroscopy. Nat Nanotechnol 7(9):583–586
14. Sun MT, Zhang ZL, Chen L, Li Q, Sheng SX, Xu HX, Song P (2014) Plasmon-driven selective reductions revealed by tip-enhanced raman spectroscopy. Adv Mater Interfaces 1:1300125 (1–6)
15. Höppener C, Novotny L (2012) Exploiting the light-metal interaction for biomolecular sensing and imaging. Q Rev Biophys 45(2):209–255
16. Bauch M, Toma K, Toma M, Zhang Q, Dostalek J (2014) Surface plasmon-enhanced fluorescence biosensors: a review. Plasmonics 9:781–799
17. Lakowicz JR (2006) Principles of fluorescence spectroscopy, 3rd edn. Springer, New York
18. Geddes CD (2011) Reviews in fluorescence. Springer, New York
19. Moskovits M (1985) Surface-enhanced spectroscopy. Rev Mod Phys 57(3):783–826
20. Purcell EM (1946) Spontaneous emission probabilities at radio frequencies. Phys Rev 69:681
21. Galloway CM, Etchegoin PG, Le Ru EC (2009) Ultrafast nonradiative decay rates on metallic surfaces by comparing surface-enhanced raman and fluorescence signals of single molecules. Phys Rev Lett 103:063003
22. Johansson P, Xu HX (2005) Surface-enhanced Raman scattering and fluorescence near metal nanoparticles. Phys Rev B 72:035427
23. Itoh T, Iga M, Tamaru H, Yoshida K, Biju V (2012) Quantitative evaluation of blinking in surface enhanced resonance scattering and fluorescence by electromagnetic mechanism. J Chem Phys 136:024703
24. Barnes WL, Dereux A, Ebbesen TW (2003) Surface plasmon subwavelength optics. Nature 424:824–830

25. Wood RW (1902) On a remarkable case of uneven distribution of light in a diffraction grating spectrum. Philos Mag 4:396–402
26. Ritchie RH, Arakawa ET, Hamm RN (1968) Surface-plasmon resonance effect in grating diffraction. Phys Rev Lett 21:1530–1533
27. Cui XQ, Tawa K, Hori H, Nishii J (2010) Tailored plasmonic gratings for enhanced fluorescence detection and microscopic imaging. Adv Func Mat 20:546–553
28. Tsuneyasu M, Sasakawa C, Naruishi N, Tanaka Y, Yoshida Y, Tawa K (2014) Sensitive detection of interleukin-6 on a plasmonic chip by grating-coupled surface-plasmon-field-enhanced fluorescence imaging. Jpn J Appl Phys 53:06JL05
29. Jiang Y, Wang HY, Wang H, Gao BR, Hao YW, Jin Y, Chen QD, Sun HB (2011) Surface plasmon enhanced fluorescence of dye molecules on metal grating films. J Phys Chem C 115:12636–12642
30. Hao YW, Wang HY, Zhang ZY, Zhang XL, Chen QD, Sun HB (2013) Time-resolved fluorescence anisotropy of surface plasmon coupled emission on metallic gratings. J Phys Chem C 117:26734–26739
31. Yuk JS, Guignon EF, Lynes MA (2014) Sensitivity enhancement of a grating-based surface plasmon-coupled emission (SPCE) biosensor chip using gold thickness. Chem Phys Lett 591:5–9
32. Zhang ZY, Wang HY, Du JL, Zhang XL, Hao YW, Chen QD, Sun HB (2015) Surface plasmon-modulated fluorescence on 2D metallic silver gratings. IEEE Photonics Technol Lett 27(8):821–823
33. Lin JH, Liou HY, Wang CD, Tseng CY, Lee CT, Ting CC, Kan HC, Hsu CC (2015) Giant enhancement of upconversion fluorescence of $NaYF_4$, Yb^{3+}, Tm^{3+} nanocrystals with resonant waveguide grating substrate. ACS Photonics 2(4):530–536
34. Lu DW, Cho SK, Ahn S, Brun L, Summers CJ, Park W (2014) Plasmon enhancement mechanism for the upconversion processes in $NaYF_4$: Yb^{3+}, Er^{3+} nanoparticles: Maxwell versus Förster. ACS Nano 8(8):7780–7792
35. Ebbesen TW (1998) Extraordinary optical transmission through sub-wavelength hole arrays. Nature 391:667–669
36. Sambles JR, Bradbery GW, Yang FZ (1991) Optical-excitation of surface-plasmons-an introduction. Contemp Phys 32:173–183
37. Ghaemi HF, Thio T, Grupp DE, Ebbesen TW, Lezec HJ (1998) Surface plasmons enhance optical transmission through subwavelength holes. Phys Rev B 58:6779–6782
38. Chen Y, Munechika K, Ginger DS (2007) Dependence of fluorescence intensity on the spectral overlap between fluorophores and plasmon resonant single silver nanoparticles. Nano Lett 7:690–696
39. Brolo AG, Kwok SC, Moffitt MG, Gordon R, Riordon J, Kavanagh KL (2005) Enhanced fluorescence from arrays of nanoholes in a gold film. J Am Chem Soc 127:14936–14941
40. Guo PF, Wu S, Ren QJ, Lu J, Chen ZH, Xiao SJ, Zhu YY (2010) Fluorescence enhancement by surface plasmon polaritons on metallic nanohole arrays. J Phys Chem Lett 1:315–318
41. Mazzotta F, Johnson TW, Dahlin AB, Shaver J, Oh SH, Höök F (2015) Influence of the evanescent field decay length on the sensitivity of plasmonic nanodisks and nanoholes. ACS Photonics 2:256–262
42. Im H, Lee SH, Wittenberg NJ, Johnson TW, Lindquist NC, Nagpal P, Norris DJ, Oh SH (2011) Template-stripped smooth Ag nanohole arrays with silica shells for surface plasmon resonance biosensing. ACS Nano 5(8):6244–6253
43. Poirier-Richard HP, Couture M, Brulea T, Masson JF (2015) Metal-enhanced fluorescence and FRET on nanohole arrays excited at angled incidence. Analyst 140:4792–4798
44. Wu L, Bai P, Zhou X, Li EP (2012) Reflection, transmission modes in nanohole-arraybased plasmonic sensors. IEEE Photonics J 4:26–33
45. Wang Y, Wu L, Zhou XD, Wong TI, Zhang JL, Bai P, Li EP, Liedberg B (2013) Incident-angle dependence of fluorescence enhancement and biomarker immunoassay on gold nanohole array. Sens Actuators B 186:205–211

46. Wu L, Zhou X, Bai P (2014) Plasmonic metals for nanohole-array surface plasmon field-enhanced fluorescence spectroscopy biosensing. Plasmonics 9:825–833
47. Saboktakin M, Ye X, Chettiar UK, Engheta N, Murray CB, Kagan CR (2013) Plasmonic enhancement of nanophosphor upconversion luminescence in Au nanohole arrays. ACS Nano 7(8):7186–7192
48. Sanchez-Gonzalez A, Corni S, Mennucci B (2011) Surface-enhanced fluorescence within a metal nanoparticle array, the role of solvent and plasmon couplings. J Phys Chem C 115:5450–5460
49. Yang BJ, Lu N, Qi D, Ma R (2010) Tuning the intensity of metal-enhanced fluorescence by engineering silver nanoparticle arrays. Small 6(9):1038–1043
50. Qiu T, Jiang J, Zhang WJ, Lang XZ, Yu XQ, Chu PK (2010) High-sensitivity and stable cellular fluorescence imaging by patterned silver nanocap arrays. ACS Appl Mater Interfaces 2 (8):2465–2470
51. Xi L, Chen M, Zhao XM, Zhang ZL, Xia JR, Xu HX, Sun MT (2014) Visualized method of chemical enhancement mechanism on SERS and TERS. J Raman Spectrosc 45:533–540
52. Yasukuni R, Ouhenia-Ouadahi K, Boubekeur-Lecaque L, Félidj N, MaureI F, Métivier R, Nakatani K, Aubard J, Grand J (2013) Silica-coated gold nanorod arrays for nanoplasmonics devices. Langmuir 29:12633–12637
53. Huang Q, Huang Z, Meng G, Fu Y, Lakowicz JR (2013) Plasmonic nanorod arrays for enhancementof single-molecule detection. Chem Commun 49:11743–11745
54. Singh DP, Singh JP (2015) Morphology dependent surface enhanced fluorescence study on silver nanorod arrays fabricated by glancing angle deposition. RSC Adv 5(40):31341–31346
55. Seal K, Sarychev AK, Noh H, Genov DA, Yamilov A, Shalaev VM, Ying ZC, Cao H (2005) Near-field intensity correlations in semicontinuous metal-dielectric films, K. Seal. Phys Rev Lett 94:226101
56. Shalaev VM (2000) Nonlinear optics of random media. Springer, Berlin
57. Sarychev AK, Shalaev VM (2000) Electromagnetic field fluctuations and optical nonlinearities in metal-dielectric composites. Phys Rep 335:275–371
58. Bozhevolnyi SI, Markel VA, Coello V, Kim W, Shalaev VM (1998) Direct observation of localized dipolar excitations on rough nanostructured surfaces. Phys Rev B 58(17):11441
59. Weitz DA, Garoff S, Hanson CD, Gramila TJ, Gersten JI (1982) Fluorescent lifetimes of molecules on silver-island films. Opt Lett 7(2):89–91
60. Zhang YX, Aslan K, Previte MJR, Geddes CD (2007) Metal-enhanced fluorescence from copper substrates. Appl Phys Lett 90:173116
61. Aslan K, McDonald K, Previte MJR, Zhang YX, Geddes CD (2008) Silver island nanodeposits to enhance surface plasmon coupled fluorescence from copper thin films. Chem Phys Lett 464:216–219
62. Shalaev VM, Poliakov EY, Markel VA (1996) Small-particle composites. II. Nonlinear optical properties. Phys Rev B 53(5):2437–2449, and reference therein
63. Karpov SV, Gerasimov VS, Isaev IL, Markel VA (2006) Spectroscopic studies of fractal aggregates of silver nanospheres undergoing local restructuring. J Chem Phys 125:111101
64. Geddes CD, Parfenov A, Roll D, Gryczynski I, Malicka J, Lakowicz JR (2003) Silver fractal-like structures for metal-enhanced fluorescence, enhanced fluorescence intensities and increased probe photostabilities. J Fluoresc 13(3):267–276
65. Shtoyko T, Matveeva EG, Chang IF, Gryczynski Z, Goldys E, Gryczynski I (2008) Enhanced fluorescent immunoassays on silver fractal-like structures. Anal Chem 80:1962–1966
66. Dong J, Li XQ, Zheng HR (2011) Surface-enhanced fluorescence from silver fractallike nanostructures decorated with silver nanoparticles. Appl Opt 50(31):G123–G126
67. Dong J, Qu SX, Zheng HR, Zhang ZL, Li JN, Huo YP, Li GA (2014) Simultaneous SEF and SERRS from silver fractal-like nanostructure. Sens Actuators B 191:595–599
68. Gu C, Zhang T (2006) Electrochemical synthesis of silver polyhedrons and dendritic films with superhydrophobic surfaces. Langmuir 24:12010–12016

69. Dong J, Zheng HR, Yan XQ (2012) Fabrication of flower-like silver nanostructure on the Al substrate for surface enhanced fluorescence. Appl Phys Lett 100:051112 (1–3)
70. Ma N, Tang F, Wang XY, He F, Li LD (2011) Tunable metal-enhanced fluorescence by stimuli-responsive polyelectrolyte interlayer films. Macromol Rapid Commun 32:587–592
71. Dulkeith E, Morteani AC, Niedereichholz T, Klar TA, Feldmann J, Levi SA, van Veggel FCJM, Reinhoudt DN, Möller M, Gittins DI (2002) Fluorescence quenching of dye molecules near gold nanoparticles, radiative and nonradiative effects. Phys Rev Lett 89(20):203002(1–4)
72. Xie F, Baker MS, Goldys EM (2006) Homogeneous silver-coated nanoparticle substrates for enhanced fluorescence detection. J Phys Chem B 110:23085–23091
73. Aslan K, Holley P, Geddes CD (2006) Metal-enhanced fluorescence from silver nanoparticle-deposited polycarbonate substrates. J Mater Chem 16:2846–2852
74. Xu S, Cao Y, Zhou J, Wang XN, Wang X, Xu WQ (2011) Plasmonic enhancement of fluorescence on silver nanoparticle films. Nanotechnology 22(27):275715
75. Bek A, Jansen R, Ringler M, Mayilo S, Klar TA, Feldmann J (2008) Fluorescence enhancement in hot spots of AFM-designed gold nanoparticle sandwiches. Nano Lett 8(2):485–490
76. Zhang ZL, Yang PF, Xu HX, Zheng HR (2013) Surface enhanced fluorescence and Raman scattering by gold nanoparticle dimers and trimers. J Appl Phys 113:033102
77. Usukura E, Shinohara S, Okamoto K, Lim J, Char K, Tamada K (2014) Highly confined, enhanced surface fluorescence imaging with two-dimensional silver nanoparticle sheets. Appl Phys Lett 104:121906
78. Dong J, Qu SX, Zhang ZL, Liu MC, Liu GN, Yan XQ, Zheng HR (2012) Surface enhanced fluorescence on three dimensional silver nanostructure substrate. J Appl Phys 111:093101 (1–4)
79. Zhang H, Li YJ, Ivanov IA, Qu YQ, Huang Y, Duan XF (2011) Plasmonic modulation of the upconversion fluorescence in $NaYF_4$,Yb/Tm hexaplate nanocrystals using gold nanoparticles or nanoshells. Angew Chem Int Ed 49:2865–2868
80. Sudheendra L, Ortalan V, Dey S, Browning ND, Kennedy IM (2011) Plasmonic enhanced emissions from cubic $NaYF_4$,Yb, Er/Tm nanophosphors. Chem Mater 23:2987–2993
81. Lee KT, Park JH, Kwon SJ (2015) Simultaneous enhancement of upconversion and downshifting luminescence via plasmonic structure. Nano Lett 15(4):2491–2497
82. He EJ, Moskovits M, Dong J, Gao W, Han QY, Zheng HR, Liu N (2015) Luminescence enhancement mechanism of lanthanide-doped hybrid nanostructures decorated by silver nanocrystals. Plasmonics 10(2):357–368
83. Cheng D, Xu QH (2007) Separation distance dependent fluorescence enhancement of fluorescein isothiocyanate by silver nanoparticles. Chem Commun 3:248–250
84. Mishra H, Mali BL, Karolin J, Dragan AI, Geddes CD (2013) Experimental and theoretical study of the distance dependence of metal-enhanced fluorescence, phosphorescence and delayed fluorescence in a single system. Phys Chem Chem Phys 15:19538
85. Luo Q (2014) Large enhancements of $NaYF_4$,Yb/Er/Gd nanorod upconversion emissions via coupling with localized surface plasmon of Au film. Nanotechnology 25:185401
86. Saboktakin M, Oh X, Ye SJ, Hong SH, Fafarman AT (2012) Metal-enhanced upconversion luminescence tunable through metal nanoparticle-nanophosphor separation. ACS Nano 6:8758–8766
87. Ray K, Badugu R, Lakowicz JR (2006) Distance-dependent metal-enhanced fluorescence from Langmuir–Blodgett monolayers of alkyl-NBD derivatives on silver island films. Langmuir 22:8374–8378
88. Chhabra R, Sharma J, Wang H, Zou S, Lin S, Yan H, Lindsay S, Liu Y (2009) Distance-dependent interactions between gold nanoparticles and fluorescent molecules with DNA as tunable spacers. Nanotechnology 20(48):485201
89. Anger P, Bharadwaj P, Novotny L (2006) Enhancement and quenching of single-molecule fluorescence. Phys Rev Lett 96:113002
90. Zhang YX, Dragan A, Geddes CD (2009) Wavelength dependence of metal-enhanced fluorescence. J Phys Chem C 113(28):12095–12100

91. Reineck P, Gómez D, Ng SH, Karg M, Bell T, Mulvaney P, Bach U (2013) Distance and wavelength dependent quenching of molecular fluorescence by Au@SiO_2 core–shell nanoparticles. ACS Nano 7(8):6636–6648
92. Abadeer NS, Brennan MR, Wilson WL, Murphy CJ (2014) Distance and plasmon wavelength dependent fluorescence of molecules bound to silica-coated gold nanorods. ACS Nano 8 (8):8392–8406
93. Feng AL, You ML, Tian L, Singamaneni S, Liu M, Duan Z, Lu TJ, Xu F, Lin M (2014) Distance-dependent plasmon-enhanced fluorescence of upconversion nanoparticles using polyelectrolyte multilayers as tunable spacers. Sci Rep 5:7779

Chapter 16
Remote Spectroscopy Below the Diffraction Limit

James A. Hutchison and Hiroshi Uji-i

Abstract The ability to confine and direct light propagation at the surface of metals via surface plasmon polaritons has been at the centre of the growth in interest in nanoscale engineering of noble metals over the past decades. This chapter reviews the recent development in our laboratories of novel techniques for remote Raman and fluorescence sensing using sub-diffraction limit diameter metallic nanowires as plasmonic waveguides. Applications of remote sensing for chemical and live cell spectroscopy are discussed as well as insights into super-resolution imaging of metallic nanostructures.

Keywords Remote excitation • Sub-diffraction limited plasmonic waveguide • Live cell spectroscopy • Super-resolution imaging

16.1 Introduction

The ability to confine and direct light propagation at the surface of metals needs little introduction and has been at the centre of the growth in interest in nanoscale engineering of gold and silver over the past two decades. Surface plasmon polaritons (SPPs), the coherent coupling of an electromagnetic field with the free electron plasma at the metal surface, are at the origin of this surface confinement [1, 2]. The optical response of silver and gold ($>$550 nm) in the visible region, that is, a negative, real permittivity, defines the SPP at an interface with a dielectric material. The intensity of the SPP is maximal at the interface and decays into both the metal and the dielectric media (1/*e* decay length ~20 nm into the metal medium and several hundreds of nm into the dielectric medium). The high field intensity

J.A. Hutchison
ISIS & icFRC, Université de Strasbourg and CNRS, Strasbourg, France

School of Chemistry and Bio21 Institute, University of Melbourne, Melbourne, VIC, Australia

H. Uji-i (✉)
Department of Chemistry, Katholieke Universiteit Leuven, Heverlee, Belgium

Research Institute for Electronic Science (RIES), Hokkaido University, Sapporo, Japan
e-mail: hiroshi.ujii@chem.kuleuven.be

C.D. Geddes (ed.), *Reviews in Plasmonics 2015*,
DOI 10.1007/978-3-319-24606-2_16

directly at the interface makes SPPs of extreme interest for spectroscopic studies of surfaces.

The SPP propagation distance along the interface is mostly determined by absorption in the metal (i.e., the magnitude of the imaginary part of the metal permittivity). The origin of this absorption is primarily Ohmic heating however losses can also occur via radiative channels, specifically the scattering of the SPP into far field radiation by surface imperfections (roughness) [3, 4]. The latter can in principle be reduced by careful growth of atomically flat metal surfaces or by using crystalline nanoparticles. Nevertheless, even for intrinsically low-loss metals like silver and gold at red-to-near-IR wavelengths, the characteristic propagation length of the SPP is typically in the micron-to-millimetre range. Though much progress has been made in the area of gain assisted SPP propagation [5, 6], and while the practical development of intrinsically low-loss negative index materials remains in its infancy [7, 8], the transport of optical signals over distances greater than millimetres will remain the prerogative of dielectric waveguides.

Nevertheless, techniques which benefit from the confinement and transport of light as SPPs over micron distances remain eminently practical, and an entire field has emerged to make use of SPPs for optical nano-circuitry (plasmonics) [9, 10]. In early 2009, Ujii-i, Hofkens, Hutchison, and co-workers pioneered another technique, the concept of remote sensing on sub-diffraction limit diameter metallic nanowires [11]. They demonstrated the detection of surface-enhanced Raman scattering (SERS) [1] from molecules adsorbed at points of nanoparticle adsorption on a silver nanowire many microns away from the point of laser excitation, with the excitation light delivered to the hotspot via SPP propagation. The technique was dubbed remote excitation SERS or RE-SERS, shown conceptually in Fig. 16.1, and was a logical connection of two areas of very active research.

These were, firstly, the enormous body of work focusing on SERS from molecules in the vicinity of gold and silver nanoparticles begun decades earlier with the first observation of anomalously intense Raman scattering from pyridine on roughened silver films by Fleishmann, Hendra and McQuillan in 1973 [12]. It is now well-established that localized SPP resonances, particularly coupled resonances between two nanoparticles, provide very large EM field enhancements in a few-nanometre region, or hotspot, in the gap between the nanoparticles, which can enable detection of few or even single molecule SERS [1, 13, 14]. These hotspots can provide a SERs enhancement factor (EF_{SERS}) as high as 10^{11} for two silver nanoparticles of 25 nm diameter and separated by a 1 nm gap when optically pumped at the coupled SPP at ~650 nm [1]. Label-free sensing of the vibrational signatures of even a single molecule makes SERS potentially the most powerful molecular sensing technique available *if* SERS hotspots can be reproducibly fabricated and brought to the point of interest effectively.

Secondly, the study of the propagation and in/out-coupling of SPPs on crystalline metallic nanowires was also well advanced. Dickson and Lyon were the first to show the launching of SPPs by focused laser excitation at one end of a silver or gold nanowire, showing that the propagation efficiency depended on the excitation wavelength (governed by the frequency dependence of the imaginary part of the

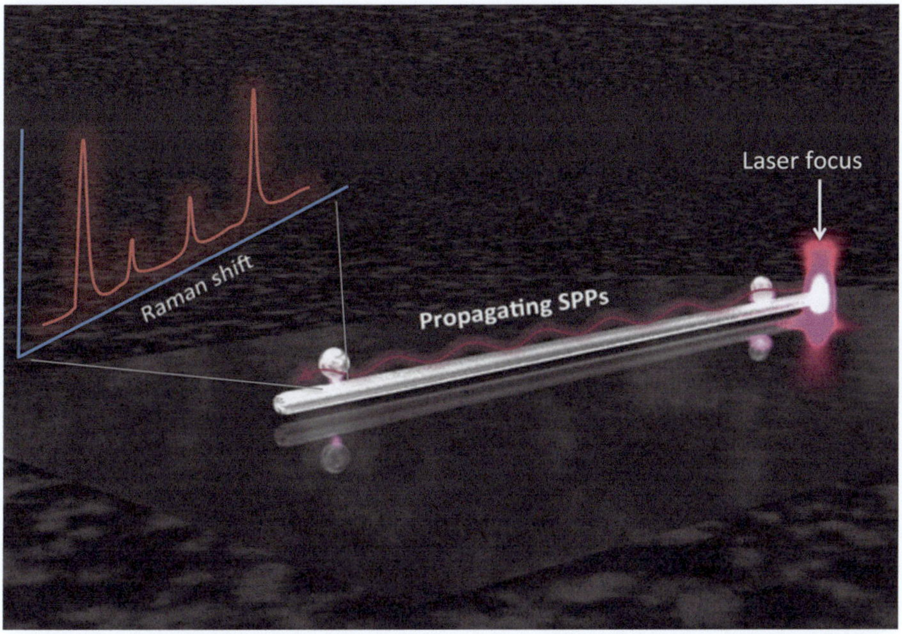

Fig. 16.1 The concept of remote spectroscopy on sub-diffraction limit diameter metallic nanowires. In this case SERS hotspots at points of nanoparticle adsorption along the nanowire are excited by SPP propagating along the wire. Separating the excitation laser focus from the detection volume reduces the spectroscopic background and photo-degradation of the sample. (Figure adapted from Ref. [11])

metal permittivity), and coupling SPPs from one wire to another [15]. Sanders et al. also showed that propagating SPPs could fan out into several nanowires at wire intersections allowing them to estimate a propagation length of 3 μm for 830 nm excitation of a 100 nm diameter nanowire [16]. Kreibig, Krenn, Ditlbacher and co-workers observed SPP decay along a silver wire using both near-field and fluorescence imaging, finding a propagation length of *ca.* 10 μm for 785 nm excitation of a 100 nm silver nanwire [4, 17]. Nordlander, Halas, and co-workers had shown that nanoparticles adsorbed on the surface of a nanowire could efficiently in-couple light into SPPs along the wire and vice versa [18], as predicted earlier by Hao and Nordlander [19]. Lukin and co-workers showed that a single quantum dot could in-couple its emission to surface plasmons along a nanowire [20].

In 2008, both Nordlander, Xu and co-workers, and Moskovits and co-workers, reported SERS from molecules at the junction between a nanoparticle and a nanowire in which excitation and detection were directly at the junction [21, 22]. Finally, in early 2009, Ihee, Kim and co-workers reported SERS from a hotspot between a metal film and a nanowire, again with excitation directly at the point of SERS detection [23].

In light of all these works on SERS sensing using localized SPPs, and nanowire plasmonics using propagating SPPs, our demonstration of *remote excitation* of a SERS hotspot between a nanoparticle and a nanowire using propagating SPPs launched far away from the hotspot was a natural next step. Just months after our publication [11], Nordlander, Xu, and co-workers published their own demonstration of remote excitation SERS [24], confirming all our conclusions. The fundamental advantages of separating by many microns the focused excitation source and the detection volume in remote spectroscopy using propagating SPPs are:

1. A dramatic improvement in signal-to-noise ratio of spectra measured by remote excitation. This is because the excitation volume is reduced from the diffraction-limited focused laser spot to the sub-diffraction limit dimensions of the SPP mode volume, reducing concomitantly background scattering and emission.
2. A reduction in photo-degradation of the sample at the point of interest. The reduced photo-thermal damage is again due to a reduction in the excitation volume, this is shown most clearly in the studies of remote spectroscopy studies of biological specimens in Sect. 16.4.
3. An increase in the spatial resolution for sensing due to the highly confined nature of the SPP mode volume.
4. A reduction in the mechanical damage done by the probe to the environment being probed. This is because metallic nanowires of diameters down to 20 nm can function as efficient SPP waveguides. The advantage of using sub-diffraction limit diameter probes is shown most clearly in Sect. 16.4 where endoscopy of live biological specimens is undertaken.

16.2 General Methods for Remote Spectroscopy on Silver Nanowires

Before discussing various studies undertaken in our laboratories employing remote spectroscopy on metallic nanowires, a brief explanation of some general methods are presented herein.

We employed chemically synthesized crystalline silver nanowires for remote spectroscopy experiments. The nanowires were synthesized using the well-known polyol method in which ethylene glycol acts as both solvent and a weak reductant [25]. Growth of long aspect ratio rods and nanowires occurs due to selective binding and passivation of the {100} side facets of the nanowire by poly(vinylpyrrolidone), PVP. The {111} facets at the end of the wire do not bind PVP as strongly and remain relatively reactive to further growth. The resulting nanowires have a 5-fold twinned structure with a pentagonal cross-section, and are typically 50–300 nm (average 100 nm) in diameter with lengths of 2–30 μm. There are always other nanoparticle types such as spheres and cubes that are also present in the final suspension and which can be partially removed by centrifugation or filtration. We modified the polyol method to improve the yield of nanowires over other particle types by irradiating the solution at the nucleation stage with a tungsten halogen lamp (wavelengths 400–800 nm). This was shown to improve the nanowire yield by fast nucleation of multiply-twinned decahedral nanoparticles, likely due to plamon-associated thermal effects [26].

Experimentally remote spectroscopy can be easily implemented on a standard confocal optical Raman/fluorescence microscope, with the addition of a piezostage recommended to aid in correct nanostructure positioning. A remote spectroscopy experiment can then be simply performed by the following steps:

- The metallic nanowire should be moved until the point of interest for spectroscopic detection along the wire is aligned with the detection optics of the microscope.
- The focused excitation laser which would normally be aligned to overlap with the detection volume for conventional spectroscopy (referred to as 'on-particle excitation' herein), should then be deliberately 'misaligned' such that it excites instead an end of the nanowire several microns away from the point of detection. This can be easily achieved by manipulating the final mirror directing the laser into the microscope.
- Once these two alignment steps have been achieved, normal imaging/spectroscopic detection (Raman, fluorescence) is undertaken.

A schematic of the relative alignment of the excitation laser spot and the detection volume for 'on-particle' and remote excitation of a silver nanowire is shown in Fig. 16.2 for the case where the nanowire serves as an endoscope for sensing inside a biological cell. The Raman signals were collected by a spectrograph equipped with an array detector/cooled electron multiplying charge-coupled device (CCD) camera combination allowing both spectroscopy and imaging to be performed.

16.3 Remote Excitation Surface Enhanced Raman Scattering (RE-SERS)

In this pioneering work [11], silver nanowires were treated with an ethanolic solution containing the small aromatic Raman reporter, 4-aminothiophenol (4-ATP). Covalent binding of the reporter molecule to the metallic wire occurred via the thiol group and the nanowires were then dispersed onto glass substrates and all non-bound 4-ATP rinsed away. Then a solution of spherical silver nanoparticles was spin-cast on the sample such that a few nanoparticles were bound to each nanowire lying on the glass substrate. In this way 4-ATP molecules would be likely positioned at the junction between the nanowire and the adsorbed nanoparticle, the optimal position for SERS. An AFM image of a typical nanoparticle-decorated silver nanowire is shown in Fig. 16.3g.

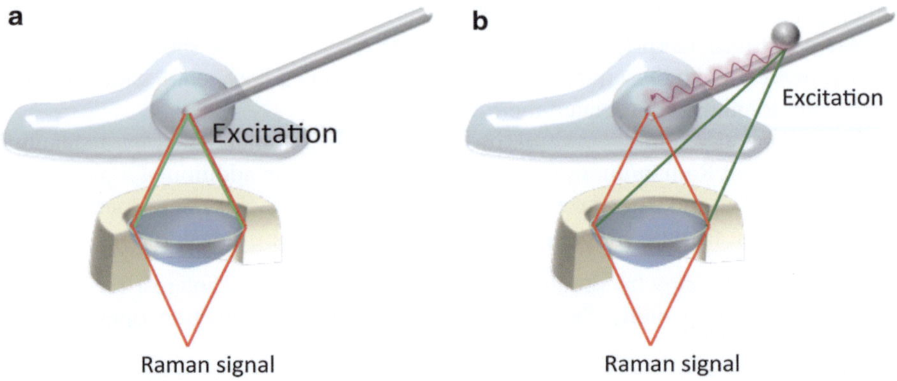

Fig. 16.2 Schematic comparing the alignment of the focused excitation laser spot and the confocal detection volume of the microscope. (**a**) For conventional ('on-particle') microspectroscopy, the laser excitation and detection volume are overlapped. (**b**) For remote microspectroscopy, the laser spot is focused many microns away from the detection volume at a launching point for SPPs on the nanowire. In these experiments the nanowire was fixed to a micromanipulator and used as an endoscope to probe inside live biological cells (see Sect. 16.4) (Figure adapted from Ref. [28])

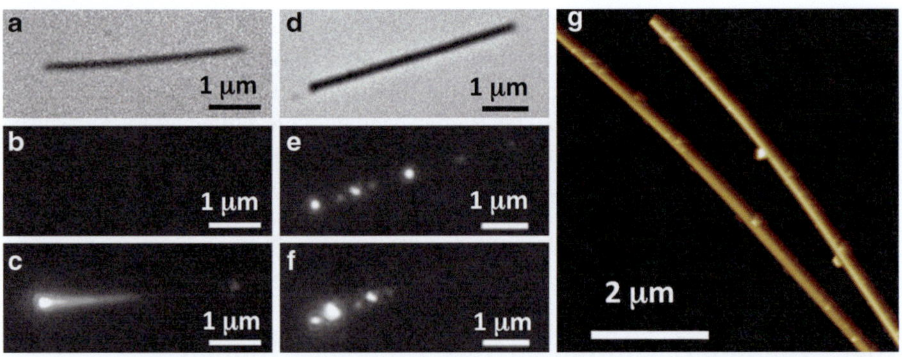

Fig. 16.3 White light transmission images of two 4-ATP-bound nanowires, without (**a**) and with (**d**) adsorbed nanoparticles. Raman images of the same nanowires under wide-field excitation, without (**b**) and with (**e**) adsorbed nanoparticles. Raman images under focused excitation at the left end of the wire, without (**c**) and with (**f**) adsorbed nanoparticles, the latter is the RE-SERS configuration. (**g**) AFM image of a typical nanoparticle-adorned silver nanowire used in experiments (Figure reproduced with permission from Ref. [11])

Two excitation configurations were employed, wide-field excitation in which the whole field of view was excited by a 632.8 nm laser, and focused excitation with the laser focused at the left end of the nanowire to launch SPPs. White light transmission images and Raman images for 4-ATP nanowires with and without adsorbed nanoparticles are shown in Fig. 16.3. For the 4-ATP bound nanowire without attached nanoparticles, no SERS due to 4-ATP is observed above the background during wide-field excitation (Fig. 16.3b). For focused excitation, scattering can be observed at the excited end and decaying away within 2 μm of the wire end, and scattering from the end of the wire can also be faintly observed, evidence that SPPs were launched from the excited end of the wire (Fig. 16.3c).

For the 4-ATP-bound nanowire with nanoparticles attached under wide-field excitation (Fig. 16.3e), bright spots of 4-ATP SERS were observed along the length of the nanowire. When the excitation light was focused at the left end of the nanowire, the same 4-ATP SERS hotspots were observed decaying in intensity with distance from the excited end (Fig. 16.3f). The latter image was good evidence that SERS hot-spots could indeed be excited when the propagating SPP on the wire is scattered by an adsorbed NP.

For SERS spectra recorded at the same hotpot for both wide-field (on particle excitation) and remote excitation, a clear difference was observed. The RE-SERS spectrum, for which the focused laser light excitation is far from the point of SERS detection, has much less background (Fig. 16.4 top). This is because the excitation volume for remote excitation is reduced to the sub-diffraction limit dimensions of the SPP mode volume and is an example of the fundamental advantages of the RE-SERS technique listed in Sect. 16.1. The background typically observed in SERS spectroscopy is due to Raman signals from carbon contamination, enhanced fluorescence or image-molecule/electronic continuum coupling [1].

It was important to establish beyond doubt experimentally that the SERS observed from 4-ATP at points of nanoparticle adsorption on the nanowire did indeed come from remote excitation of SPPs at the end of the wire, and subsequent SPP propagation along the wire. We conducted several experiments to prove this beyond doubt. Firstly, we established that intense Rayleigh scattering of the excitation light during wire-end excitation was not responsible for the SERS excitation at hotspots along the wire (i.e., a through-space excitation mechanism). Two nanowires in close proximity were excited independently by focused laser light at the wire ends (Fig. 16.5). Even if a SERS 'hotspot' on one wire was very close to the end of another wire being excited, no obvious through-space excitation could be observed. SERS was only observed on the wire being directly excited, strongly suggesting propagating SPPs were the excitation source.

The second important confirmation was the distance dependence of RE-SERS intensity during wire-end excitation (see Fig. 16.6). The SPP field intensity decays as a function of distance from the nanowire end excited due to radiative and Ohmic losses as already discussed in Sect. 16.1. The characteristic $1/e$ decay length of SPPs on silver nanowires of ~100 nm diameter is in the range of several microns for 785–800 nm excitation [16, 17], thus the intensity of SERS at various hotpots along a nanowire would be expected to decay similarly as a function of distance from the excited end. We studied RE-SERS on nanowires with multiple nanoparticles attached, to measure the exponential decay of SERS intensity as a function of distance from the wire end. The experiments were complicated by the well-known SERS blinking phenomenon [1]. Fluctuations in SERS intensity as a function of time from a single hot-spot are thought to occur due to orientation changes or diffusion of molecules in or out of the hot-spot region over the time of measurement

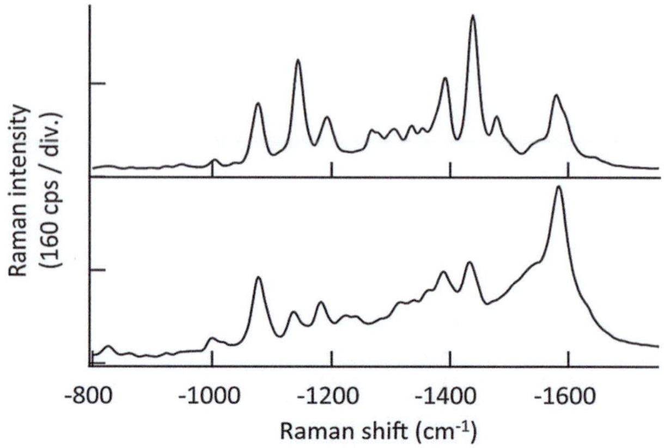

Fig. 16.4 SERS spectra due to 4-ATP at a nanowire/nanoparticle junction during remote excitation (*upper spectrum*) and during 'on-particle' excitation (*lower spectrum*)
(Figure reproduced with permission from Ref. [11])

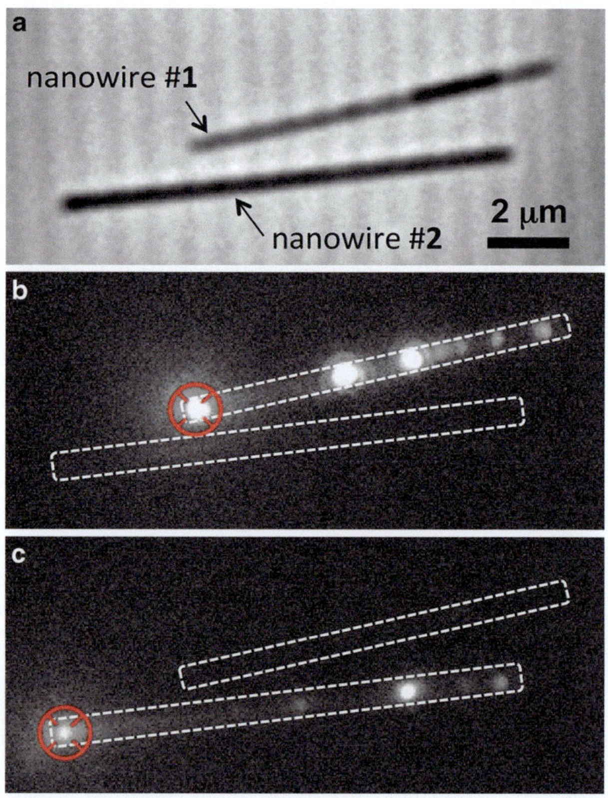

Fig. 16.5 (**a**) White light transmission image of two closely spaced nanowires. (**b**, **c**) Raman images of the same two nanowires separately excited by focused laser light at the wire end (Figure reproduced with permission from Ref. [11])

(typical image integration times were 1–5 s). Nevertheless, studying a single nanowire over time, and several different nanowires, we could estimate an average characteristic 1/*e* decay length for RE-SERS intensity of ~2.7 μm, in good agreement with that observed previously for SPP decay (Fig. 16.6e).

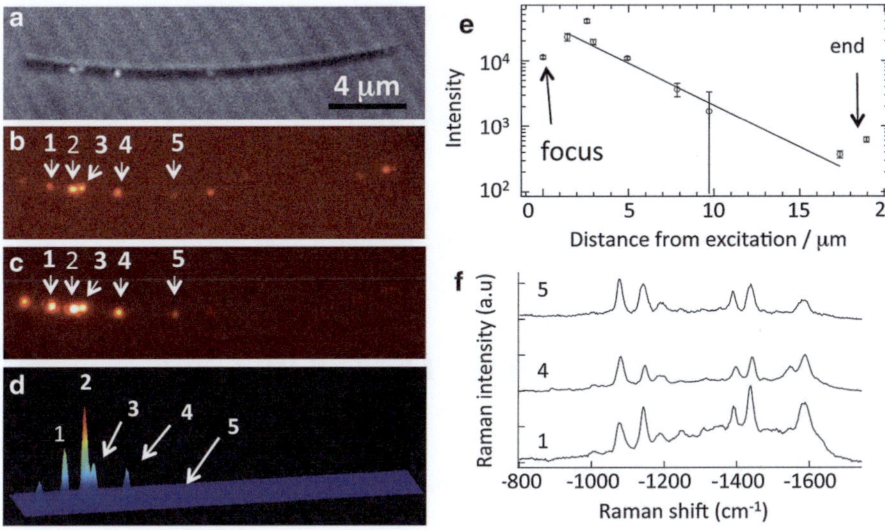

Fig. 16.6 A 4-ATP-bound nanowire with multiple nanoparticles adsorbed. (**a**) White light transmission image, (**b**) Raman image under wide-field excitation, and Raman image under focused excitation at left end of the wire in (**c**) 2-D, and (**d**) 3-D. Multiple hotspots are numbered 1–5 as indicated by *arrows*. (**e**) Semi-log plot of RE-SERS intensity at each hotspot as a function of distance from the excited end of the wire. (**f**) 4-ATP spectra at the hotspots indicated, the background is reduced as the distance between the hotspot and the point of excitation increases (Figure reproduced from Ref. [11])

Finally both the excitation polarization and wavelength dependence of RE-SERS were also consistent with the known efficiencies of SPP launching and propagation on nanowires. RE-SERS was more efficient during 633 nm laser excitation than during 488 nm laser excitation, which reflects the wavelength dependence of the imaginary part of the permittivity of silver and thus the higher absorption of Ag at near-UV frequencies. RE-SERS signal intensity was also much greater when the focused laser was polarized along the long axis of the nanowire compared to when the polarization was parallel. This agrees with the known polarization dependence of in-coupling efficiency for incident light focused at the end of a metallic nanowire [17].

Just months after the publication of our work, Xu, Nordlander and co-workers published a confirmation of the RE-SERS technique [24], measuring RE-SERS of malachite green isothiocyanate (MGITC) on silver nanowires. They employed an MGITC concentration such that, on average, less than one molecule was likely to be located at any single nanoparticle/nanowire junction along the wire, thus the work was an important demonstration of the sensitivity of the technique. Their theoretical studies confirmed our conclusions that SERS hotspots between a nanoparticle and a nanowire could be efficiently excited by propagating SPPs running along the wire. They estimated a $1/e$ decay length of the SPP field to be 4.4 μm, very similar to our experimental result, and an enhancement factor, EF_{SERS}, of 10^7–10^8. Finally their

SEM studies corroborated our AFM studies proving that the RE-SERS hotspots along the nanowire did indeed exactly correspond to points of nanoparticle adsorption.

Since our initial report in 2009, the use of metallic nanowires for remote spectroscopy has expanded dramatically, and a very recent review has detailed the experiments showing dual remote excitation of RE-SERS on coupled nanowires, remote sensing of Raman optical activity, remote excitation of SPP-induced catalysis, live cell endoscopy, and much more [27].

16.4 Applications of RE-SERS: Live Cell RE-SERS Endoscopy

For remote spectroscopy using sub-diffraction limit nanostructures to fulfil its ultimate promise, the probe must be able to be controllably placed on or near a region of interest with high precision. In this section the application of metallic nanowires as probes for remote Raman sensing in live cells is presented [28]. The work clearly demonstrates the advantages of remote spectroscopy on sub-diffraction limit nanostructures listed in Sect. 16.1. A sub-diffraction limit diameter probe is required to penetrate the cell membrane causing as little stress as possible, while the separation of the excitation and detection regions avoids photo-thermal damage to the cell during RE-SERS and improves the signal-to-noise ratio of the SERS spectra.

Non-invasive, high spatial resolution, live cell spectroscopy using nanowires has enormous potential application for in vitro disease diagnostics and for sensing the intracellular distribution and chemistry of bioactive compounds [29]. SERS spectroscopy is well suited to this task as it is label free, it emanates from a hotspot of a few nanometers diameter, and it can provide well-resolved spectral signatures for identification of compounds with near-single molecule sensitivity. Live cell SERS endoscopy, in which a SERS probe is precisely positioned at intracellular regions of interest, is thus a powerful technique *if* it does not alter normal cell function. To this end we developed a cost efficient, high throughput method for attaching silver nanowires of 50–150 nm diameters to a tungsten tip, as shown in Fig. 16.7, for use in SERS endoscopy of live cells.

Briefly the nanowires were attached to an electrochemically etched tungsten tip using the alternating current dielectrophoresis (AC-DEP) method (shown schematically in Fig. 16.7a). High throughput was ensured by adhering many tungsten wires to an ITO glass electrode using silver paste (Fig. 16.7b). The tungsten tips were immersed in a solution of silver nanowires in a water/ethanol mixture and an AC voltage applied (frequency 1 MHz, amplitude 8 V) between the tungsten tip and a ring-shaped platinum wire electrode for ~10 s. The silver nanowires were thus attached in a bundle to the end of the tungsten tip, with a single nanowire normally protruding from the bundle to serve as the SERS endoscope. By chance, a few small

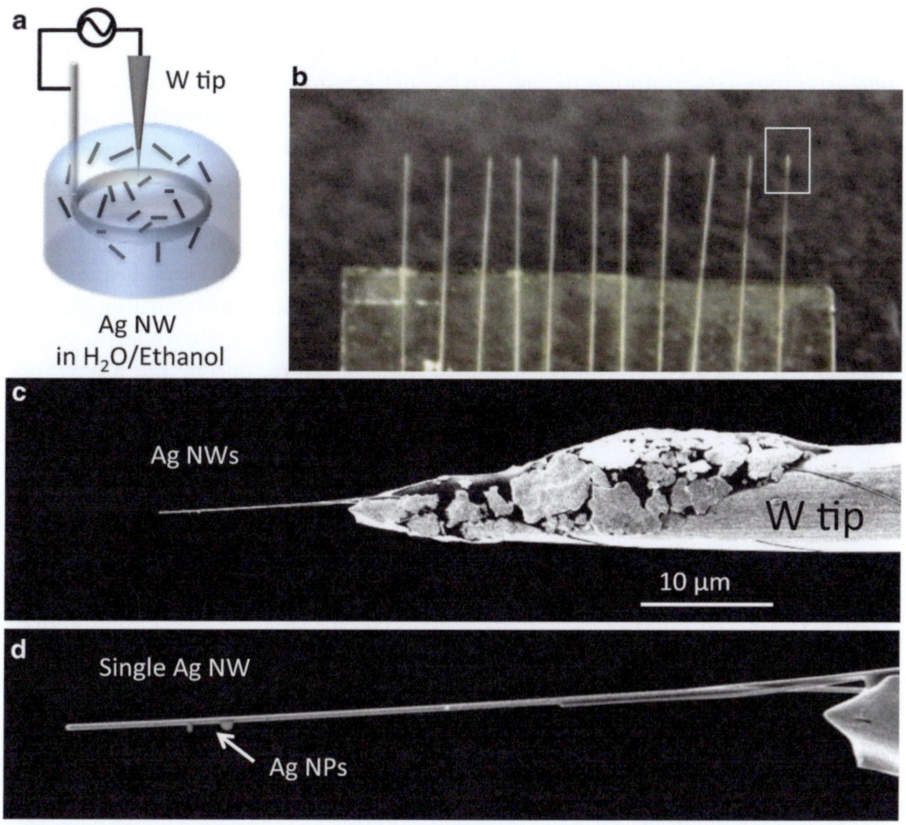

Fig. 16.7 (**a**) Schematic of the alternating current dielectrophoresis (AC-DEP) method for attaching silver nanowires to an electrochemically etched tungsten tip. (**b**) Multiple tunsgsten tips arrayed on an ITO support to aid throughput. (**c**, **d**) SEM images of a SERS endoscope protruding out from a bundle of silver nanowires glued to a tungsten tip, with some silver nanoparticles attached (Figure adapted from Ref. [28])

nanoparticles adsorbed to the nanowire during the AC-DEP process, which could then act as hotspots for RE-SERS spectroscopy (see SEM images, Fig. 16.7c, d). Following annealing and gluing conductive epoxy on the silver nanowire bundle/tungsten tip, the probe was ready for use in live cell endoscopy experiments.

The target for the endoscopy measurements were living HeLa cells grown in a glass bottom dish. The dish could be placed on an inverted fluorescence microscope such that both SERS excitation and imaging/spectroscopy could be undertaken in epi-configuration using an ordinary high NA objective (see schematic in Fig. 16.2). The cell growth solution was replaced by Hank's balanced salt solution (HBSS) just before endoscopy commenced. The live cell was approached from above by the silver nanowire/tungsten probe attached to a micromanipulator and live endoscopy studies undertaken.

As already mentioned, a challenge for live cell endoscopy is to be able to penetrate the cell and undertake analyses without unduly stressing the cell. An increase in the concentration of Ca^{2+} in the cytoplasm is an indicator that the cell membrane has been damaged. Here we assessed cell membrane stress using the fluorescent dye Calcium Green^{TM}-1, which exhibits a strong increase in fluorescence quantum yield when bound to Ca^{2+} (the Calcium Green-1 was added to the growth solution earlier together with Chremaphore EL to enhance uptake).

The nanowire probe was brought to the cell at an angle of 30° to the horizontal with the micromanipulator and gently inserted through the cell membrane. For a comparison stress test, a typical conical probe used in tip-enhanced Raman scattering studies (TERS), i.e., the electrochemically etched tungsten wire, was also inserted into HeLa cells. The fluorescence images before, during and after probe penetration in each case are shown in Fig. 16.8. For conical probe insertion, Calcium Green-1 fluorescence was increased ~20 times at the point of insertion. For the silver nanowire probe however, insertion did not lead to any significant increase in Calcium Green-1 emission during insertion or following withdrawal. Nanowires with diameters between 50 and 150 nm showed similar results, suggesting endoscopy with sub-diffraction limit diameter probes does not unduly modify the normal behaviour of the cell.

Live HeLa cell endoscopy was then undertaken using the two configurations shown in Fig. 16.2. Conventional ('on-particle') SERS spectroscopy/imaging was undertaken, in which the excitation laser (632.8 nm) and the detection volume were overlapped. The circularly polarized laser excitation was focused either at the wire end or at positions of nanoparticle adsorption along the nanowire within the cell, and SERS detected from the same position. For the RE-SERS spectroscopy/imaging configuration, excitation was focused on a part of the nanowire outside the cell, launching SPPs that then propagated into the cell for SERS hotspot excitation. Laser excitation was focused either at an adsorbed nanoparticle on the nanowire but outside the cell, at the nanowire end outside the cell, or at a junction between the nanowire probe and another nanowire outside the cell.

SERS spectra at hotposts along the nanowire in different cell compartments during 'on-particle' excitation are detailed in Fig. 16.9a–d. Spectra taken in the cell cytoplasm showed very little signal, similar to spectra taken from hotspots in the HBSS carrier solution. This is not surprising since the cytoplasm contains a dilute mixture of salts and other species with low Raman cross-sections. SERS spectra from hotspots positioned in the nucleus showed much stronger signals in the fingerprint region from 900 to 1700 cm^{-1}, likely due to aromatic species with large Raman cross-sections associated with proteins or DNA. Exact identification is impossible however due to the high background of the spectrum when taken using 'on-particle' excitation.

A more serious consequence of conventional SERS endoscopy, with direct laser excitation of the probe inside the cell, was that following a SERS experiment parts of the cell typically become strongly adhered to the probe. Continuing to withdraw the nanowire then destroyed the cell, pulling its contents into the buffer solution (Fig. 16.10a–c). The cell then typically became round in shape and floated to the top

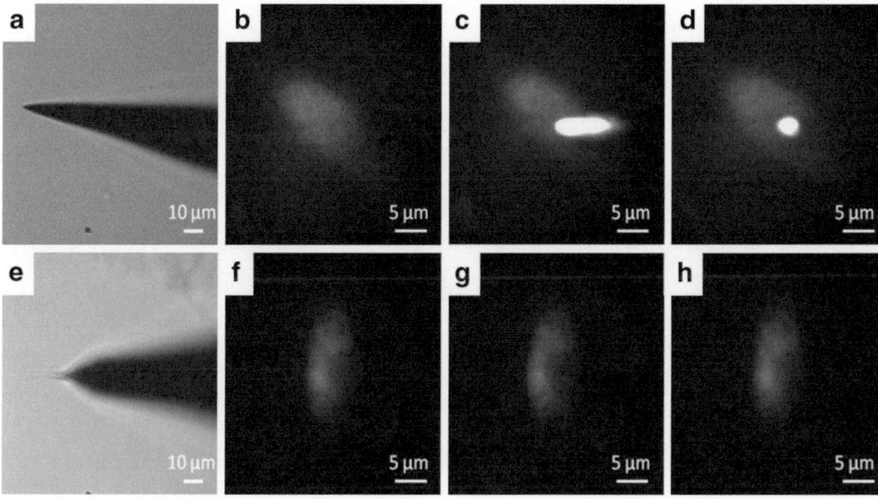

Fig. 16.8 White light transmission image of the tungsten probe (**a**) and fluorescence images of a Calcium Green-1-stained HeLa cell before (**b**), during (**c**), and after (**d**) probe insertion. (**e–h**) The corresponding images but using a 50 nm diameter nanowire probe (Figure adapted from Ref. [28])

of the buffer solution indicating cell death. This is likely due to photo-thermal degradation of the cell due to direct irradiation by the focused laser on the probe and surrounds (typical power density employed was 1–10 kW/cm^2, and the image integration time was ~5 s).

RE-SERS endoscopy overcomes both of the issues exemplified in Figs. 16.9 and 16.10. Transmission and RE-SERS images of a live Hela cell are shown in Fig. 16.11a–c. Laser excitation is focused at the probe outside the cell, launching SPPs which travel down the wire and into the cell, exciting SERS at an adsorbed nanoparticle inside the cell. Transmission and RE-SERS images show that point of excitation and point of SERS detection are clearly separated (Fig. 16.11c). In this case SERS spectra can be taken and the probe removed typically without damaging or destroying the cell. Furthermore the spectra have much better signal-to-noise ratio compared to 'on-particle' excitation. An example is shown in Fig. 16.11d, for a RE-SERS spectrum taken from a hotspot inside the nucleus. The peaks are clearly distinguishable which could allow identification species, though this task is not undertaken here.

These experiments are an excellent demonstration of the unique advantages of remote spectroscopy on sub-diffraction limit diameter nanostructures. Low-background SERS spectra can be taken from different compartments of a cell with minimal damage to the cell either mechanically during probe penetration/removal, or by photo-thermal effects during the SERS measurements. The great limitation of RE-SERS endoscopy is the limitation of all SERS experiments, that is, that the hotspot for SERS enhancement is highly spatially localized within a few nm^2 right in the nanogap between the nanoparticle and the nanowire [1]. Thus this hotspot is not always easily accessible to analytes in solution. Future work to

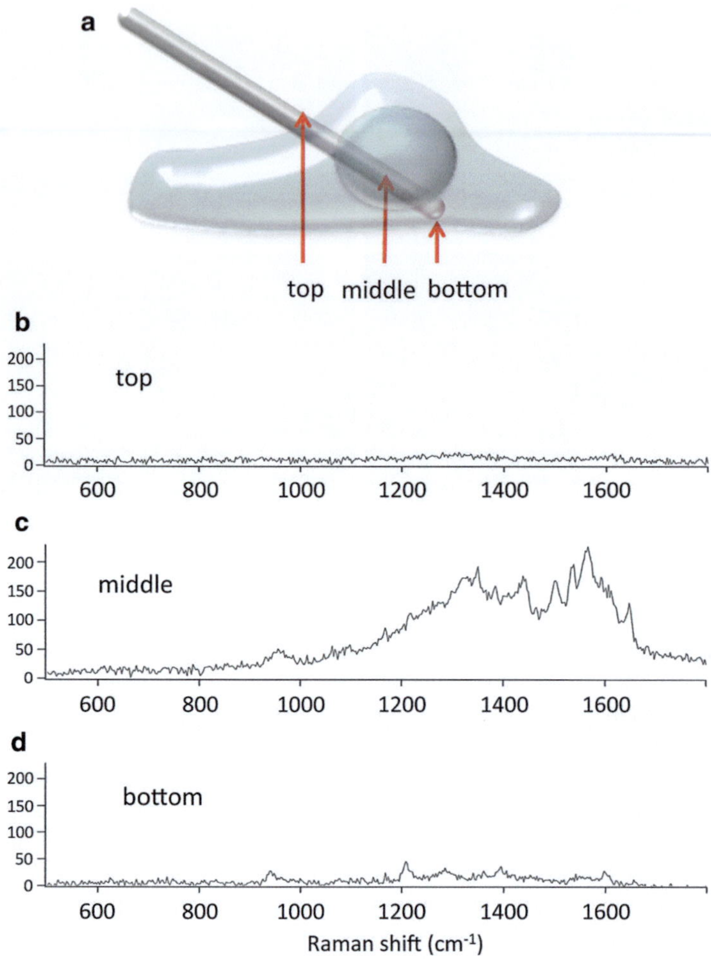

Fig. 16.9 (**a**) Schematic of the different regions (buffer solution, HeLa cell nucleus, HeLa cell cytoplasm) from which were recorded the conventional 'on-particle' SERS spectra shown in (**b**, **c**, and **d**), respectively (Figure adapted from Ref. [28])

improve these RE-SERS endoscopy experiments would focus on making the SERS hotspot more reproducible, perhaps by on-wire etching, and also making it more specific for biomolecules of interest by binding a receptor at the hotspot. Finally, we recently extended remote spectroscopy using silver nanowire probes by using them in Tip-enhanced Raman Scattering (TERS) experiments where they reduced tip-sample convolution effects compared to conventional conical tips [30].

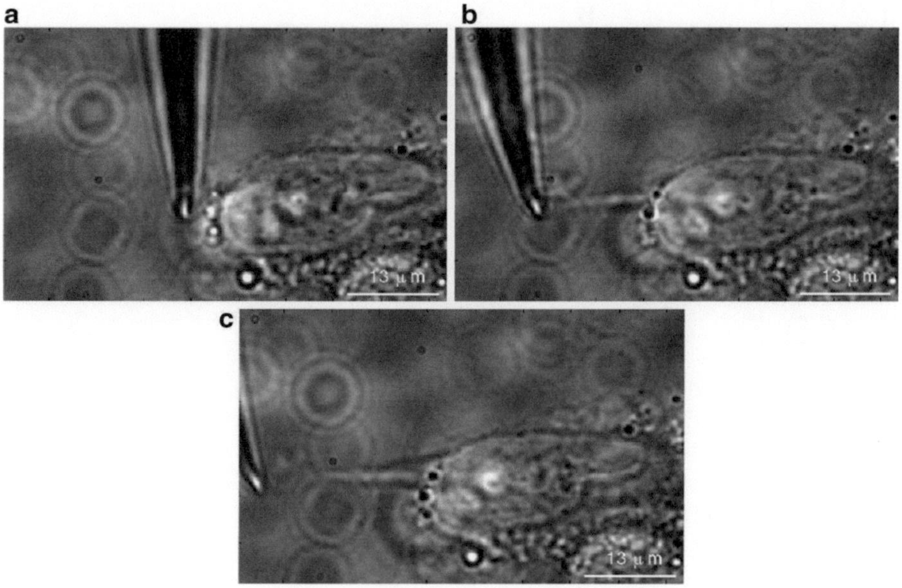

Fig. 16.10 (**a–c**) White light transmission images during retraction of a Ag nanowire after conventional 'on-particle' excitation of SERS inside a HeLa cell (Figure adapted from Ref. [28])

16.5 Remote Excitation of Single Molecule Fluorescence

The perturbation of atomic and molecular photophysics within the near field of a metal surface has been a topic of interest for more than 70 years. Purcell was the first to predict that the lifetime of emission depends on the photonic mode density of the metallic structure in the vicinity of the molecule, confirmed by Drexhage when he observed oscillations in the lifetime of a europium complex's luminescence as it was brought close to a metal mirror [31]. Viewed classically, the modification of the luminescence lifetime is due to the emitted field being reflected and re-interacting with the molecule, either damping or driving the transition dipole depending on the relative phase of the reflected field. From the quantum mechanical viewpoint the presence of the metal perturbs the continuum of vacuum states into which the excited molecule can transfer energy, thus the probability of emission fluctuates proportionally with that modulated density of vacuum states.

Drexhage also showed that when the molecule approaches within 10 nm of the surface, its emission is strongly quenched, either by energy transfer to surface plasmons, or to non-optical modes in the metal [31]. Thus the optimum distance for molecular fluorescence enhancement is typically in the region 10–20 nm from the metal surface. This surface-enhanced fluorescence (SEF) near metallic interfaces, whether planar or a metallic nanoparticle, has seen intense research [32, 33]. SEF is most effective where non-radiative processes dominate a molecules

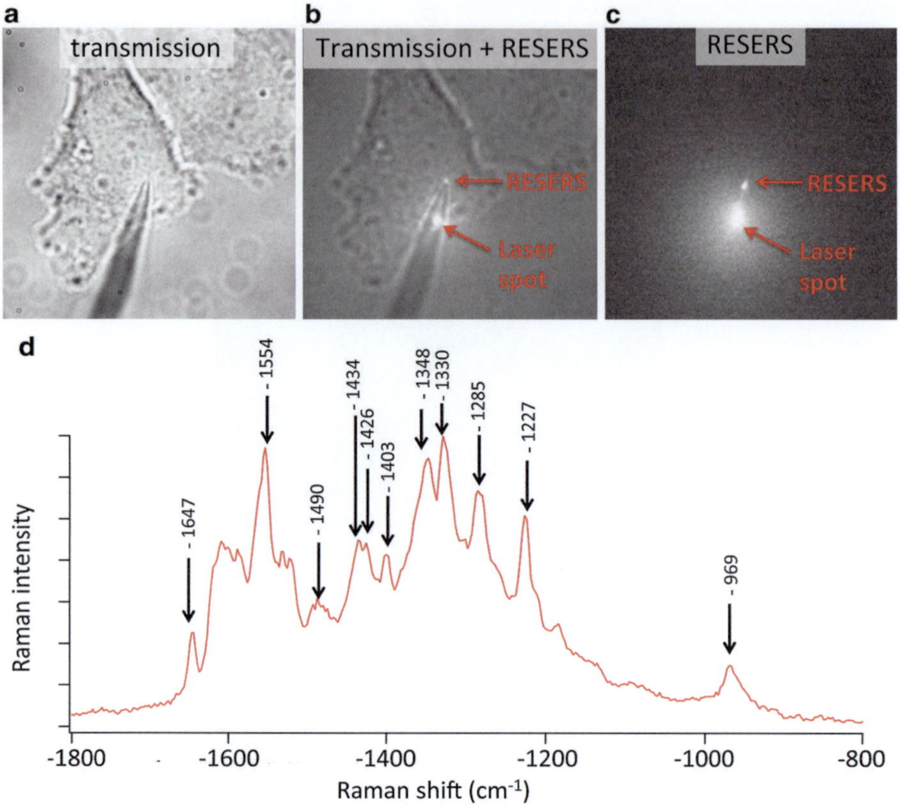

Fig. 16.11 (**a–c**) White light transmission and Raman images of RE-SERS endoscopy and (**d**) a RE-SERS spectrum from a hotspot in the HeLa cell nucleus (Figure adapted from Ref. [28])

photophysics such that the enhancement in the radiative rate has the strongest effect on the fluorescence quantum yield.

The enhancement or damping of the radiative rate of a molecule is not the only perturbation that occurs near a metal interface. The typical free space $\sin^2\theta$ angular distribution of emission about the transition dipole is also modified due to scattering off the metal surface/particle, a process that depends strongly on the orientation of the molecule at the interface [31]. The result of this angular distribution when imaged in the far-field, the point spread function of emission, is particularly important in light of recent advances in super-resolution optical imaging, in which a molecule is localized to sub-diffraction limit certainty by fitting the point spread function of its emission [34]. Super-resolution imaging involves the construction of an image by localizing the emission of many single molecules separately over time if conditions are adjusted such that only one molecule in a sample is emitting within a diffraction-limited radius at any time.

Therefore super-resolution always requires an assumption about the fluorescence point spread function of a molecule in order that it may be adequately fit and the emitter localized. As mentioned already, if the emitter is immersed in an environment of homogenous permittivity, the distribution is radially symmetrical and the point spread function can be well fit by a 2-D Gaussian for localization. However, given the strong scattering properties of metallic nanoparticles and their potential effect on the emission point spread function of molecules [31], it is not obvious that super-resolution imaging could be successfully applied to image sub-diffraction limit diameter metallic nanostructures.

In 2012 we published the first studies that attempted to use fluorescence localization of emitters in the vicinity of metallic nanostructures for imaging of metallic nanoparticles [35]. Silver nanowires were imaged as well as arrays of polycrystalline gold nanotriangles fabricated by the well-known nanosphere lithography technique [36]. The metallic particles were then decorated with the photo-switchable green fluorescent protein derivative DRONPA via a biotin linker. DRONPA can be photo-switched reversibly from a fluorescent state to a dark state by alternating excitation at 488 and 405 nm. Thus only one protein could be switched on in isolation from other emitters adsorbed within a diffraction limit-radius around it. Each frame containing single protein fluorescence point spread functions was fit to a 2D-Gaussian function and localised to build up a reconstructed super-resolution image of the nanopartcles.

It was immediately apparent during these studies that the super-resolution reconstructed image did not perfectly reproduce the dimensions of the nanostructures as shown by electron microscopy. As the laser excitation overlapped the detection volume, scattering and broad emission from the metallic nanostructure itself made it difficult to resolve the single molecule fluorescence spot above the background. Some of the point spread functions appeared elongated rather than Gaussian, but due to the high background it was difficult to make a more accurate assessment of the true form of the point spread function. Following this early work, others have also noticed that super-resolution reconstruction imaging can fail to give a faithful representation of metallic nanoparticle dimensions [37, 38].

To gain a clearer understanding of the nature of molecular fluorescence point spread functions in the vicinity of metallic nanoparticles we turned once again to remote spectroscopy on metallic nanowires to avoid the problems of high background due to direct excitation [39]. Firstly we modified the technique for super-resolution imaging, employing stochastic switching for fluorescence localization, the so-called DSTORM technique [40, 41]. The latter relies on the random switching on of dye emission due to its (photo)chemical environment. In this case we employed Alexa647 whose intersystem crossing rate to a non-emissive triplet state is excitation intensity dependent. The Alexa647 was bound by biotin linker to silver nanowires (Fig. 16.12a). The lifetime of the Alexa647 on the nanowires was around 1.2 ns, slightly shorter than that measured away from the metal. The biotin linkage holds the Alexa647 molecules on average ~15 nm from the silver surface, thus it is not surprising that the relaxation rate of the molecule is perturbed as discussed above.

Excitation either directly on the nanowire (wide field excitation) or by remote excitation focusing the laser at one end of the nanowire to launch SPPs, which then

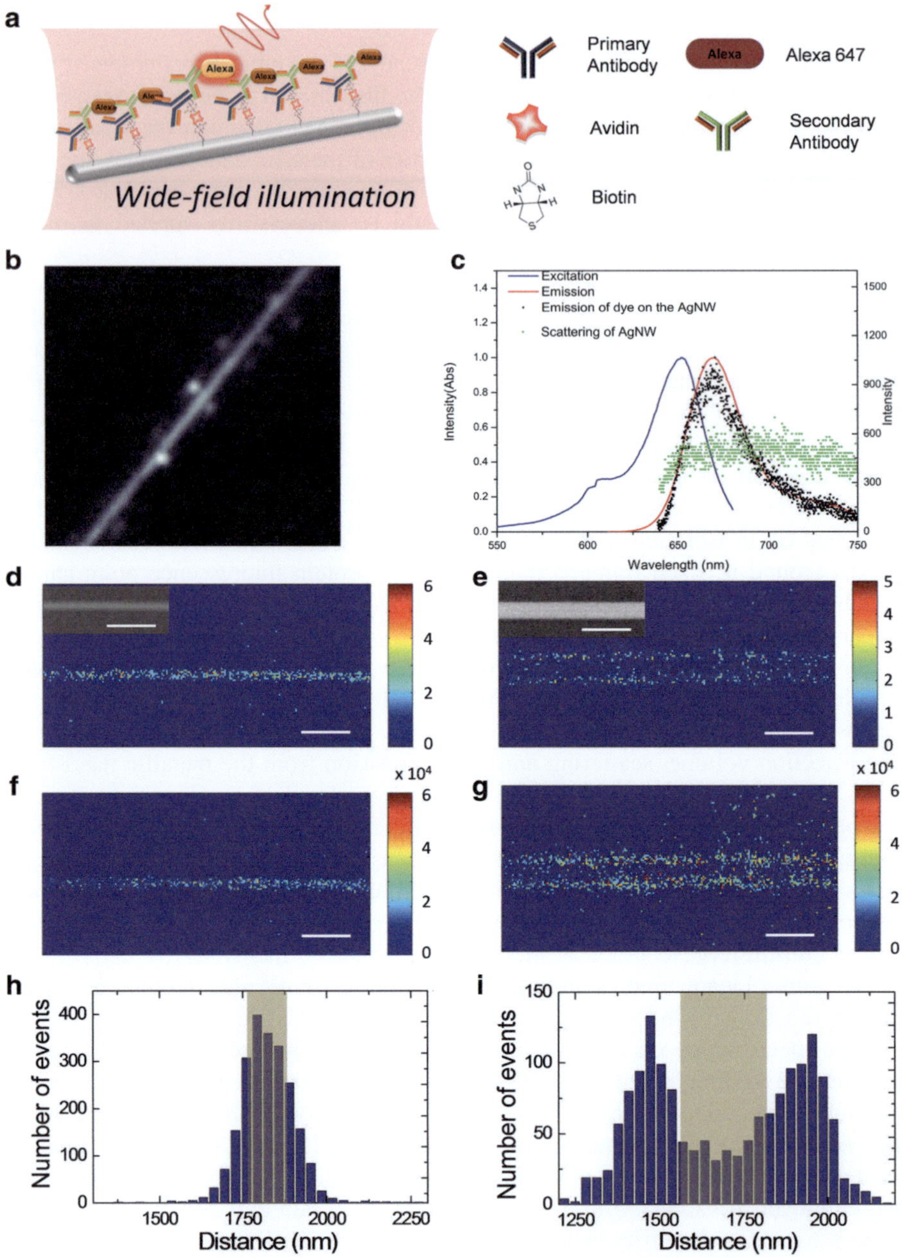

Fig. 16.12 (**a**) Schematic of the binding of Alexa647 to silver nanowires. (**b**) Fluorescence image of an Alexa647-bound nanowire under wide-field excitation. (**c**) Absorption (*blue curve*) and emission (*red curve*) of Alexa647 in bulk solution. The fluorescence spectrum of an Alexa647-bound nanowire (*black dots*) and a bare silver nanowire (*greens dots*) is also included. (**d–g**) High resolution reconstructed images of silver nanowires following wide-field excitation assuming that each emission spot is due to a single Alexa647 molecule localised at its centroid. (**d, e**) Localised single molecule density maps for 110 and 250 nm diameter nanowires respectively as shown in SEM images inset. (**f, g**) Localised single molecule intensity maps. (**h, i**) Cross sections of the single molecule density maps perpendicular to the nanowire long axis, the *grey shading* indicates the true diameter of the nanowires in each case (Figure reproduced with permission from Ref. [39])

excited molecules in the SPP near field further along the nanowire, was used to image the fluorescence point spread functions of the bound Alexa647. In each case the excitation intensity was adjusted such that only a few Alexa647 molecules were emitting at any one time, thus allowing single molecule localization to be performed.

In Fig. 16.12b is shown a fluorescence image of an Alexa647-decorated silver nanowire under wide-field excitation. The bright spots that appear as satellites along the nanowire are spots of single molecule emission along the wire. What is also clear is that the nanowire itself shows strong emission (see Fig. 16.12c). The large background obscured the single molecule emission such that it had to be assumed such that each fluorescent spot was due to a single molecule at its centroid. High resolution reconstructed images built up in this way are shown in Fig. 16.12d–g for nanowires of diameters 110 and 250 nm respectively. The reconstruction imaging suggests the diameter of the 110 nm nanowire is 205 nm, and completely fails to represent the true form of the nanowire with 250 nm diameter (see cross-sections of the density maps perpendicular to the nanowire long axis in Fig. 16.12h, i in which grey shading indicates the true diameter of the nanowire in each case).

When the same experiment is done utilizing remote excitation, much clearer images of the fluorescence point spread functions emerge (Fig. 16.13). They appear as bright spots clearly discernable against the weak background emission, demonstrating once again the benefit of using remote excitation for reducing spectroscopic background. Importantly, some groups of spots clearly blink on and blink off again at the same time. That is to say, they are actually multi-spot emission point spread functions due to a single Alexa647 molecule.

The point spread functions of single molecules of Alexa647 thus identified form a family of patterns including: symmetrical, single 2-D Gaussian spots located directly on the wire; elongated spots on the side of the nanowire (running along the long axis); two-spot 2-D Gaussian patterns in which the spots are on one side of the wire, two-spot 2-D Gaussian patterns in which the spots are on opposite sides of the wire; and finally four spot 2-D Gaussian patterns with two spots on either side of the wire (see Fig. 16.14 for examples).

Having collected fluorescence images of the single molecule point spread functions consisting of complex, multi-spot patterns, finite-difference time-domain (FDTD) calculations were undertaken to model the far field emission patterns expected if a fluorescent molecule was located on the nanowire at a certain facet or corner, and with a certain orientation [39]. The single molecular emitter was modelled as an oscillating point dipole. Figure 16.15 shows a selection of simulated far-field emission patterns, all of which were found experimentally (Fig. 16.14). The angular distribution of emission is also shown in radial plots next to each simulated point spread function. The close agreement between experiment and simulation suggests that fast identification of a molecule's binding orientation and position on a nanoparticle could be determined directly from the emission pattern, if the size and shape of the particle is known. Some binding positions are predicted to be particularly sensitive, for instance when the molecule is located at the apex of the

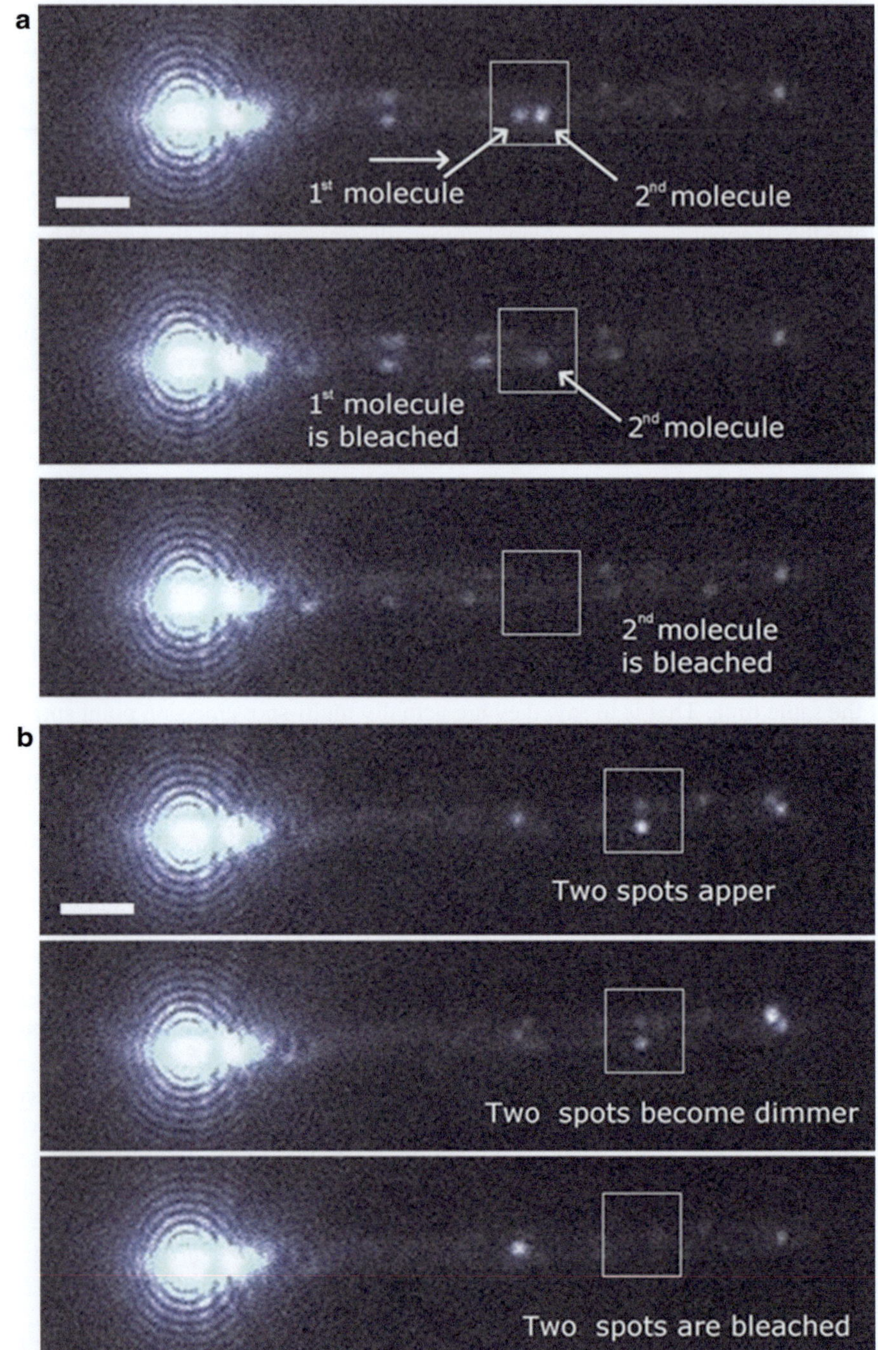

Fig. 16.13 (**a**,**b**) Fluorescence images of Alexa647-bound nanowires under remote excitation at the left end of the wire at different times. The synchronous blinking on and off of multiple emission spots are indicated (Figure reproduced with permission from Ref. [39])

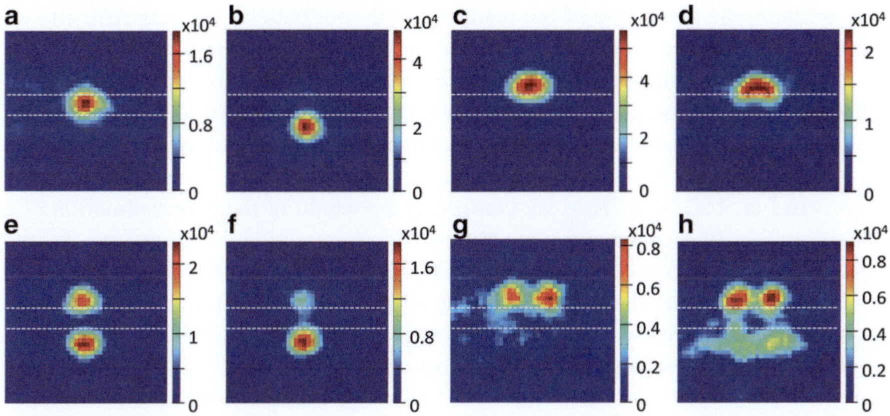

Fig. 16.14 (**a–h**) Examples of emission point spread functions for single molecules on a silver nanowire under remote excitation (wire position indicated by *white dashed lines*) (Figure reproduced with permission from Ref. [39])

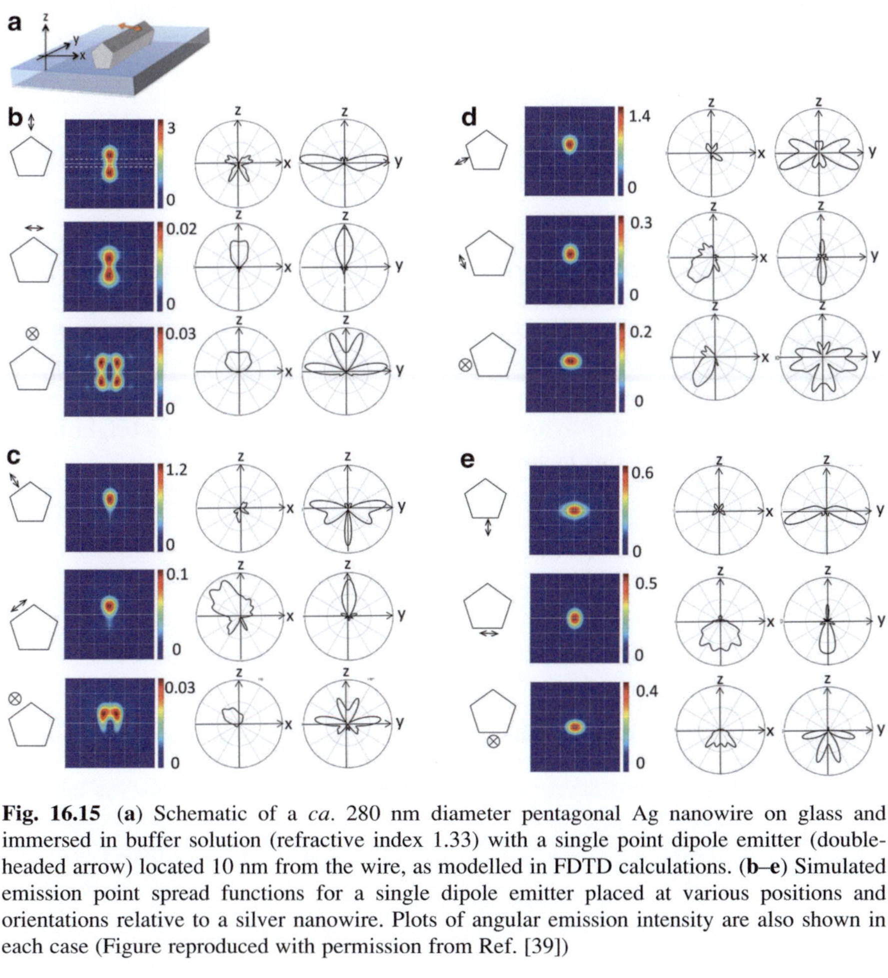

Fig. 16.15 (**a**) Schematic of a *ca.* 280 nm diameter pentagonal Ag nanowire on glass and immersed in buffer solution (refractive index 1.33) with a single point dipole emitter (double-headed arrow) located 10 nm from the wire, as modelled in FDTD calculations. (**b–e**) Simulated emission point spread functions for a single dipole emitter placed at various positions and orientations relative to a silver nanowire. Plots of angular emission intensity are also shown in each case (Figure reproduced with permission from Ref. [39])

nanoparticle, the symmetry of the resulting four-spot pattern depends critically on the orientation of the molecules transition dipole [39].

By deliberately incorrectly assuming that even complex multi-spot fluorescence point spread functions due to single molecules were due to several molecules localized at the centroid of each fluorescent spot, one can build up a reconstructed image that fails to represent the true structure of the nanowire in exactly the same way as in Fig. 16.12d, e. Thus we confirmed the origin of the poor estimation of a nanoparticle's true dimensions by optical super-resolution imaging, that is, the assumption that each bright emission spot is due to a single molecule at its centroid.

These experiments show once again the advantage of using remote spectroscopy on metallic nanowires for minimizing background emission/scattering. In this case remote excitation revealed a complex array of multi-spot emission point spread functions for single molecules that were otherwise obscured during direct excitation. The results also suggest a fundamental paradox for the goal of taking super-resolution optical images of metallic nanoparticles. Normally the technique is used to assess the size of an unknown object of interest. Unfortunately, for molecules near metallic nanoparticles it is not possible to know the form of the point spread function without some a priori knowledge of the nanoparticle dimensions! This work has consequences also for counting (for instance) single molecule catalytic turn over events on metallic nanoparticles using fluorescence imaging.

16.6 Conclusions

This short review has retraced work in our laboratories over the last 6 years, from our first report of remote excitation SERS on silver nanowires [11], to recent applications including live cell SERS endoscopy and super-resolution fluorescence imaging of metal nanowires [28, 39]. The combination of fluorescence and Raman sensing with metallic nanowire plasmonics is clearly a powerful one, and the many new applications continuing to appear in the literature [27] suggest a bright future for this remote spectroscopy technique.

References

1. Ru EL, Etchegoin P (2009) Principles of surface-enhanced raman spectroscopy and related plasmonic effects. Elsevier, Amsterdam
2. Genet C, Ebbesen TW (2007) Light in tiny holes. Nature 445:39–46
3. Kuttge M et al (2008) Loss mechanisms of surface plasmon polaritons on gold probed by cathodoluminescence imaging spectroscopy. Appl Phys Lett 93:113110
4. Ditlbacher H et al (2005) Silver nanowires as surface plasmon resonators. Phys Rev Lett 95:257403
5. Berini P, De Leon I (2012) Surface plasmon-polariton amplifiers and lasers. Nat Photonics 6:16–24

6. Paul A et al (2014) Dye-assisted gain of strongly confined surface plasmon polaritons in silver nanowires. Nano Lett 14:3628–3633
7. Shalaev VM (2007) Optical negative-index metamaterials. Nat Photonics 1:41–48
8. Paniagua-Domínguez R, Abujetas DR, Sánchez-Gil JA (2013) Ultra low-loss, isotropic optical negative-index metamaterial based on hybrid metal-semiconductor nanowires. Sci Rep 3: 1507
9. Fang Y, Sun M (2015) Nanoplasmonic waveguides: towards applications in integrated nanophotonic circuits. Light Sci Appl 4, e294
10. Barnes WL, Dereux A, Ebbesen TW (2003) Surface plasmon subwavelength optics. Nature 424:824–830
11. Hutchison JA et al (2009) Subdiffraction limited remote excitation of surface enhanced Raman scattering. Nano Lett 9:995–1001
12. McQuillan AJ (2009) The discovery of surface-enhanced Raman scattering. Notes Rec R Soc 63:105–109
13. Nie S, Emory SR (1997) Probing single molecules and single nanoparticles by surface-enhanced Raman scattering. Science 275:1102–1106
14. Kneipp K et al (1997) Single molecule detection using surface-enhanced Raman scattering (SERS). Phys Rev Lett 78:1667–1670
15. Dickson RM, Lyon LA (2000) Unidirectional plasmon propagation in metallic nanowires. J Phys Chem B 104:6095–6098
16. Sanders AW et al (2006) Observation of plasmon propagation, redirection, and fan-out in silver nanowires. Nano Lett 6:1822–1826
17. Graff A, Wagner D, Ditlbacher H, Kreibig U (2005) Silver nanowires. Eur Phys J – At Mol Opt Plasma Phys 34:263–269
18. Knight MW et al (2007) Nanoparticle-mediated coupling of light into a nanowire. Nano Lett 7:2346–2350
19. Hao F, Nordlander P (2006) Plasmonic coupling between a metallic nanosphere and a thin metallic wire. Appl Phys Lett 89:103101
20. Akimov AV et al (2007) Generation of single optical plasmons in metallic nanowires coupled to quantum dots. Nature 450:402–406
21. Wei H et al (2008) Polarization dependence of surface-enhanced Raman scattering in gold nanoparticle−nanowire systems. Nano Lett 8:2497–2502
22. Lee SJ, Baik JM, Moskovits M (2008) Polarization-dependent surface-enhanced Raman scattering from a silver-nanoparticle-decorated single silver nanowire. Nano Lett 8:3244–3247
23. Yoon I et al (2009) Single nanowire on a film as an efficient SERS-active platform. J Am Chem Soc 131:758–762
24. Fang Y, Wei H, Hao F, Nordlander P, Xu H (2009) Remote-excitation surface-enhanced Raman scattering using propagating Ag nanowire plasmons. Nano Lett 9:2049–2053
25. Sun Y, Mayers B, Herricks T, Xia Y (2003) Polyol synthesis of uniform silver nanowires: a plausible growth mechanism and the supporting evidence. Nano Lett 3:955–960
26. Lin H et al (2011) Light-assisted nucleation of silver nanowires during polyol synthesis. J Photochem Photobiol Chem 221:220–223
27. Huang Y, Fang Y, Zhang Z, Zhu L, Sun M (2014) Nanowire-supported plasmonic waveguide for remote excitation of surface-enhanced Raman scattering. Light Sci Appl 3, e199
28. Lu G et al (2014) Live-cell SERS endoscopy using plasmonic nanowire waveguides. Adv Mater 26:5124–5128
29. Yan R et al (2012) Nanowire-based single-cell endoscopy. Nat Nanotechnol 7:191–196
30. Fujita Y et al (2014) A silver nanowire-based tip suitable for STM tip-enhanced Raman scattering. Chem Commun 50:9839–9841
31. Barnes WL (1998) Fluorescence near interfaces: the role of photonic mode density. J Mod Opt 45:661–699

32. Geddes CD, Aslan K, Gryczynski I, Malicka J, Lakowicz JR (2004) Noble-metal surfaces for metal-enhanced fluorescence. In: Geddes CD, Lakowicz JR (eds) Reviews in fluorescence 2004. Springer, Boston, pp 365–401
33. Fort E, Grésillon S (2008) Surface enhanced fluorescence. J Phys Appl Phys 41:013001
34. Orrit M (2014) Nobel prize in chemistry: celebrating optical nanoscopy. Nat Photonics 8:887–888
35. Lin H et al (2012) Mapping of surface-enhanced fluorescence on metal nanoparticles using super-resolution photoactivation localization microscopy. ChemPhysChem 13:973–981
36. Haynes CL, Van Duyne RP (2001) Nanosphere lithography: a versatile nanofabrication tool for studies of size-dependent nanoparticle optics. J Phys Chem B 105:5599–5611
37. Blythe KL, Titus EJ, Willets KA (2014) Triplet-state-mediated super-resolution imaging of fluorophore-labeled gold nanorods. ChemPhysChem 15:784–793
38. Titus EJ, Willets KA (2013) Accuracy of superlocalization imaging using gaussian and dipole emission point-spread functions for modeling gold nanorod luminescence. ACS Nano 7:6258–6267
39. Su L et al (2015) Visualization of molecular fluorescence point spread functions via remote excitation switching fluorescence microscopy. Nat Commun 6: 6287
40. Rust MJ, Bates M, Zhuang X (2006) Sub-diffraction-limit imaging by stochastic optical reconstruction microscopy (STORM). Nat Methods 3:793–796
41. Schüttpelz M, Wolter S, van de Linde S, Heilemann M, Sauer M (2010) dSTORM: real-time subdiffraction-resolution fluorescence imaging with organic fluorophores. In: Proceedings of SPIE 7571, 75710V–75710V–7

Index

C.D. Geddes (ed.), *Reviews in Plasmonics 2015*,
DOI 10.1007/978-3-319-24606-2

N

MIX
Papier aus verantwortungsvollen Quellen
Paper from responsible sources
FSC® C105338

FSC
www.fsc.org

If you have any concerns about our products,
you can contact us on
ProductSafety@springernature.com

In case Publisher is established outside the EU,
the EU authorized representative is:
Springer Nature Customer Service Center GmbH
Europaplatz 3, 69115 Heidelberg, Germany

Printed by Libri Plureos GmbH
in Hamburg, Germany